# Green Energy and Technology



For further volumes:
http://www.springer.com/series/8059

Jahangir Hossain · Apel Mahmud
Editors

# Renewable Energy Integration

## Challenges and Solutions

Springer

*Editors*
Jahangir Hossain
Griffith School of Engineering
Griffith University
Gold Coast, QLD
Australia

Apel Mahmud
Electrical and Electronics Engineering
Swinburne University of Technology
Hawthorn, VIC
Australia

ISSN 1865-3529 ISSN 1865-3537 (electronic)
ISBN 978-981-10-1190-0 ISBN 978-981-4585-27-9 (eBook)
DOI 10.1007/978-981-4585-27-9
Springer Singapore Heidelberg New York Dordrecht London



Printed on acid-free paper

Springer is part of Springer Science+Business Media (www.springer.com)

# Editorial Advisory Board



## Reviewers

# Preface

Recent concerns regarding the environmental protection and sustainable development have resulted in there being a critical need for cleaner energy technologies. Some potential solutions have evolved including energy conservation through improved energy efficiency, reductions in the use of fossil fuels, and increases in the supply of environmental-friendly energy sources which has led to the use of intermittent renewable energy sources (RESs). These RESs are connected close to loads in the distribution network to reduce transmission losses and delay in the upgrade of transmission systems. The inclusion of renewable sources gives rise to a new set of problems which are due to the intermittency of the sources and the dynamics of interfacing equipments. Therefore, it is essential to investigate the potential challenges of renewable energy integration and to find out the effective and innovative solutions. This book includes different aspects of renewable energy integration—from the current trends of renewable energy integration to the current development of smart grids.

Chapter 1 of this book discusses the importance of green energy which is structured into two parts: (i) the available knowledge with regard to the general decision-making processes is described, followed by a critical perspective about today's decision making and (ii) a review of three enhanced approaches using Real Options Theory, Multi-Criteria Decision Analysis, and Multi-Criteria Cost Benefit Analysis, which are applied to RES decision making from the personal or investment point of view as well to the policy and the latter pan-European point of view.

Various aspects, such as classification and specifications of the grid codes, the anomalies that exist between the grid codes developed and standards used in conventional power plants are discussed in Chap. 2 and a fault-ride-through criteria by satisfying these grid codes are developed in Chap. 3 where the criteria is tested on New Zealand power systems.

Chapter 4 presents a voltage imbalance sensitivity analysis and stochastic evaluation based on Monte Carlo method carried out based on the ratings and locations of single-phase grid-connected rooftop PVs in a residential low voltage distribution network. On the other hand Chap. 5 includes comparative studies on the performance evaluation of grid-connected photovoltaic systems with different maximum power point tracking techniques.

One of the most important tasks in renewable energy integration is to determine optimal size and location of renewable energy sources which is discussed in Chap. 6 in which wind energy is considered as a renewable energy sources (RESs). After determining the optimal size and location, it is essential to investigate the characteristics of RESs and the steady state characteristics of wind energy conversion systems (WECSs) is presented in Chap. 7 from where it can be seen that WECSs affect the performance of power systems. A detailed study in which the effects of variable-speed wind generators to frequency regulation and oscillation damping is discussed elaborately in Chap. 8. The behaviors of power systems change with the penetration of RESs and Chap. 9 discusses some power management approaches for low and medium voltage distribution networks.

The negative impacts of RESs need to be minimized for stable and reliable system operation. Keeping this in mind, a new control methodology is proposed in Chaps. 10 and 11 which incorporates a review study on a new load, plug-in hybrid electric vehicles in power distribution networks. The coordination and aggregation of RESs during emergency conditions are discussed in Chaps. 12 and 13, respectively. Since the cost is an important issue for power system operation, this aspect of study for a residential application is presented in Chap. 14.

The latest trend in the area of renewable energy integration is the operation of power system in a smarter way. The operation of interconnected smart grids with self-healing capability is addressed in Chap. 15 and an agent-based scheme for smart-grid protection and security is presented in Chap. 16. In the last two chapters (Chaps. 15 and 17), the vulnerability analysis of complex smart grids is discussed from the cyber attacks and renewable energy integration points of view.

# Contents

# Chapter 1
# Green Energy and Technology: Choosing Among Alternatives

**Brian Azzopardi**

**Abstract** The primary renewable energy system (RES) investment decision-making criteria are economics. These criteria are focused on the RES and its support ancillary infrastructure technical superiorities, such as efficiency and cost, which is reasonable in the context of generous financial support schemes. However, when financial supports are phased out the energy market becomes technologically diversified environmental, political and social concerns, which include both quantitative as well as qualitative criteria, become significant. The technical superiorities may fail to describe RES or the relevant technology properly. This chapter is structured in two parts. Firstly, the available knowledge with regards to the general decision making processes is described, followed by a critical perspective about today's decision making. The second part presents a review of three enhanced approaches using Real Options Theory, Multi-Criteria Decision Analysis and Multi-Criteria Cost Benefit Analysis which are applied to RES decision making both from the personal or investment point of view as well to the policy and the latter pan-European point of view. Finally, the society challenges are discussed within this context.

**Keywords** Decision making · Renewable energy system (RES) · Real options (RO) theory · Multi-criteria analysis (MCA)

---

B. Azzopardi (✉)
Faculty of Electrical and Electronics, Department of Electrical Energy Systems and Renewable Energy Centre, Kaunas University of Technology (KTU), Studentų g. 48, LT-51367 Kaunas, Lithuania
e-mail: brian.azzopardi@ktu.lt

J. Hossain and A. Mahmud (eds.), *Renewable Energy Integration*, Green Energy and Technology, DOI: 10.1007/978-981-4585-27-9_1,

## 1.1 Introduction

Historically, the choice of energy has been economics and local conditions. Our society has been driven to choose inexpensive energy. However nowadays, the technical superiorities of energy systems may fail to describe for instance renewable energy systems (RESs) or its technology properly.

In this chapter, the RESs decision making process is examined with a number of analytical lenses that may not give priority to their technological superiorities. Although the real life decision-making process is far from the aspired process in this chapter, this chapter will provide the understanding of the complex multi-disciplinary decision making approach that today's experts and policy makers are faced to address the challenges that society will dictate in the future.

The large scale definition in this chapter is not limited to large system size magnitudes such as in Mega Watt equivalent but also considers the high deployment rates of micro-generation which when collected together may provide a large scale renewable power generation potential.

## 1.2 The Decision-Process Complexity

The decision process starts when there is more than one alternative to a solution. As will be described later on in the second part of the chapter, with the use of Real Options (RO) Theory decision-process kicks-off even between two simple alternatives do nothing or do something. There is no fixed framework or a single sequential approach how to achieve to the best decision.

Figure 1.1 tries to depict the complexity of choosing among alternatives.The first phase is the Problem Definition which is the crucial stage in decision-making. It is usually difficult or impossible to fully complete one component in the process without reflecting on the other components within a decision making process. This first phase groups the components into more malleable and therefore more realistic manner. This phase involve data processing that filters the relevant data and the feasible alternatives by their attributes leading to objective mapping. The criteria description helps the evaluation which is subjectively mapped to quantitative of qualitative attributes. While several components of the decision making process may be considered in parallel, there is no dictated or standardised procedure of what component comes before the other. Hence, Fig. 1.1 to some difficulty tries to draw the interrelationships between all components. It also includes a very important component in today's world, the perspective of the decision maker. Information and data can be gathered to understand more holistically the perspective of the decision maker being at political level, business level or individual customer level. Perspective change depending on the decision maker position and during the next phase due importance may be provided. This will lead to the most

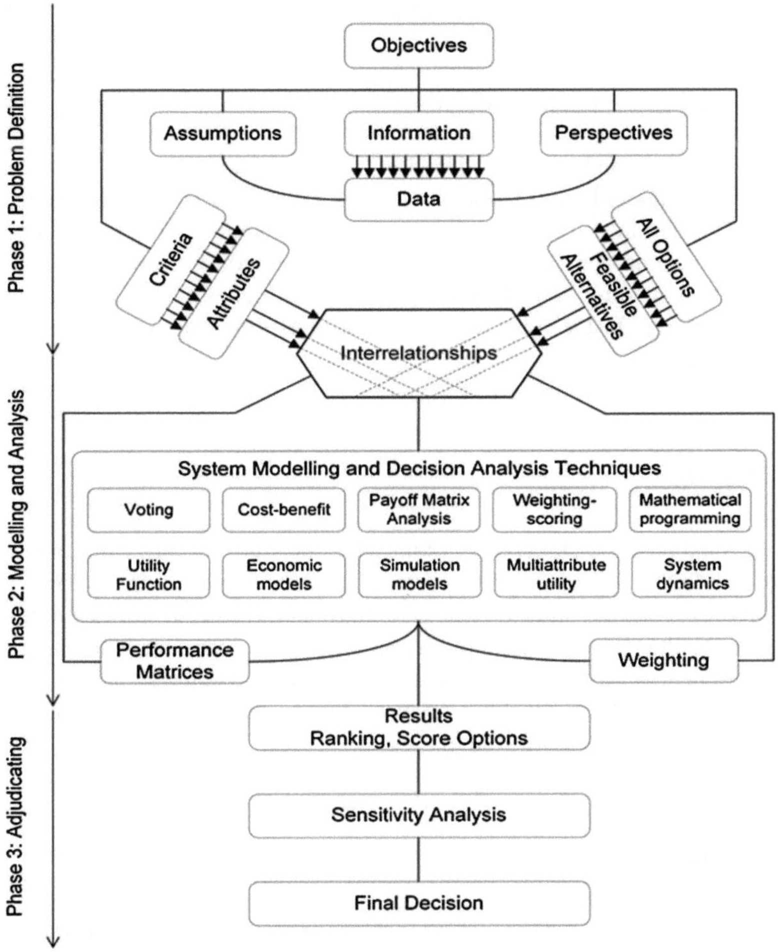


**Fig. 1.1** The complexity of choosing among alternatives

missing aspect in decision making which is usually referred to 'feedback' and will help finding the trade-off between different decision makers' perspectives.

The objective of a decision making process is the focus of the problem definition. The objective may have a multi-dimension perspective and several impacts. This is illustrated by the following example for decision making:

- To integrate nationwide renewable energy systems, an example using multi-criteria decision making technique is further illustrated in second part of this chapter.
- To inform stakeholders including policy, business and end-user on diversified renewable energy technology for example in case of photovoltaic (PV) technologies. An example is also illustrated in second part of the chapter.

- To evaluate production high throughput practices in producing renewable energy systems or components.

The second phase is the modelling and decision analysis. Figure 1.1 shows a wide range of techniques that may be used, while the importance of results can differ for decision makers. The decision analytics methods may be undertaken combined such as with simulation models which are fed to multi-attribute utility analysis. There is an extensive classic works on decision making techniques [1, 2] which goes beyond the scope for this chapter, including aspects related to the behavioural viewpoint in Kahneman et al. [3] which aspect has already been highlighted above. In the next subsections, a brief description is given to all the named techniques.

### *1.2.1 Voting*

In voting methods, the stakeholders that have the right to vote can voice up and choose democratically an alternative over another. However this allows for the possibility of political interference and may not result in complete unbiased solution to be the right choice. Voting can take many forms such as using simple voting system or preferential voting system. Voting technique may also inform decision makers to form the "big picture" that is the holistic approach and fed back to the decision support system.

### *1.2.2 Cost-Benefit*

Cost-Benefit Analysis (CBA) computes the "net present value" (NPV) which is usually monetary value based on one time snap-shot all the benefits and costs of a project, decision or policy. CBA has been widely used in RES projects to justify investment or compare projects and sometimes even coupled by other economic theories such as the RO Theory which will be further described in the second part of this chapter.

Related formal techniques include cost-effectiveness analysis, cost–utility analysis, economic impact analysis, fiscal impact analysis, and social return on investment (SROI) analysis. When Life Cycle Costing (LCC) is incorporated within CBA, the technique finds out the total cost of ownership. It is a structured methodology which deals with all the elements of this total cost of ownership. Hence an expenditure profile of a system over its anticipated life-span can be formed. The results of an LCC study can be employed in the decision-making process over a number of products or systems. The accuracy of LCC analysis diminishes, as the project's finances are more into the future. Hence, long term assumptions are preferred on all alternatives.

Investment appraisal may be assessed with varies LCC metrics. The NPV which is the basis for CBA, or worth present value, is the most common technique [4, 5], representing the investment wealth level. The NPV is calculated on net annual cumulative present value cash flows, that is annual inflows less annual outflows. The benefit of an investment is indicated by a positive NPV. All net annual cash flows are discounted over the lifetime of the investment. This incorporates the time preference, which reflects the investor's preference of having money today versus future revenues.

Similarly, the Internal Rate of Return (IRR) indicates the rate of return generated by the investment and is the discount rate by which the NPV equals zero. The selection over alternative investments is based on the highest IRR. IRR entails more complex calculations than NPV and does not always provide a single answer [5]. In fact, IRR does not provide an indication of NPV sensitivity to cost of capital.

Another LCC metric for investment appraisal is the Profitability Index (PI), which represents the present value of future cash flows generated by the project per unit of invested capital. The viability is indicated by PI greater than one.

Some other investment appraisal tools are Payback Time (PBT), defined as the period it takes for a project to recover cost outlays. Feasibility by payback period is predetermined by a period which is always significantly less than the project lifetime. Last but not least is the Annualised Life-Cycle Cost (ALCC) appraisal metric which averages upfront present value of the life cycle project cost over the investment lifetime [5].

The basis for CBA based on LCC was recently used to determine the cost of electricity from emerging plastic-based PV systems [6] and the impact and affordability of PV RES financial support schemes such as the feed-in-tariffs (FIT) in the UK [7].

### *1.2.3 Payoff Matrix Analysis*

Originating from Game Theory, a payoff matrix analysis tabulates the advantages and disadvantages of a choice. This includes the uncertainty of a possible outcome, however the technique is limited to a set of alternatives and outcomes under the decision maker's control. Payoff Table Analysis is widely used in the energy field such as in energy markets [8].

### *1.2.4 Weighing Scoring*

A less political dependent but powerful and flexible tool is the weighted-scoring method. These kind of methods compare and prioritise the attributes of alternatives against a fixed set of "must have" (high importance) to "no need" (low importance) criteria. It is a commonly used tool when we normally buy things to make an objective impression to our decision from a subjective process. The final score will rank the alternatives for example on submitted RES projects for a public tender, product selection by a bidder within a RES project tender call, risk analysis and design.

### 1.2.5 Mathematical Optimisation

Mathematical optimisation also known as mathematical programming technique is the selection of best available variables to an objective function within a domain defined by a number of equalities and inequalities constraints. The objective function, also described as cost function and indirect utility function in case of minimsation, utility function in case of maximisation and energy function incase of other fields, represent the decision maker's objective such as maximising profits, or minimising environmental impacts. In complex scenarios where there exits conflicting objectives, multi-objective optimisation provides trade-offs between the objective functions, and a possible compromised solution may be suggested. Optimisation may also be used in combination with modelling and simulation especially in large complex problems where uncertainty would require to analysis the system dynamically-such attempts are called stochastic optimisation. Optimisation has been used widely in scholarly studies such as in recent studies [9] and has been useful in many power systems scheduling examples. The methodology may become useful in the case for real-time optimisation for smart grid application. However this application would have to see novel mathematical solutions to achieve fast-response solutions as optimisation problems are usually iterative and require processing time and machine power to solve.

### 1.2.6 Utility Function

This analysis is the process for structuring a logical framework which may even include the use of other decision making methods and techniques, while informing the decision maker and other stakeholders. Graphical representations are commonly used in decision analysis such as influence diagrams and decision trees. Uncertainty may also be incorporated with the use of probability distributions. The objective is represented by the 'utility' functions maximisation, hence the name.

### 1.2.7 Economic Models

Complex models were developed based on 'free market' approach which basically is surrounded by the assumption that prices dictate choices and that the economy may be modeled as a system. The assumption leads that if there is any change in prices, a general equilibrium is reached and that self-interested individuals in a free market economy would collectively produce benefits for the society. However these models have been discussed in depth that assumptions taken may be too drastic and not real-life scenarios. Hence micro-economists learned that these models are not necessary stable. Later on in this chapter we shall discuss further theories to enhance such models.

### *1.2.8 Simulation Models*

This technique usually models a time-series historical collected data to assess the real life behavior of a physical model based on a soft prototype. Simulation modelling is an inexpensive process to predict the performance of engineering systems, including for example grid control with RES penetration or simply RES generation capacity for the next day. Uncertainties are usually addressed through a number of case scenario studies.

### *1.2.9 Multi-Attribute Utility Analysis*

Multi-Attribute Utility Analysis, also called Multi-Criteria Analysis (MCA) is an extension to the utility function technique in which multiple criteria and set of defined alternatives are evaluated in a hierarchical decision that allows a mathematical solution to be found. A family of methods has recently seen an increase interest in the energy field. Typical used methodologies such as SMART, PROMETHEE and ELECTRE outranking methods rank alternatives from best preferred to worse according to the decision makers preferences. In the second part of this chapter the ELECTRE III is applied on PV technology selection.

### *1.2.10 System Dynamics Modelling*

This technique models complex systems, which incorporate feedback loops and time delays that are present in a system and effects the holistic performance. This technique although might look as describing a simple system the influence on these mathematical formulation to represent the system dynamics may lead to a computational time and power demanding problem.

The third and final phase of a decision making process is adjudicating. Here results are plotted, listed or tabled, and a final decision can be taken after testing the robustness of the decision analysis framework through sensitivity analysis.

In the end, the choice for an alternative over another or the decision is yours. The general techniques, which have been basically developed in mid-twentieth century have been described above. Further details on recently enhanced methodologies to address the shortcomings of these techniques is described in the second part of the chapter.

## 1.3 How are Decisions Taken?

Unfortunately even though there exist multiple techniques to analyse the most suitable or optimal or feasible solution, it is still early days that complex techniques are truly being exploited. As individual concerns for example we face decisions on a daily basis that would require time and money to analyse each decision properly, even though some techniques discussed are quire straightforward these techniques are unlikely being used on individual basis.

On the political aspect, politicians would like to keep protecting re-election chances. While the use of regulators which may balance these political interference situations in many cases in EU regulators, sometimes the regulators themselves have been over passed by the political will of politicians.

On the other hand decision analysis is important in business and planning. However at the end these dimensions are still at risk due to any political interference which may change the scenario completely within especially an emerging market. A case in point is the generous FIT support schemes that emerged in the EU during the last decades. While initially, the purpose of these tariffs has been achieved by early adaptors that is to increase RES deployment these schemes are nowadays obsolete. Many companies that invested a lot in research and development in new promising technologies, such as in the case of PV technologies, most often even supported by government funds, had to file bankruptcy in view that mature technology dominated the market with the old engineering decision analysis that is biased towards cost and efficiency. This basically let to downgrading innovation.

## 1.4 Engineering Decision Analysis

Engineering decision making is based on the same components highlighted in Fig. 1.1 and Sect. 1.2, plus one final important phase, phase four, that is the post audit of results. The post audit of results is a verification stage that decisions taken had the right course of action and if required amendments may be considered. This verification process is performed in some cases at regular intervals such as in the manufacturing process of products for RES or even in real-time such as for grid monitoring.

Engineering decision analysis has often focused on technical superiorities such as costs and efficiency. Hence the economic decision analysis using criteria such as

NPV, IRR and PBT has dominated the decision approach within the industry in general for many decades.

However the social responsibility adopted by major entities and the 'greeness' criteria that reflect the impact or benefits one makes towards the environmental criteria are taking shape within decision making frameworks. In addition qualitative criteria which most often are difficult to quantify cannot be missed out either. These added complexity to decision making will be described more in the next section.

## 1.5 Decision Support Systems

This second part of the chapter provides a brief overview of selected three approaches that apply a combination of techniques described in Sect. 1.2, and further developed more robust frameworks for decision support systems to choose among a number of green energy and/or technology solutions.

The first is using the enhanced economic theory of RO Thoery used in recent studies exploring the investment decision on PV technologies and systems [10]. Although the work was when the FIT in UK was not yet tracked with an IRR, the framework provides a degree of understanding on the uncertainty, which can be used in many scenarios worldwide as well as on many different technologies and energy systems. RO Theory has recently increased its popularity within academia on energy systems. A comprehensive review is given in Martinez-Cesena [11].

The second approach uses Weighed in system planning: ENTSO-E's cost/benefit method, a new European method developed by the European Network of Transmission System Operators for Electricity (ENTSO-E) [12]. The methodology falls under the Ten Year Network Development Projects (TYNDP) and includes network and market modelling for a harmonised energy system-wide cost-benefit analysis at Union-wide level for projects of common interest.

The third approach is a complex decision support system that integrates the technical, economic and environment qualitative and quantitative criteria to help with a diversified technology decision choice. A number of quantitative criteria are pre-calculated through simulation and modelling. This example also contains different perspective of the decision maker being political, business or end-customer.

### *1.5.1 The Real Options (RO) Theory Approach*

As discussed earlier, the failure and dissatisfaction in economic models using discounted cash flow methods (DCF) [13], and the introduction of the Black–Scholes equation [14] gave rise to the creation of RO theory. In fact what RO theory does is enhance DCF methods to better address uncertainty for example with technology development and derive its options.

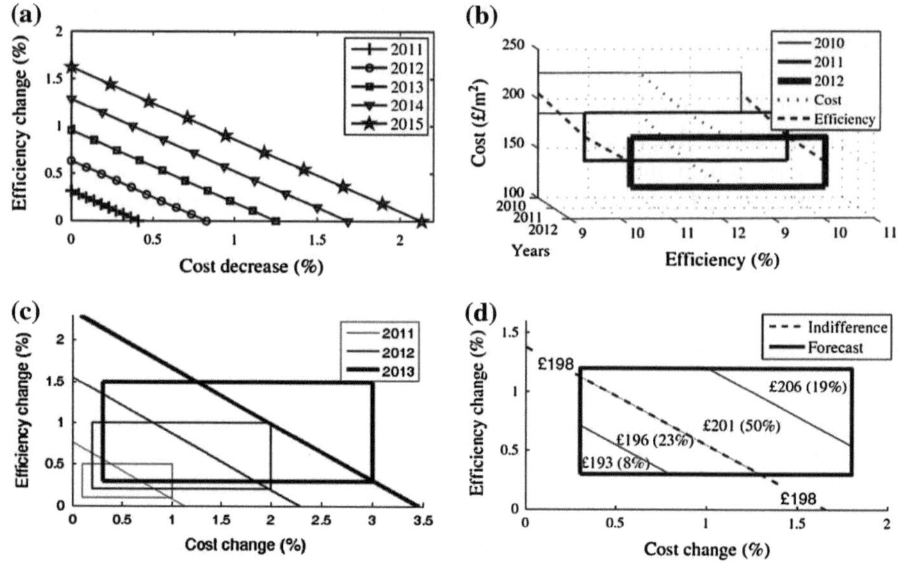


**Fig. 1.2** **a** Mono-Si PV system indifference *curves*, **b** CdTe PV system efficiency and cost forecasts, **c** a-Si PV system forecasts indifference, and **d** CIGS PV system delay assessment example

RES projects and their associated investment projects from manufacturing to site installations possess 'real' options that if identified and implemented may increase the investment worth [15]. These 'real' options may be defined as projects adjustments that the decision maker seem fit in response to the evolution of uncertainty such as delaying or rushing investment and building additional infrastructure [16].

Today, the RO theory has been applied widely to RES, and a comprehensive literature review is available [11]. The applications have also differed from assessing investment time [17], examining the investment indifference [18] and addressing uncertainty in the LCC of RES systems due to the Emissions Trading market [19].

In this study, RO theory was based on indifference, and cost and efficiency forecasts [20] for a diversified range of PV technologies to assess investment, while considering solely the option for delay.

The indifference within this developed RO theory model was based on the assumption that PV module prices will not increase and efficiencies will not decrease in the future for a particular technology. However if the price for PV module is higher, efficiency increase is expected to achieve the indifference. The normalised result of this process is shown in Fig. 1.2a showing the combinations of Mono-Si PV type modules' cost reduction and efficiency increase.

Meanwhile the forecasted technology development is based on the targeted efficiency values [20], while those for costs are based on PV module predictions

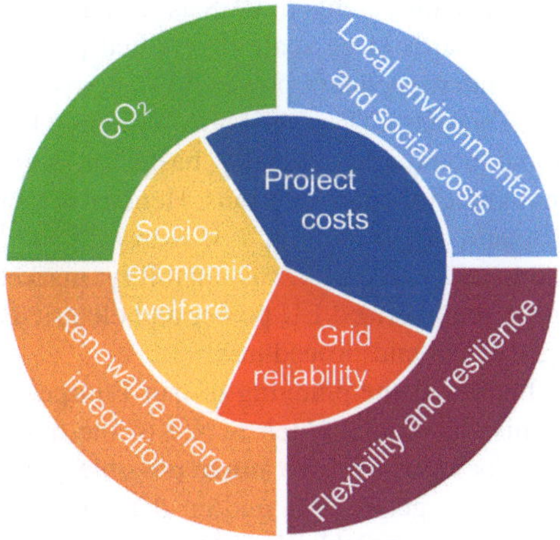


*(Author's Compilation)*

**Fig. 1.3** The scope of the multi-criteria cost benefit analysis [22]

[21] and experience curves. Figure 1.2b shows the forecast for CdTe module costs based on available efficiency predictions and learning rates of 15 and 22 %.

These indifference and forecast curves are then normalised by their current average values and contrasted on top of each other as shown in Fig. 1.2c for a-Si PV and (d) CIS PV systems delay assessment examples. Further from the origin means higher efficiency improvements and cost reductions while the area above the indifference curve represent a convenient delay, below is vice versa.

In summary this approach has simplified the assessment of the economic convenience of delaying projects such as in this case investments in PV technology within a domestic grid-connected PV system. It has been noticed that the option to delay investment is mainly affected by the time-value of money, type of loan and technology forecasts. An interesting aspect is that a delay option may become a driver for emerging technologies exhibiting inferior technical superiorities over mature technologies to penetrate the market.

### *1.5.2 Weighed in System Planning Approach*

The EU Regulation 347/2013 requires the European Network of Transmission System Operators for Electricity (ENTSO-E) to establish a methodology, including on network and market modelling, for a harmonised energy system-wide cost-benefit analysis at Union-wide level for projects of common interest [12]. The developed framework is a Multi-Criteria Cost Benefit Analysis for candidate projects (feasible alternatives) of common trans-European interest and projects which fall within the scope of Ten Year Network Development Plan (TYNDP).

Fast developments and deployments in RES as well as the EU electricity market liberisation has moved towards large scale diversified power generation portfolio offering more and more interdependent power flows across Europe, with large and correlated variations. Therefore, Transmission System Operators (TSOs) must look beyond traditional boundaries, usually based on national conditions, and move towards regional and European solutions. Hence the scope of developing this Multi-Criteria Cost Benefit Analysis, Fig. 1.3, with mid and long term objectives which addresses system safety, security, sustainability, market access and increase energy efficiency within the liberised EU power and electricity market regulations, policies, national legislations and procedures.

The methodology facilitates the planning of pan-European transmission system and assess new transmission assets which is one of the many number of feasible solutions. However it does not exclude other possible alternatives in particular energy storage, as well as generation and/or demand side management. The quantitative assessment is performed both by translating the inner ring attributes in Fig. 1.3 to a monitised value in € for example as well as through the measured physical units such as $kgCO_2$-eq or kWh for the outer ring attributes in $CO_2$ emissions reduction would have benefits on production cost savings given an adequate carbon price. Considering emissions as a separate attribute and an adequate carbon price may lead to double counting. Similarly, integration of renewable energy would have benefits on production cost savings. On the other hand, local and environmental costs contribute to the infrastructure costs. Such impacts includes the impact on human beings, on the local fauna and flora, on material assets, and on cultural heritage, as described in the Environmental Impact Assessment (Directive 85/337/EEC) [22].

This set of common European-wide indicators will form a complete and solid basis, both for project evaluation within the TYNDP, and coherent project portfolio development for the common interest selection process. Initially, the planning process consists of the definition of scenarios, which represent a coherent, comprehensive and internally consistent description of a plausible future. The aim of scenario analysis is to depict uncertainties on future system developments on both the production and demand sides. In order to incorporate these uncertainties in the planning process, a number of planning cases are built, taking into account forecasted future demand level and location, dispatch and location of generating units, power exchange patterns, as well as planned transmission assets.

### *1.5.3 Multi-Criteria Analysis (MCA) Approach*

During the past decade, decision support systems using multi-criteria analysis (MCA) models and techniques has been widely applied for energy systems in a multitude scenario on political and national level to technology specific and manufacturing processes. MCA techniques such as utility theory, hierarchy process, weighted sum and others are available in literature [23, 24].

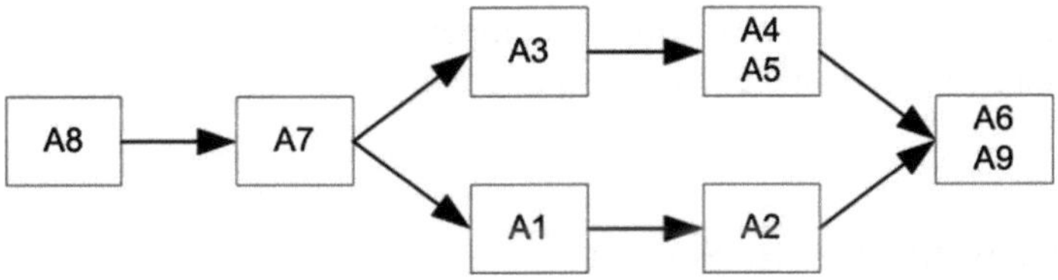


**Fig. 1.4** Final ranking under all three perspectives with 0.80 PR (Performance Ratio), where *A1* refer to mono-Si, *A2* refer to multi-Si, *A3* refer to a-Si, *A4* refer to CIS, *A5* refer to CdTe and *A6*, *A7*, *A8* and *A9* refers to stable/fixed, reference, optimistic and pessimistic technology developments respectively in organic-based PV (OPV) technologies with respect to efficiency, lifetime and price

In this approach, the outranking ELECTRE III method was chosen as it makes use of the discordance concept and do not hold structural properties in outranking relations. In addition the ELECTRE III has other superior features of partially having non-compensatory treatment of the problem and proportional thresholds for imprecise data from other outranking methods such as SMART and PROMETHEE, which may turn out to be a difficult task [25–27].

The purpose of this study was to elaborate and demonstrate the first multi-criteria decision support tool for an investment in PV systems [28]. Currently the PV micro-generation market, which is the largest PV market, is dominated by mainly five technologies namely mono-crystalline silicon (mono-Si), multi-cyrstalline silicon (multi-Si), amorphous silicon (a-Si), copper indium gallium diselenide (CIGS) and cadmium telluride (CdTe). Additionally, emerging polymeric organic-based PV (OPV) technologies have the potential to enter this market [6].

In the future, the use of PV technologies will have to be diversified to ensure a sustainable PV energy supply to the market. This implies that the overall assessment and selection of PV technologies will have to consider several criteria to address the particular technical, environmental, and economic factors associated to a wide range of available PV technologies [29].

The application of MCA can combine the various viewpoints into a standardised evaluation procedure. Figure 1.4 illustrates the final ranking considering all technical, environmental and economic quantitative and qualitative criteria such as PV contribution to local load, Module Design, $CO_2$ impact, aesthetic aspects, NPV and technology maturity. Of interest is the result in the ranking of crystalline silicon based technology alternatives (A1 and A2) and thin film technologies (a-Si A3, CIGS A4 and CdTe A5) which illustrates their incomparability.

In summary the study demonstrated that reduction in cost and enhanced technology developments are required for preferential ranking. Today the thin-film (TF) PV technology and the crystalline (c-Si) PV technology have a competitive edge in the market, and this study shows certain insuperabilities. In this context, crystalline technologies are more expensive. However, this PV technology offers a much better $CO_2$ benefit than emerging technologies due to high efficiency levels. MCA study as demonstrated in this study can be used as a preferential indicator for any PV technology integration within a competitive market under three categories namely technical, environmental and economic.

## 1.6 The Challenges to Society

Several international cooperative programme frameworks, protocols and summits have taken shape to address global concerns such as population growth, greenhouse gas emissions and climate change. These frameworks such as the Framework Convention on Climate Change (FCCC) and the Kyoto Protocol led by the United Nations are a first step to enhance the conservative engineering decision making focused on purely efficiency and cost economic analysis and provide achievable targets to enhance the future sustainability of the planet and the current living population.

The energy debate is open but decision making to the society is only reserved to the big players. Better access to the energy markets by consumer will provide more stability and improve the functionality of markets, as well as their transparency and information on the demand and supply side. This includes the removal of harmful subsidies both for long standing support to fossil fuels and the generous financial supports for renewable that distorts the markets.

On the other hand energy poverty needs to be elevated and improve access to reliable, affordable, economically viable, social acceptable and environmentally sound energy services and resources. In this chapter we have seen that not all criteria are quantitative, some are qualitative ones which brings about the problem of measuring their attributes. Using decision making techniques can help also provide more knowledgeable decisions and may act as a key to resolve syndromes such as the NIMBY (Not In My Back Yard) or BANANA (Build Absolutely Nothing Anywhere Near Anything) ones.

## 1.7 Summary

This chapter has assessed the decision making of technologies in RES. The defined large scale renewable power generation in this chapter is not limited to large size system magnitudes of Mega Watt equivalent but also considers the high deployment rates of micro-generation which when collected together may provide a large scale renewable power generation potential. Therefore the chapter has considered key decision making techniques across a range of applications in industry from manufacturing RES components or products for large scale generation, operations decision within an energy grid, investment decisions as well as policy guidance.

We took off with the thesis that the technical superiorities of efficiency and cost may not describe well all the alternatives of green energy and technology and therefore we need to move forward from the traditional engineering decision making. Unfortunately, today most real-life decision making within traditional industries such as the utilities are still based on conservative old style of thinking. In addition traditional decision making techniques may have also failed to take into account the likely technology developments.

A case in point is the lack of uncomfortable use of net demand planning that is considering also all distributed and intermittent generating sources within the demand profiles, over conservative demand planning that is considering only dispatched generation sources regulated by the utilities. In addition there exists conflicts in the context of generous financial support schemes or even to some extent support of fossil fuels, mostly in terms of subsidies, which is leading to dominant mature players take up the market while emerging technologies end up investing but hard to integrate.

The chapter therefore described three main enhanced approaches:

- RO Theory, where the technology developments and the scenario changes are taken into consideration for an investment decision in case of RES for instance-making a wiser and knowledgeable investor decision,
- Weighed in system planning: the Multi-Criteria Cost Benefit Approach being developed by ENSTO-E for transmission system planning, and
- Multi-Criteria Decision Analysis, where multiple criteria and several alternatives may be brought together under many different techniques, even in some cases considering or combining both quantitative and qualitative matrices to help with decision support under a respective perspective.

Beyond the economic costs and benefits green energy and technology solutions offer, there are other considerations to be taken into account when choosing truly from alternatives. Here we tried to highlight some of these criteria which may stand well, but not limited, under one of these four streams technical, environment, economic and social.

All energy systems and their technologies have both positive and negative impacts under these streams be it termed 'Green' or 'Dark' energy. The task that engineers, politicians and basically human kind face as we transit to a sustainable energy future is to weigh up all relevant criteria and decide more transparent and structurally correct way.

## References

1. Raiffa H (1968) Decision analysis: introductory lectures on choices under uncertainty. McGraw-Hill, New York
2. Keeney RL (1992) Value-focused Thinking: a path to creative decisionmaking. Harvard University Press, Cambridge, Massachusetts
3. Kahneman D, Slovic P, Tversky A (1982) Judgment under uncertainty: heuristics and biases. Cambridge University Press, Cambridge
4. Lumby S (1988) Investment appraisal and financing decisions. International Thomson Business Press, London

5. McLaney E (2009) Business finance: theory and practice. Pearson Education, Essex
6. Azzopardi B, Emmott CJM, Urbina A, Krebs FC, Mutale J, Nelson J (2011) Economic assessment of solar electricity production from organic-based photovoltaic modules in a domestic environment. Energ Environ Sci 4(10):3741
7. Candelise C (2009) Technical and regulatory developments needed to foster gridconnected photovoltaic (PV) within the UK electrcity sector. Imperial College London, London
8. Ferrero RW, Shahidehpour SM, Ramesh VC (1997) Transaction analysis in deregulated power systems using game theory. IEEE Trans Power Syst 12(3):1340–1347
9. Azzopardi B, Mutale J (2009) Optimal integration of grid connected PV systems using emerging technologies. In: 24th European photovoltaic solar energy conference. Hamburg, pp 3161–3166
10. Martinez-Cesena EA, Azzopardi B, Mutale J (2013) Assessment of domestic photovoltaic systems based on real options theory. Prog Photovoltaics Res Appl 21(2):250–262
11. Martinez-Cesena EA (2012) Real options theory applied to renewable energy generation projects planning. University of Manchester
12. ENTSO-E (2013) ENTSO-E guideline for cost benefit analysis of grid development projects, European Network of Transmission System Operators for Electricity (ENTSO-E), Belgium
13. Hastie KL (1974) One businessman's view of capital budgeting. Financ Manag 3(4):36–44
14. Black F, Scholes M (1973) The pricing of options and corporate liabilities. J Polit Econ 81(3):637–654
15. Dixit RK, Pindyck RS (2012) Investment under uncertainty. Princeton University Press, Princeton, New Jersey
16. Trigeorgis L (1996) Real options: managerial flexibility and strategy in resource allocation. MIT Press, Cambridge, Massachusetts
17. Hoff TE, Margolis R, Herig C (2003) A simple method for consumers to address uncertainty when purchasing photovoltaics, S Consulting, cleanpower.com
18. Clean Power Estimator®—Clean Power Research. Available http://www.cleanpower.com/products/clean-power-estimator/. Accessed 08 Aug 2013
19. Sarkis J, Tamarkin M (2008) Real options analysis for renewable energy technologies in a GHG emissions trading environment. In: Antes R, Hansjürgens B, Letmathe P (eds) Emissions trading. Springer, New York, pp 103–119
20. Frankl P, Nowak S, Gutschner M, Gnos S, Rinke T (2010) Technology roadmap: solar photovoltaic energy. France, International Energy Agency (IEA)
21. Hoffman SW, Pietruszko W, Viaud M (2004) Towards an effective european industrial policy for PV solar electricity. In: Presented at the 19th European photovoltaic science and engineering conference and exhibition, Paris
22. Home—THINK—European University Institute (2013) Available http://www.eui.eu/Projects/THINK/Home.aspx. Accessed 08 Aug 2013
23. Bana e Costa CA (1990) Readings in multiple criteria decision aid. Springer, Heidelberg
24. Bragge J, Korhonen P, Wallenius H, Wallenius J (2010) Bibliometric analysis of multiple criteria decision making/multiattribute utility theory. In: Ehrgott M, Naujoks B, Stewart TJ, Wallenius J (eds) Multiple criteria decision making for sustainable energy and transportation systems, vol 634. Springer, Berlin, pp 259–268
25. Bouyssou D (1996) Outranking relations: do they have special properties? J Multi-Criteria Decis Anal 5(2):99–111
26. Simpson L (1996) Do decision makers know what they prefer?: MAVT and ELECTRE II. J Oper Res Soc 47(7):919–929
27. Salminen P, Hokkanen J, Lahdelma R (1998) Comparing multicriteria methods in the context of environmental problems. Eur J Oper Res 104(3):485–496
28. Azzopardi B, Martínez-Ceseña E-A, Mutale J (2013) Decision support systems for ranking photovoltaic technologies. IET Renew Power Gener 7(6):669–679. doi:10.1049/iet-rpg.2012.0174
29. Azzopardi B (2010) Integration of hybrid organic-based solar cells for micro-generation. PhD, The University of Manchester, Manchester

# Chapter 2
# Grid Codes: Goals and Challenges

**Pradeep Kumar and Asheesh K. Singh**

**Abstract** Electric power has always been one of the driving forces for progress in human life. This has popularized electric energy as the most utilized form of energy. However, dispersed locations of energy resources and continuously increasing demand of electricity have led to a large electric power transmission network across the landscape. To operate the system effectively, a large number of components, such as protection systems, monitoring systems, operational procedures, etc. are required to work in a synchronized and efficient manner; otherwise, contingencies may arise in the system. Development and integration of renewable energy sources into the existing power system has enhanced the complexity in the network. Efficient operation of this complex network is a tedious task for the authorities. Thus, to simplify the planning, operational, and other tasks, grid codes have been developed. Grid codes are the rules laid by the authorities for all its stakeholders, i.e., the users and power generating stations for connecting to the network and operate as per the standards. These grid codes implement the regulations for smooth operation of the grid and its connected components. It implies to the existing and future plants. This chapter gives an overview of the grid codes, its various components and their development considering integration of renewable energy into the grid. Various aspects, such as classification and specifications of the grid codes, the anomalies that exist between the grid codes developed and standards used in conventional power plants are discussed in this chapter.

**Keywords** Grid codes · Renewable energy sources · Thermal power plants · Nuclear power plants

P. Kumar (✉) · A. K. Singh
Electrical Engineering Department,
Motilal Nehru National Institute of Technology Allahabad, Uttar Pradesh 211004, India
e-mail: pradeepkumar@ieee.org

A. K. Singh
e-mail: asheesh@mnnit.ac.in

J. Hossain and A. Mahmud (eds.), *Renewable Energy Integration*,
Green Energy and Technology, DOI: 10.1007/978-981-4585-27-9_2,

## 2.1 Introduction

Electric power grid is the largest man-made system in the world. A variety of different components, such as synchronous generators, transmission lines, switches, relays, active/reactive compensators, and controllers etc., are the main constituents of this system [1]. Owing to its complex nature, proper planning, and design become essential for the operation of the power system. Traditionally, a single authorised-body designs and plans the type, location, capacity of new power stations and their connection requirements to the grid.

Recent trend of penetration of the renewable energy sources to the grid has further enhanced the complexity of the network. Due to their stochastic nature, their integration has added more uncertainty to the grid, leading to increased concerns related to the accurate prediction of generation and control of power flow [2]. In addition, the deregulation (privatisation and liberalisation) in the electricity sector have invited more generation and supplier entities to meet up the ever-growing demand of electricity [3]. This addition of the new generation schemes along with a large number of competitors has led to following new challenges for the proper planning and operation of the present and future power systems [4]:

- Increased level of complexity,
- Stochastic nature of power transfer capabilities, and
- Bidirectional power flow across the system.

Inclusion of these renewable energy sources and stringent requirements by the sensitive loads has raised serious concerns related to the security, reliability, stability and efficiency, viz., adequate dimensioning, establishment, connection, and operation [5]. To safeguard, the electric power systems, against failures and address the issues raised, rules and regulations are required. These rules act as standard procedures and requirements for including or prohibiting connection of the generation plants and loads to the grid. The rules should be applicable to the both, new and existing, generation plants and users, who are interested to connect to the grid. Grid codes are attempts in this direction to ensure supply quality for the consumers.

Grid codes, also known as the 'interconnecting guidelines', are the instructions, which specify technical and operational characteristic requirements of power plants and different parties involved in the production, transportation and utilization of electric power. In other words, these are the technical requirements to interconnect new generating plants to the local or bulk systems [3, 6, 7]. With the generation deregulation in USA in early 1990s, these guidelines started appearing in USA and other countries [7]. These rules are laid by experts in the area of transmission operation and control from the very early time. As transmission acts as the link between the generation and distribution, handling bulk of the power, grid codes give prime importance to it. The main goal for development of these codes was to formalize the criteria, earlier used by the predecessor utility organizations, to specify the details of the power generation equipment and connection

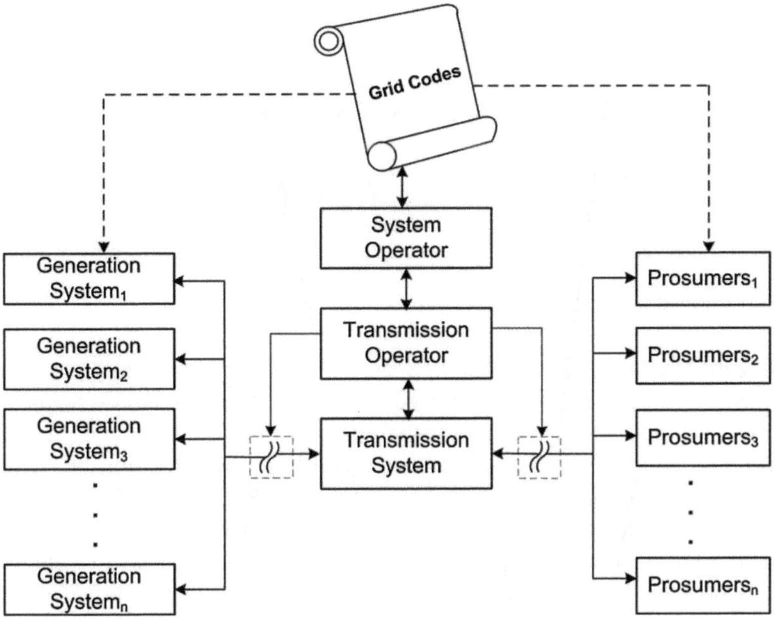


**Fig. 2.1** Diagram showing the importance of grid codes in power system [9]

requirements for their connection to transmission system. Also, it attempts to establish and maintain compliance of the new generation plants, with recognized industry standards, except in some cases, where some sites requires stringent standards [7]. As formation of these codes is controlled by the local regulatory authorities, under different legal and technical environment, the requirements and specifications mentioned in the document may vary. The reason for variation in the grid code may be due to different system types or grid characteristics in different countries [3].

The importance of the grid codes in any power system can be easily explained with the help of Fig. 2.1. It depicts that grid codes act as a standard document for all the entities interacting in power systems. It is the responsibility of the system operator to check whether the codes are being obeyed at every level or not. The next component is the transmission operators, who have direct control over the transmission system. Activities such as increasing awareness regarding following the codes and taking necessary steps for not obeying it are the duties of the transmission operator. Based on the compliance report provided by the operators, the generating station or prosumer is permitted to make a connection with the transmission system. The term 'prosumer' is referred in the context of the smart grids, as in smart grids, the consumer is likely to *pro*duce as well as consume electricity [8].

Although, grid codes may appear as a simple document, they address all significant concerns related to the grid. This chapter attempts to discuss the main

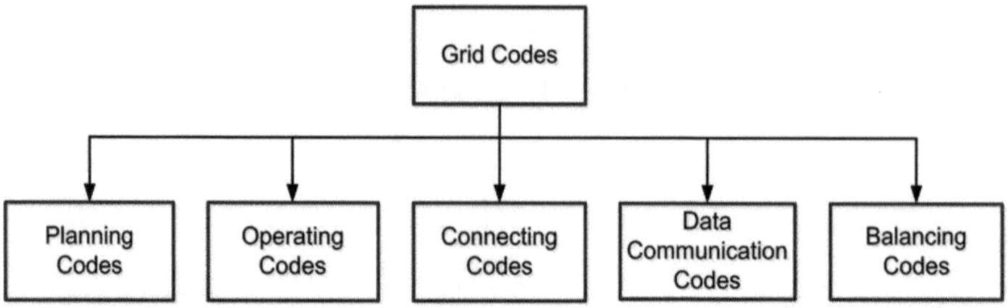


**Fig. 2.2** Classification of the grid codes [6]

areas covered under the grid codes. In Sect. 2.2, classification of the grid codes and their various components are discussed in detail. Relevant issues regarding the frequency response, reactive power capabilities, safety, security and efficiency of the grid are addressed. The challenges regarding incorporation of renewable energy sources are discussed in Sect. 2.3. Section 2.4 presents existing grid codes, and the trend for development of grid codes and certification procedure for the renewable energy sources, especially wind power generation and its integration, etc. In Sect. 2.5, the issues of compliance between the conventional generation schemes and grid codes are discussed. The discussions made in this chapter are based on the following grid codes:

(1) USA: Interconnection for Wind Energy [10]
(2) India: Indian Electricity Grid Code [11]
(3) Great Britain : The Grid Code [6].

However, in general, it is based on the inferences drawn from [6], being a recent grid code, and clearly representing the requirements and classifications.

## 2.2 Specifications of the Grid Codes

Grid Codes are instructions that describe the technical and operational characteristic requirements of the power plants. These codes are the guidelines, to be followed by any user from, installing new power plants or making any alteration in the existing power plants to the, connection of the different components of the power system to the grid. Since, these instructions involve a variety of components to be dealt with the grid codes; they are classified into different components [7], as shown in Fig. 2.2. This classification simplifies the understanding and requirements for implementation and development of the new grid codes.

Amongst the various components, planning codes (PC) deal with dimensioning, planning, and development, etc. of the power plant and other equipments that are, required to be connected between the user/generation plant and the transmission system. The connecting code (CC) discusses the connection requirements and conditions to remain connected to the grid. The operational requirements of various equipments in the power systems are dealt by the operating codes (OC). Data communication codes (DCC) deals with the requirement for data storage, amount of data storage and quantities to be stored. Balancing codes (BC) discuss the steps taken by authorities, to maintain the power balance between load and generation. Detailed analysis on these different classes of grid codes is presented in the following sub-sections.

### *2.2.1 Planning Codes (PC)*

Planning codes (PC) are the vital component of the grid codes. These codes are implemented at the planning or modifying stages of power plants. It may involve substation or connection site, transmission lines or other facilities connecting the connection site to remainder of the transmission network. It discusses the technical specifications and procedures for planning of the system. At initial stages of the project, transmission operator checks abidance of the PC by user. It may accept or decline the project, depending upon the fulfilment of the grid code criteria. In most cases, the codes are specified either as per the established standard, or in form of a bilateral agreement, between the authorities.

As per the grid code [6], details (generally required by the authorities at initial stages) can be given as:

- Description of the plant or apparatus being installed or modified,
- Standard planning data,
- Desired completion date of the proposed development, and
- Connection entry capacity and transmission entry capacity.

At initial stages, all the information exists in form of data. Utilizing this, data interaction between transmission operators and the power generation plant takes place. However, for lucid understanding and uniformity, the data should be in the format, as prescribed in the PC. Only then, a bilateral agreement between the generation plant and operator may take place. The offer is made for a fixed period of time, which may vary as per the licence standard (varies from country to country). The time for development may vary depending upon magnitude, complexity, nature, and location of the transmission system project.

The objectives of the PC can be summarized as:

(a) Promotion of interaction between different entities involved in the transmission system for discussion on proposals affecting, directly or indirectly, and performance of the transmission system.
(b) Collection of information for planning and development in relevance to the standards and existing facilities.
(c) To increase awareness about licensing standards in planning and development.

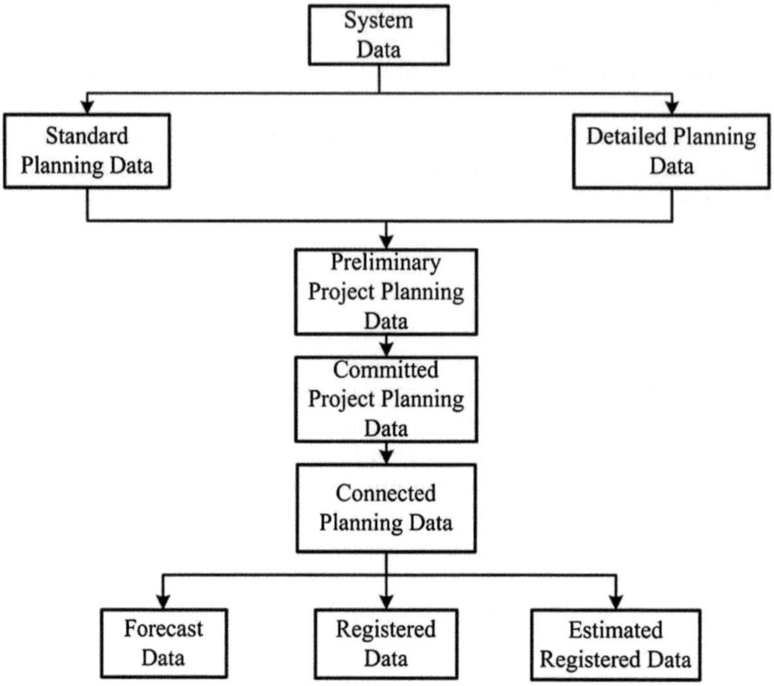


**Fig. 2.3** Classification of data under planning codes [6]

(d) Spreading knowledge about various duties, under the licences, for proper operation of transmission system.

To understand PC, the data requirement is divided into two parts, namely [6]:

(a) Standard planning data, and
(b) Detailed planning data.

Standard planning data is general-purpose data, required at initial stages of the project for, having information about the planning stages. Thus, some data may slightly change due to implementation issues. On the other hand, detailed planning data is specific data, satisfying a particular requirement of the PC. This type of data is required at advanced stages on the normal time scale, to enable transmission operation. Figure 2.3 shows further classification of these two forms of data, as following:

- *Preliminary Project Planning Data*: It is required at initial stages of the project, involving quantities like, plant name, details, ratings, etc. Due to its initial stage requirements, it is termed as preliminary data.
- *Committed Project Planning Data*: After planning stages, the data required, is termed as committed project planning data. Generally, it involves the updated and new data, obtained after initialization of the project.

**Table 2.1** Specifications of different components represented in SLD

| Components | Specifications |
|---|---|
| Transmission circuitry | Rated voltage (kV), operating voltage (kV), positive phase sequence reactance, positive phase sequence resistance, positive phase sequence susceptance, and zero sequence impedance, both self and mutual |
| Transformers | Rated voltage (kV), voltage ratio, winding arrangement, positive sequence reactance (maximum, minimum and nominal tap), positive sequence resistance (maximum, minimum and nominal tap), and zero sequence reactance |
| Interconnecting transformers | Tap changer range, tap change step size, tap changer type (on load or off load), earthing method, and direct resistance and reactance and Impedance (if not directly earthed) |
| Reactive compensation | Connection node, voltage rating, power loss, tap range, connection arrangement, mathematical representation in block diagram format to control any dynamic compensation plant, HV node, LV node, control node, nominal voltage (kV), target voltage (kV), voltage dependent Q limit, and normal running mode |

- *Connected Planning Data*: After initialization of the project, network operator updates the value estimations for planning purpose with actual values after reasonable practicing.

The connected planning data is further divided into three types namely:

(i) *Forecast Data*: Those items of standard planning data and detailed planning data which will always be forecasted are known as forecast data.
(ii) *Registered Data*: Standard planning data and detailed planning data, which become fixed upon connection, are termed as registered data.
(iii) *Estimated Registered Data*: Data estimated for a particular term are termed as the estimated registered data.

It is necessary to specify the dates for these data [6].

Standard planning data that is, required at initial stages of the project, include the following components:

- Single line diagram (*SLD*),
- Current carrying capacity of the apparatus, and
- Connection point.

(a) *Single line diagram* (*SLD*): For a particular portion of transmission system, an SLD should clearly depict its circuitry. General information, such as, substation names, circuit breakers, phasing arrangements, rated voltage (kV), operating voltage (kV) etc. should be specified. However, the amount of data required under various categories may vary from country to country. The other various specifications needed to be specified in the transmission system are summarized in Table 2.1 [6].

**Table 2.2** Specifications of different components in SLD at connection point

| Category | Description |
|---|---|
| Short-circuit contribution | Type of connection (direct or indirect) and bus bar arrangement |
| Generator | Registered capacity, output usable, system constrained capacity, and minimum generation |
| Generation performance chart | Specified in the charts |
| Rated parameter data | Ratings of different components connected |
| For all generating units | Rated MVA, rated MW |
| Synchronous generating units | Short circuit capacity, direct axis reactance, inertia constant—MWs/MVA; |
| Transformer units | Step up transformer, rated MVA, positive sequence reactance (at maximum, minimum and nominal tap) |
| Forecasted data | Peak day on each of the user's user system, day of peak (date and time), day of minimum (date and time) |
| Unit control | Maximum droop (%), normal droop (%), minimum droop (%), maximum frequency deadband(±Hz), minimum frequency deadband(±Hz), maximum output deadband(±Hz), normal output deadband(±MW), minimum output deadband(±MW) |
| Frequency setting of unit droop controller | Maximum Hz, normal Hz, and minimum Hz |

(b) *Load/current carrying capacity of the apparatus*: Under abnormal operating conditions, the different components connected to the transmission line may have to bear the extra amount of current. Therefore, to protect the components, it is necessary to specify the current carrying capacity of the equipments. The current carrying capacity of the components is an indication of the short-circuit capacity of the planned facility. Thus, it may be specified either as short-circuit capacity or Ampere ratings.

(c) *Connection point*: The connection point is decided according to the connecting capacity of the transmission system. Based on this capacity, the location for plant connection is decided.

In detailed planning data, the data related to the operations, viz. usable capacity during the operation, operating range of the connected equipment, etc. of the grid is provided. Other different components, usually specified under the detailed planning data, are shown in Table 2.2 [6].

### *2.2.2 Connection Codes (CC)*

Completion of plant as per the planning codes does not ensure the eligibility of the plant/user for connection to the grid. To connect to the grid, it is essential for the plant authorities to obtain clearance as per the connection code (CC). These codes specify the minimum technical, design and operational requirements, for the plant to obey and connect to the connection site/interface point prior and start operating. These requirements are to be compiled and checked by the transmission operator under the supervision of the system operator. Objective of the CC is to ensure that the specifications of system comply with its statutory and transmission licence obligations.

These codes are applicable to the generators/power stations, network operators, different converter station (acting as an interface between the generation and transmission systems), and other externally connected operators. To test the compliance for their connection to grid, certain parameters are checked to ensure proper operation of the transmission system. Conditions laid before permitting connections to the grid are specified as per the standards or bilateral agreement. Only those users/power generation plants are allowed to connect to the grid who complies with these norms, considered in terms of following parameters:

**(a) Grid Frequency Variations**
Different components connected in a grid are connected synchronously, i.e., at same frequency. Any variation in frequency will lead to loss of this link and separation of the different components. Thus, it is crucial to maintain frequency within the prescribed limits. The bands for frequency described in the grid codes may vary with region; for example, for a 50 Hz transmission system operation band is 49.5–50.5 Hz, which may be allowed to extend from 52 to 47 Hz, in exceptional circumstances [6]. This wide range under exceptional circumstances is further divided into smaller bands as shown in Table 2.3.

**(b) Grid Voltage Variations**
For efficient operation of electrical equipment, it is essential to maintain the voltage levels. The permissible voltage levels vary as per the agreement. However, the allowable voltage variation varies with the voltage level. At higher voltage levels, the voltage is not allowed to vary as much as at the lower levels. For example, at higher voltages, say 400 kV, a variation of ±5 % is permissible unless abnormal conditions prevail. Under abnormal conditions, a voltage variation of ±10 % is permissible, but voltage variation from ±5 to ±10 % is permissible for about 15 min, only. For lower voltages, like 275 and 132 kV at connection site a variation of ±10 % of the nominal value is permissible unless abnormal conditions prevail. Under fault conditions, voltage may collapse transiently to zero at the point of fault until the fault is cleared. The permissible variation of voltages for different voltage levels [6] is presented in Table 2.4.

**(c) Voltage Waveform Quality**
Most of the components present in the power system are non-linear in nature. However, the equipment, specially meters and protection devices, are designed for their use with linear equipment. The generators, power station equipment, and loads, connected to the transmission system may operate erroneously, under the influence of waveform distortion. Thus, it becomes critical to put a limit on the waveform distortion at the connection site. The most common distortions encountered are discussed below:

**Table 2.3** Permissible frequency variation under exceptional circumstances [6]

| Frequency range (Hz) | Requirement |
|---|---|
| 51.5–52 | Operation for a period of at least 15 min is required each time the frequency is above 51.5 Hz |
| 51–51.5 | Operation for a period of at least 90 min is required each time the frequency is above 51 Hz |
| 49.0–51 | Continuous operation is required |
| 47.5–49.0 | Operation for a period of at least 90 min is required each time the frequency is below 49.0 Hz |
| 47–47.5 | Operation for a period of at least 20 s is required each time the frequency is below 47.5 Hz |

**Table 2.4** Permissible voltage band at different voltage levels [6]

| Nominal voltage (kV) | Normal operating range |
|---|---|
| 400 | 400 kV ± 5 % |
| 275 | 275 kV ± 10 % |
| 132 | 132 kV ± 10 % |

*(i) Harmonic content*

In the grid, the switching of components and non-linear loads, connected to the system, induce harmonics in the system. The different generation, transmission and distribution equipment connected under the influence of harmonic distortion operate erroneously. However, for proper operation of the transmission system, the electromagnetic compatibility levels of harmonic distortion should be within the prescribed limits. These limits are to be either based on the IEEE/IEC standards or described clearly in the bilateral agreement. To obey these limits, it is vital to limit the harmonic content at a particular connection point and its emission to the rest of the network. Thus, while considering the harmonic limits in a system, it is important to consider:

(1) Location of new connections,
(2) Position of existing connections, and
(3) Number of connections at the node.

*(ii) Phase unbalance*

Phase unbalance is a major common problem in the grid. Due to inherent structural and loading properties of the transmission system, it is nearly impossible to remove it. It severely affects the components connected to the transmission system, especially the induction machines [12]. Therefore, to maintain the unbalance within the tolerable limits, certain limitations are imposed on the power plants and users. For example, in England and Wales, the limit for unbalance is 1 %, whereas it should be below 2 % in Scotland, unless abnormal conditions prevail. For off-shore connections, the limits are described in the bilateral agreement, in general.

*(iii) Voltage fluctuations*

Voltage fluctuations are the variation in the voltage magnitude. It arises mainly due to the switching of heavy loads and different components connected directly to the point of common coupling. However, to protect the equipment and limit the voltage fluctuations, limit is defined in the bilateral agreement. Design of the connections between transmission system and generating units, network operator's user system, etc. should be consistent with the licence standards. For example, in England and Wales, for onshore transmission system, it shall not exceed 1 % of the voltage level for repetitive step changes. In case of abnormal conditions, voltage excursions up to a level of 3 % may be allowed; however, it should be safe for the transmission system.

For offshore transmission system, the limits are set out in the agreement. In relation to the connection point, the plant and apparatus must comply with the requirements specified in the bilateral agreement.

**(d) Protection System**

Protection system is a vital component in operation of the grid. It separates the faulty components of the system from the healthy part. Since, the transmission system is complex in nature; the protection devices should have co-ordinated control. The co-ordination should be with minimum fault clearing time, in accordance with the bilateral agreement. At higher voltage system, the fault clearing time is minimum, which may increase with lesser voltage levels. For example, at 400 kV the permissible limit is 80 ms, whereas, at 132 kV, it is 120 ms.

In case of the converter stations, the owner is required to install the circuit breaker fail equipments with settings in accordance with the bilateral agreement. The important protection schemes to be considered in the protection system are:

(1) Loss of excitation,
(2) Pole-slipping protection,
(3) Signals for tariff metering,
(4) Work on protection equipment, and
(5) Relay settings.

**(e) Plant Performance Requirements**

The plant, either thermal or renewable energy is required to follow the plant performance norms to remain connected to the transmission system. The plant should maintain performance in terms of frequency, voltage and power level, as per the requirements of the agreement.

For onshore installed synchronous generator units, the units must be capable to operate continuously, between the limits of 0.85 power factor lagging to 0.95 power factor leading at the generator terminals. However, at operating conditions other than the rated value, the units must supply continuous power at reactive power capabilities identified on the generator performance chart. In addition, the generator connected should have the connection entry capacity of above rated value, during which:

(i) the reactive power capabilities must be at a minimum of 0.9 power factor,
(ii) all active power output levels in excess of rated value or as per the bilateral agreement.

For such a generator with apparent power rating of less than 1600 MVA the short circuit ratio should not be less than 0.5, whereas above 1600 MVA, it should not be less than 0.4.

For all generators, at offshore grid entry point of the LV side, the active power output levels should be maintained at zero transfer of reactive power. The steady state tolerance on reactive power transfer, to and from an off shore transmission system, should not be greater than 5 % of the rated value, or as per the reactive power capability specified in the bilateral agreement.

Further, the different generation units, converter stations, etc. must be capable of continuously maintaining constant active power output for system frequency changes.

**(f) Excitation and Voltage Control Performance Requirements**
The generators connected to the transmission system shall maintain the terminal voltage as per the grid voltage variation specified in the agreement. To implement this, it is necessary to implement continuously acting automatic excitation control system. This provides the necessary control, to generator, for maintaining constant terminal voltage without causing instability over the entire operational range.

The continuously acting control system should be capable of providing the reactive power control or an alternative reactive capability, as specified in the agreement.

**(g) Steady State Load Inaccuracies**
To regulate the frequency variation, it is important to have error free characteristics of the load. The major application of load data is in forecasting the load. In [6], at steady state the upper limit on the standard deviation in the load, over a period of 30 min, is fixed at 2.5 % of the generation capacity.

**(h) Negative Phase Sequence Loadings**
Due to faults and unbalanced load connected etc., negative sequence current flows in the generator. This affects the operation of the synchronous generator. Therefore, it is important for the generator to withstand the negative phase sequence voltages/currents, without tripping of the circuit breaker.

**(i) Communication Plant**
In a certain situation, communication between the operators and the plant is required. To achieve this, a communication link is required between the two. The system should be planned in such a way that both calling and receiving parties can signify the priority and prioritise disconnect tones.

### 2.2.3 *Operating Code (OC)*

Operational phase of the power plant begins after obtaining clearances for the planning and connection codes from the authorities. Operation is a vital part of the power system. Any improper operational event may lead to failure of the equipment and damage to the connected systems. Thus, to keep the system working, while maintaining the supply quality without any contingencies, it is essential for the plant operator to follow the operating codes. Operating codes are a part of the grid code, dealing with the operation of the plants connected to the transmission system. The operating code is further divided into following components [6]:

(a) Demand forecasting,
(b) Operational planning and data provisions,
(c) Testing and monitoring,
(d) Demand control,
(e) Operational liaison,
(f) Safety coordination,
(g) Contingency planning,
(h) Event information supply,
(i) Numbering and nomenclature of high voltage apparatus at sites, and
(j) System tests.

Since most of the operating codes involve time duration, equipment specification and other related requirements that, change with the local authorities, the details are not discussed here. A generalized discussion is rather presented to make clear understanding. For more details, readers can refer to the codes of their country.

**(a) Demand Forecasting**
Demand forecasting is an important aspect for operational purposes. It deals with the prediction of future demand in for a fixed duration of time as, specified by the grid codes. This information helps utilities in scheduling the generation to maintain the balance between generation and demand of power. It is usually performed for the active power, but the reactive power demand forecasting may also be required, for some authorities in their grid codes. The forecasting can also be used by different network operators and power suppliers for planning purposes.

Forecasting information is also required during the planning phase of the plant. But, the time durations for the demand forecasting and planning forecasting are different. In case of demand forecasting, the duration is in hours, whereas, for planning purposes, forecasting is done in years [6]. When the arrangement at a grid supply point is expected to be operated in separate sections, separate sets of forecast information are required for each section. The information required by the transmission operator, for a period specified in the grid code, is given as:

(a) Half-hourly annual peak transmission system demand,

(b) Output of the power station, and
(c) Forecasts of demand to be relieved by demand control.

Generally, the data is provided at a time-resolution of half-hourly basis along with the following information:

(i) Proposed date, time and duration of implementation of demand management,
(ii) Proposed reduction in demand by use of demand management, and
(iii) Proposed switching times, on daily basis.

The following factors are taken into account by National Grid Electricity Transmission (NGET) for daily demand forecasting [6], in programming phase and control phase as:

(a) Historic demand data including transmission losses.
(b) Weather forecasts with current and historical weather conditions.
(c) The incidence of significant events or activities, known to NGET in advance.
(d) Anticipated interconnection flows across external interconnections.
(e) Proposed demand control, equal to or greater than the demand control notification level, to be exercised by Network Operators with due information to NGET.
(f) Customer demand management equal to or greater than the proposed customer demand management notification level, to be exercised by suppliers with due information to NGET.
(g) Other information supplied by users.
(h) Anticipated pumped storage unit demand.
(i) Sensitivity of demand to anticipated market prices, for electricity.
(j) Demand taken up by station transformers.

**(b) Operational Planning and Data Provision**
Operation of the plant requires transmission of data, i.e., information at various levels. It includes information regarding the construction, maintenance, repair of the plant, and surplus generation available, etc. Each generator and interconnector operator has to provide this data in writing periodically, say weekly. This time scale may vary as per the grid codes. It is important to perform the operational planning of the plant to maintain the generation reserves for emergency conditions, like outages, maintenance schedule of plants, and financial planning, etc.

**(c) Testing and Monitoring**
The components of the generation plants and transmission systems require continuous monitoring and testing. The monitoring is essential to observe the operation of the equipment connected. The input and output of the plant should be according to the voltage, frequency and other quantities, as specified by the operational code. Testing is required to check the compatibility amongst the different components for input and output limits, and to obtain the dynamic parameters. If any failure in components is expected by obtaining this code, necessary measures are taken, to keep the system operational. The procedures to be

followed and duration after which the component has to be tested are described in the bilateral agreement.

**(d) Demand Control**

Demand control is the reduction of demand in the event of insufficient availability of active power generation to meet the demand, or in emergency conditions. Primarily, it deals with following events [6]:

(a) Customer voltage reduction,
(b) Customer demand reduction by disconnection,
(c) Demand reduction,
(d) Automatic low frequency demand disconnection, and
(e) Emergency manual demand disconnection.

Demand control is essential for operation of the power station. It keeps the power station in a state of proper start up and shut down. The authorities authorized to perform the demand control, and the measures to be taken, are described in the agreement. Since, the forecasting data is required at a resolution of half an hour or less, the actions/implementations are also executed for the same time interval.

**(e) Operational Liaison**

Operational liaison is the requirement laid for information exchange, in operations or events, having an operational effect on the transmission system, generators or the users connected to the system. Monitoring the terminal voltage, frequency and their violation outside the statutory limits is important to check the status of the system.

Based on the monitoring and analysis, alerts are issued to the connected plants and users of the system about the shortages, problems or demand reductions. Also, the duration, as described in the data, allows users to be in a state of readiness, to react properly.

**(f) Safety Coordination**

Safety co-ordination is a major concern for the plant whether online or offline. It is, concerned with the safety of the people working in the plant and equipment. It prescribes the steps or procedures to be followed in order to keep the safety. Disregarding these instructions may lead to cancel/suspend the license for lifetime or certain duration.

Thus, the plant has to maintain a logbook for, recording the safety violations happened in the past and the measures taken to prevent their future occurrence.

## *2.2.4 Data Communication Code (DCC)*

From planning to the operation of the power plant, exchange of information between the authorities, in form of data is important. The data from one end is communicated to other plants and to the transmission and system operator in the grid. The data communication refers to the

(a) Generators,
(b) Network operators,
(c) Converter station owners,
(d) Suppliers, and
(e) Users.

It is important to maintain the prescribed format, and security of the data.

### *2.2.5 Balancing Code (BC)*

To keep the voltage and frequency variations within the statutory limits, it is important to control the reactive power flow and maintain the generation-demand balance, respectively. BC specifies the requirements to maintain the terminal voltage and frequency, under normal and emergency conditions.

In case of conventional power plants, such as, thermal and hydro, it specifies the opening and closing of the valves and gates to control the rotation of the turbines. Also, it specifies the time required by the mechanical devices, to respond to changes in the system. For voltage control, the amount of reactive power flow required, and time constant of the mechanical devices, depending upon the forecasting data is described in this code. Also, it specifies the permissible changes/rate of power transfer to the transmission system.

For renewable energy systems, such as wind or solar, the input to the system cannot be controlled. Therefore, it describes the control requirements, such as, rate of power generation, actions required to control the amount of power flow, and specifications of the controllers, etc.

## 2.3 Grid Code Challenges

The specifications of the grid code, as discussed in the Sect. 13.2, deal with the conventional generation schemes, such as, thermal or hydro power plants, etc. Inputs to the plants, i.e., steam and water, in case of thermal and hydro power plants, respectively, are controllable. However, in case of renewable generation schemes, the input is uncontrollable. For example, for wind energy generation the wind flowing cannot be controlled, and for solar energy, the presence/absence of sun cannot be controlled. Thus, the grid codes for such a system shall contain following information:

(a) Forecasting the availability of the renewable energy sources.
(b) Timing requirements for the decision-making and taking the necessary actions.
(c) Voltage level to be maintained at different input conditions.
(d) Frequency band to be followed.

(e) Characteristics, power ratings, and operational requirements of the converter stations.
(f) Protection system for these variable operating conditions.
(g) Characteristics and specifications of the controllers to control the amount of active and reactive power to be generated.

## 2.4 Grid Codes for Integration and Operation of Renewable Energy Sources

Nowadays, share of renewable energy generation in the total generated power is becoming more and more prominent. Because of the intermittent generation, improvements are important for the generation plant [13]. The discussions made in this section are based on the wind generation, primarily.

Sometimes, the power generation using the renewable sources can be a costlier affair. Therefore, governments provide incentive based schemes for its promotion. They help on the components of the generation system, especially the technology related to the development of controllers to generate the specified amount of voltage and frequency [14]. Also, it helps in training the personnel for the operation.

Operation of the wind generators, in line with the grid codes, is the prime concern for the plant owner. Thus, analyzing technical requirements for wind generators, due to their different physical characteristics from the synchronous generators becomes essential.

Along with the components of the grid codes, as discussed in Sect. 13.2, the major concerns for the wind energy systems can be described as [5, 15]:

(1) Low voltage ride-through capability,
(2) Reactive power control, and
(3) Power quality, etc.

*(a) Low voltage ride-through capability*: For any utility, connecting the wind energy systems to the transmission line is the prime concern. The wind technology is sensitive to the changes in voltage and power levels of the transmission system, which may lead to plant shut down. This sudden shutdown is a significant concern for the reliability of system, along with wind generators connected to the transmission system, as each transmission provider designs its own low voltage ride-through requirement, which affects the design and operation cost of the wind turbines. As a result, both wind generators and most transmission authorities prescribe the low voltage limits with time duration, or in other words, the characteristics that a wind farm, should have in order to connect to the transmission system. Wind generators demonstrating these characteristics are allowed to connect to transmission lines. In comparison to the synchronous generators, these

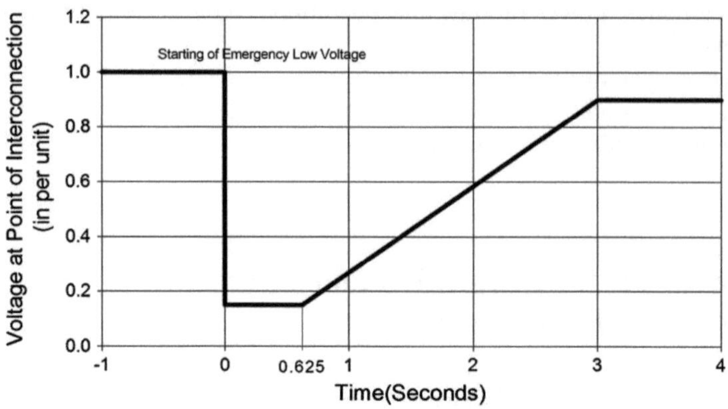


**Fig. 2.4** Low Voltage ride through characteristics as per USA grid code [5]

characteristics are more important for wind energy system. The synchronous generators are connected to the automatic voltage regulators, to maintain the constant voltage, but wind energy systems do not have any such arrangements. The characteristics of the wind energy systems can be depicted, using the example of USA grid code [5] as shown in Fig. 2.4.

Figure 2.4 shows the time duration (in second) with the voltage (in per unit) at the point of interconnection. The characteristics show that the voltage dip occurs during 0 to 0.625 s, and then reaches to 90 % of the actual value in 3 s. The wind energy system, to be connected to the transmission system, should operate from 0 to 0.625 s, whereas the wind energy system is allowed to get disconnected from the system, when the system starts regaining the voltage level.

*(b) Reactive power Control*: For small wind energy system, the utilities are not much concerned for reactive power control, due to small requirement of reactive power. However, for larger wind generators, it is mandatory to maintain the reactive power levels. Such plants must be capable to provide the dynamic voltage support for interconnection to the transmission systems. Also, it is important for the wind energy service provider to maintain the power factor within the range of 0.95 capacitive to 0.95 inductive. To add the reactive power handling capabilities, the service provider can use the fixed/switched capacitors with consent of the transmission operator.

*(c) Power quality*: Amongst the various power quality problems, voltage sags are the major problems for the wind energy system. However, the transmission system operators demand the low voltage ride-through capability in these sources. Thus, huge investment has been made in developing devices, to support the wind system during voltage sags.

Doubly-Fed Induction Generators (DFIG) is one of the most popular technologies for wind energy systems, in which rotor and stator are connected to the transmission system through the converters. Voltage sags in the transmission side cause large currents to flow in the rotor and rotor side converter. If the required

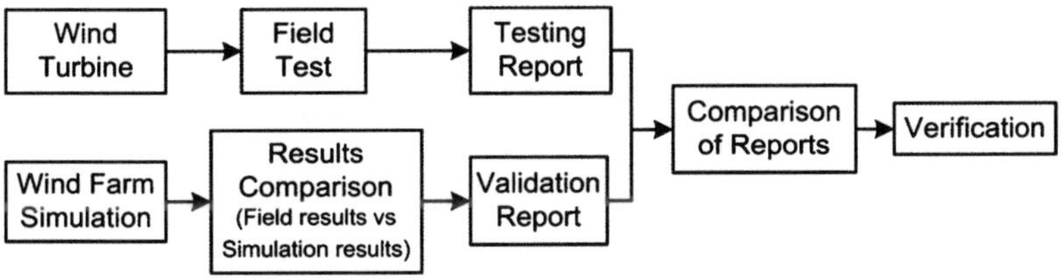


**Fig. 2.5** General verification process [5]

voltage exceeds the maximum voltage of the converter, current control becomes tough [16]. To avoid damage to the converter switches, DFIG wind turbines are equipped with crowbars. When the rotor currents become high, it is bypassed through the crowbars. Then generator operates as an induction machine, with a high rotor resistance. Thus, it can support the grid during voltage sags [16, 17]. In case of short-circuits in the power system, large reactive power is required to recover the air gap flux. If not supplied, the induction generator may become unstable, leading to its disconnection from the transmission system [18, 19].

The other practical solutions, to deal with the voltages sags are:

- Static Var Compensator (SVC)
- Static Compensator (STATCOM).
- Dynamic Voltage Restorer (DVR), and
- Unified Power-Quality Conditioner (UPQC).

It is important for the wind energy system authorities, to select the techniques having capability to make a grid code complied system. The usefulness of these technologies has to be proven before their implementation in the system. The general verification process, adopted to verify that the wind system does not get disconnected during voltage sags, is shown in Fig. 2.5.

The general verification is performed in two steps, the field test and simulation test. If the reports of these tests are found in compliance with the grid codes, the construction, and operation of the wind farm is approved.

## 2.5 Grid Codes and Conventional Generation Schemes

The major concerns of the grid codes are the transmission system voltage and frequency variations. The transmission system operators demand the reliable compatibility, with all statutory warnings.

Synchronous generators are the most important component of the power system. It is vital for them to follow the specifications, and requirements, laid by the grid codes. However, in many cases the grid codes do not comply with the established international standards, such as IEC and ANSI [7].

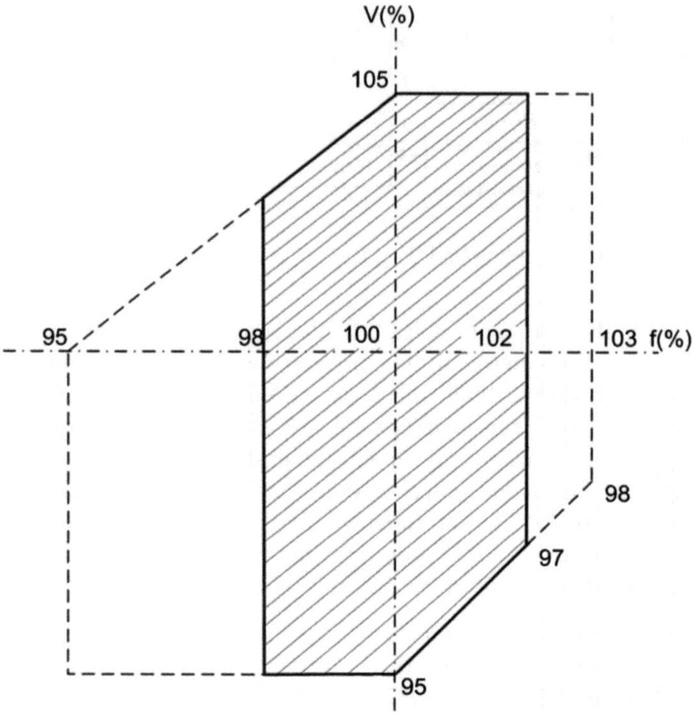


**Fig. 2.6** Voltage and frequency variation [3]

The grid codes make no mention of the relevant generator or turbine industry standards and require capabilities that significantly fulfil industry norms. IEC 60034-3 [21] spells out the requirements for the design of turbines and the synchronous generators. For a turbine system, having synchronous generator systems connected to it, the voltage and frequency variation is shown in Fig. 2.6, where shaded portion represents the voltage-frequency characteristics for continuous operation, while unshaded portion shows the inoperable region.

As per the industry norms, synchronous generator terminal voltage can vary from 95 to 105 % of its nominal values. However, the foremost concern is frequency rather than the voltage. The permissible frequency variation is 5 % of nominal value, but the generator standards limit it from 98 to 102 %. Several other discrepancies can be clearly seen while observing the grid codes and industry norms for the synchronous generator systems [3].

Another major concern is the variation of cost involved, for changing the generator performance as per changes in the grid code. The short-circuit ratio or SCR, defined as the inverse of the value of its saturated direct axis reactance, helps in determining the generator leading reactive capability, with direct impact on the static stability [7].

For same apparent power, a higher value of the SCR depicts more ampere-turns in the field winding. However, the operation is limited by the temperature. A change of mere 0.1 % in the SCR increases the total machine volume by 5–10 %, which increases the cost of the system, depending upon the type of generator. Therefore, any minor change in the requirements of the grid code, may lead to typical changes in the design, and increased cost of the whole system.

## 2.6 Grid Codes and Nuclear Generation

Safety is a major concern for the nuclear power plants. Cooling of these plants is essential requirement. Thus, reliable power supply is required for the cooling purpose; otherwise, catastrophic situations, like FUKUSHIMA, may arise. This event showed the importance of maintaining auxiliary electric power to the nuclear power plants for safety.

The power supply to the nuclear power plants is divided into two components, i.e., mains or normal (unclassified, on a conventional island) and the emergency (safety classified, located in the nuclear island) power supply systems. Under normal conditions, the normal supply is used to cool the nuclear plant, whereas in case of emergencies, the emergency supply system is required to cool the plant. Thus, it is important to have continuous supply for the nuclear power plant.

Also, in case of blackout, nuclear power plants are unable to start of their own, due to the unavailability of power. It always requires a small power source to start with. Thus, reliability and availability of power are important in view of safety.

Usually, diesel oil based emergency power plants, producing limited amount of power are not reliable. To solve this problem, the nuclear power plants are supplied with dual supply, independent of each other, from the grid [20]. The standby connection should be able to provide the spinning reserve for operating the safety loads for cooling.

Supply quality is a major concern for the power plant. Voltage variation is an important factor for the operation of components connected in the nuclear power plant. Normally, a plant is designed to handle the slow variations in the voltage. To handle fast variations plants have to be specifically designed. Thus, it is important to maintain the voltage levels within the limit. The limits should be implemented for both, the normal and emergency power plants. Frequency variation is another important factor of power quality. It is a difficult component to be compensated. For a nuclear power plant, frequency variation is of utmost importance, as most of the motors in the plant are directly connected to the grid.

Thus, any variation in the frequency may affect the speed of the motor, indirectly related to the safety of the plant [20]. Thus, considering these factors, the grid code considering nuclear power plant should have:

- Strict rules for voltage variation,
- Strict rules for frequency variation, and
- Effective protection system, to maintain supply to the plant.

These requirements should be applicable for normal and emergency power plants, both.

## 2.7 Conclusion

In this chapter, a discussion on the grid codes is presented. Initially, the main classification of the grid codes is presented. Further, its classification into planning codes, connecting codes, operational codes, data communication codes and balancing codes, allows clear understanding of the grid codes. It signifies the various stages in the development of the plant and their requirements. Since, most of the codes are designed presuming conventional generation schemes, the requirements for developing grid codes for renewable sources are presented.

Further, the important components for the renewable energy plants are discussed. It shows that, power quality, low voltage ride-through capabilities, etc. are important component that needs to be addressed in the grid codes. In the end, a discussion on the conventional generation schemes and the grid codes is presented. It shows that a conflict may arise due to differences in the standards being followed by the plants. Further, a procedure on the requirement of the nuclear power plants and the grid codes is presented, to discuss the requirement of the strict rules for voltage and frequency variations, along with the protection system, for proper operation of the nuclear power plants. In summary, the chapter presents in-depth classification of the grid codes with its present status, compliance dilemma between the grid codes and conventional generation schemes, and requirements for their developments considering renewable energy sources and nuclear power plants.

## References

1. Valle Y, Venayagamoorthy GK, Mohaghegi S, Hernandez JC, Harley RG (2008) Particle swarm optimization: basic concepts, variants and applications in power systems. IEEE Trans Evol Comput 12(2):171–195
2. Bae Y, Vu TK, Kim RY (2013) Implemental control strategy for grid stabilization of grid-connected PV system based on German grid code in symmetrical low-to-medium voltage network. IEEE Trans Energy Convers 28(3):619–631
3. Stephan CE, Baba Z (2001) Specifying a turbogeneraor's electrical parameters guided by standards and grid codes. IEEE international conference on electric machines and drives, pp 63–68
4. Katiraei F, Iravani R, Hatziargyriou N, Dimeas A (2008) Microgrids management. IEEE Power Energ Mag 6(3):54–65
5. Comech MP, Gracia MG, Susana MA, Guillen MAM (2011) Wind farms and grid codes. Turbines to wind farms, ISBN 978-953-307-237-1
6. The Grid Code (April 2013) Issue 5, Revision 3

7. Nelson RJ (2001) Conflicting requirements for turbogenerators from grid codes and relevant generator standards. IEEE international conference on electric machines and drives conference, pp 57–62
8. Larsen GKH, Foreest ND, Scherpen JMA (2013) Distributed control of the power supply-demand balance. IEEE Trans Smart Grid 4(2):828–836
9. Joseph DM, Haigh P, McCullagh J (2012) Ensuring grid code harmonic compliance of wind farms. 10th IET international conference AC and DC transmission, pp 1–6
10. USA FERC (2005) Interconnection for wind energy. 18 CFR Part 35 (Docket No. RM05-4-001; Order No. 661-A)
11. Indian Electricity Grid Code (IEGC) (2010)
12. Lee CY (1999) Effects of unbalanced voltage on the operation performance of a three-phase induction motor. IEEE Trans Energy Convers 14(2):202–208
13. Erlich I, Bachmann U (2005) Grid code requirements concerning connection and operation of wind turbines in Germany. IEEE power engineering society general meeting, pp 1253–1257
14. Zavadil RM, Smith JC (2005) Status of wind-related US national and regional grid code activities. Power engineering society general meeting, pp 1258–1261
15. Armenakis A (2012) Grid code compliance test for small wind farms connected to the distribution grid in Cyprus. 8th Mediterranean conference on power generation, transmission, distribution and energy conversion, pp 1–6
16. Morren J, de Haan SWH (2007) Short-circuit current of wind turbines with doubly fed induction generator. IEEE Trans Energy convers 22(1):174–180
17. Lopez J, Gubía E, Olea E, Ruiz J, Marroyo LL (2009) Ride through of wind turbines with doubly fed induction generator under symmetrical voltage dips. IEEE Trans Industr Electron 56(10):4246–4254
18. Muyeen SM, Takahashi R, Murata T, Tamura J, Ali MH, Matsumura Y, Kuwayama A, Matsumoto T (2009) Low voltage ride through capability enhancement of wind turbine generator system during network disturbance. IET Renew Power Gener 3(1):65–74
19. Muyeen SM, Takahashi R (2010) A variable speed wind turbine control strategy to meet wind farm grid code requirements. IEEE Trans Power Syst 25(1):331–340
20. Sobott O (2012) White paper: grid code and nuclear safety. IEEE power and energy society general meeting, pp 1–3
21. IEC 60034-3. Rotating electrical machines—part 3: specific requirements for synchronous generators driven by steam turbines or combustion gas turbines

# Chapter 3
# Fault Ride-Through Criteria Development

**Nirmal-Kumar C. Nair and Waqar A. Qureshi**

**Abstract** Wind energy integration into transmission systems has been accelerated by establishing robust grid codes relevant to the network. Fault Ride-Through (FRT) criteria is one of them, which requires generation to stay connected to transmission system under certain operational voltage envelope. Development of the criterion is a specialized task which will take into account existing grid characteristics, network history and generation response towards voltage and frequency disturbances. This chapter will outline a detailed systematic methodology for achieving FRT criteria through a set of analysis involving contingency analysis, static and dynamic scenario based system assessment. The uniqueness of this chapter is to present a case study for a grid having a high percentage of the generation through renewable energy and has HVDC connected island based transmission grid. Each island sub-system separated by HVDC link is pre-dominant with non-uniform distribution of generation units and load centres. Therefore, two different FRT criteria are found to be effective for this power systems network. None of the reported criteria in prevailing networks have introduced similar approach before. New Zealand power system has similar characteristics and has been utilized for this case study. FRT criteria development has been in practice but the current context has not been published in articles or books and hence the motivation for this book chapter.

**Keywords** Wind generators · Low-voltage ride-through

N.-K. C. Nair (✉) · W. A. Qureshi
Department of Electrical and Computer Engineering, The University of Auckland,
38 Princes Street, Building 303, Science Centre 1142 Auckland, New Zealand
e-mail: n.nair@auckland.ac.nz

W. A. Qureshi
e-mail: w.qureshi@auckland.ac.nz

J. Hossain and A. Mahmud (eds.), *Renewable Energy Integration*,
Green Energy and Technology, DOI: 10.1007/978-981-4585-27-9_3,

## 3.1 Introduction

Increased wind generation and its integration on transmission system dictate the development of stricter grid code requirements to maintain the security of supply. Large wind generation connected to transmission system raises many concerns for not being able to support the system during disturbances. The response of a large wind farm during a voltage disturbance can possibly affect system stability as discussed in [1–4]. One of the major requirements is Fault Ride-Through (FRT) requirement. FRT is subdivided into Low Voltage Ride-Through (LVRT) or High Voltage Ride-Through (HVRT). LVRT capability means that generators should be able to remain connected during any fault event. In some countries, grid codes require all new installed wind turbines to have FRT or LVRT capability [5–7]. Most Conventional generators are normally capable enough to meet these requirements while wind generators vary in capability depending on the technology used. Wind generator capabilities and different techniques used to satisfy LVRT criteria are also discussed in next few paragraphs. A typical FRT criteria or voltage duration profile is shown as in Fig. 3.1.

All generators are required to stay connected to satisfy the system security. They may disconnect if the voltage enters into the region below the line. These voltage profile or LVRT requirements are part of grid code or grid connection guidelines established by transmission system operator. Based on the grid code requirements manufacturers have designed various solutions for different wind generator such as Fixed Speed Induction Generator (FSIG), Doubly Fed Induction Generator (DFIG) and Full Convertor Synchronous Generator (FCSG) in order to meet ride through requirements.

In [8–10] reactive power injection during the fault to support and improve voltage ride through capability has been discussed. A comparison of performance of uncompensated large scale wind farm having FSIG units have been provided with wind farm having similar units with Static VAR Compensation (SVC) and wind farm having DFIGs in [9]. Reactive power support is apparently the only technique for FSIG to ride through the fault at the Point of Common Coupling (PCC). However, various techniques using convertors and controllers are available for DFIGs.

A controller technique applied to grid-side converter to keep DFIG connected to the system during the fault is presented in [11]. Series-connected voltage sources converter has been proposed for enhancing the FRT capability in [12]. Crowbar resistor which is a bypass resistor-set connected to rotor terminal is also proposed as technique to improve FRT capability for DFIG in publications [13–17]. Crowbar improves the FRT capability along with protecting the rotor from damage but it converts a DFIG into a conventional induction generator thus degrading the DFIG performance [18, 19]. However, in [17] it is suggested that DFIG performance can be improved if appropriate value of crowbar resistor is selected. Kasem et al. [20] proposes that both grid-side and rotor-side converter inject reactive power to the grid during the fault along with improved timing algorithms for crowbar resistor, thus improving the FRT capability even more.

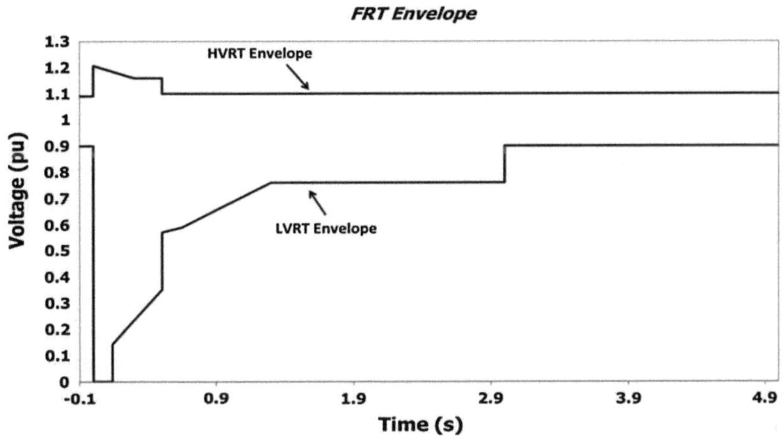


**Fig. 3.1** Typical limit curve for LVRT requirements

FCSG has a totally different arrangement than above two technologies. FCSG is a relatively new practice for large wind farms connecting into transmission system. It has been found to have better FRT capability and grid support during the fault. Sufficient documents and grid codes have been found stating LVRT criteria for various countries [21]. However, no significant literature could be found discussing FRT development from a transmission network's perspective.

## 3.2 FRT Criteria Development for New Zealand

The increasing amount of wind generation that is proposed for connection to the New Zealand transmission system has prompted the need for tougher grid code requirements for generators to maintain the security of supply. The technological variations in wind generator technology available for large scale wind farms raise concerns around their ability to support the grid during system events. Most conventional synchronous generators in New Zealand have the capability to meet these requirements while wind generators vary in capability depending on the Wind Turbine Generator (WTG) technology employed. Section 3.3.2 discusses WTGs types and their capabilities and different techniques used to satisfy FRT criteria. Sufficient documentation and grid codes has been researched which cites FRT criteria for various countries [21]. However, no significant publication has been found discussing FRT criteria development in detail from a network's perspective.

At present wind farms have a total installed capacity of 5 % in New Zealand. An increase in this percentage is expected as there are on-going projects yet to be commissioned [22]. There are about fifteen small and large wind farms operating in New Zealand. These include turbines of various manufacturers and sizes with a total operating capacity of over 600 MW. Five of these are connected at high voltage or medium voltage levels. Security of supply is a concern when large wind farms are connected at the high voltage transmission level with little or no demographic separation, thus requiring stricter grid code requirements. The response of conventional generators or plants are well known but wind farms having different technologies may differ in responses and may affect the Transmission System Operator's ability to manage and operate the grid in a secure manner.

A variety of investigations have been carried out by the New Zealand Electricity Authority (EA) for large wind integration scenarios. These include the impacts on the market, transient stability, small signal stability, power quality, and dynamic response of wind farms during disturbances [23]. FRT capability is done of the major concerns of all types of generation, specifically large wind farms, in order to ensure security of supply in New Zealand. Wind generator technologies in New Zealand and their FRT capabilities are discussed in Sect. 3.3.2.

By definition, FRT requirements bind generators to operate between allowable voltage limits and stay connected to the grid within the same voltage envelope to maintain the security of supply. Normally, these FRT requirements become the part of grid code requirements established by the Transmission System Operator (TSO). Grid operators define a 'ride-through' profile in order to avoid situations where generators are disconnected during grid faults. Some grid operators not only require participation from generators during the fault but also post-fault period towards voltage stabilization and system recovery [24].

In this chapter, the process for designing a FRT criterion is outlined from a TSO perspective. Next section discusses fundamental considerations followed by the methodology used for development of FRT criteria. It is followed by a section discussing assumptions for the FRT studies which have been taken into account for analysis and a New Zealand grid case study detailed in Sect. 3.5. This section also briefly discusses some of the large wind farms integrated in the New Zealand transmission system categorized by Wind Turbine Generator (WTG) Type. Following that, the processes of developing voltage envelopes for FRT criteria are given, and this proposed criterion is then compared with existing Transpower and international criteria. Finally, recommendations for future development and conclusions are presented.

## 3.3 Fundamental Considerations

### *3.3.1 System Characteristics*

The first and foremost consideration towards developing a FRT criterion is to understand the characteristics of the transmission network. For instance if we consider the New Zealand Network then we have a two island network i.e. the North Island (NI) and South Island (SI) networks interconnected through a 2 pole

HVDC link currently. The total network is comprised of 173 substations and over 1,200 power transformers. The transfer capacity of the HVDC link is 1040 MW with two cables on pole-1. Another pole is under commission process to increase the transfer capacity between two islands. The transmission voltages in both networks are 220 and 110 kV, with some 66 kV transmissions in the SI. The generation sources are scattered through the NI that include hydro, thermal, combined cycle gas, co-generation, geothermal and wind. However, the SI is mostly hydro generation which serves as the base load and under normal system conditions energy is typically transferred from SI to NI via the HVDC link to meet the demand of major load centres in NI such as Auckland City.

Due to the isolated nature of the New Zealand power system it means that it prone to experiencing frequent voltage and frequency disturbances. To avoid cascade failure, it is critical that generation remains connected and the system may require a specific fault ride through criteria that differ from those specified in large well-interconnected continental power grids [25, 26]. There are few obligations towards generators side as per existing requirements in New Zealand. The requirements are as follows:

- Maintain a certain level of reactive power output under steady state conditions
- Plant must continuously operate in a manner to support voltage and frequency stability.

FRT is part of the latter requirement to a generators capability for voltage and is a key factor in assisting the System Operator to manage the system to avoid black outs.

#### 3.3.1.1 Fault Types

In order to make the criteria more robust, a history of network faults which may have direct impact on system stability is considered. Fault types vary from network to network but their classifications are universally accepted. Grid planning guidelines in New Zealand state that power system should remain stable for fault types such as loss of generation, three phase, single and double phase faults transmission faults [25].

The most common faults in power system are short-circuits. Three phase balanced short-circuit or a balanced three-phase to ground short-circuit are referred to as symmetrical faults. They cause highest fault currents but are less frequent power systems. Machine short-circuit ratings are designed based on balanced or symmetrical fault current calculations. The most common short-circuits in any system are unbalanced or unsymmetrical in nature. These can be single line to ground, double line to ground, or line to line faults. Unbalanced faults to ground are affected by fault resistance, soil resistivity, fault position, and grounding technique used: thus related issues also need to be considered while addressing unbalanced or unsymmetrical faults. Other types of faults may include the sudden loss of any generating unit or plant which causes severe voltage drops thus requiring other generation to support the system through frequency and power control. Apart from short-circuits such outages are also included in the possible list of contingencies.

#### 3.3.1.2 Protection and Fault Clearance Times

Voltage envelopes are influenced by fault clearance times for any network or part of the network. Standard distance or differential protection schemes protect transmission lines. The transmission protection coordination with neighbouring sub- transmission circuits, bus bars and other DC/AC links plays a significant role. High voltage transmission line protection normally has very short clearance time in order to maintain system security and avoid catastrophic failure of the system. In New Zealand transmission level protection schemes target 120 ms clearance times for critical parts of the grid to avoid blackouts. Without aided signal schemes some parts of the 110 kV transmission network are subject to prolonged clearance times. Protection philosophies and clearance times in practice have direct part in development of FRT criteria for any network. Protection performance assumed in the FRT studies is discussed in Sect. 3.4.5. Typically, In NZ power system 110 kV voltage, sub-transmission is aimed to achieve 10–12 cycles, while transmission protection scheme is targeted to achieve 5–7 clearance cycles.

Along with standard protection schemes some special protection schemes designed for specific contingencies may also be considered for FRT studies. These are generally either runbacks or overload schemes, and operate within the period of voltage recovery following a fault removal. These schemes only be active during specific grid conditions. In case of New Zealand FRT studies, there are a few special protection schemes but they have not been considered for this analysis.

### *3.3.2 Protection Issues*

Longer clearance times may also be expected due to other protection issues within the system and need to be considered during FRT analysis. These issues are specific to networks and could be identified from history of fault events from available data. Some of these issues might be high resistance faults, weak in-feeds, weak back-feeds, system configuration changes, grounding issues, circuit breaker failure, loss of signalling for transmission protection, loss of reactive compensation etc. Not many of these issues have been identified for New Zealand; thus any additional issues have not been included in the analysis.

### *3.3.3 Technology*

Modern wind farms employ different types of wind generators and these can be categorized into five main types. These types vary in inherent FRT capabilities. In

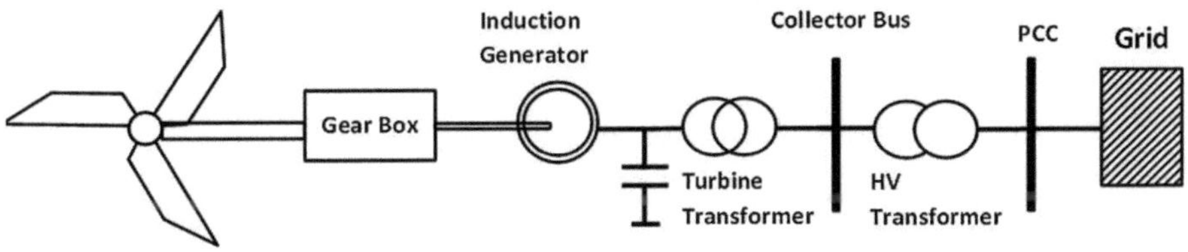


**Fig. 3.2** Squirrel Cage Induction Generator (SCIG) wind turbine (Type-1)

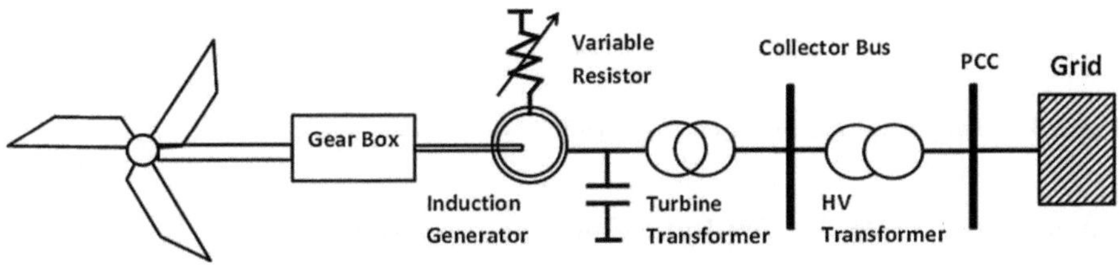


**Fig. 3.3** Wound Rotor Induction Generator (WRIG) wind turbine (Type-2)

the context of the case study for this work; New Zealand has five different wind generator technology based farms [22].

#### 3.3.3.1 Type-1 and Type-2 (Fixed and Variable Speed Induction Generators)

Type-1 and Type-2 utilize conventional induction generators as shown in Figs. 3.2 and 3.3. They are basically constant speed machines with the minor fluctuation of speed during change of load. Type-1 has a squirrel cage induction generator. However, wound rotor induction generators for Type-2 have some speed control through rotor winding access. In order to avoid an adverse impact on system performance and voltage, these generators use soft starters for grid connection. They absorb reactive power from the system to maintaining the rotating field in the air gap between rotor and stator winding. Naturally they do not have much ride-through capability for not supplying reactive power in case of a voltage event. Earlier designs of these types were connected to distribution scale without any binding to stay connected to the grid. Thus, the probability of disconnection for any fault in the vicinity was always high. In order to improve FRT capability turbine manufacturers offer FRT packages to comply with grid codes requirements. SVC or STATCOM support is used to improve FRT capability of these types. One of the very first grid connected wind farms in New Zealand was the Te Apiti (TAP) Wind Farm. The capacity of this wind farm is around 90 MW and all 55 turbines are Fixed Speed Induction Generators (FSIG) i.e. Type-1. Type-2 Wind technology is also available in New Zealand but not on significant scale.

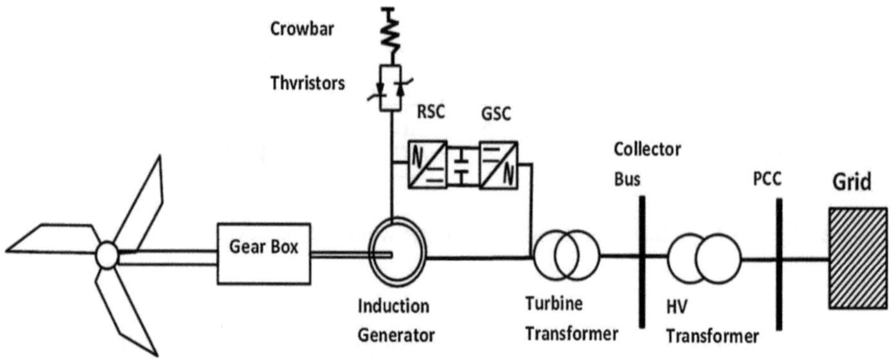


**Fig. 3.4** Doubly Fed Induction Generator (DFIG) wind turbine (Type-3)

#### 3.3.3.2 Type-3 (Doubly-fed Induction Generator (DFIG))

One of the most widely used variable speed WTG is the Type-3. This WTG arrangement is shown in Fig. 3.4. This is wound rotor machine, fed through a series back to back frequency converter at rotor side, having a rating of 30 % of the maximum stator power rating. The total power is the arithmetic summation of powers from stator and rotor. The DFIG operates either in super-synchronous, synchronous or sub-synchronous modes. Power is injected from the rotor, through the converter, into the system when the DFIG operates at super-synchronous speed. The real power is absorbed through the converter from the system by the rotor when the DFIG operates at sub-synchronous speeds. There is no power exchange through rotor at synchronous speed as voltage at the rotor is essentially dc.

Reactive power is supplied to the system through d-axis excitation control on rotor in most cases. The convertor could also be used as STATCOM for dynamic reactive compensation even when the turbine generator is not operational. In earlier designs the Type-3 has been more sensitive to disturbances and would disconnect from the network in a much shorter duration than conventional generators. The safety of DC convertors was the main cause. Now DFIG machines employ more controls than FSIG.

The first proposed solution for the above mentioned issue is known as crowbar protection. This technique short-circuit's the rotor side converter with or without additional resistance and leaves the DFIG as standard induction generator during the disturbance or fault and brings the convertor back after pre-defined time period. This technique is not acceptable anymore in some countries like Germany. Thus turbine manufacturers are coming up with more advanced controls to protect the rotor and provide P and Q control at the same time. One of the techniques is called advanced grid option (AGO) by the Vestas Ltd. This is an additional control which is only activated in case of a ride-through requirement.

Most of the wind farms commissioned in New Zealand after 2004 employ Doubly Fed Technology. The Tararua-3 was commissioned in 2005 has the

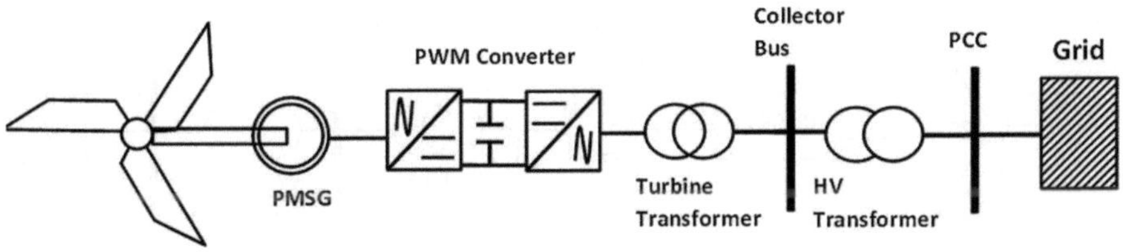


**Fig. 3.5** Full Scale Frequency Convertor (FSFC) Permanent Magnet Synchronous Generator (PMSG) wind turbine (Type-4)

capacity of 93 MW. This is wound rotor machine which is fed through a series voltage-source converter.

#### 3.3.3.3 Type-4 (Full Scale Frequency Converters (FSFC))

In Type-4 the wind generator is decoupled through full back-to-back convertor. These could be conventional generators, dc field or permanent magnet generators. The generator spins at any available rotational speed through direct coupling to the turbine. The frequency may not be 50 Hz at the generator end but electrical power is converted through a back-to-back converter to the required grid frequency, thus giving generator a wide range of speeds because of full frequency convertors. The arrangements of Type-4 are shown in Fig. 3.5. The grid side convertor has the ability to independently control real and reactive power to improve FRT capability, voltage regulation, and reactive power control of the electrical generator.

In New Zealand, the West-Wind farm was first connected in 2009 and employs the Type-4 WTG arrangement. This wind farm has a total capacity of around 140 MW. Another large wind farm, the Te Uku wind farm employs the same technology but this wind farm is connected at a sub-transmission level of 33 kV.

#### 3.3.3.4 Type-5 (Synchronous Generator Technology)

The Te Rere Hau Wind Farm employs Type-5 WTG arrangement in New Zealand. Type-5 WTGs have locally been developed and commissioned by a local New Zealand company [27]. It is based on a gear box technology, converting the variable wind speed to a fixed shaft speed for synchronous generator as shown in Fig. 3.6. The synchronous generator then generates the electricity at grid frequency. Currently, these turbines are smaller and are not widely used but give much better advantages during integration. FRT capability of this type is as good as conventional generators.

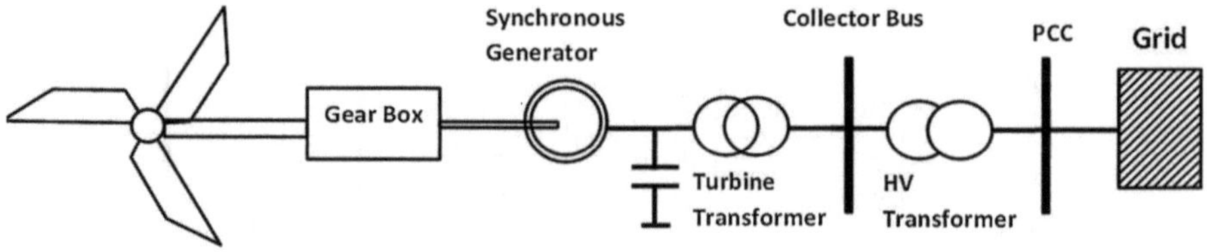


**Fig. 3.6** Type-5 wind turbine generator

### *3.3.4 Frequency Ride-Through*

Apart from voltage ride-through requirements, smaller grids like New Zealand and Ireland are also concerned about frequency ride-through of the modern wind farms. In New Zealand the North Island (NI) the lower frequency limit is 47 Hz and upper frequency limit is 52 Hz. However, In South Island (SI) Frequency limits are slightly extended due to presence of most of hydro plants in South Island and a positive flow of real power from SI to NI. The SI lower frequency is 45 Hz and upper limit is 55 Hz. Normally, the modern wind farms have arbitrary frequency settings and design capability of 47–52 Hz. This configuration may suit the under frequency protection operation of wind farms in NI but their suitability in SI can be an issue [28].

### *3.3.5 International Practices*

The consideration of existing criteria gives a better understanding towards development of new criteria. FRT criterion has already been established and is mandatory in many countries. International experiences have also been considered while developing criteria for the New Zealand system and have been used as a comparison in Sect. 3.5 as shown in Fig. 3.7 [21].

## 3.4 Methodology and Assumptions

The proposed methodology towards development of ride-through envelope could be summarized as shown in the Fig. 3.8.

The above illustration shows the six steps towards developing a FRT criterion. Data collection about the network, faults, protection schemes and other network related information is the first step towards the criteria. Based on the data available a full list of credible contingencies is formed. Normally, the network models are available with Transmission System Operator (TSO), if not then required validated models need to be developed. Electrical load could be static or dynamic in nature,

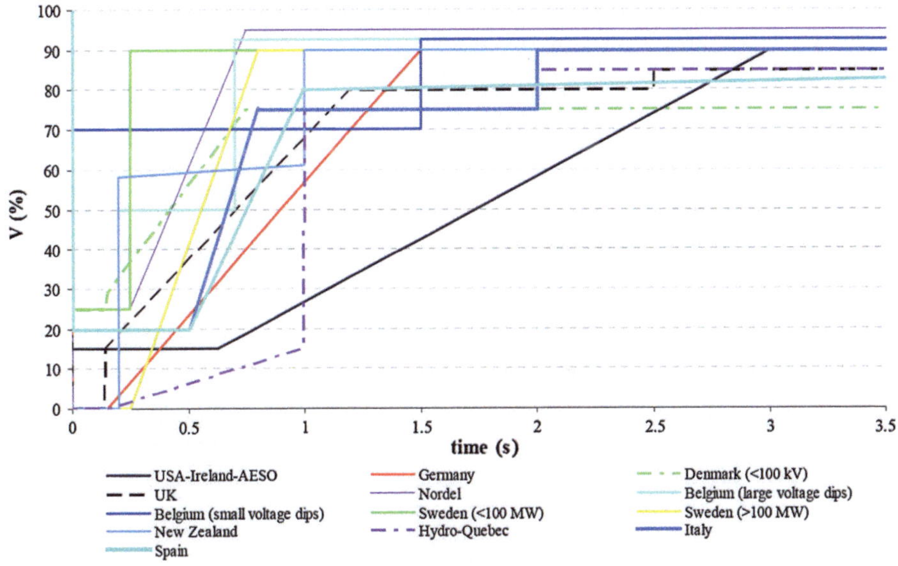


**Fig. 3.7** International FRT criteria in practice

thus basic load modelling assumptions are to be made as well. Static and Dynamic network analysis are carried out to determine, short-circuit strengths at different buses and also obtaining voltage profiles at various buses. Through statistical analysis of the weakest nodes the final voltage profile is achieved for most affected bus bars.

The purpose of this study is to formulate FRT requirements for all generation, wishing to connect to the New Zealand power system through dynamic studies. The outcome of this study is determination of FRT criteria for the New Zealand power system. This criterion is under process of approval and planned to be implemented into the transmission code to assist the SO in managing system security; it will provide a benchmark for manufacturer type testing and commissioning testing [26, 29].

Transpower has existing FRT requirements for the HVDC upgrade project and these and other international criteria are also considered and compared as part of this study as shown in Fig 3.7 [25, 26, 29]. The existing HVDC design requirements of System Operator have been used as inputs into the development of a FRT standard for the transmission system.

The New Zealand transmission system is dispersed into its respective electrical regions. The assumptions discussed below from Sect. 3.4.1 to 3.4.5 are applied and bus voltages recorded to determine the worst case system response or performance to an N-1 scenario. Performance and existing criteria are then analysed towards development of suitable criteria.

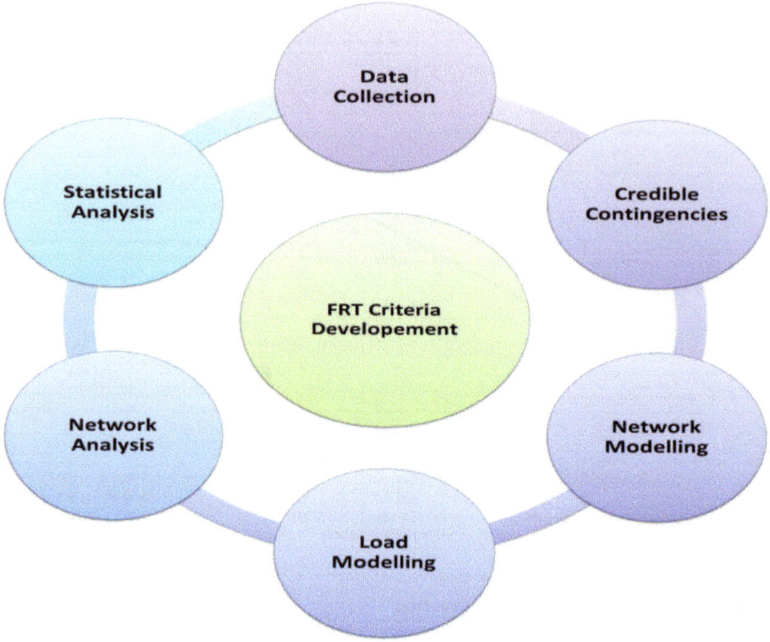


**Fig. 3.8** Proposed methodology for FRT criteria development

### *3.4.1 Load Modelling Assumptions*

For the New Zealand case study a composite load model has been assumed based on similar studies to determine reactive reserve requirements for the grid. For the dynamic studies aggregated load model at Grid Exist Point (GXP) has been assumed [26, 29].

The aggregated load model includes dynamic load such as motor load, static load, and a distribution system model. The motor load model has three different protection groups (Group I, II and III). These groups are determined based on protection type and motor control each group has further subdivisions into groups based on motor sizes i.e. large and small [30]. A middle ground approach has been adopted for this study, where it is assumed that 25 % of motor loads may trip under low voltage conditions and up to 50 % for high voltage conditions. The distribution system is modelled as an 8–10 % impedance transformer based on the forecasted summer or winter MW load respectively [26, 29].

The proportion of each load type at each GXP is derived through survey data for different regions in the system [30–32]. These load surveys were conducted for a peak winter and the extreme summer period.

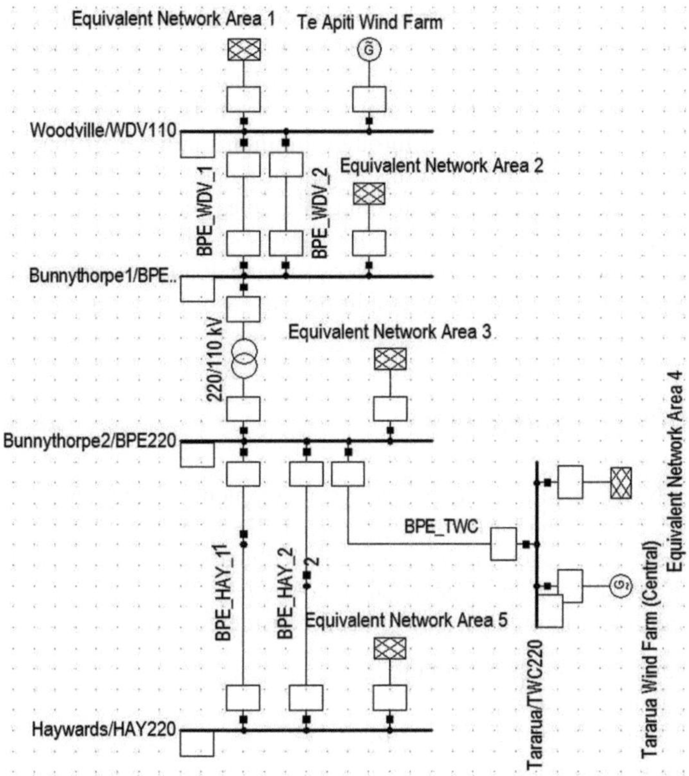


**Fig. 3.9** New Zealand test case study

## *3.4.2 Additional Controllers*

The study assumes that the transient period is well within the operating times for shunt connected reactive equipment and tap changer controllers This equipment is therefore assumed to be fixed at the pre-contingent state for the duration of the dynamic study. An exception to this are the HVDC filters that are controlled by the fast acting Reactive Power Controller (RPC) [26, 33].

## *3.4.3 Load and Generation Scenarios*

A basic assumption was made that all existing wind farms are unavailable. This assumption is based on the fact that existing wind farms may not remain connected for close in faults, with the exception of Type-4 wind farm in the North Island, where their unavailability is the worst case condition for the scenarios considered.

The load and generation scenarios are based on historical data collected through SCADA. High seasonal loads are assumed for LVRT that corresponded with high HVDC north transfer with respect to the North Island studies, and low north or south transfer for the South Island and lower North Island regions. The lowest short-circuit capability is assumed through minimum number of machines available, due to the level of HVDC transfer offsetting the dispatch of synchronous generators. In order to reduce the amount of studies required, the seasonal peak loads are tested for worst case response for various system faults, i.e. high summer load with more motor load or higher overall winter load but with a higher static load percentage. The summer load condition was also a limiting factor [26].

### 3.4.4 Credible Contingencies

Considering N-1(loss of a single power system element) contingency level, all the contingencies are chosen based on their impacts they have on a regional and sub-regional transmission level. Faults at the Low Voltage (LV) level are not modelled as these are considered an issue local to a single generating station only. The following credible contingencies have been considered [26]:

- Loss of single transmission circuit
- Loss of single generating unit
- Loss of a single dynamic reactive plant (SVC)
- Loss of an HVDC pole.

The complete list of credible contingencies in North and South Islands are given in Appendix Table 3A-1 and 3A-2 respectively.

### 3.4.5 Protection Performance

Three-phases, zero impedance faults resulting in the loss of a single element are assumed for all studies. Clearance times assumed are actual operating times; if this information is unavailable then standard operating times are used. With protection signaling, the fault is cleared almost simultaneously at both ends of the circuit. Where signaling is unavailable, zone based protection timings are assumed for Zone-1 and Zone-2. For some of the HVRT contingencies, it is assumed that the equipment is disconnected from the system without any fault at the transmission level. The targeted protection clearance times of various level of protection in New Zealand transmission system as follows:

- Main protection for 220 kV (Transmission) circuits : 120 ms (6 cycles)
- Main protection for 110 kV (Transmission) circuits : 200 ms (10 cycles)
- Main protection for 66 kV (Sub-Transmission) circuits : 200 ms (10 cycles)

**Table 3.1** Study Scenarios

| | Contingency type | Protection type |
|---|---|---|
| **1** | 3-phase local fault on BPE_WDV_2 line near Woodville 110 bus bar | Zone based protection (Zone 1:200 ms, Zone 2:600 ms) |
| **2** | 3-phase local fault on BPE_WDV_1 near Bunnythorpe110 bus bar | Zone based protection (Zone 1:200 ms, Zone 2:600 ms) |
| **3** | 3-phase local fault on BPE_HAY_1 near Bunnythorpe220 bus bar | Main protection with signalling (Main: 120 ms, Back up: 350 ms) |

- Circuit Breaker Failure Time : 350 ms (17.5 cycles)
- Auto-reclosing Time : 1.5 S (75 cycles).

#### 3.4.5.1 Significance of Clearance time on Voltage–time Profile

In a voltage–time profile (LVRT), the magnitude of the voltage is determined by the system characteristic and network configuration. However, targeted protection clearance times are a simple guideline for achieving duration in voltage–time profile, and actual protection clearance times dictate the performance of those criteria during the fault. This ride-through study is a simple extension of investigation of protection performance of WTG system in large scale wind integration context. In order to achieve results closer to realistic network conditions; approximately similar protection clearance times have been used for the network under study as shown in Table 3.1.

## 3.5 Analyses and Results

### *3.5.1 Test Case Study*

Based on the discussed methodology a brief laboratory based contingency analysis was carried out with the help of New Zealand North Island Power Systems (NIPS) as presented in Appendix B. The system diagrams of South Island Power System Networks are also presented in Appendix B.

In this section some contingencies are analyzed on two different voltage levels i.e. 110 and 220 kV. Woodville (WDV110), Bunnythorpe (BPE110, BPE220), and Haywards (HAY220) are the bus bars under the scope of this test case study. WDV110, BPE110, BPE220 are busbars very close to the region where most of the wind generation of New Zealand is currently installed, including Te Apiti Wind Farm, Tararua Wind Farm etc. as shown in Fig. 3.10. The network under consideration is as shown in Figs 3.9 and 3.10. The study includes three different contingencies each for one unique scenario as shown in Table 3.1. Table 3.1 also illustrates the targeted protection clearance times of each bus bar. These protection clearance times are realistically close to actual targets by Transpower.

**Fig. 3.10** Grid section under study

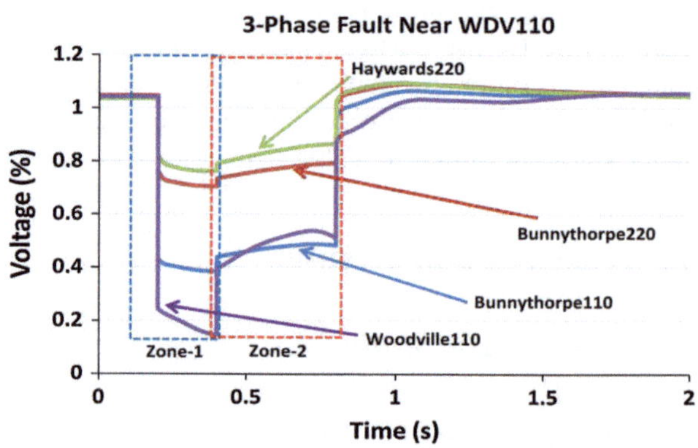


**Fig. 3.11** A 3-phase symmetrical fault near WDV110

In order to assess the ride-through capability of each of the wind farm we need to collect the voltage profiles of neighbouring buses under worst fault conditions. The voltage profiles for these busbars i.e. Woodville110, Bunnythorpe110, Bunnythorpe220, and Haywards220 are presented for all three scenarios.

### 3.5.1.1 Scenario 1

A 3-phase fault was created on BPE_WDV_2 (110 kV) transmission line at a 10 % distance from Woodville Substation. This line is protected using zone based protection scheme and clearance times are mentioned in Table 3.1. The voltage response of bus bars under study is shown in Fig. 3.11.

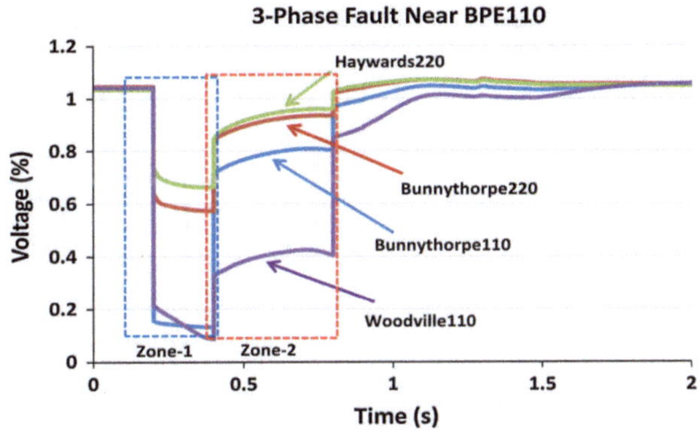


**Fig. 3.12** A 3-phase symmetrical fault near BPE110

The worst voltage is observed on WDV110 before operation of Zone1 protection after 200 ms. The voltage drops down to a value of 15 % of the nominal voltage. The second bus affected is certainly the other end of the line i.e. BPE110 which has a dropped voltage of 40 % of the nominal value. After Zone1 protection operates the voltage at WDV110 and BPE110 returns back to 45–47 % of the nominal value. After Zone 2 protection operates and fault is completely cleared, the voltage quickly recovers and settles back to the nominal value within 1.6 s.

#### 3.5.1.2 Scenario 2

A 3-phase fault was created on BPE_WDV_1 (110 kV) transmission line at a 10 % distance from Bunnythorpe substation. This line is protected using zone based protection scheme and clearance time is around 200 ms. The voltage response of bus bars under study is shown in Fig. 3.12.

The worst voltage in scenario 2 is also observed on WDV110 before operation of Zone1 protection. It shows that this part of the network is very weak and there is not much voltage support at this busbar from any generation nearby. There is a Wind Farm connected to this busbar which is FSIG wind farm and cannot contribute much towards voltage support at this busbar. The voltage drops down to a value of 10 % of the nominal voltage. The second bus affected is certainly BPE110 as fault was closest to this busbar, which has a dropped voltage of 15 % of the nominal value. After Zone1 protection operates the voltage at BPE110 recovers back to 75 % of the nominal value while the voltage at WDV110 could still get back to 40 % of the nominal value. After Zone 2 protection operates and fault is completely cleared the voltage quickly recovers and settles back to the nominal value within 1.75 s.

#### 3.5.1.3 Scenario 3

A 3-phase fault was created on BPE_HAY_1 (220 kV) transmission line at a 10 % distance from Bunnythorpe substation. This line is protected using main and back

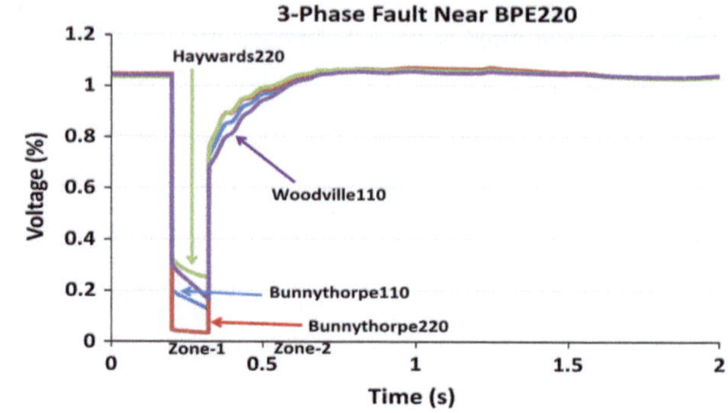


**Fig. 3.13** A 3-phase symmetrical fault near BPE220

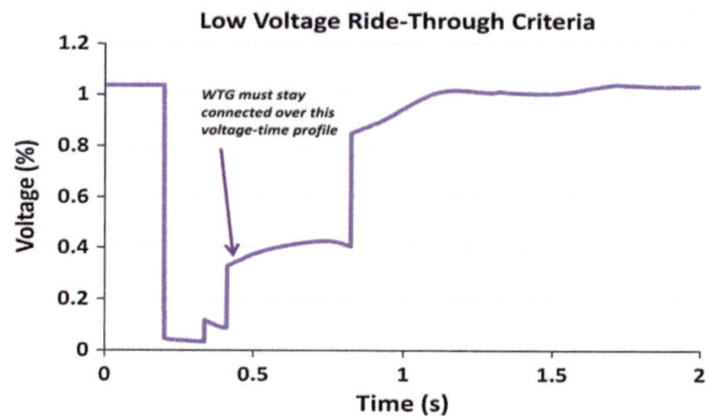


**Fig. 3.14** LVRT criteria for network under study

up protection scheme and clearance times are mentioned in Table 3.1. The voltage response of bus bars under study is shown in Fig. 3.13.

The worst voltage in scenario 3 is also observed on BPE220 during the fault. The voltage drops down to 5 % of the nominal value. After Main protection operates the voltage at all buses recovers back to 70 % of the nominal value and gradually in duration of 500 ms it reaches the nominal value.

Three scenarios have been presented here and four different bus bars of two different levels of transmission voltages and protection philosophies were selected. Figures 3.11– 3.13 presents different voltage envelopes which could be combined together to achieve one single voltage duration profile for the network under study. The accumulative LVRT voltage–time profile is presented in Fig. 3.14.

A detailed contingency analysis was carried out at Transpower to suggest voltage profiles for New Zealand Power System. The author spent a few months at Transpower to observe and be part of this process.

Section 3.5.2 describes the analysis and final results of that collaborative study with Transpower. These envelopes and ride-through criteria have been developed under similar methodology; considerations and assumptions as discussed above.

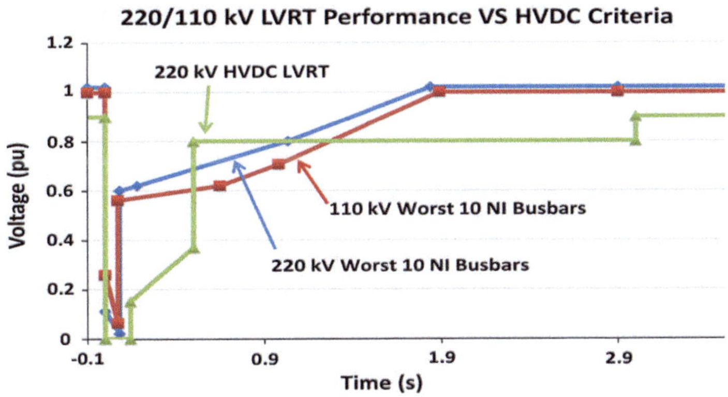


**Fig. 3.15** Summary and existing HVDC transpower comparison—LVRT

## *3.5.2 New Zealand LVRT Summary and Comparison*

Based on the methodology, the system response for the North Island is different than that for the South Island, due to the difference in load composition and contingencies for the region, a separate LVRT criterion for each island is required.

The worst 10 bus results are recorded and averaged on this basis for the LVRT system performance and a safety margin is applied to these profiles to allow for the condition where less motor load is tripped. These results are discussed in the next section.

It was found that the North Island system LVRT performance is worse than that of the South Island. The results for the North Island are compared with existing system operator HVDC and international LVRT criteria in the Figs. 3.15 and 3.16 below.

As shown in Fig. 3.15, the North Island performance violated the System Operator HVDC criteria after 500 ms where the voltage recovery is outside the 0.8 pu voltage requirements. The HVDC criterion is used as a benchmark for these studies because it considers the limitations on distance protection under faulted conditions, the LVRT should not be prolonged so that Zone 3 and Zone 4 distance relays do not operate inadvertently [25, 33].

As can be seen from Fig. 3.15 the LVRT performance and System Operator HVDC criteria are within the boundaries of many of the international FRT criteria. It is also visible that the LVRT curves are within the AGO2 wind turbine criteria with the exception of the zero voltage periods. It is worth noting that many of the international criteria are applied to highly interconnected transmission systems. This comparison gives us some confidence towards existing Wind Turbine Generator technology to be connected to New Zealand Grid without any major ride-through issues.

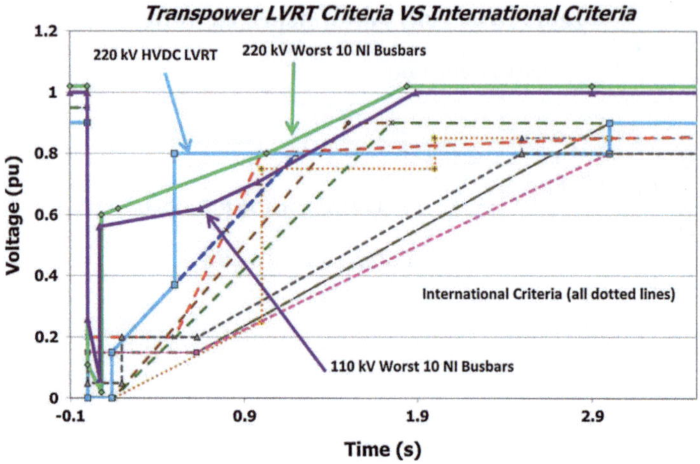


**Fig. 3.16** International comparison-LVRT

## 3.5.3 *Final Proposed FRT Envelopes*

Based on the considerations, discussed methodology, results and analysis above, a separate FRT criteria or envelope can be developed for North and South Islands due to the difference in LVRT system performance [26].

### 3.5.3.1 Proposed North Island LVRT Requirement

The FRT envelope proposed for the North Island system is the combination of the existing Transpower HVDC criteria and system performance for the LVRT criteria with margin applied. The profiles are shown in Fig. 3.17 [26].

### 3.5.3.2 Proposed South Island LVRT Requirement

The envelope proposed for the South Island system is the existing Transpower HVDC LVRT criteria with margin. The envelope is shown in Fig. 3.18 [26].

## 3.6 Compliance Tests

Existing Wind farms and generators need to be tested against proposed criteria for compliance; thus it is proposed that the Generator Asset Owner are required to conduct dynamic simulations using the curves and observe the modeled response of the generator. It is also proposed that type testing be performed using laboratory

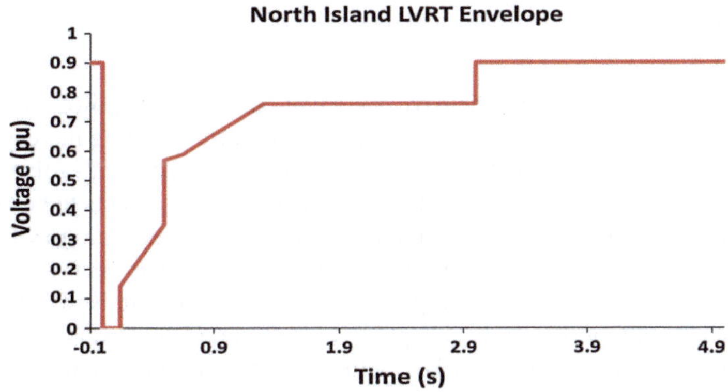


**Fig. 3.17** North Island LVRT Envelope

**Fig. 3.18** South Island LVRT Envelope

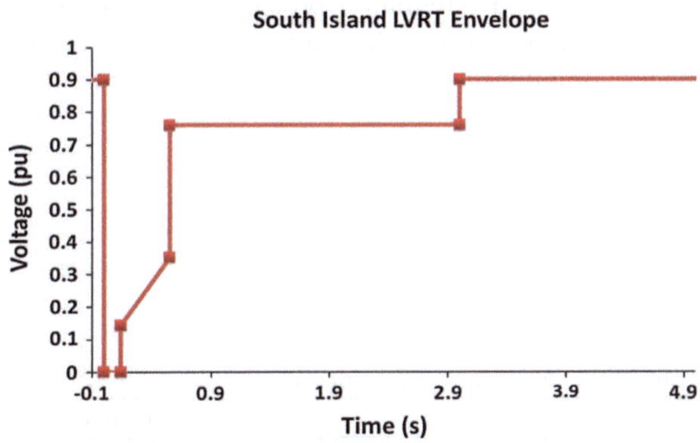


testing on the actual generator or turbine under faulted conditions using a simulated HV bus fault. Necessary and proper data recording is also suggested to be installed on generator site to enable the System Operator to review generator performance for actual system disturbances. The System Operator is in progress of drafting process documentation for the commissioning and testing of large scale wind farms.

## 3.7 Recommendations and Future Direction

In future, country level criteria may be proposed but it is recommended at this stage that an island based FRT envelope is proposed. It is also proposed that the FRT envelope be reviewed once additional reactive plants are commissioned as it is expected that a common FRT envelope for both islands can be implemented once this has been achieved.

## 3.8 Summary

This chapter provides a comprehensive review of New Zealand power system from wind integration viewpoint. FRT criteria are required for enabling increasing wind integration scenario for New Zealand network. The establishment of these criteria requires special analysis, data set and other network related aspects to be taken into account. This chapter discussed available wind generator technologies from FRT capability viewpoint and system aspects. A methodology has been proposed and adopted to develop a voltage envelope for the grid. A complete New Zealand grid example has been reported in this chapter which explains the development of FRT criteria for a section of a grid. Further, actual proposed ride-through envelopes for New Zealand North Island and South Island grids have also been presented towards the end of this chapter. This work was carried out in collaboration with Transpower; the Transmission System Operator. The motivation of this work was to explore wind integration protection aspects where FRT criteria development plays an important role for protection clearance times and coordination management. Compliance testing, future improvements, directions and recommendations are also identified.

## Appendix A

**Table 3A-1** North Island contingencies

| Region | Contingent events | Condition |
|---|---|---|
| Northland | HEN-SWN-1 | TUV |
| | MDN-MPE-1 | TUV |
| Auckland | OTA-PEN-5 or 6 | TUV |
| Upper North Island | OTC G1 | TUV |
| | HLY G5 | TUV |
| | HLY-OTA-2 | TUV |
| | OHW-OTA-1 | TUV |
| | HLY-DRY-1 (committed upgrade) | TUV |
| | WKM-BRH (committed upgrade) | TUV |
| Waikato | HAM-WKM-1 | TUV |
| | ARI-HAM-1 or 2 | TUV |
| | BOB-HAM-1 or 2 | TUV |
| | HAM-WHU-1 or 2 | TUV |
| | HLY-TWH-1 | TOV |
| BOP | OHK-WRK-1 | TUV |
| | ARI-KIN1 or 2 | TUV |
| Hawke's Bay | RDF-WHI-1 | TUV |
| | FHL-RDF-1 or 2 | TUV |

(continued)

**Table 3A-1** (continued)

| Region | Contingent events | Condition |
|---|---|---|
| Taranaki | SPL G1 | TOV |
| | BRK-SFD-3 | TUV |
| | HWA-WVY-1 | TUV |
| Central North Island (Bunnythorpe) | BPE-MTR-OKN-ONG | TUV |
| | TKU-WKM-1 or 2 | TUV |
| | BPE-WDV | TUV |
| | MGM-WDV-1 | TUV |
| | SFD-TMN-1 | TUV |
| | BPE-TKU-1 or 2 | TUV |
| Wellington | BPE-TWC-LTN-1 | TUV |
| | Pole 2 | TOV |
| | HAY-TKR-1 or 2 | TUV |
| | HAY-UHT-1 or 2 | TUV |

**Table 3A-2** South Island contingencies

| Region | Contingent events | Condition |
|---|---|---|
| Nelson/Marlborough | ISL-KIK-2 or 3 | TUV |
| | BLN-KIK-1 | TUV |
| | KIK Statcom (committed upgrade) | TOV |
| West Coast | IGH-RFN-ATU | TUV |
| | HOR-ISL-1 or 2 | TUV |
| | COL-OTI-2 | TUV |
| Upper South Island | ASB-TIM-TWZ-1 or 2 | TUV |
| | ISL-TKB | TUV |
| Otago/Southland | CYD-CML-TWZ-1 or 2 | TUV |
| | CYD-CML-TWZ-1 or 2 -CB fail event | TUV |
| | Pole 2 | TOV |
| | TWI pot line | TOV |
| | INV-ROX | TUV |
| | GOR-ROX-1 | TUV |
| | BAL-BWK-HWB-1 | TUV |

# Appendix B

North Island Power Systems Diagram

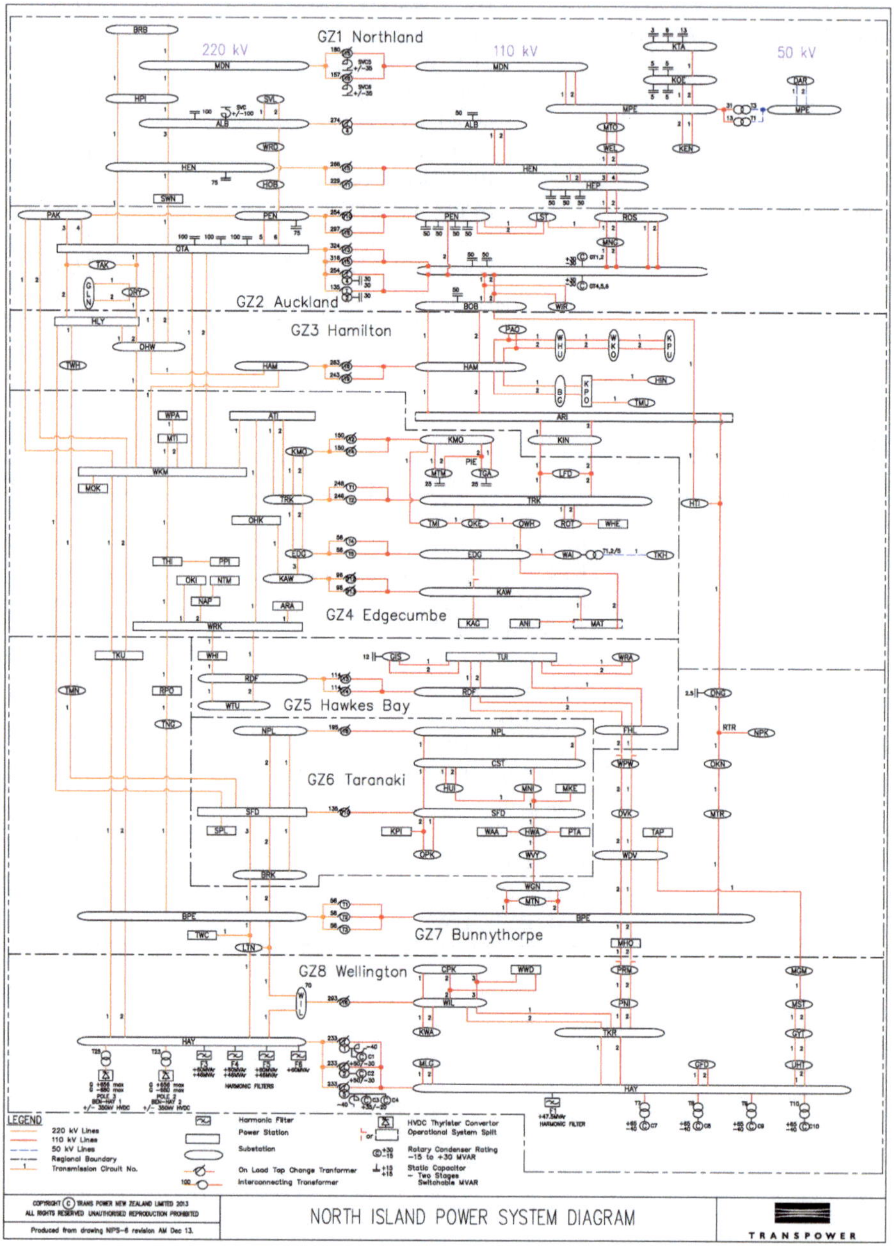

## South Island Power Systems Diagram

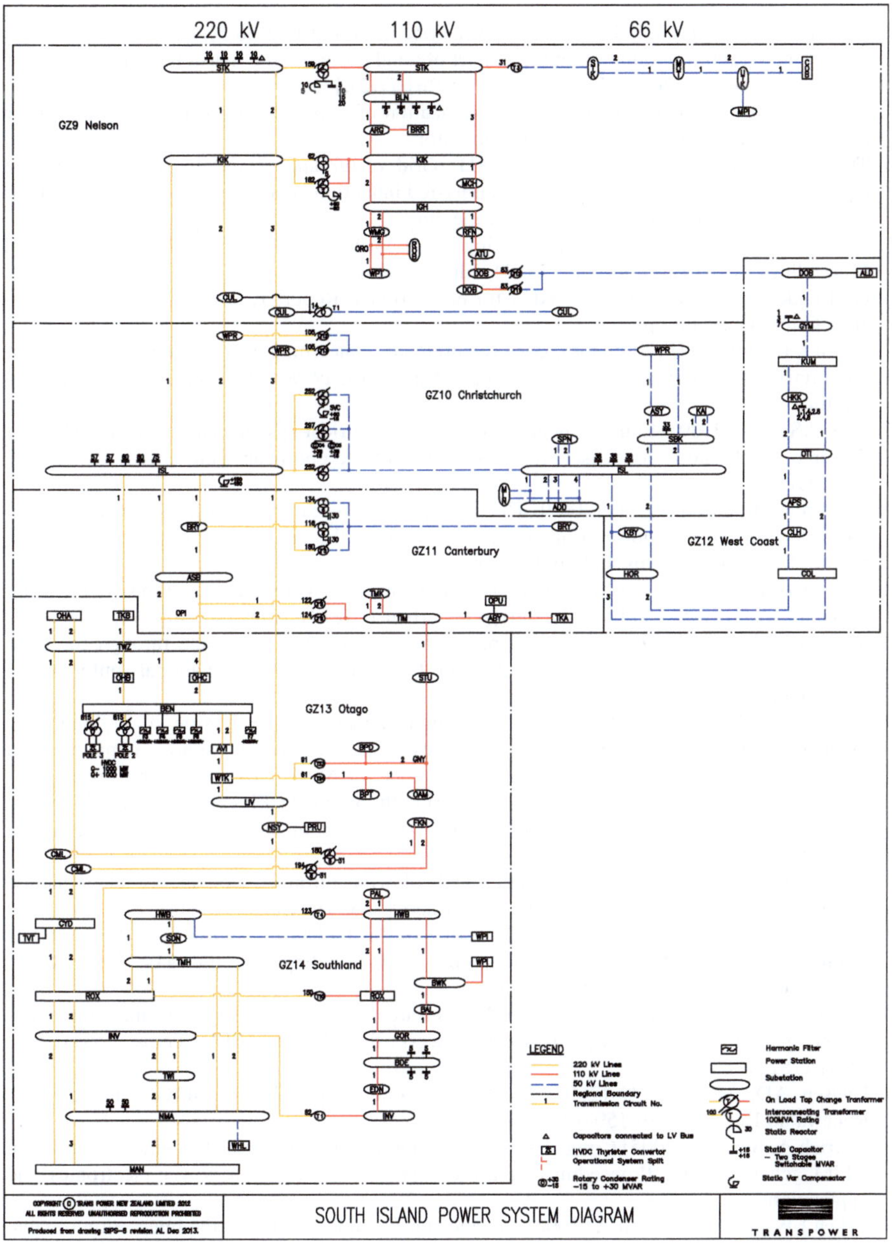

## References

1. Ackermann T (2006) Wind power in power systems. Wind Eng 30(5):447–449
2. Jauch C et al (2007) Simulation of the impact of wind power on the transient fault behavior of the Nordic power system. Electr Power Syst Res 77(2):135–144
3. Sørensen P et al (2003) Simulation and verification of transient events in large wind power installations. Denmark, Risø National Laboratory
4. Thiringer T, Petersson A, Petru T (2003) Grid disturbance response of wind turbines equipped with induction generator and doubly-fed induction generator. In: Proceddings of the IEEE power engineering society general meeting
5. Eltra (2000) Specification for connecting wind farms to the transmission network, In: Proceedings of the Eltra doc. no. 74174, Eltra, Denmark
6. Grid code regulations for high and extra high voltage Report ENENARHS (2006) E.ON. Netz Gmbh Germany. p 46
7. Erlich I, Bachmann U (2005) Grid code requirements concerning connection and operation of wind turbines in Germany. In: Proceedings of the IEEE power engineering society general meeting
8. Kalsi SS et al (2009). Enhancement of wind farm electrical system with a superconducting dynamic synchronous condenser. In: Proceedings of the european wind energy conference (EWEC)
9. Huan-ping L, Jin-ming Y (2009) The performance research of large scale wind farm connected to external power grid. In: Proceedings of the 3rd international conference on power electronics systems and applications
10. Chompoo-inwai C et al (2005) Reactive compensation techniques to improve the ride-through capability of wind turbine during disturbance. IEEE Trans Ind Appl 41(3):666–672
11. Rathi MR, Mohan N (2005) A novel robust low voltage and fault ride through for wind turbine application operating in weak grids. In: Proceedings of the 31st annual conference of IEEE industrial electronics society
12. Zhan C, Barker CD (2006) Fault ride-through capability investigation of a doubly-fed induction generator with an additional series-connected voltage source converter. In: Proceedings of the 8th IEE international conference on AC and DC power transmission
13. Bing X, Fox B, Flynn D (2004) Study of fault ride-through for DFIG based wind turbines. In: Proceedings of the 2004 IEEE international conference on electric utility deregulation, restructuring and power technologies
14. Morren J, de Haan SWH (2005) Ridethrough of wind turbines with doubly-fed induction generator during a voltage dip. IEEE Trans Energy Convers 20(2):435–441
15. Jayanti NG et al (2006) Optimising the rating of the UPQC for applying to the fault ride through enhancement of wind generation. In: Proceedings of the 41st international universities power engineering conference
16. Seman S, Niiranen J, Arkkio A (2006) Ride-through analysis of doubly fed induction wind-power generator under unsymmetrical network disturbance. IEEE Trans Power Syst 21(4):1782–1789
17. Hansen AD, Michalke G (2007) Fault ride-through capability of DFIG wind turbines. Renewable Energy 32(9):1594–1610
18. Dittrich A, Stoev A (2005) Comparison of fault ride-through strategies for wind turbines with DFIM generators. In: Proceedings of the european conference on power electronics and applications
19. Erlich I,Winter W, Dittrich A (2006) Advanced grid requirements for the integration of wind turbines into the German transmission system. In: Proceedings of the IEEE power engineering society general meeting
20. Kasem AH et al (2008) An improved fault ride-through strategy for doubly fed induction generator-based wind turbines. IET Renew Power Gener 2(4):201–214

21. Tsili M, Papathanassiou S (2009) A review of grid code technical requirements for wind farms. IET Renew Power Gener 3(3):308–332
22. Wind Farms in New Zealand (2012) Available from http://www.windenergy.org.nz/. Accessed 30 Oct 2012
23. Electricity Authority New Zealand (2012) Available from http://www.ea.govt.nz. Accessed 1 Nov 2012
24. Bublat T (2008) Comparison of high technical demands on grid connected wind turbines defined in international grid codes. In: Proceedings of the european wind energy conference, Brussels, Belgium
25. Generator Fault Ride Through (FRT) Investigation—Stage 1: Literature Review (2009) Feburary 2009, System Operator, Transpower NZ Ltd
26. Demler GL (2009) Generator Fault Ride Through (FRT) Investigation—Stage 2. May 2009, System Operator, Transpower New Zealand Ltd
27. Wind Turbine Manufacturer Wind Flow (2013) Available from http://www.windflow.co.nz/. Accesssed 26 Jan 2013
28. Nutt S (2010) Wind generation compliance with South Island frequency limits In: Proceedings of the New Zealand wind energy conference, Palmerston North, New Zealand
29. Upper South Island Voltage Stability Study—Information Collation (2008) April 2008, Sinclair Knight Mertz
30. Transmission Code-TP.DG 25.01, Section 6.7 (2009) Transpower NZ Ltd, December 2009
31. Auckland Voltage Stability Study—Information Collation (2005) March 2005, Sinclair Knight Mertz
32. Wellington Voltage Stability Study—Information Collation (2008) October 2008, Sinclair Knight Mertz
33. New Zealand Inter Island HVDC Pole 3-Project Contract Document (2009) Transpower NZ Ltd, October 2009. Vol 2(Chapter 5): p Clauses 6.1.4 and 8.3.1.4

# Chapter 4
# High Penetration of Rooftop Photovoltaic Cells in Low Voltage Distribution Networks: Voltage Imbalance and Improvement

**Farhad Shahnia and Arindam Ghosh**

**Abstract** Installation of domestic rooftop photovoltaic cells (PVs) is increasing due to feed–in tariff and motivation driven by environmental concerns. Even though the increase in the PV installation is gradual, their locations and ratings are often random. Therefore, such single–phase bi–directional power flow caused by the residential customers can have adverse effect on the voltage imbalance of a three–phase distribution network. In this chapter, a voltage imbalance sensitivity analysis and stochastic evaluation are carried out based on the ratings and locations of single–phase grid–connected rooftop PVs in a residential low voltage distribution network. The stochastic evaluation, based on Monte Carlo method, predicts a failure index of non–standard voltage imbalance in the network in presence of PVs. Later, the application of series and parallel custom power devices are investigated to improve voltage imbalance problem in these feeders. In this regard, first, the effectiveness of these two custom power devices is demonstrated vis–à–vis the voltage imbalance reduction in feeders containing rooftop PVs. Their effectiveness is investigated from the installation location and rating points of view. Later, a Monte Carlo based stochastic analysis is utilized to investigate their efficacy for different uncertainties of load and PV rating and location in the network. This is followed by demonstrating the dynamic feasibility and stability issues of applying these devices in the network.

F. Shahnia (✉)
Center of Smart Grid and Sustainable Power Systems,
Department of Electrical and Computer Engineering,
Curtin University, Perth, Australia
e-mail: farhad.shahnia@curtin.edu.au

A. Ghosh
School of Electrical Engineering & Computer Science,
Queensland University of Technology, Brisbane, Australia
e-mail: a.ghosh@qut.edu.au

J. Hossain and A. Mahmud (eds.), *Renewable Energy Integration*,
Green Energy and Technology, DOI: 10.1007/978-981-4585-27-9_4,

**Keywords** Distribution network · Single–phase rooftop PV · Voltage imbalance · Sensitivity analysis · Stochastic evaluation · DSTATCOM · DVR

## 4.1 Introduction

Current and voltage imbalance is one of the major power quality problems in low voltage distribution networks [1]. Voltage imbalance is more common in individual customer loads due to phase load imbalances, especially where large single–phase power loads are used [2]. Although voltages are well balanced at the supply side, the voltages at the customer level can become unbalanced due to the unequal system impedances, unequal distribution of single–phase loads or large number of single–phase transformers [2]. Usually, the electric utilities aim to distribute the residential loads equally among the three phases of distribution feeders [3].

An increase in the voltage imbalance can result in overheating and de-rating of all induction motor types of loads [4]. Voltage imbalance can also cause network problems such as mal-operation of protection relays and voltage regulation equipment, and generation of non–characteristic harmonics from power electronic loads [3].

In recent years, there is a growing interest in the residential customers in installing of single-phase grid-connected rooftop Photovoltaic cells (PVs) due to new energy and incentive polices in several countries [5]. The most important characteristic of these PVs is that their output power being fed to the grid is not controlled and is dependent on the instantaneous power from the sun. Several technical problems of these systems such as harmonics, voltage profile and power losses are already studied and investigated in [6–8].

The residential rooftop PVs are currently installed randomly across distribution systems. This may lead to an increase in the imbalance index of the network [9]. This will increasingly cause problems for three-phase loads (e.g. motors for pumps and elevators). Therefore, there is a need for investigating the sensitivity analysis of voltage imbalance in these networks.

A deterministic analysis may not be suitable due to the randomness of PV installations and their intermittent nature of power generation. Monte Carlo method is already applied for analysis of uncertainties in the network in order to study load flow, voltage sag, fault and reliability [10]. Therefore, a stochastic evaluation based on Monte Carlo method is carried out in this chapter to investigate and predict the network voltage imbalance for the uncertainties arising due to rooftop PV power ratings and locations [11].

Some voltage imbalance improvement methods are already proposed and investigated in [12–14]. However, parallel and series converter–based Custom Power Devices (CPD) can also be used for power quality improvement [1, 15]. In this chapter, the application of CPDs, in particular, Distribution Static Compensator

(DSTATCOM) and Dynamic Voltage Restorer (DVR) are demonstrated for voltage imbalance reduction and voltage profile correction within these networks. Their optimum installation location, efficacy and ratings will be discussed. However, due to uncertainty in load demand and PV generation in addition to PV ratings and locations along the feeder, a Monte Carlo based stochastic analysis is carried out later to investigate the efficacy of the proposed CPDs for many different uncertainties within the network. Finally, the dynamic feasibility of the system in the presence of the proposed improvement methods will be demonstrated.

## 4.2 Voltage Imbalance in Low Voltage Distribution Networks

The Electricity supplies are nominally 110 V (in US and Canada) or 220–240 V. National standards specify that the nominal voltage at the source should be in a narrow tolerance range. Based on Australian Standard Voltages, AS60038–2000, Australian low voltage network is 230 V with a tolerance between +10 % and −6 % [16].

Voltage imbalance in three–phase systems is a condition in which the three phase voltages differ in magnitude and/or do not have normal 120° phase differences. IEEE Recommended Practice for Monitoring Electric Power Quality defines voltage imbalance as [17]

$$VUF\,\% = \left|\frac{V_-}{V_+}\right| \times 100 \tag{4.1}$$

where $V_-$ and $V_+$ are the negative and positive sequence of the voltage, respectively. This will be referred to as percentage voltage imbalance in the chapter. According to [17], the allowable limit for voltage imbalance is limited to 2 % in low voltage networks. Engineering Recommendation P29 in UK not only limits voltage imbalance of the network to 2 %, but also limits the voltage imbalance to 1.3 % at the load point [18]. In this chapter, we assume that the standard limit is 2 %.

The utilities try to minimize the imbalance index in their network by distributing single–phase loads equally across all three phases. Probabilistic studies have shown that it is very rare that the residential and small business loads can result in higher values of the voltage imbalance in the network. The measurements done in a low voltage distribution network in US [19], Brazil [20] and Iran [21] conclude that the probability of the voltage imbalance to be more than 3 % in the network is about 2–5 %. However, this can only be achieved if the engineering judgements have been applied for selecting the appropriate size of the conductors and cables, transformer ratings and also if the load dispatch among the three phases are recorrected by later observations and measurements. If the appropriate designs have not been done or there is a non–standard voltage drop in the network, it is highly probable that the network also suffers from higher voltage imbalance.

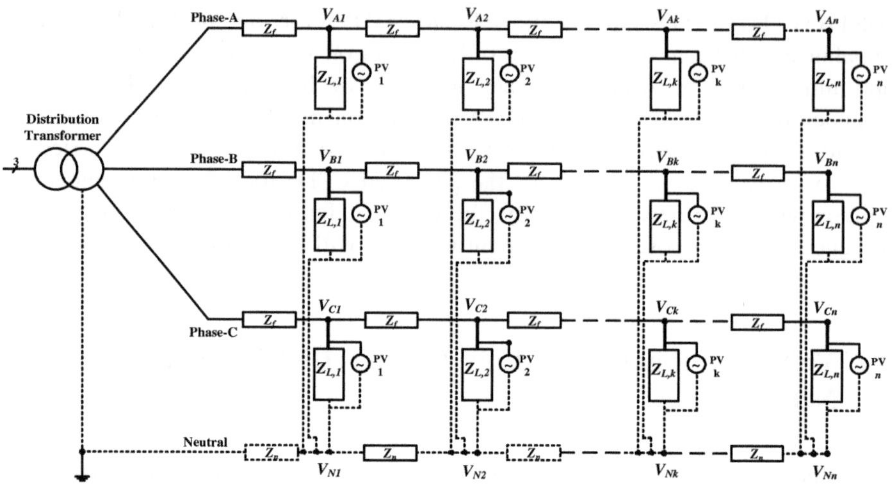


**Fig. 4.1** Schematic single line diagram of the considered low voltage distribution network

As mentioned before, rooftop PVs that are currently being installed depend on outlook and financial condition of householders. Therefore, it is not unexpected that these installations are randomly placed along distribution feeders. For example, it is possible that 80 % of customers on a phase have installed PVs, while the other two phases have only 50 and 10 % installed. In such a condition, even if the voltage imbalance of the network was within the standard limits without any PVs, it is not guaranteed to remain so. Therefore, the possible PV installation number or rating on such systems must be investigated in a way that the voltage imbalance is still kept within the standard limit.

### *4.2.1 Network Structure*

A sample radial low voltage residential urban distribution network is considered for voltage imbalance investigations. The simplified equivalent single line diagram of one feeder of the network is illustrated in Fig. 4.1. In this model, the PVs are all grid–connected such that the surplus of the electricity generated will flow to the grid. It has been assumed that all the PVs work at unity power factor based on IEEE Recommended Practice for Utility Interface of Photovoltaic Systems [22]. It is also assumed that the neutral conductor is making the path for unbalance current circulation and the analysis is based on the mutual effect of three phases.

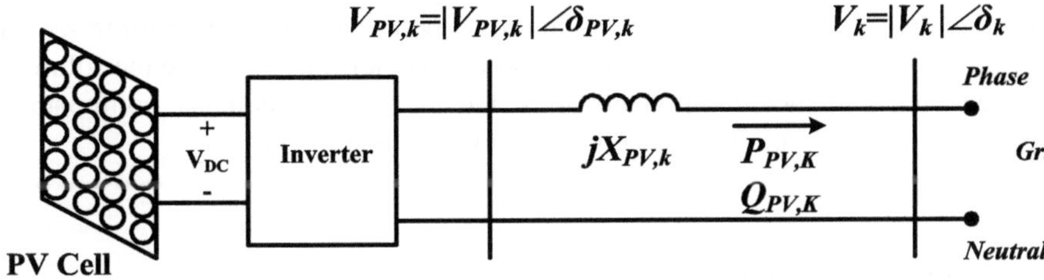


**Fig. 4.2** Schematic diagram of a PV connection to grid

### 4.2.2 Power Flow Analysis

For calculating the voltage imbalance, it is necessary that the network to be analysed and the voltages at the desired nodes to be calculated. Based on the KCL on $k$th node of phase $A$, we have

$$\frac{\beta\left(V_{A,PV,k}-V_{A,k}\right)}{X_{A,PV,k}}+\frac{V_{A,k-1}-V_{A,k}}{Z_f}+\frac{V_{A,k+1}-V_{A,k}}{Z_f}+\frac{V_{N,k}-V_{A,k}}{Z_{A,L,k}}=0 \tag{4.2}$$

where $Z_f$ is the feeder impedance between two adjacent nodes in phase lines, $V_{A,i,}$ $i=1,\ldots,n$ is the single–phase voltage of the $i$th node of phase $A$, $Z_{A,L,k}$ is the load impedance connected to $k$th node of phase $A$ and and $V_{N,k}$ is the voltage of the neutral wire connected to $k$th node. $V_{A,PV,k}$ and $X_{A,PV,k}$ are the PV voltage and impedance connected to $k$th node of phase $A$. Similar equations are valid for phases $B$ and $C$. In (4.2), the controlling constant $\beta$ is equal to 1 when there is a PV connected to $k$th node, otherwise, it is zero.

KCL for each node on the neutral line is

$$\frac{V_{N,k-1}-V_{N,k}}{Z_n}+\frac{V_{N,k+1}-V_{N,k}}{Z_n}+\frac{V_{N,k}-V_{A,k}}{Z_{A,L,k}}+\frac{V_{N,k}-V_{B,k}}{Z_{B,L,k}}+\frac{V_{N,k}-V_{C,k}}{Z_{C,L,k}}=0 \tag{4.3}$$

where $Z_n$ is the feeder impedance between two adjacent nodes in neutral line.

The simplified diagram of the PV connection to the grid is shown in Fig. 4.2. Based on this figure, we have

$$P_{PV,k}=\frac{\left|V_{PV,k}\right|\left|V_k\right|}{X_{PV,k}}\sin\left(\delta_{PV,k}-\delta_k\right) \tag{4.4}$$

$$Q_{PV,k}=\frac{\left|V_k\right|}{X_{PV,k}}\left(\left|V_{PV,k}\right|\cos\left(\delta_{PV,k}-\delta_k\right)-\left|V_k\right|\right) \tag{4.5}$$

where $P_{PV,k}$ and $Q_{PV,k}$ are respectively the active and reactive power output of the PV connected to $k$th node. Assuming $P_{PV,k}$ and $Q_{PV,k}$ to be constant and $|V_k|$ and $\delta_k$ are known, $|V_{PV,k}|$ and $\delta_{PV,k}$ can be calculated. Please note that $Q_{PV,k}$ will be zero if the PV operates in Unity Power Factor (UPF).

To calculate $V_k$ from (4.2)–(4.5), an iterative method is required. Starting with a set of initial values, the entire network is solved to determine $V_k$. Once the solution converges, the sequence components are calculated. These sequence components are later used to voltage imbalance calculation given in (4.1).

### 4.2.3 Sensitivity Analysis

The voltage at any node can be considered as a function of the location and rating of PV. Therefore the sensitivity of network voltage imbalance to a PV location and rating is expressed as

$$S_k = \frac{\partial VUF}{\partial P_{PV,k}} \cdot \frac{P_{PV,k}}{VUF} \tag{4.6}$$

Since voltages at each node are calculated iteratively, the sensitivity is calculated numerically once the iterations converge as

$$S_k = \frac{VUF(\gamma+1) - VUF(\gamma)}{P_{PV,k}(\gamma+1) - P_{PV,k}(\gamma)} \tag{4.7}$$

where $0 \le \gamma \le 4$ defines the rating of the PV (i.e. 0, 1, 2, 3, 4, 5 kW).

### 4.2.4 Stochastic Evaluation

Monte Carlo simulation is a powerful numerical method of stochastic evaluation based on random input variables [17].

The inherent characteristic of low voltage distribution networks includes random variation of residential load demand and PV power generation at different time periods. This variation is based on load demand in about 12 h sunshine daily pattern and summer–winter periods. In addition to this, the random location and nominal power of PVs increase the randomness of the network.

For investigating the voltage imbalance in the network when there is a random combination of rooftop PVs with different ratings and at different location on all phases and feeders on the network, a stochastic evaluation based on Monte Carlo method has been carried out. Three random inputs of the stochastic evaluation are the number of householders with installed rooftop PVs, the ratings of the PVs and their location in the different phases and feeders. The flowchart of Monte Carlo method is shown in Fig. 4.3. In this figure, the inputs are network (load, feeder and transformer) data. The random number generation and selection of the other parameters for the Monte Carlo method is explained in Sect. 4.3.2. For each set of data, the load flow is carried out and voltage imbalance is calculated.

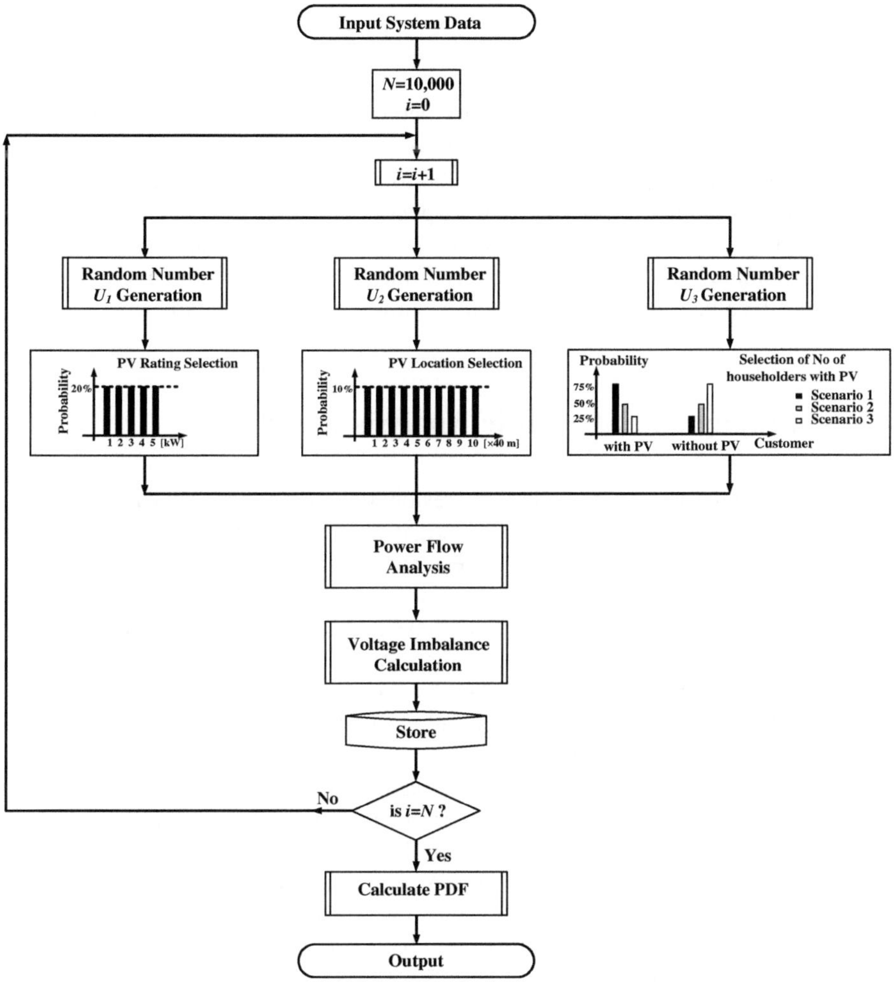


**Fig. 4.3** Monte Carlo flowchart for stochastic evaluation

The expected voltage imbalance at the calculation node $\overline{VUF_j}$ from each trial $1 \leq k \leq N$, is calculated by

$$\overline{VUF_j} = \frac{1}{N}\sum_{k=1}^{N} VUF_k \tag{4.8}$$

The unbiased sample variance for voltage imbalance at the calculated nodes (beginning or end of feeder) is as follows:

$$Var(VUF_j) = \frac{1}{N-1}\sum_{k=1}^{N}\left(VUF_k - \overline{VUF_j}\right)^2 \quad (4.9)$$

The stopping rule of the Monte Carlo method is chosen based on achieving an acceptable convergence for $\overline{VUF}$ and *Var(VUF)*. In this study, the number of Monte Carlo trials is chosen as N = 10,000 to achieve an acceptable convergence. This is more explained in the table in Appendix A.

The voltage imbalance results as the output of the Monte Carlo simulations are used to calculate the Probability Density Function (PDF) and the average (mean value) of all voltage imbalances which is shown as $\lambda$ in the chapter.

## 4.3 Sensitivity and Stochastic Analysis Results

A sample radial low voltage (415 V) residential urban distribution network is considered for voltage imbalance investigations. This network is supposed to supply electricity to a combination of residential and small business customers. It has three feeders, each three-phase and 4-wire system with equal length (400 m) and equal number of customers on each phase and feeder. The poles are located at a distance of almost 40 m from each other. At each pole, 2 houses are supplied from each phase. The feeders and their cross–section are also designed appropriately based on the amount of power and the voltage drop. The technical data of the network is given in the Appendix B.

It is assumed that the total electric demand of the network is almost 1 MVA including the low voltage network under consideration that is supplying a total demand of 360 kW. It is also assumed that during the period of study the loads of phase *A*, *B* and *C* are 60, 120 and 180 kW, respectively. The rest of the network load (the portion not included in this study) is considered as a lumped load. The rooftop PVs installed by the householders have an output power in the range of 1–5 kW working in unity power factor. Several studies are performed [11] and their results are discussed below.

The utilities usually measure and monitor the voltage imbalance at the beginning of the feeder (i.e. secondary side of the distribution transformer). As described previously, the probabilistic studies of the measurement results show that there is low chance of having a voltage imbalance beyond 2 % at this location. Let us assume that the length of the three phases and number of customers per phase are the same. Even then, since the power consumption of the loads are different, there will be different voltage drops along the feeder. This will result in different voltage amplitudes and angles at different locations along the feeder. This phenomenon can result in high voltage imbalance especially at the end of the feeders. To keep the voltage drop within the limit all along the feeder, the utilities install single–phase pole mounted capacitors or increase the cross–section of the feeder. However, voltage imbalance at the end of the feeder still remains higher

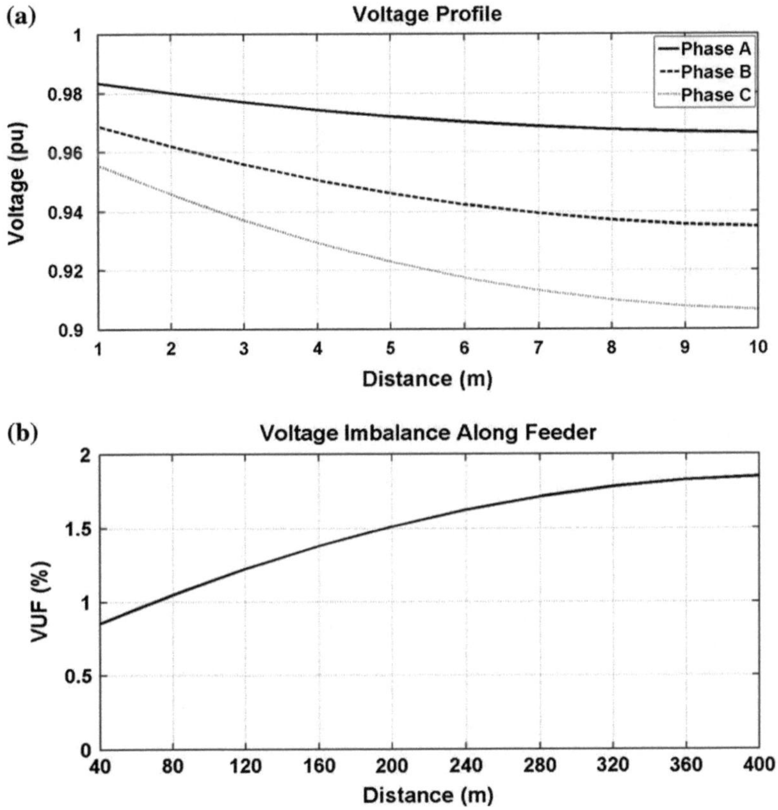


**Fig. 4.4** **a** Network voltage profile along the feeder, **b** Voltage imbalance profile along the feeder

than the beginning of the feeder. This might cause a problem if some three–phase induction motors or power electronics based equipment are located far from the beginning of the feeder. The voltage profile and voltage unbalance profile along the feeder in the system under consideration is shown in Figs. 4.4a and b, respectively. From this figure, it can be seen that the voltage amplitude at the beginning of the feeder is 0.98, 0.96 and 0.95 pu for phases *A*, *B* and *C* respectively. These values are decreased to 0.96, 0.93 and 0.90 pu at the end of the feeder, respectively. Therefore, voltage imbalance at the beginning of the feeder has increased from 0.85 to 1.84 % at the end.

Due to time varying characteristic of residential load demand and PV output power, a study is included to investigate the voltage imbalance variation during a specific time period. Let us assume some PVs with the power output profile as shown in Fig. 4.5a and the residential loads with load profile as shown in Fig. 4.5b are considered [23, 24]. Time varying characteristic of voltage imbalance at the beginning and end of the feeder is shown in Fig. 4.5c. Voltage imbalance has a time varying characteristic and it may increase depending on the PVs.

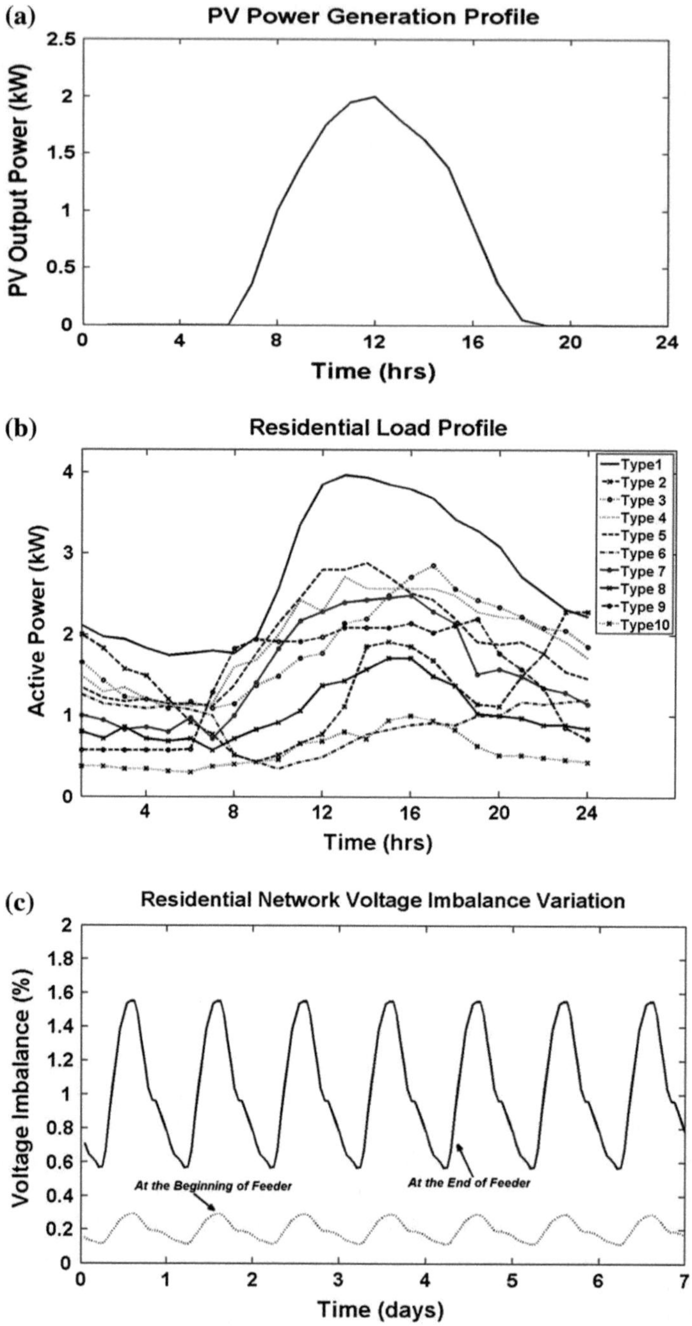


**Fig. 4.5** **a** Power generation profile of a 2 kW rooftop PV, **b** 10 different types of residential loads profiles, **c** Time varying characteristic of voltage imbalance with constant location for PVs in the network

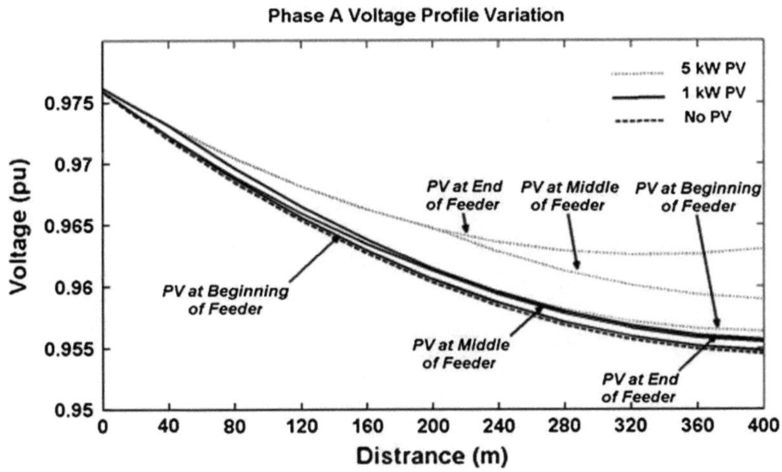


**Fig. 4.6** Variation of phase *A* voltage profile versus the location and rating of the PV in phase *A*

### *4.3.1 Sensitivity Analysis of a Single PV on Voltage Imbalance*

The voltage imbalance variation due to the location of one PV with a constant rating will depend on the total load of the phase in which it is installed. Usually rooftop PVs can have ratings up to 5 kW for urban residential customers. It is also important to define the point at which the voltage imbalance will be measured.

It is expected that the voltage profile will improve in the phase the PV is installed. Let us consider either a 1 or a 5 kW PV installation at the beginning, middle and end of a feeder. In Fig. 4.6, the voltage profile of phase *A* is shown. As expected, the voltage amplitude increases with PV of higher ratings or when it is installed at the end.

The installation of a PV in a low load phase (phase *A* in this case) results in the increase in voltage difference and hence voltage imbalance at the end of the feeder while having minor effect at the beginning of the feeder. This voltage imbalance is more if the PV is installed at the end of the feeder or if the rating of the PV is high. The sensitivity analysis of voltage imbalance (calculated at the end of the feeder) versus the location and rating of rooftop PV installed in low load phase *A* is shown in Fig. 4.7a.

The voltage difference between the phases reduces if the PV is installed in high load phase (phase *C*). The decrease is more pronounced at the end of the feeder than at the beginning. This decrease at the end of the feeder is more if the PV is installed at the end of the feeder or if the rating of the PV is high. The sensitivity analysis of voltage imbalance (calculated at the end of the feeder) versus the location and rating of rooftop PV installed in high load phase *C* is illustrated in Fig. 4.7b.

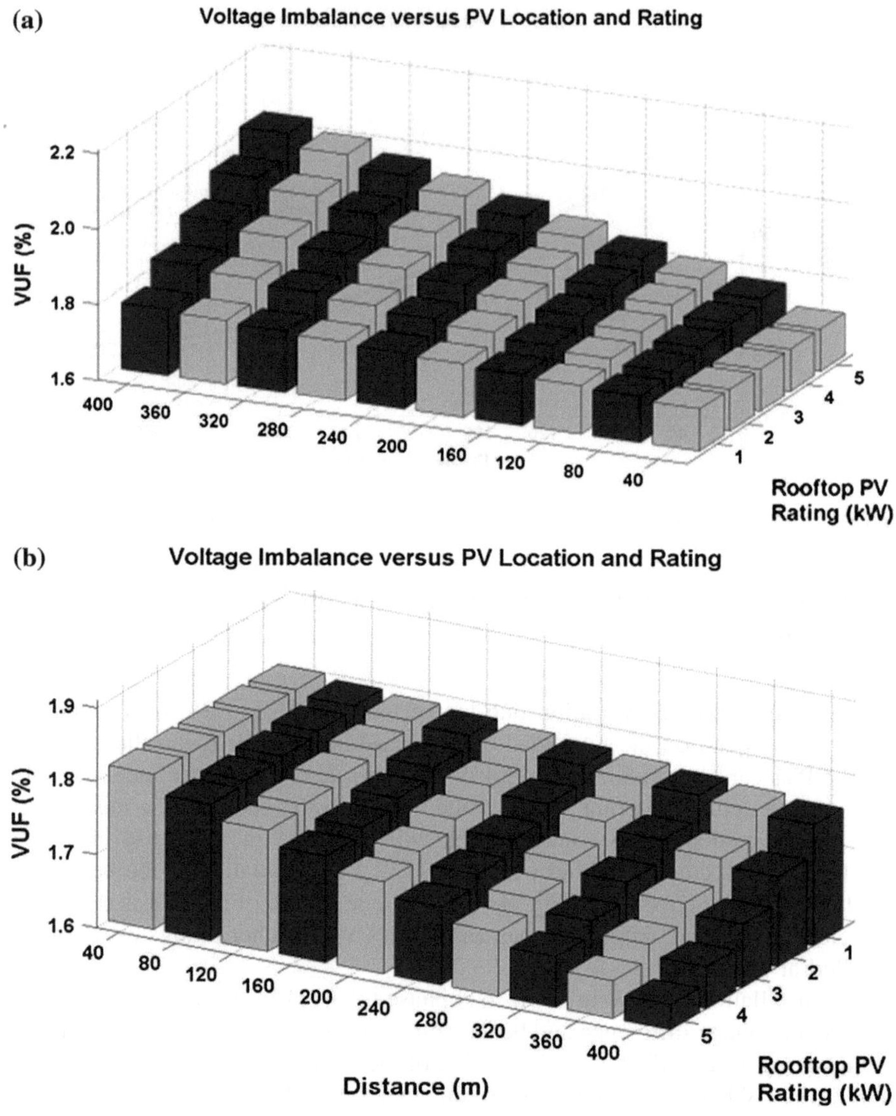


**Fig. 4.7** Voltage imbalance sensitivity analysis versus PV location and rating in **a** low load phase—Phase *A*, **b** high load phase—Phase *C*

These results prove that a rooftop PV (with a rating of less than 5 kW) can cause network voltage imbalance to increase by 0.1 % when installed at the beginning of the feeder and by 25 % when installed at the end of the feeder, specifically when the feeder supplies up to 1 MW load. For example, in the worst case, the voltage imbalance of 1.84 % without any PV increased to 2.02 % (i.e., a 25 % rise) when a 5 kW PV was installed at the end of the feeder. Even in this worst case, the voltage imbalance, at the end of the feeder, is not significant since it still does not lead to non-standard voltage imbalance. However, this may not be true when more than one PV get installed in the network.

### 4.3.2 Stochastic Evaluation of Voltage Imbalance

A stochastic evaluation is carried out for investigating the uncertainties in the network [11]. In this study, it is assumed the PVs (with rating of 1, 2, 3, 4, 5 kW) have equal probability of 20 % each, as shown in Fig. 4.3. The uncertainty in the PV rating is modelled by drawing a random number $U_1$ distributed uniformly under [0, 1]. Using $U_1$, the instantaneous output power of each PV is selected in 0–5 kW range during day–night period. The uncertainty of PV location along the feeder, [0–400 m], is modelled by drawing a random number $U_2$ distributed uniformly under [0, 1]. This is done for all phases and feeders of the studied network independently.

The number of the householders with installed rooftop PVs is assumed to be ¼, ½ and ¾ of the total number of householders as 3 different scenarios, shown in Fig. 4.3. For selecting 1 out of these 3 scenarios, a random number $U_3$ distributed uniformly under [0, 1] is used. If $U_3 < 0.33$ then scenario 1 is selected, if $0.33 \leq U_3 \leq 0.66$ then scenario 2 is selected and if $U_3 > 0.66$ then scenario 3 is selected.

Another study was also carried out with normal distributions for $U_1$ with ($\mu = 0.5$ and $\sigma = 0.2$) and also $U_3$ with ($\mu = 0.3$ and $\sigma = 0.08$) that due to similarity in the results are not shown here.

This study was carried out for several times with minimum number of N = 10,000 trials. A sample result for the scenario of ½ householders having PV are shown in Fig. 4.8a. It can be seen that the voltage imbalance calculated at the beginning of the feeder always remain about 0.8 %. This value for the end of the feeder varies between 1 and 3 %.

The PDFs for the cases when ½ of the householders have PVs is shown in Fig. 4.8b. The PDFs for all the three cases have mean value ($\lambda$) equal to 0.61 % at the beginning of the feeder and 1.80 % at the end of the feeder.

From Fig. 4.8b, it can be seen that, there is a high probability that the voltage imbalance at the end of the feeder is more than the 2 % standard limit. This probability is referred to as the failure index ($F_I$ %) which is the frequency of the cases in the shaded area in probability density function and is calculated as

$$F_I\% = shaded\,area \times 100 \qquad (4.10)$$

While voltage imbalance failure index is zero at the beginning of the feeder, it is about 30.19 % at the end of the feeder.

The customer load consumption is different at different times. Therefore, the residential loads also have an effect on the voltage imbalance. This phenomenon is

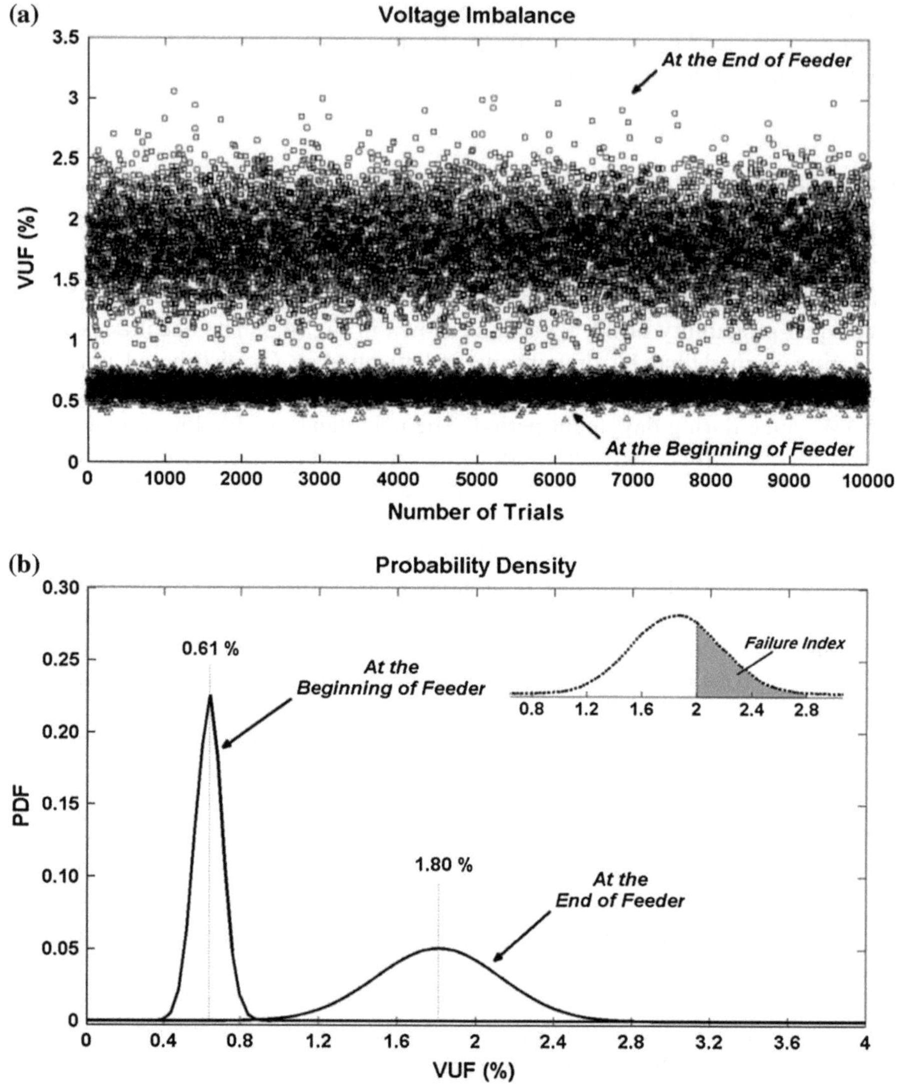


**Fig. 4.8** **a** Voltage imbalance for 10,000 scenarios of random location and ratings of PVs. **b** Probability density function of voltage imbalance

included as the fourth uncertainty condition for Monte Carlo method. The results of this analysis are given in Table 4.1 for different load consumption levels in the network during different times. It can be seen that when the loads are almost balanced, $\lambda$ and failure index decrease while they increase if the loads are highly unbalanced.

**Table 4.1** $\lambda$ and failure indices of voltage imbalance of the studied low voltage distribution network for different residential load levels

| Residential load status | Highly unbalanced | Lightly unbalanced | Almost balanced |
|---|---|---|---|
| $\lambda$ at beginning of feeder | 0.66 | 0.58 | 0.46 |
| $\lambda$ at end of feeder | 1.92 | 1.70 | 1.37 |
| Failure index ($F_I$ %) | 46.0 | 22.3 | 6.0 |

**Table 4.2** $\lambda$ and failure indices of voltage imbalance of the studied network considering majority of pvs installed at beginning or end of the feeder

| Majority of PVs installed at | Beginning of feeder | Middle of feeder | End of feeder |
|---|---|---|---|
| $\lambda$ at beginning of feeder | 0.61 | 0.61 | 0.62 |
| $\lambda$ at end of feeder | 1.88 | 1.97 | 2.06 |
| Failure Index ($F_I$ %) | 16.2 | 51.8 | 67.9 |

The influence of the location of the PVs (at the beginning or end of the feeder) on voltage imbalance is discussed before. In the previous studies, it was assumed the PVs are distributed randomly along the feeder. It is of high interest to investigate the case when the majority of the PVs are installed at the beginning or at the end of the feeder. Therefore, another Monte Carlo study is carried out to investigate this phenomenon. The results of this study given are in Table 4.2. It can be seen that generally when the majority of the PVs are installed at the end of the feeder, the failure index and $\lambda$ increase at the end points of the network.

## 4.4 Voltage Imbalance Improvement using Custom Power Devices

In this section, the series and parallel custom power devices are applied to fix the voltage at Point of Common Coupling (PCC) by forcing the three phase voltages magnitudes to be the same as desired, while their phases are separated by 120° [25].

### 4.4.1 DSTATCOM

A DSTATCOM is connected in parallel to the network, as shown in Fig. 4.9a. It will fix the voltage magnitude at PCC to desired value of $0.94 \leq E_{DSTAT} \leq 1$ pu by injecting/absorbing a required amount of reactive power. Hence the DSTATCOM can be supplied by only a DC capacitor.

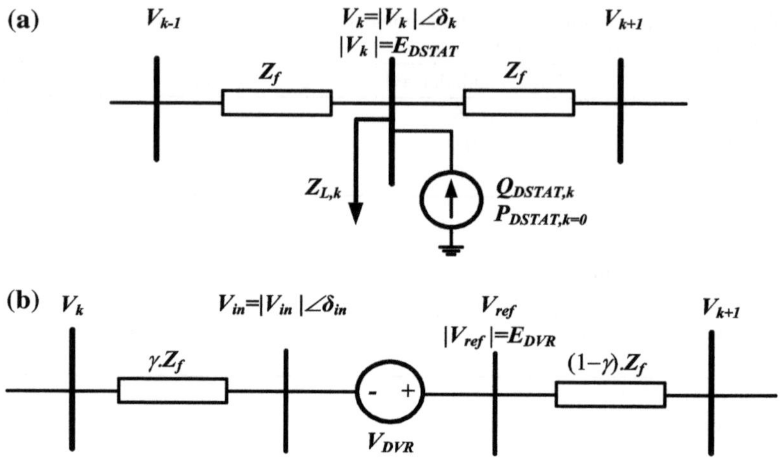


**Fig. 4.9** **a** Single line diagram of DSTATCOM connection, **b** Single line diagram of DVR connection

Let us assume that a three–phase DSTATCOM is installed at the *k*th node, which will be assumed to be a voltage controlled node (i.e., with constant active power and voltage magnitude). The sum of active power injected by DSTATCOM in three phases is set to be zero ($P_{DSTAT,k} = 0$) and the voltage magnitude for all three phases is set to be $E_{DSTAT}$.

The amount of reactive power to be injected/absorbed by DSTATCOM is

$$Q_{DSTAT,k} = -\text{Im}\left\{V_k^* \times \left[\left(V_k \times \left(\frac{2}{Z_f} + \frac{1}{Z_{L,k}}\right)\right) + \left(\frac{V_{k-1} + V_{k+1}}{Z_f} + \frac{V_{N,k}}{Z_{L,k}}\right)\right]\right\} \quad (4.11)$$

Based on the calculated $Q_{DSTAT,k}$, PCC voltage ($V_k$) will be modified as

$$V_k = \frac{1}{\frac{2}{Z_f} + \frac{1}{Z_{L,k}}}\left[\frac{P_{DSTAT,k} - jQ_{DSTAT,k}}{V_k^*} - \left(\frac{V_{k-1} + V_{k+1}}{Z_f} + \frac{V_{N,k}}{Z_{L,k}}\right)\right] \quad (4.12)$$

Equations (4.11)–(4.12) are used in the network analysis iterative method along with equations (4.2)–(4.5) for the node in which the DSTATCOM is installed.

When a DSTATCOM is installed in a node, it will inject/absorb reactive power to fix the voltage in that node to the desired value. By changing the PCC voltage, the voltages of all the nodes will be improved.

### 4.4.2 DVR

A DVR is connected in series within the network as shown in Fig. 4.9b, where the DVR buses are indicated with voltages of $V_{in}$ and $V_{ref}$. The DVR adds/subtracts a small amount of voltage in series such that the voltage magnitude such that the magnitude of $V_{ref}$ is equal to a desired value $0.94 \leq E_{DVR} \leq 1$ pu. In Fig. 4.9b, $0 \leq \gamma \leq 1$ represents the location of DVR between two adjacent buses $k$ and $k + 1$. Unlike a DSTATCOM, a DVR needs to inject/absorb both active and reactive power. However, as will be shown later, its rating is much smaller than that of a DSTATCOM for the same network.

The amount of necessary voltage to be added by the DVR to phase–*A* of the network is

$$V_{DVR,A} = V_{ref,A} - V_{in,A} \tag{4.13}$$

The desired voltages at the output of DVR for all three phases are based on same desired magnitude ($E_{DVR}$) and are displaced 120° from each other. These reference voltages are set based on the angle of one of the phases of the voltage $V_{in}$ as

$$\begin{aligned} V_{ref,A} &= E_{DVR}\angle\delta_{in,A} \\ V_{ref,B} &= E_{DVR}\angle\left(\delta_{in,A} - 2\pi/3\right) \\ V_{ref,C} &= E_{DVR}\angle\left(\delta_{in,A} + 2\pi/3\right) \end{aligned} \tag{4.14}$$

The selection of $E_{DVR}$ is based on the location of the DVR along the feeder and will have a significant effect on the rating of the DVR. To optimize the rating while satisfying the voltage and voltage imbalance conditions, $E_{DVR}$ needs to have a higher value if the DVR is installed close to the beginning of feeder and have a lower value if it is installed at far end of the feeder. For network analysis, (4.13) is used in the iterative method of equations (4.2)–(4.5) at the DVR connection point. Unlike a DSTSTCOM, which improves the voltages of all nodes, a DVR installation will improve the voltages of all the nodes downstream of the DVR.

## 4.5 Application of CPDs: Steady–State Results

Let us consider one 11 kV overhead line is feeding several 11 kV/415 V distribution transformers. Only one radial low voltage (415 V) residential feeder is considered with a total load demand of 120 kW. The feeder length is taken as 400 meters. The poles are located at a distance of 40 meters from each other. At each pole, 2 houses are supplied from each phase. The feeders and their cross–section are also designed appropriately based on the nominal power drawn and voltage drop. The technical data of the network is given in Table B.1. in the Appendix B.

It is assumed that during the period of study, the loads of phase *A*, *B* and *C* are 20, 40 and 60 kW, respectively. Other distribution transformers with their loads in the 11 kV network are considered as one single lumped load. The rooftop PVs installed by the householders have an output power in the range of 1–5 kW working in UPF. Several studies are performed [25, 26], some of which are discussed below.

### 4.5.1 Nominal Case

Let us first assume there is no rooftop PV installed in the network. The voltage magnitude at the beginning of the feeder is 0.98, 0.97 and 0.97 pu for phases *A*, *B* and *C*, respectively. However, these values decrease to 0.96, 0.94 and 0.92 pu, respectively, at the end of the feeder. Voltage imbalance along the feeder has increased from 0.32 % at the beginning of feeder to 1.31 % at the end.

Let us now consider the case in which rooftop PVs are installed in such a way that very high voltage imbalance results at the end of the feeder. For this purpose, the total power generation of rooftop PVs are assumed to be 40, 5 and 1 kW in phases *A*, *B* and *C*, respectively. Each PV location and rating is given in Table B.1 in the Appendix B. Now, in phase *A*, the power generation of PVs is 40 kW while the load demand is 20 kW. This results in reverse active power flow in phase *A* from PVs to the transformer. Therefore, the voltage profile of phase *A* increases from the beginning of the feeder to the end of the feeder. Hence, voltage imbalance profile along the feeder increases to 2.56 % at the end of the feeder due to unequal distribution of PVs in the network.

### 4.5.2 DSTATCOM Application

Now, let us now assume a DSTATCOM is installed at 280 m (2/3rd) from the beginning of the feeder. The DSTATCOM is controlled to fix the PCC voltage magnitude to $E_{DSTAT} = 0.98$ pu. The voltage profile of the three phases of the network before and after DSTATCOM installation is shown in Fig. 4.10a. In this figure, the dashed lines show the voltage profile when there was no DSTATCOM installed, while the solid lines show that with the DSTATCOM installed. It is clearly evident that the DSTATCOM is capable of fixing the magnitude of all three phases to $E_{DSTAT}$ by injecting reactive power to phases *B* and *C* and absorbing reactive power from phase *A*. In this case, the DSTATCOM has a rating of 80 kVA.

For investigating the effect of DSTATCOM location in voltage imbalance reduction, another study is carried out [26]. In this study, the DSTATCOM is installed in different nodes along the feeder and the voltage imbalance profiles are compared. In Fig. 4.10b, voltage imbalance profile along the feeder is shown before and after DSTATCOM installation in 4 different locations along the feeder: at 1/3rd of feeder from the transformer, midpoint, 2/3rd of feeder and at the end. Comparing the voltage imbalance profiles for these four cases, it can be concluded that DSTATCOM installation is not effective near the beginning of the feeder. Also, when the DSTATCOM is installed exactly at the end of the feeder, the nodes around the midpoint can suffer from higher voltage imbalance values. While if the DSTATCOM is installed somewhere around 2/3rd of feeder, it will have the best result all along the feeder. From this figure, it can be concluded that DSTATCOM will have better results when installed anywhere between the midpoint and 2/3rd of the feeder. Maximum value of voltage imbalance in the network after DSTATCOM

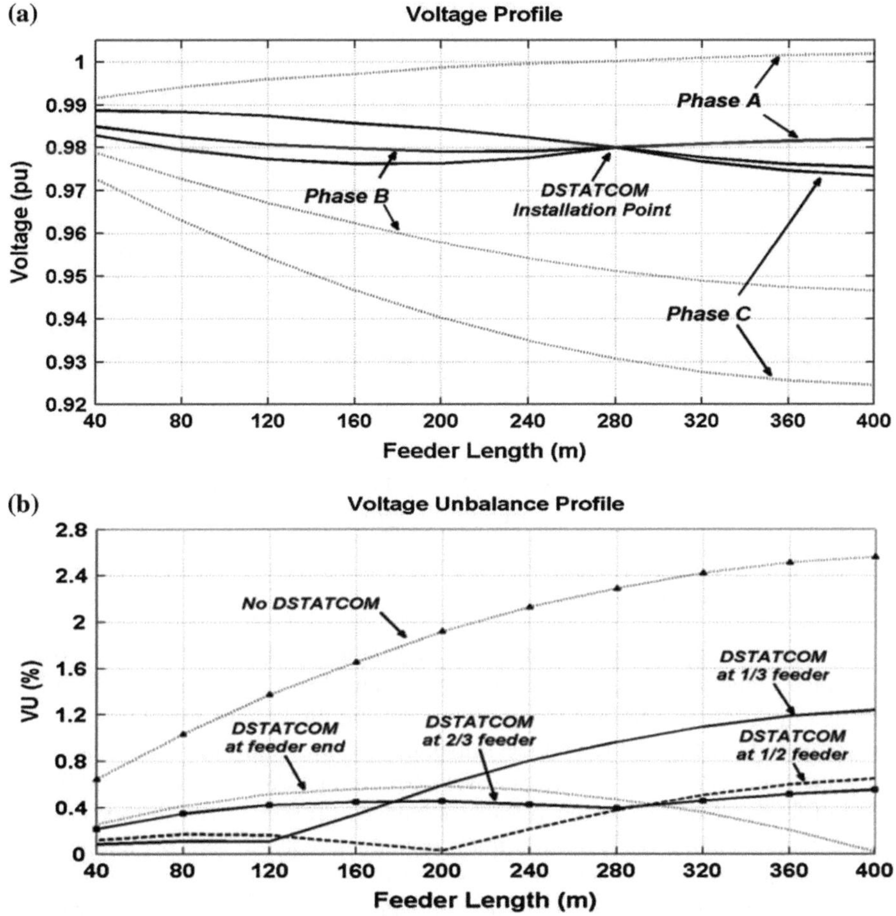


**Fig. 4.10** **a** Low voltage feeder voltage profile before and after DSTATCOM installation at 2/3 of feeder beginning, **b** Low voltage feeder voltage imbalance profile before and after DSTATCOM installation in four different locations along the feeder

installation at 2/3rd of the feeder is 0.55 %, calculated at the end of the feeder. This value is even smaller than the case when no PVs were installed in the network. This verifies the high effectiveness of DSTATCOM application for voltage imbalance reduction and voltage profile improvement in these networks.

### 4.5.3 DVR Application

Instead of a DSTATCOM, let us now assume a DVR is installed in series with the low voltage feeder, at 120 m (1/3rd) from the feeder beginning. The DVR is applied to fix its output voltage ($V_{ref}$) magnitude to $E_{DVR} = 0.975$ pu. For this case, the DVR will add some positive voltage to phases $B$ and $C$ and some negative voltage to phase $A$.

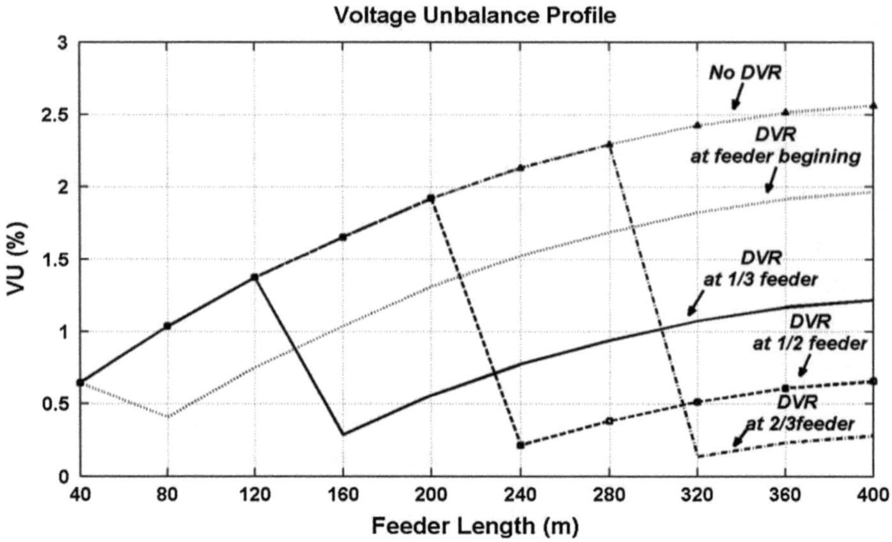


**Fig. 4.11** Low voltage feeder voltage imbalance profile before and after DVR installation in four different locations along the feeder

For investigating the effect of DVR installation and its location in voltage imbalance reduction, another study is carried out [25]. In this study, the DVR is installed in different locations along the feeder and the voltage imbalance profiles are compared. In Fig. 4.11, voltage imbalance profile along the feeder is shown when the DVR is installed in series at very beginning, 1/3rd of feeder, middle and 2/3rd of feeder. Comparing the voltage imbalance profiles for these four cases, it can be concluded that DVR installation is not effective at the very beginning or far end of the feeder. Also, when the DVR is installed exactly at the middle of the feeder, there is a high value of voltage imbalance in the middle of feeder (DVR input side). However, when the DVR is installed at 1/3rd of feeder beginning, voltage imbalance value is smaller all along the feeder compared to other locations. Maximum value of voltage imbalance in the network after DVR installation at 1/3rd of feeder is 1.21 %, calculated at the end of the feeder. Although DVR has reduced voltage imbalance even at the end of the feeder; it is not as successful as DSTATCOM. Nevertheless, it must be noted that the applied DVR has a 3 kVA rating which is very smaller compared to the DSTATCOM rating.

### *4.5.4 Stochastic Analysis Results*

Now, the stochastic analysis explained in Sect. 4.2 is carried out. In this case, first, let us assume that a DSTATCOM is installed at 2/3rd distance of feeder beginning. Voltage imbalance is only calculated at the end of the feeder since it was seen that it has its highest value at this point. The PDF of voltage imbalance at feeder end is

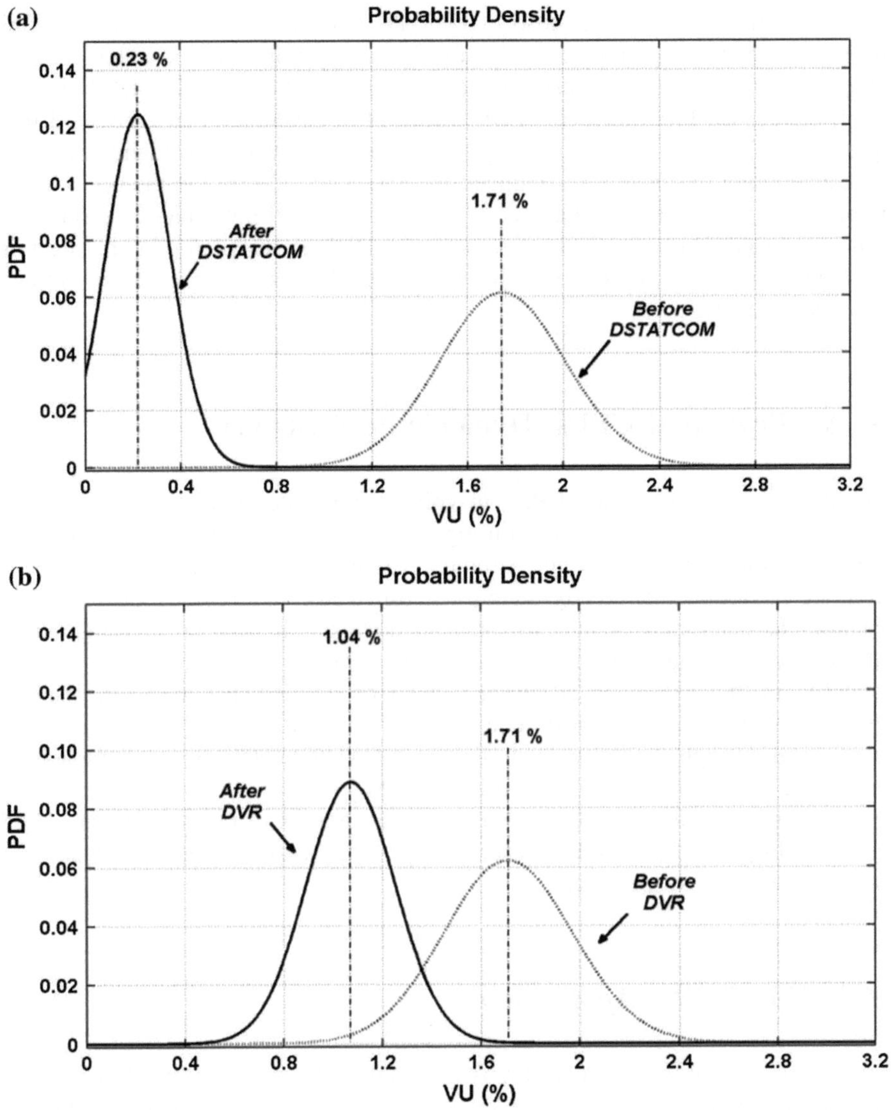


**Fig. 4.12** **a** Comparing PDF of voltage imbalance at feeder end before and after DSTATCOM installation, **b** Comparing PDF of highest voltage imbalance all along the feeder before and after DVR installation

shown in Fig. 4.12a. In this figure, the dashed line represents PDF before DSTATCOM installation and the solid line represents PDF after DSTATCOM installation. The average value of voltage imbalance at feeder end has reduced from 1.71 to 0.23 %. On the other hand, the probability of voltage imbalance at feeder end to be more than 2 % standard limit is reduced from 33.5 % to zero with the DSTATCOM is installed.

Now, let us assume that instead of DSTATCOM, a DVR is installed at 1/3rd distance of feeder beginning. Maximum of voltage imbalance is calculated all along the feeder. PDF of highest voltage imbalance is shown in Fig. 4.12b. In this figure, the dashed line represents PDF before DVR installation and the solid line represents PDF after DVR installation. The average value of highest voltage imbalance along the feeder has reduced from 1.71 to 1.04 %. On the other hand, the probability of voltage imbalance at feeder end to be more than 2 % standard limit is reduced from 33.5 % to zero with the DVR is installed. Comparing this figure with Fig. 4.12a, it is shown that DSTATCOM has better results in voltage imbalance reduction.

## 4.6 Application of CPDs: Dynamic Performance

The efficacy of the proposed CPDs in voltage imbalance reduction and voltage profile improvement was verified through steady–state load flow and stochastic analyses in the previous section. Although voltage imbalance and voltage regulation are predominantly steady–state (or quasi steady–state) issues, however, the dynamic characteristics of the proposed methods should also be investigated. It must be verified that these CPDs and their control strategies do not lead to instability in the system. In addition, it must be verified that they can respond effectively and promptly to load demand and PV output variations. This has been investigated through several case studies in PSCAD/EMTDC. Several studies are performed, some of which discussed below [25].

Let us assume that initially the total power generation of rooftop PVs are 40, 5 and 1 kW in phases *A*, *B* and *C*, respectively, while the loads in these phases are 20, 40 and 60 kW, respectively.

### 4.6.1 DSTATCOM Application

Now, let us assume a DSTATCOM is installed at the 2/3rd point of the feeder, however it is not connected to the feeder. Then at $t = 1$ s the DSTATCOM is connected to the network. The PCC instantaneous and RMS voltages are shown in Fig. 4.13a and b. It can be seen from these figures that the three–phase voltage waveform seems more balanced and the RMS values of the three phases are shifted up close to the desired value ($E_{DSTAT} = 0.98$ pu).

With the system operating in the steady state, with the DSTATCOM being connected, the PV generation is increased by 13 and 1 kW in phases *A* and *B*, respectively at $t_1 = 0.05$ s. Subsequently at $t_2 = 0.35$ s, a load change is created in the network. A total 4 kW load is reduced from phase *A*, 8 kW increased in Phase *B* and 12 kW increased in phase *C*. Furthermore at $t_3 = 0.55$ s, another PV generation and load variation occur. A total 8 kW of PV output is reduced from

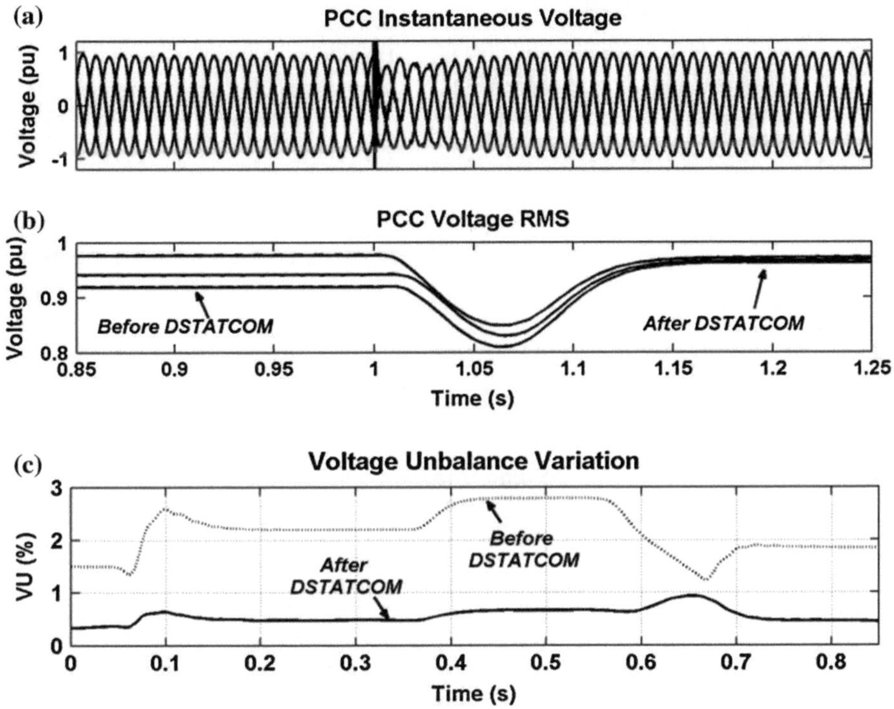


**Fig. 4.13** **a** PCC instantaneous voltage before and after DSTATCOM connection, **b** RMS voltage of PCC before and after DSTATCOM connection, **c** Voltage imbalance variation at low voltage feeder end before and after DSTATCOM installation

phase *A* while 2 kW and 6 kW PV output is increased in phases *B* and *C*, respectively. At the same time, the load in phase *A* in increased by 2 kW while the load in phases *B* and *C* is decreased by 4 and 6 kW, respectively.

The variation in voltage imbalance at the feeder end before and after DSTATCOM installation is shown in Fig. 4.13c. Comparing the voltage imbalance results in this figure, the efficacy of DSTATCOM application is verified. As seen from this figure, the DSTATCOM will vary the amount of its injected reactive power to the network based on network load and power parameters to fix the PCC voltage magnitude to the desired value.

### *4.6.2 DVR Application*

A similar study is carried out to investigate DVR dynamic performance. Let us assume a DVR is installed at 1/3rd point along the feeder. In Fig. 4.14a and b, the PCC instantaneous and RMS voltages are shown for cases before and after DVR connection. As it can be seen from these figures, the three–phase voltage

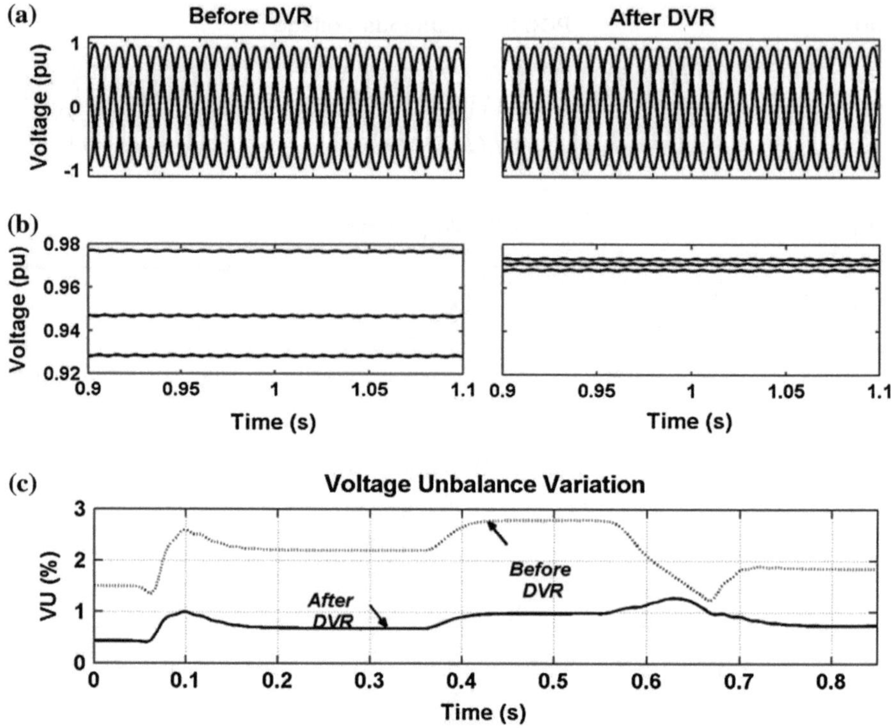


**Fig. 4.14** **a** PCC instantaneous voltage before and after DVR application, **b** PCC voltage RMS before and after DVR application, **c** Voltage imbalance variation at low voltage feeder end before and after DVR installation

waveform seems more balanced and the RMS values of the three phases are shifted up to the desired value of $E_{DVR} = 0.975$ pu.

The voltage imbalance at the end of the network before and after DVR installation is shown in Fig. 4.14c for the same PV generation and load variation discussed in the previous sub–section.

## 4.7 Conclusions

Random location and rating of single–phase rooftop PVs installed by householders may result in high voltage imbalance in low voltage feeders especially at far end nodes. For investigating this issue, a voltage imbalance sensitivity analysis and stochastic evaluation based on the rating and location of single–phase grid–connected rooftop PVs in a residential low voltage distribution network were presented in this chapter. Through the studies, it is proved that rooftop PV installation will have minor effect on the voltage imbalance at the beginning of a low voltage feeder designed with engineering judgments. However, it might increase at the end of the feeder to more than the standard limit. It is also proved that depending on the load of the phase in which the PV is installed, the voltage imbalance will increase or decrease based on the location and rating of the PVs. The stochastic simulation demonstrated that the failure index of non–standard voltage imbalance in these networks is high (30.19 %). Later, the applications of DSTATCOM and DVR were investigated for voltage imbalance reduction. Based on numerical analyses in steady–state condition, it was shown that a DSTATCOM has better results for voltage profile improvement and voltage imbalance reduction in comparison with a DVR. The Monte Carlo based stochastic analyses proved that for any random load and PV rating and location scenarios, the discussed CPDs are successful in reducing voltage imbalance to below the standard limits.

## Appendix A

The stopping rule of the Monte Carlo method is chosen based on achieving an acceptable convergence for $\overline{VUF}$ and *Var(VUF)*. In this study, the program was rerun for several trial numbers. The mean ($\lambda$) and *Var(VUF)* at the beginning and end of feeder in addition to Failure Index ($F_I$ %)for different trial numbers is listed in Table A.1. From this table, it can be seen that the mean, variance and failure index values do not vary much after N = 10,000 trials. The error value in *Var(-VUF)* is given in the last column of the table assuming the base case of 10,000 trials. It can be seen that an increase in trial number from 10,000 does not increase the error in variance significantly. Therefore this value is chosen as the stopping rule.

**Table A.1** Convergence of Monte Carlo method for different trial numbers

| N (Trial number) | 1,000 | 5,000 | 10,000 | 20,000 | 30,000 | 50,000 | 75,000 | 100,000 |
|---|---|---|---|---|---|---|---|---|
| Failure index (*FI*) (%) | 4.9 | 4.34 | 6.89 | 6.99 | 7.03 | 7.07 | 6.96 | 7.00 |
| $\lambda$ at beginning of feeder | 0.38 | 0.38 | 0.38 | 0.38 | 0.38 | 0.38 | 0.38 | 0.38 |
| $\lambda$ at end of feeder | 1.54 | 1.53 | 1.53 | 1.53 | 1.53 | 1.53 | 1.53 | 1.53 |
| *Var(VUF)* at beginning of the feeder (%) | 0.16 | 0.17 | 0.19 | 0.18 | 0.19 | 0.18 | 0.18 | 0.18 |
| *Var(VUF)* at the end of feeder (%) | 5.92 | 6.0554 | 8.04 | 7.95 | 8.01 | 8.12 | 8.07 | 8.05 |
| Error (%) of end *Var (VUF)* to other trial numbers | 26.28 | 24.71 | 0 | 1.12 | 0.32 | 0.98 | 0.39 | 0.15 |

# Appendix B

The technical parameters of the considered network within the chapter are provided below (Table B.1).

**Table B.1** Technical parameters of the studied low voltage distribution network

| | |
|---|---|
| Transformer | 11 kV/415 V, 250 kVA, $\Delta/Y_{grounded}$, $Z_l = 4~\%$ |
| Feeder | $3 \times 70 + 35$ mm$^2$AAC overhead line for low voltage feeder, |
| | $3 \times 50$ mm$^2$ ACSR, 2 km overhead line |
| Rooftop PV | 1–5 kW, unity power factor, $L = 5$ mH |
| Loads | 1 kW, $\cos\varphi = 0.95$, $z = 51.9840 + j \times 17.0863~\Omega$ |
| | 2 kW, $\cos\varphi = 0.95$, $z = 25.9920 + j \times 8.5432~\Omega$ |
| | 3 kW, $\cos\varphi = 0.95$, $z = 17.3280 + j \times 5.6954~\Omega$ |
| PV location | 1 kW PV at node 7 of phase *A* and node 9 of phase *C* |
| | 2 kW PV at nodes 1 (2 PVs), 6 (2 PVs), 9 (2 PVs), 10 (2 PVs) of phase *A* and node 1of phase *B* |
| | 3 kW PV at nodes 2 (2 PVs), 3 (2 PVs), 5 (2 PVs), 8 of phase *A* and node 4 of phase *B* |

# References


1. Ghosh A, Ledwich G (2002) Power quality enhancement using custom power devices. Kluwer Academic Publishers, Boston
2. Short TA (2004) Electric power distribution handbook. CRC Press, Boca Raton
3. Jouanne AV, Banerjee B (2001) Assessment of voltage imbalance. IEEE Trans Power Delivery 16(4):782–790
4. Gnacinski P (2008) Windings temperature and loss of life of an induction machine under voltage unbalance combined with over—or undervoltages. IEEE Trans Energy Convers 23(2):363–371
5. International Energy Agency (IEA) (2008) PVPS annual report—implementing agreement on photovoltaic power systems,—photovoltaic power systems programme
6. Eltawil MA, Zhao Z (2010) Grid–connected photovoltaic power systems: technical and potential problems—a review. Renew Sustain Energy Rev 14(1):112–129
7. Papathanassiou SA (2007) A technical evaluation framework for the connection of DG to the distribution network. Electr Power Syst Res 77:24–34
8. Lopes JAP, Hatziargyriou N, Mutale J, Djapic P, Jenkins N (2007) Integrating distributed generation into electric power systems: a review of drivers, challenges and opportunities. Electr Power Syst Res 77:1189–1203
9. Shahnia F, Majumder R, Ghosh A, Ledwich G, Zare F (2010) Sensitivity analysis of voltage imbalance in distribution networks with rooftop PVs. IEEE Power Energy Soc Gen Meet 80:1–8
10. Li W (2005) Risk assessment of power systems: models, methods, and applications. Wiley Publishers, New york
11. Shahnia F, Majumder R, Ghosh A, Ledwich G, Zare F (2011) Voltage imbalance analysis in residential low voltage distribution networks with rooftop PVs. Electr Power Syst Res 81(9):1805–1814

12. Mazumder S, Ghosh A, Shahnia F, Zare F, Ledwich G (2012) Excess power circulation in distribution networks containing distributed energy resources. IEEE Power Energy Soc Gen Meet, 1–8
13. Shahnia F, Ghosh A, Ledwich G, Zare F (2012) An approach for current balancing in distribution networks with rooftop PVs. IEEE Power Energy Soc Gen Meet, 1–6
14. Shahnia F, Wolfs P, Ghosh A (2013) Voltage unbalance reduction in low voltage feeders by dynamic switching of residential customers among three phases. IEEE Power Energy Soc Gen Meet, 1–5
15. Ghosh A (2005) Performance study of two different compensating devices in a custom power park. IEE Gener Transm Distrib 152(4):521–528
16. Australian Standard Voltage, AS60038–2000
17. IEEE recommended practice for monitoring electric power quality, IEEE Standard 1159–1995
18. Planning Limits for Voltage Unbalance in the United Kingdom (1990) The Electricity Council, Engineering Recommendation P29
19. Lee K, Venkataramanan G, Jahns T (2006) Source current harmonic analysis of adjustable speed drives under input voltage unbalance and sag conditions. IEEE Trans Power Delivery 21(2):567–576
20. Valois PVS, Tahan CMV, Kagan N, Arango H (2001) Voltage unbalance in low voltage distribution networks. In: Proceeding of 16th International Conference on Electricity Distribution (CIRED)
21. Power quality measurement results in 120 points in Eastern Azarbayjan Electric Power Distribution Co., Technical report, 2008 (in Persian)
22. IEEE Standard 929–2000. IEEE recommended practice for utility interface of photovoltaic (PV) systems
23. Makrides G, Zinsser B, Norton M, Georghiou GE, Schubert M, Werner JH (2010) Potential of photovoltaic systems in countries with high solar irradiation. Renew Sustain Energy Rev 14:754–762
24. Parker D, Mazzara M, Sherwin J (1996) Monitored energy use patterns in low–income housing in a hot and humid climate. In: Proceeding of 10th symposium on improving building systems in hot humid climates
25. Shahnia F, Ghosh A, Ledwich G, Zare F (2011) Voltage correction in low voltage distribution networks with rooftop PVs using custom power devices. In: 37th Annual Conference on IEEE Industrial Electronics Society (IECON), pp 991–996, Nov 2011
26. Shahnia F, Ghosh A, Ledwich G, Zare F (2010) Voltage unbalance reduction in low voltage distribution networks with rooftop PVs. In: 20th Australasian universities power engineering conference (AUPEC), pp. 1–5, Dec 2010

# Chapter 5
# Performance Evaluation of Grid-Connected Solar Photovoltaic (SPV) System with Different MPPT Controllers

**R. Singh and B. S. Rajpurohit**

**Abstract** Renewable energy plays an important role in electric power generation. Solar energy is one of them. It has the advantage of no pollution, low maintenance cost, no installation area limitation, and no noise due to the absence of the moving parts. However, high initial cost and low conversion efficiency have deterred its popularity. Due to the non-linear relation between the voltage and current of the PV cell, it is observed that there is unique Maximum Power Point (MPP) at particular environmental conditions, and this peak power point keeps changing with solar irradiations and ambient temperature. To achieve high efficiency in SPV power generation it is required to match the SPV source and load impedance properly for any weather conditions, thus obtaining maximum power generation. The technique process of MPP is being tracking which is called Maximum Power Point Tracking (MPPT). In recent years, a large number of techniques have been proposed for MPPT and some based on Computational Intelligence (CI) techniques. In this chapter performance evaluation of DC–DC boost converter based on P&O and INC has been compared. The scope of the work is to first give the detailed mathematical model of grid connected three-phase SPV system. A parametric model of SPV cell is presented. Second, thermal modeling and switching loss calculation of switching devices has been discussed and then the performance evaluation will be carried out for P&O and INC based MPPT algorithms for various operating conditions of the SPV array, in terms of energy injected to grid, switching losses, junction temperature and sink temperature, for switching in the DC–DC boost converter. Application of an Adaptive Neuro-Fuzzy Inference Systems (ANFIS) based intelligent controller has been described and

R. Singh (✉) · B. S. Rajpurohit
School of Computing and Electrical Engineering, Indian Institute of Technology, Mandi, Himachal Pradesh, India
e-mail: er.ranjitsingh2010@gmail.com

B. S. Rajpurohit
e-mail: bsr@iitmandi.ac.in

J. Hossain and A. Mahmud (eds.), *Renewable Energy Integration*, Green Energy and Technology, DOI: 10.1007/978-981-4585-27-9_5,

applied for fast, accurate, and efficient control of DC–DC boost converter used for SPV system, in place of conventional (PI) controllers.

**Keywords** Computational intelligence solar photovoltaic system · Maximum power point tracking · Perturb and observe · Incremental conductance · Switching losses and thermal model

## 5.1 Introduction

In the current century, the world is increasingly experiencing a great need for additional energy resources so as to reduce the dependency on conventional sources, and photovoltaic energy could be an answer to that need. Generally a Solar Photovoltaic (SPV) system can be divided into three categories: stand alone, grid-connection and hybrid system. For places that are far away from a conventional power generation system, standalone power generation systems have been considered a good alternative. These systems can be seen as a well-established and reliable economic source of electricity in rural areas, especially where the grid power supply is not fully extended. Solar energy has the advantage of no pollution, low maintenance cost, no installation area limitation, and no noise due to the absence of the moving parts. However, high initial cost and low conversion efficiency have deterred its popularity. Therefore, it is important to reduce the installation cost and to increase the energy conversion efficiency of SPV arrays and the power conversion efficiency of SPV system. Due to the non-linear relation between the voltage and current of the PV cell, it is observed that there is unique Maximum Power Point (MPP) at particular environmental conditions, and this peak power point keeps changing with solar irradiations and ambient temperature. To achieve high efficiency in SPV power generation it is required to match the SPV source and load impedance properly for any weather conditions, thus obtaining maximum power generation. Hence, tracking the MPP of a SPV array is usually an essential part of a SPV system. Maximum power extraction can be obtained by a method called as Maximum Power Point Tracking (MPPT). As such, many MPPT methods have been developed and implemented. The methods vary in complexity, sensors required, convergence speed, cost, range of effectiveness, implementation hardware, popularity, and in other respects. They range from the almost obvious (but not necessarily ineffective) to the most creative (not necessarily most effective). In fact, so many methods have been developed that it has become difficult to adequately determine which method, newly proposed or existing, is most appropriate for a given PV system. Given the large number of methods for MPPT, a survey of the methods would be very beneficial to researchers and practitioners in PV systems [1]. In recent years, a large number of techniques have been proposed for MPPT such as Constant Voltage Tracking (CVT), the Perturb-and-Observe (P&O) method, the Incremental Conductance

(INC) method, and some based on Computational Intelligence (CI) techniques. Computational intelligence (CI) techniques, such as fuzzy logic (FL), artificial neural network (ANN), and evolutionary computation (EC), are recently promoted for the control of systems. Overall, the dynamic performance of a system can be substantially improved by the introduction of CI based techniques for the intelligent control. The broad categories for different MPPT are given bellow:

**Conventional Algorithms**

- Curve Fitting Method
- Perturb and Observe
- Incremental Conductance
- Fractional Open-Circuit Voltage
- Fractional Short-Circuit Current
- Ripple Correlation Control (RCC)
- Current Sweep
- DC Link Capacitor Droop Control.

**Computational Intelligence Based Techniques**

- Fuzzy Logic Control (FLC)
- Artificial Neural Network (ANN)
- Genetic Algorithm (GA)
- Hybrid methods (such as ANFIS).

This chapter presents a performance analysis of grid connected SPV system for different MPPT algorithms. The performance evaluation of DC–DC boost converter based on P&O and INC has been compared in this chapter. The scope of the work is to first give the detailed mathematical model of grid connected three-phase SPV system. A parametric model of SPV cell is presented. Second, thermal modeling and switching loss calculation of switching devices has been discussed and then the performance evaluation will be carried out for different MPPT algorithms for various operating conditions of the SPV array, in terms of energy injected to grid, switching losses, junction temperature and sink temperature, for switching in the DC–DC boost converter. In this work, a computational strategy directed more towards intelligent behavior is employed as a tool for controlling DC–DC converter employed in SPV system. The conventional proportional-integral (PI) controller is replaced with a nonlinear adaptive neuro-fuzzy inference system (ANFIS) based controller, that is applied for fast, accurate, and efficient control of DC–DC boost converter used for SPV system. The design and procedure for selection of parameters and training of ANFIS are described. The performance of the conventional PI and ANFIS based controllers is compared using simulation results on a test system.

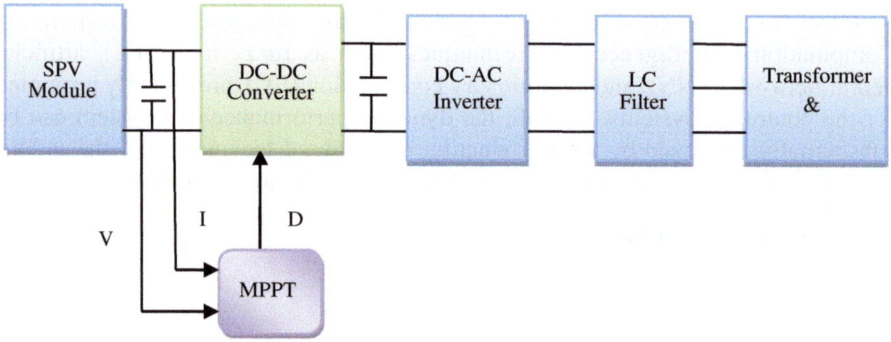


**Fig. 5.1** Schematic diagram of SPV system blocks

## 5.2 Three-Phase Solar Photovoltaic System

The configuration of the solar power generation system is as shown in Fig. 5.1. In this system sunlight is captured by SPV array. The SPV array is connected a DC–DC converter to increase the voltage level and to operate at the desired current and desired voltage to match the maximum available power from the PV module. This MPPT DC–DC converter is followed by a DC–AC inverter for grid connection or to supply power to the AC loads in stand-alone applications. The grid connected SPV array system is thus composed of the following blocks:

- Solar Photovoltaic Array
- DC–DC Boost Converter and Controller
- DC Link Capacitor
- DC–AC Three Phase Inverter and Controller
- LC Filter
- Transformer
- Grid.

## 5.3 Solar PV Cell

The equivalent circuit of the solar PV cell is given below in Fig. 5.2. $I_{ph}$ is the cell photocurrent that is proportional to solar irradiation, $I_{rs}$ is the cell reverse saturation current that mainly depends upon the temperature, $K_o$ is a constant, $N_s$ and $N_p$ are the number of series and parallel strings in the PV array respectively, $R_{sh}$ and $R_p$ is the series and parallel resistance of the PV array. Generally, a PV module comprises of a number of PV cells connected in either series or parallel and its mathematical model can be simply expressed as given below. The equation describing the I–V characteristics of the solar array are as follows [2]:

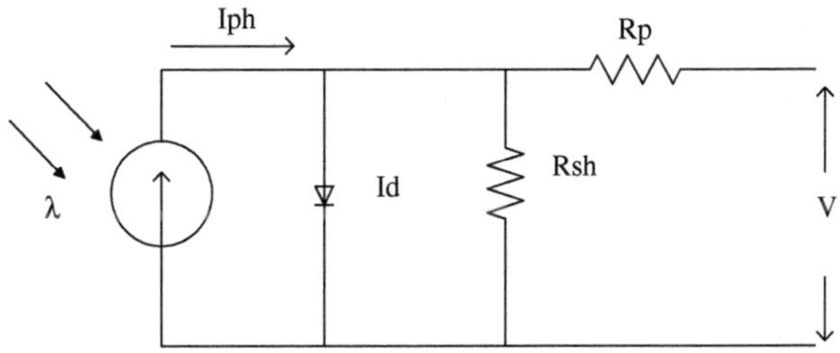


**Fig. 5.2** Equivalent circuit of solar cell

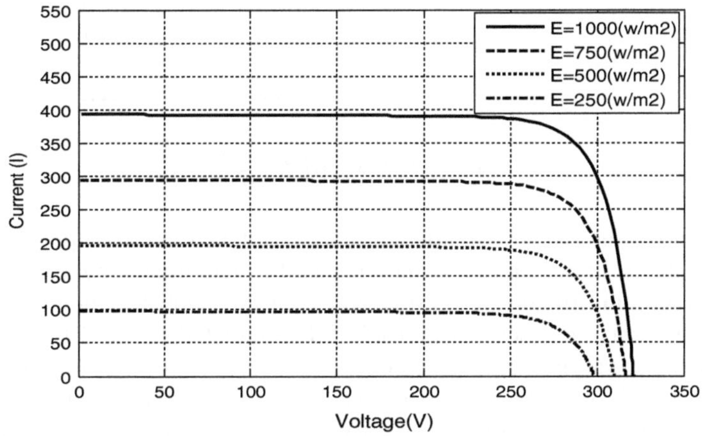


**Fig. 5.3** Simulated *I–V* curve of a SPV module for varying irradiance condition at 25 °C temperature

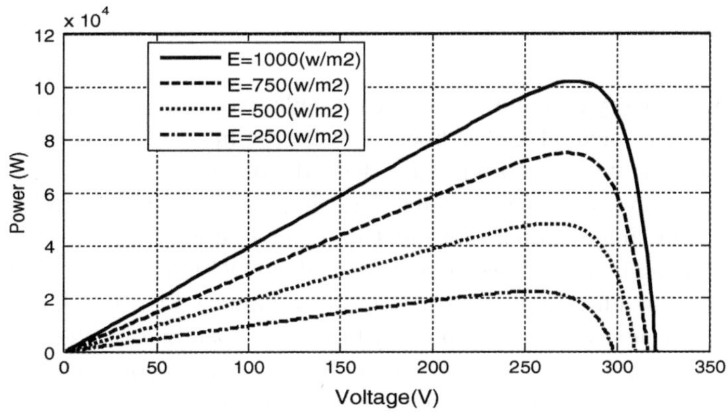


**Fig. 5.4** Simulated *P–V* curve of a SPV module for varying irradiance condition at 25 °C temperature

**Table 5.1** Specification from sun power module (SPR-305) data sheet

| | |
|---|---|
| No of series connected cells | 96 |
| Open circuit voltage ($V_{oc}$) | 64.2 V |
| Short circuit current ($I_{sc}$) | 5.96 A |
| Maximum power | 100.7 kW |
| Voltage at maximum power ($V_{mpp}$) | 54.7 V |
| Current at maximum power ($I_{mpp}$) | 5.58 A |

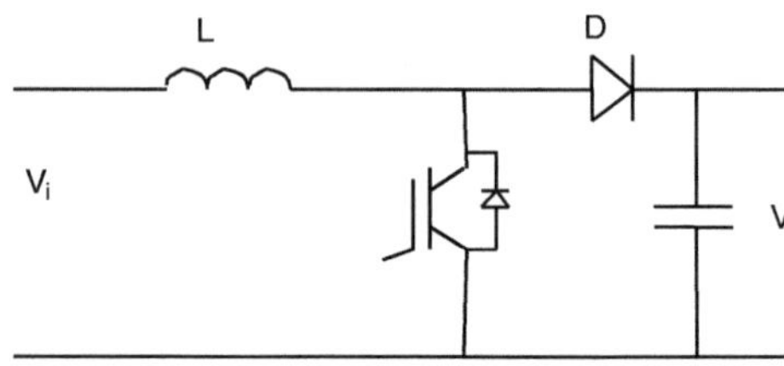


**Fig. 5.5** Schematic of a DC–DC boost converter

$$I = N_p I_{ph} - N_p I_{rs}\left[\exp\left(K_o \frac{V}{N_s}\right) - 1\right] \tag{5.1}$$

where, I denotes the PV array output current, V is the PV array output voltage. All of the constants in the above equation can be determined by examining the manufacturer rating of the SPV array and then the published or measured I–V curves of the array as described in Table 5.1. Simulated *I–V* and *P-V* curve of a SPV module for varying irradiance condition at 25 °C temperature are shown in Figs. 5.3 and 5.4, respectively. As a typical case, the Sun Power modules (SPR-305) array is used to illustrate and verify the model. The model parameters are given in Table 5.1 and can be found in the datasheet [3].

## 5.4 Voltage Boost DC–DC Converter and MPPT Algorithms

For maximum energy exploitation it is reasonable to operate SPV at the MPP. The SPV array is connected a DC–DC boost converter to increase the voltage level and to operate at the desired current and desired voltage to match the maximum available power from the PV module. The simplest form of DC–DC boost converter based on single switch and input inductor is used. The boost topology is capable of raising input voltage to the intermediate DC-link voltage, with the only limitation due to efficiency drop at very low voltage [4]. The DC–DC boost converter equivalent circuit is shown in Fig. 5.5, depending on the load and the circuit parameters, the inductor current can be either continuous or discontinuous. The inductor value, L, required to operate the converter in continuous conduction mode is calculated such that the peak inductor current at maximum output power does not exceed the power switch current rating. Thus, L and output capacitor value, C, to give the desired peak-to-peak output ripple is calculated as:

$$L = \frac{(1 - D)^2}{2f} \tag{5.2}$$

$$C \geq \frac{DV_0}{V_r Rf} \tag{5.3}$$

where f is switching frequency, D is duty cycle of the IGBT switch, R is the load resistance, $V_o$ is output voltage and $V_r$ is peak-to-peak ripple voltage. The DC–DC boost converter has the following simplified input-output equation.

$$V_i = (1 - D)V_o \tag{5.4}$$

where $V_o$ is the DC-link voltage regulated to be constant by the DC-link PI controller based voltage control. So D is the degree of freedom to change the work point of the SPV cells.

The Fig. 5.4 shows the char *P–V* characteristics of solar cells. The problem considered by MPPT techniques is to automatically find the voltage and current of at which a SPV array should operate to obtain the maximum power output under a given temperature and irradiance. It is noted that under partial shading conditions, in some cases it is possible to have multiple local maxima, but overall there is still only one true MPP. Most techniques respond to changes in both irradiance and temperature, but some are specifically more useful if temperature is approximately constant. Most techniques would automatically respond to changes in the array due to aging, though some are open-loop and would require periodic fine-tuning [1]. Systems composed of various PV modules located at different positions should have individual power conditioning units to ensure the MPPT for each module. The various MPPT algorithms are briefly described as given below.

### *5.4.1 Fractional Open-Circuit Voltage Based MPPT*

The near linear relationship between voltage at maximum power ($V_{MPP}$) and Open Circuit Voltage ($V_{OC}$) of the PV array, under varying irradiance and temperature levels, has given rise to the fractional $V_{OC}$ method [5–12].

$$V_{MPP} \approx k_1 V_{OC} \tag{5.5}$$

where $k_1$ is a constant of proportionality. Since $k_1$ is dependent on the characteristics of the PV array being used, it usually has to be computed beforehand by empirically determining $V_{MPP}$ and $V_{OC}$ for the specific PV array at different irradiance and temperature levels. The factor $k_1$ has been reported to be between 0.71 and 0.78. Once $k_1$ is known, $V_{MPP}$ can be computed using (5.5) with $V_{OC}$ measured periodically by momentarily shutting down the power converter. Figure 5.6 shows the implementation of fractional open-circuit based MPPT.

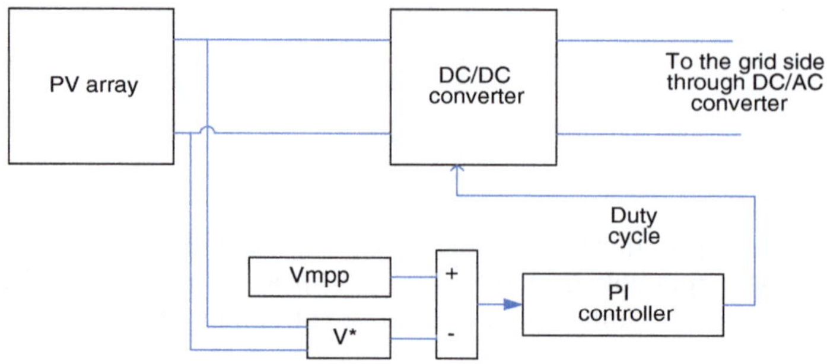


**Fig. 5.6** Fractional open-circuit based MPPT

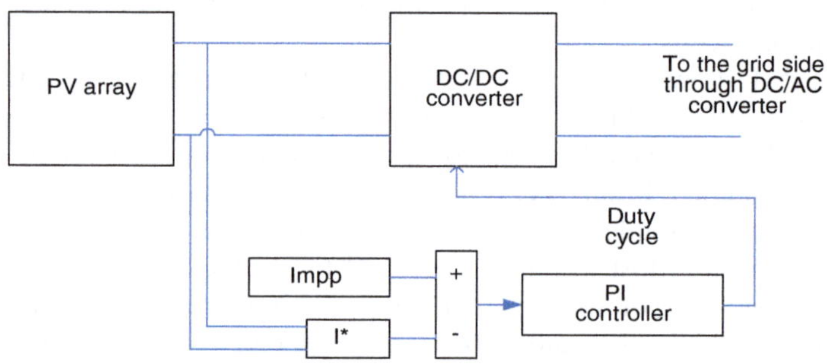


**Fig. 5.7** Fractional short-circuit based MPPT

### *5.4.2 Fractional Short-Circuit Current Based MPPT*

Fractional $I_{SC}$ results from the fact that, under varying atmospheric conditions, $I_{MPP}$ is approximately linearly related to the $I_{SC}$ of the PV array [13–15].

$$I_{MPP} \approx k_2 I_{SC} \tag{5.6}$$

where $k_2$ is a proportionality constant. Just like in the fractional $V_{OC}$ technique, $k_2$ has to be determined according to the PV array in use. The constant $k_2$ is generally found to be between 0.78 and 0.92. Figure 5.7 shows the implementation of fractional short-circuit based MPPT.

### *5.4.3 Perturb and Observe Based MPPT*

Perturb and Observe (P&O) involves a perturbation in the duty ratio of the power converter, i.e. a perturbation in the operating voltage of the PV array. In the case of a PV array connected to a power converter, perturbing the duty ratio of power converter perturbs the PV array current and consequently perturbs the PV array voltage [16–25]. From Fig. 5.3, it can be seen that incrementing the voltage

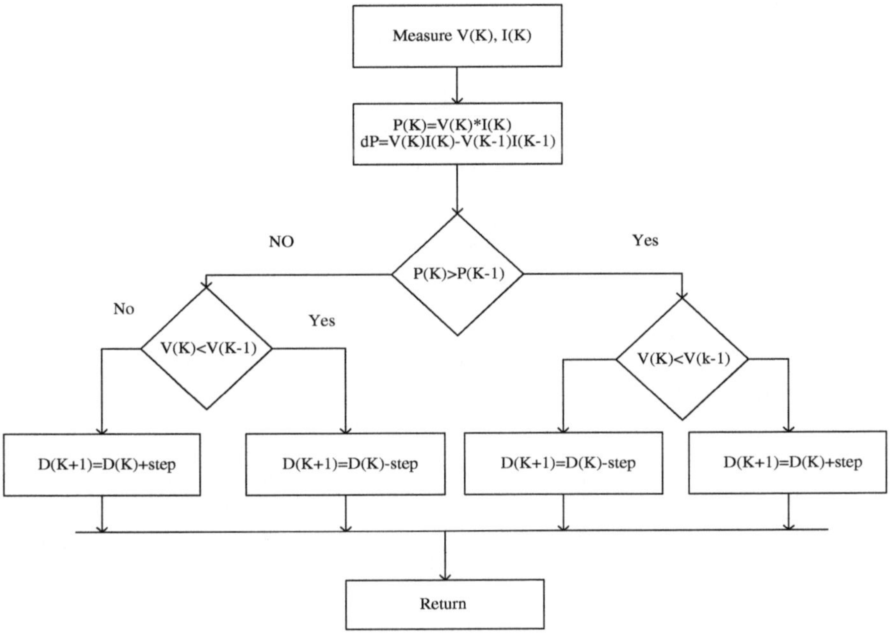


**Fig. 5.8** Flow chart of P&O based MPPT

increases the power when operating on the left of the MPP and decreases the power when on the right of the MPP. Therefore, if there is an increase in power, the subsequent perturbation should be kept the same to reach the MPP and if there is a decrease in power, the perturbation should be reversed. The flow chart of the P&O based method is given in Fig. 5.8.

### *5.4.4 Incremental Conductance Based MPPT*

The incremental conductance is based on the fact that the slope of the PV array power curve is zero at the MPP, positive on the left of the MPP and negative on the right [26–35]. The MPP can be thus be tracked by comparing the instantaneous conductance (I/V) to the incremental conductance ($\Delta I/\Delta V$).

So,

$$\begin{cases} dP/dV = 0 \\ dP/dV > 0\,. \\ dP/dV < 0 \end{cases} \tag{5.7}$$

The flow chart of the INC algorithm is given in Fig. 5.9.

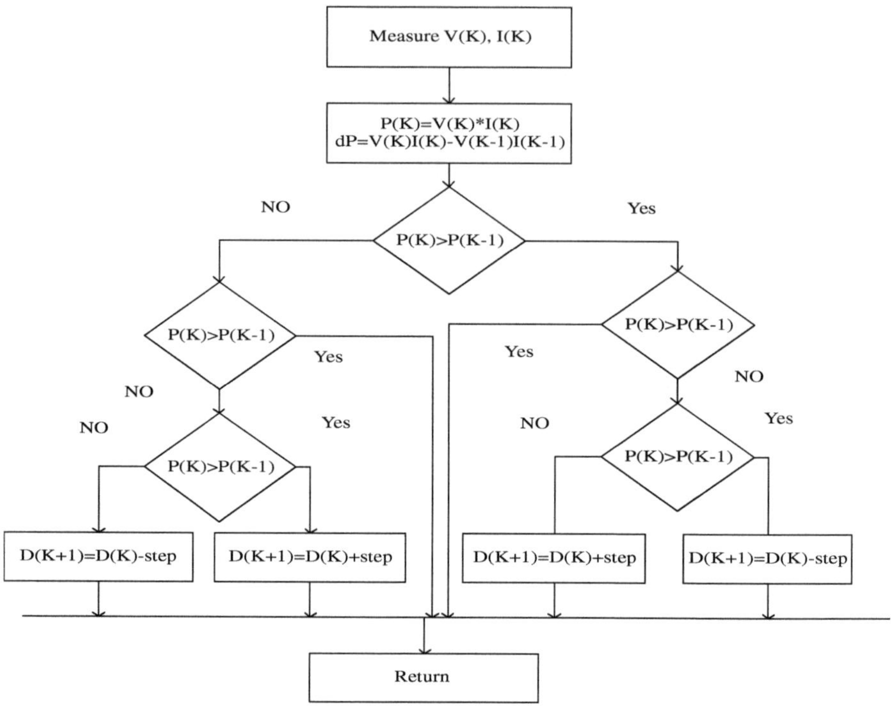


**Fig. 5.9** Flow chart of INC based MPPT

## *5.4.5 Fuzzy Logic Control Based MPPT*

Fuzzy logic controllers (FLC) have the advantages of working with imprecise inputs, not needing an accurate mathematical model, and handling nonlinearity. FLC generally consists of three stages: fuzzification, rule base lookup table, and defuzzification. During fuzzification, numerical input variables are converted into linguistic variables based on a membership function. In this case, rule base are given in Table 5.2. The inputs to a MPPT fuzzy logic controller are usually an error E and a change in error ΔE. The user has the flexibility of choosing how to compute E and ΔE. Since dP/dV vanishes at the MPP [36–40]. By calculate the following

$$E(n) = \frac{P(n) - P(n-1)}{V(n) - V(n-1)} \tag{5.8}$$

$$\Delta E(n) = E(n) - E(n-1). \tag{5.9}$$

Then the error signal can be calculated as

**Table 5.2** Fuzzy rule base table [40]

| ΔE | NB | NS | ZE | PS | PB |
|---|---|---|---|---|---|
| E | | | | | |
| NB | ZE | ZE | NB | NB | NB |
| NS | ZE | ZE | NS | NS | NS |
| ZE | NS | ZE | ZE | ZE | ZE |
| PS | PS | PS | PS | ZE | ZE |
| PB | PB | PB | PB | ZE | ZE |

$$E(n) = \frac{I}{V} + \frac{dI}{dV}.$$

Once E and ΔE are calculated and converted to the linguistic variables, the fuzzy logic controller output, which is typically a change in duty ratio ΔD of the power converter, can be looked up in a rule base Table 5.2. Implementation of a fuzzy logic controller based MPPT is shown in Fig. 5.10.

### *5.4.6 Artificial Neural Network*

Another intelligent technique is the artificial neural network. Neural networks commonly have three layers: input, hidden, and output layers. The number of nodes in each layer vary and are user-dependent. The input variables can be PV array parameters like $V_{OC}$ and $I_{SC}$, atmospheric data like irradiance and temperature, or any combination of these. The output is usually one or several reference signal(s) like a duty cycle signal used to drive the power converter to operate at or close to the MPP [39–42]. The most commonly used neural network in the MPPT is feed forward neural network.

Since most PV arrays have different characteristics, a neural network has to be specifically trained for the PV array with which it will be used. The characteristics of a PV array also change with time, implying that the neural network has to be periodically trained to guarantee accurate MPPT. Implementation of ANN based MPPT is shown in Fig. 5.11.

### *5.4.7 Genetic Algorithm*

A genetic algorithm (GA) is a procedure used to find approximate solutions to search problems through application of the principles of evolutionary biology. The evolutionary process of a GA is a highly simplified and stylized simulation of the biological version. It starts from a population of individuals randomly generated according to some probability distribution, usually uniform and updates this population in steps called generations. Each generation, multiple individuals are randomly selected from the current population based upon some application of

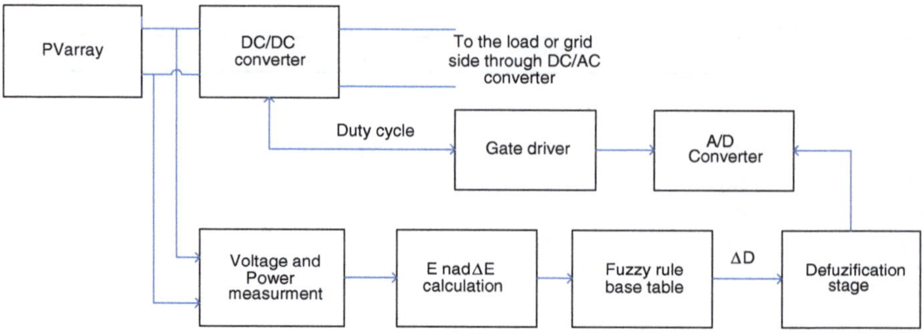


**Fig. 5.10** Implementation of a fuzzy logic controller based MPPT

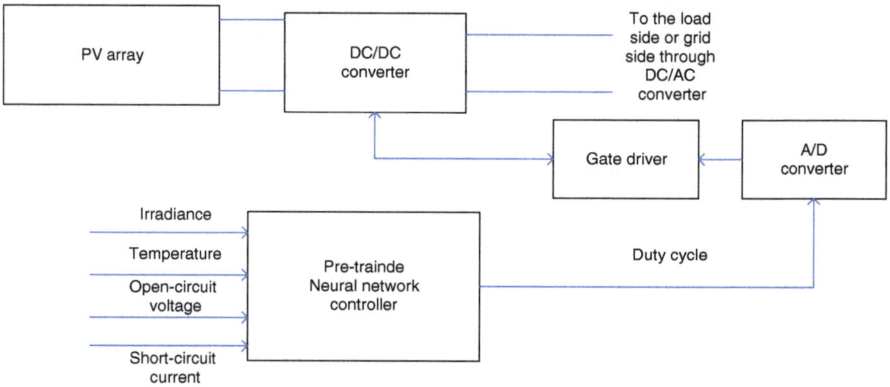


**Fig. 5.11** Implementation of ANN based MPPT

fitness, bred using crossover, and modified through mutation to form a new population.

## 5.5 Three-Phase Inverter for Grid Current SPV System

This MPPT DC–DC converter is followed by a DC–AC inverter for grid connection or to supply power to the AC loads in stand-alone applications. The basic operation principle of the DC–AC inverter is to keep the dc-link voltage at a reference value meanwhile keep the frequency and phase of output current are same as grid voltage. The error signal generated from voltage comparison is adjusted by voltage adjuster and it decides the value of reference current, then it's used to switch ON and OFF the values of the inverter. The load of the grid-connected inverter is power grid, grid power is controlled by grid current [43]. Figure 5.12 shows the schematic of three-phase gird connected inverter. Assume

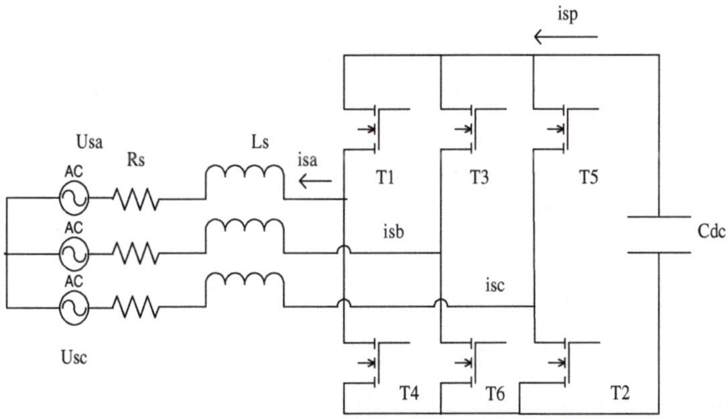


**Fig. 5.12** Schematic of three-phase grid connected inverter

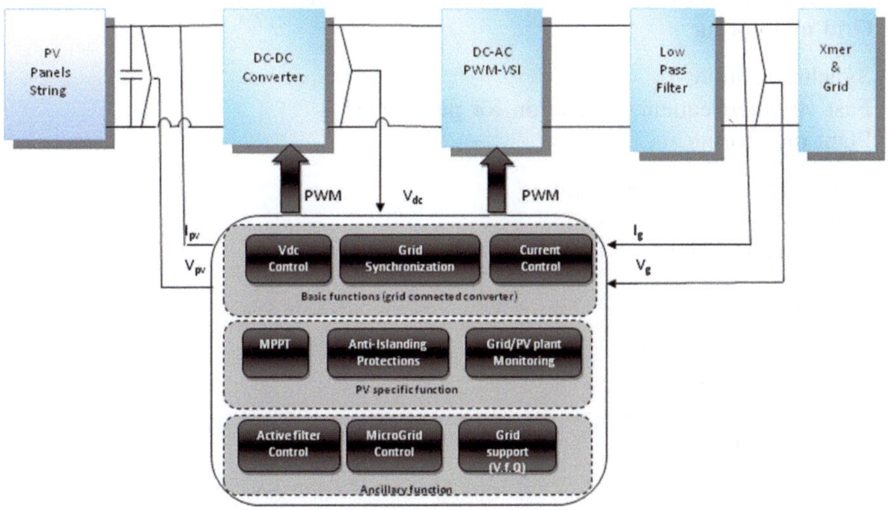


**Fig. 5.13** Generalized control structure of the SPV system

that three-phase grid voltage is symmetrical, stable and internal resistance is zero; three phase loop resistance $R_S$ and $L_S$ are of the same value respectively; switching loss and on-state voltage is neglectable; affection of distribution parameter is neglectable; switching frequency of the rectifier is high enough.

The generalized control structure of SPV system is shown in Fig. 5.13.

Following are the three different classes of control functions of power electronics converter of SPV system.

1. Basic functions-common for all grid connected inverters

- Grid current control

- THD limits imposed by standards
- Stability in the case of large grid impedance variations
- Ride through grid voltage disturbances

• DC voltage control

- Adaptation to grid voltage variation
- Ride-through grid voltage disturbance

• Grid synchronization

- Operation at the unity power factor as required by standards
- Ride through grid voltage disturbances

2. PV specific functions-common for all PV inverters

• Maximum power point tracking (MPPT)

- Very fast MPPT efficiency during steady state (typically >99 %)
- Stable operation at very low irradiance levels

• Anti-islanding (AI), as required by standards (VDE 0126, IEEE 1574, etc.)
• Grid monitoring

- Synchronization
- Fast voltage/frequency detection for passive AI

• Plant monitoring

- Diagnostic of PV panel array
- Partial shading detection

3. Ancillary functions

• Grid support

- Local voltage control
- Q compensation
- Harmonic compensation
- Fault ride through.

## 5.6 Power Losses and Junction Temperature Estimation of Semiconductor Devices Used in Power Electronics Convertor of SPV System

As with the increased application and usage of semi-conductor devices the estimation of the power loss and temperature of the junction and thermal model (case and sink) has become a major issue with the increase of the capacity and switching frequency of devices. One method for estimation of power loss of devices is based on the exact current and voltage waveforms of the devices. But, it is very difficult to get the waveforms from simulating each pulse of PWM exactly, with the variation of current and voltage. Usually the power loss is calculated under the

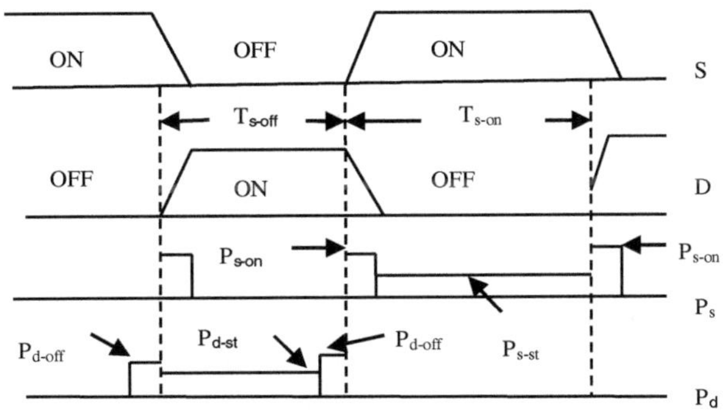


**Fig. 5.14** Power loss estimation

constant junction temperature. However, the power loss does depend on the junction temperature, not only the loss of saturation, but also the loss of transient switching operation. Therefore, the power loss estimation and the junction temperature calculation should be combined to find out the working point of devices [43]. The power loss of each switching operation of semiconductor device (IGBT) is divided into three main portions, which are illustrated in Fig. 5.14. Total power loss during each pulse of the IGBT is the sum of turn-on loss, turn-off loss, and saturation loss. Also, the losses of the anti-paralleled diode are included, if any.

It can be assumed that the IGBT power loss of turn-on or turn-off depends on the dc-link voltage and collector current of the IGBT. From application, it was found that the transient switching waveforms change with increase of junction temperature (50). It should be pointed out that the turn-on loss and turn-off loss are also functions of junction temperature which are expressed in (5.10), (5.11), (5.13), and (5.14). The saturation voltage of the IGBT and its antiparallel diode is usually defined as the function of junction temperature and collector current which are shown in (5.12) and (5.15).

$$P_{s-on} = f_{s-on}(V_d, i, T_j) \tag{5.10}$$

$$P_{s-off} = f_{s-off}(V_d, i, T_j) \tag{5.11}$$

$$V_{s-st} = f_{s-st}(i, T_j) \tag{5.12}$$

$$P_{d-on} = f_{d-on}(V_d, i, T_j) \tag{5.13}$$

$$P_{d-off} = f_{d-on}(V_d, i, T_j) \tag{5.14}$$

$$V_{d-st} = f_{d-st}(i, T_j) \tag{5.15}$$

where

$P_{s-on}$ = Power Losses during on time of the IGBT
$P_{s-off}$ = Power Losses during off time of the IGBT
$P_{d-on}$ = Power Losses during on state of the Diode
$P_{d-off}$ = reverses recovery losses of the Diode.

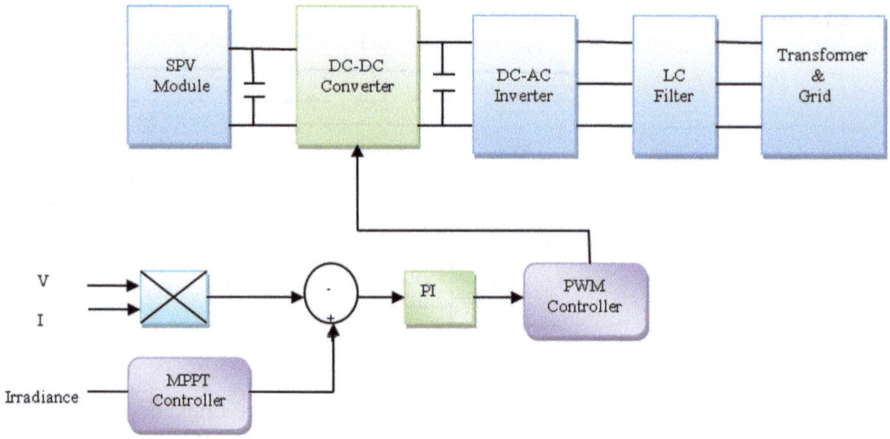


**Fig. 5.15** PI controller based INC MPPT

## 5.7 Thermal Model

A state-space block is used to build a one-cell Cauer network modeling the thermal capacitance of the device junction as well as its junction-to-case thermal resistance. The state space equations are given below:

$$x' = \left[\frac{-1}{R_{th}C_{th}}\right]x + \left[\begin{matrix}\frac{1}{R_{th}C_{th}} & \frac{1}{C_{th}}\end{matrix}\right]\left[\begin{matrix}T_c \\ P_l\end{matrix}\right] \tag{5.16}$$

$$\left[\begin{matrix}T_j \\ P_c\end{matrix}\right] = \left[\begin{matrix}1 \\ \frac{1}{R_{th}}\end{matrix}\right]x + \left[\begin{matrix}0 & 0 \\ -\frac{1}{R_{th}} & 0\end{matrix}\right]\left[\begin{matrix}T_c \\ P_l\end{matrix}\right] \tag{5.17}$$

where, $T_j$ is junction temp of IGBT, $P_l$ is power loss across IGBT, $T_c$ is case temperature of IGBT, $R_{th}$ is junction to case thermal resistance. $C_{th}$ is thermal capacitance of junction; $P_c$ is heat flow from junction to case. Now by calculating the junction temperature we can calculate the power losses of the IGBT. The same analysis is extended for power losses and junction temperature calculation of anti-parallel diode.

## 5.8 ANFIS Based Controllers

Classical PI and PID controllers that are used in conventional control are mainly tuned using specific methods. The design of the PI controller based INC MPPT is as shown in Fig. 5.15. Several methods provide initial values of the controller parameters. The most commonly used methods are based on the Ziegler-Nichols approach. However, these methods can be time consuming and fixed controllers

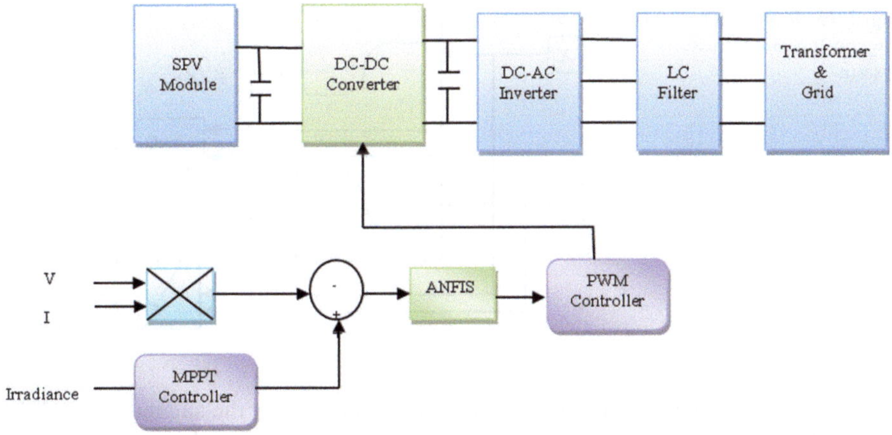


**Fig. 5.16** ANFIS controller based INC MPPT

cannot necessarily provide acceptable dynamic performance over the complete operating range of the SPV system. Performance will degrade mainly because of factors such as system non-linearities and parameter variations. Adaptive controllers can be used to overcome these problems. Alternatively, performance-index based optimal control techniques can be adopted, but these may suffer from convergence related problems.

The purpose of using a computational intelligence based controller is to reduce the tuning efforts for improved response and to remove the shortcomings of conventional controllers. The design of the ANFIS controller is shown in Fig. 5.16. There are various possibilities to obtain the training data from the classical PI controlled transient simulation of the SPV system. The ANFIS controller is trained with the input and output data obtained from the transient simulations of the conventional PI controller with a wide range of operating conditions. The ANFIS controller acts like the conventional PI controller without the need to design and tune for different operating conditions repeatedly. The fuzzy logic toolbox in MATLAB™ has been used for designing and testing the ANFIS controllers [41].

The Adaptive-Neuro Fuzzy Inference system is a hybrid system that combines the potential benefits of both the methods ANN and FL. This technique has been employed in numerous modeling and forecasting problems. ANFIS starts its functionality with the fuzzyfication of input parameters defining the membership function and design of fuzzy IF-THEN rules, by employing the learning capability of ANN for automatic fuzzy rules generation and self adjustment of membership functions [44].

In this work, the Sugeno method or Takagi-Kang method of fuzzy inference has been used. The Sugeno method was introduced in 1985 [45]. It is similar to the Mamdani method in many aspects. The first two parts of the fuzzy inference process, fuzzifying the inputs and applying the fuzzy operator are exactly the

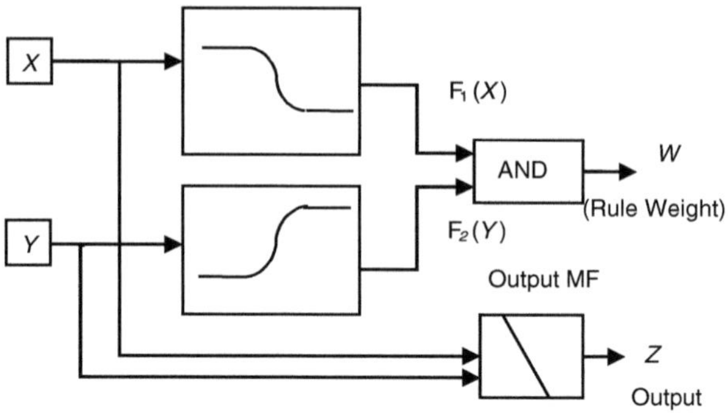


**Fig. 5.17** First order Sugeno-type inference system

same. The difference is that unlike the mamdani method, in the sugeno method the output MFs are only constant or have linear relationship to the inputs. With a constant output MF, this method is known as the Zero-order Sugeno method, whereas with a linear relation, it is known as the first-order Sugeno method.

A typical rule in a Sugeno fuzzy model has the following form:

If Input-1 = x, and Input-2 = y, then, Output z = ax + by + c.

For a Zero-order Sugeno model, the output level z is a constant (a = b = 0). The output level zi of each rule is weighted by the firing strength $w_i$ of the rule. For example, for an AND rule with Input-1 = x, and Input-2 = y, the firing strength is

$w_i$ = AND (F1(x), F2(y)),

where F1 (.) and F2 (.) are the inputs for 1 and 2.

The final output of the system is the weighted average of the output of all the rules, computed as

$$\text{Final output} = \frac{\sum_{i=1}^{N} w_i z_i}{\sum_{i=1}^{N} w_i} = \sum_{i=1}^{N} g_i z_i, g_i = \frac{w_i}{\sum_{i=1}^{N} w_i} \tag{5.18}$$

A sugeno rule operates as shown in Fig. 5.17.

The basic structure of fuzzy inference system seen, so far, is a model that maps input characteristics to input membership functions, input membership function to rules, rules to a set of output characteristics, output characteristics to output membership functions, and the output membership function to a single-valued output or a decision associated with the output. In both Mamdani and Sugeno type of inference systems, when used for data modeling, membership functions and rule structure are essentially predetermined by the human interpretation of the characteristics of the variables of the data model.

The shape of the membership functions depends on the values of the parameters. Instead of just looking at the data to choose the membership function parameters, by using ANFIS membership function, the parameters can be chosen automatically. The basic idea behind neuro-adaptive learning techniques is very simple. These techniques provide a method for the fuzzy modeling procedure to learn information about a data set, in order to compute the membership function parameters that best allow the associated fuzzy inference system to track the given

**Fig. 5.18** Typical ANFIS structure

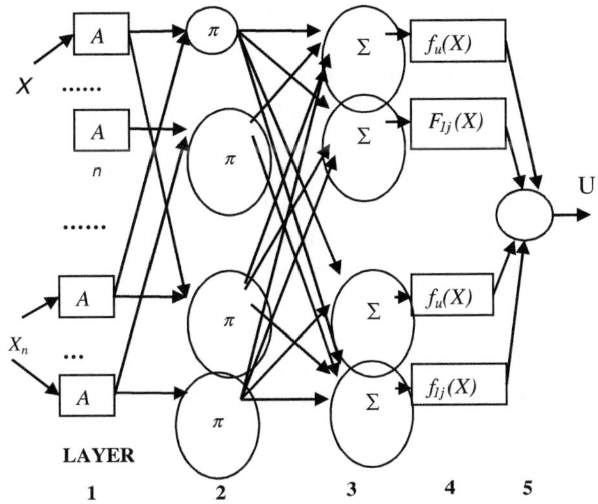


input/output data. This learning method works similar to the neural networks. In an adaptive neuro-fuzzy inference technique, using a given input/output data set, a fuzzy inference system (FIS) is constructed, whose membership function parameters are tuned (adjusted) using either a back-propagation algorithm alone, or in combination with a least squares type of method. This allows fuzzy systems to learn from the data. A network-type structure, similar to that of a neural network, which maps inputs through input membership functions and associated parameters, and then through output membership functions and associated parameters to outputs, can be used to interpret the input/output map.

Figure 5.18 shows the basic structure of the ANFIS algorithm for a first order Sugeno-type fuzzy system. The various layers shown in Fig. 5.18 are explained below [46].

*Layer 1*

Every node i, in this layer, is a square node with a node function

$$O_i^1 = \mu_{A_i}(x)$$

where, x is the input to node i, and $A_i$ is the linguistic label (small, large, etc.,) associated with this node function. In other words, $O_i^1$ is the membership function of $A_i$ and it specifies the degree to which the given x satisfies the quantifier $A_i$. Usually $\mu_{Ai}(x)$ is selected to be bell shaped with maximum value equal to 1, and minimum value equal to 0, such as

$$\mu_{A_i}(x) = \exp\left\{-\left(\frac{x - c_i}{a_i}\right)^2\right\}$$

where, $\{a_i, b_i, c_i\}$ is the parameter set. As the values of these parameters change, the bell-shaped functions vary accordingly, thus exhibiting various forms of membership functions on linguistic label $A_i$. In fact any piecewise differentiable function, such as commonly used trapezoidal or triangular-shaped membership function, is also qualified candidates for node functions in this layer. Parameters in this layer are referred to as premise parameters.

*Layer 2*

Every node in this layer is a circle node, labeled $\prod$, which multiplies the incoming signals and sends the product out. For example $w_i = \mu_{A_i}(x) X \mu_{B_i}(y)$, i = 1, 2. Each node output represents the firing strength of a rule. In fact, other T-norm operators that performs generalized AND can be used as the node function in this layer.

*Layer 3*

Every node in this layer is a circle node, labeled N. The *i*th node calculates the ratio of the *i*th rule's firing strength to the sum of all rule's firing strengths, as given below.

$\overline{w_i} = \frac{w_i}{w_1 + w_2}, i = 1, 2$ Outputs of this layer are known as normalized firing strengths.

*Layer 4*

Every node i in this layer is a square node with a node function

$$O_i^4 = w_i f_i = w_i(p_i x + q_i y + r_i)$$

where, $\overline{w_i}$ is the output of layer 3, and $\{p_i, q_i, r_i\}$is the parameter set. Parameters in this layer will be referred to as consequent parameters.

*Layer 5*

The single node in this layer is a circle node labeled Σ that computes overall output as the summation of all incoming signals, i.e.

$$O_i^5 =_i overall\ output = \sum_i \overline{w_i} f = \frac{\sum_i w_i f_i}{\sum_i w_i} \tag{5.19}$$

The adjustment of modifiable parameters is a two-step process. First, information are propagated forward in the network until Layer-4, where the parameters are identified by a least-squares estimator. Then, the parameters in Layer-2 are modified using gradient descent. The only user specified information is the number of membership functions in the universe of discourse for each input and output as training information. ANFIS uses back propagation learning to learn the parameters related to membership functions and least mean square estimation to determine the consequent parameters. Every step in the learning procedure includes two parts. The input patterns are propagated, and the optimal consequent parameters are estimated by an iterative least mean square procedure. The premise parameters are assumed fixed for the current cycle through the training set. The pattern is propagated again, and in this epoch (iterations), back propagation is used to modify the premise parameters, while the consequent parameters remain fixed.

The parameters associated with the membership functions will change through the learning process. The computation of these parameters (or their adjustment) is facilitated by a gradient vector, which provides a measure of how well the fuzzy

Table 5.3 System parameters

| $T_{PV}$ | Temperature of PV cells | 25 °C |
|---|---|---|
| G | Irradiance of PV cells | 1 kW/m$^2$ |
| $C_{PV}$ | Boost capacitance | 100 μF |
| $C_{DC}$ | DC link capacitance | 6 mF |
| $L_{LC}$ | Inductance of the LC filter + resistance | 250 μH + 2 mΩ |
| $F_{req}$ | Inverter switching frequency | 1.65 kHz |
| $V_{grid}$ | Grid RMS voltage | 25 kV |

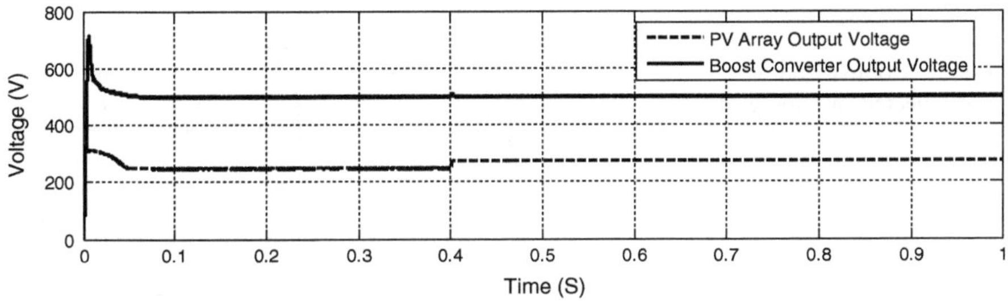


**Fig. 5.19** DC output voltage of PV array DC–DC boost converter

inference system is modeling the input/output data for a given set of parameters. Once the gradient vector is obtained, some of the available optimization routines can be applied to adjust the parameters so as to reduce some error measure (usually defined by the sum of the squared difference between actual and desired outputs).

The big advantage of the Sugeno-type FIS, is that it avoids the use of time consuming defuzzification, since it is a more compact and computationally efficient representation than the Mamdani system, the Sugeno system lend itself to the use of adaptive technique for construction fuzzy models. Theses adaptive technique can be used to customize the MFs so that fuzzy system accurately models the data. Some of the advantages of the Sugeno-type method are that it is computationally efficient; it works well with linear techniques (e.g., PID control).

## 5.9 Performance Comparisons

The schematic diagram of the main system of 100.7 kW is shown in Fig. 5.1. The modeling and simulation has been done using Matlab/Simulink software. The main system parameters are given in Table 5.3. The system is simulated with zero initial conditions hence results are settling down to steady-state values after transient period. The MPPT algorithms have been activated at the 0.4 s instant. The SPV array has been simulated and the steady state voltage output is 240 V and the boost converter has been used to boost the voltage level at steady-state without MPPT as shown in Fig. 5.19. The operating voltage of the SPV array increases after using

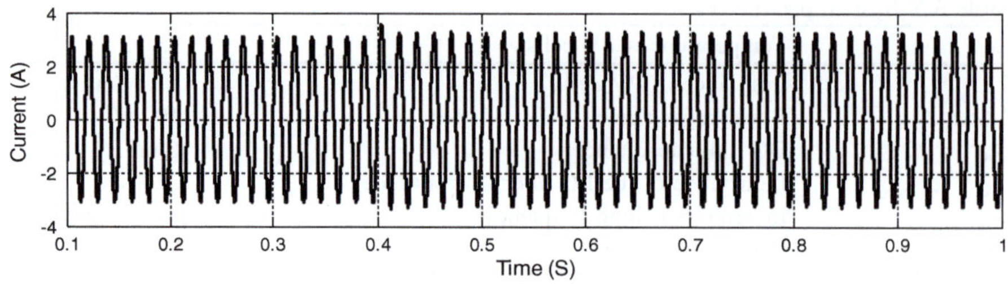


Fig. 5.20 Grid injected current

Fig. 5.21 Comparison of energy injected into the grid

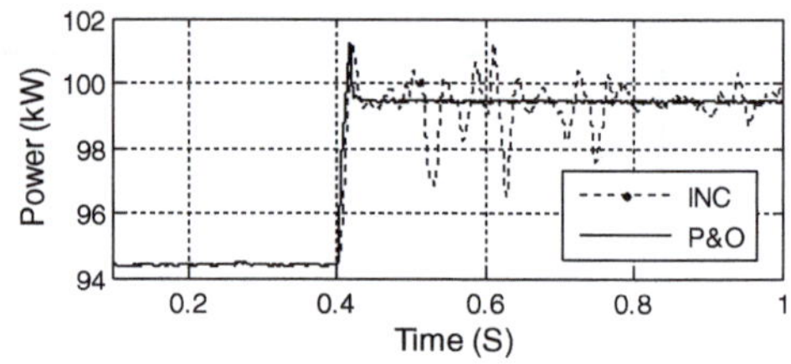


Fig. 5.22 Comparison of junction temperature of IGBT and diode

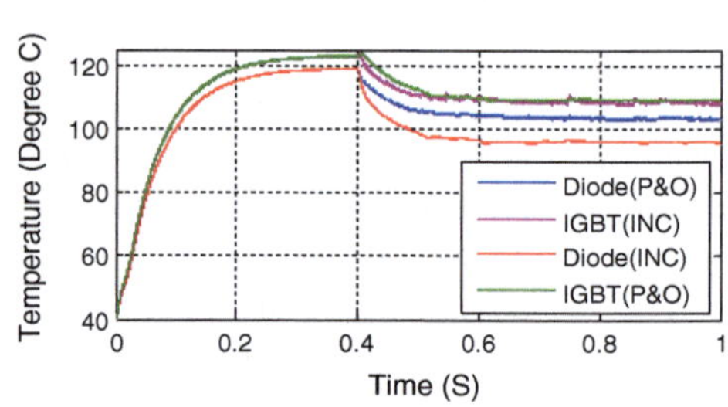


MPPT algorithms (MPPT algorithm starts at 0.4 s). The grid current follows the sinusoidal waveform as shown in Fig. 5.20 and the Total Harmonic Distortion (THD) is 1.66 % which is acceptable as per the IEEE-519 standard.

The energy injected into the grid using INC method is better than P&O method and the efficiency of the SPV system is 99.11 % using INC MPPT as compared to P&O MPPT having efficiency of 99.06 %. The Fig. 5.21 clearly show the advantage of using MPPT algorithms, as energy injected into the grid has increased from 95.6 kW to almost 100.7 kW after the instant 0.4 s when MPPT algorithms activated. The Fig. 5.22 shows the comparison of junction temperature of IGBT and DIODE using INC and P&O MPPT. It clearly indicates that without MPPT the junction temperature is high and the junction temperature of IGBT and DIODE are less using INC MPPT as compared to P&O MPPT. The switching losses of the DC–DC boost converter are higher in absence of MPPT algorithms as shown in Fig. 5.23 initially during transient state the switching losses are increasing exponentially and after some time they become constant. When the

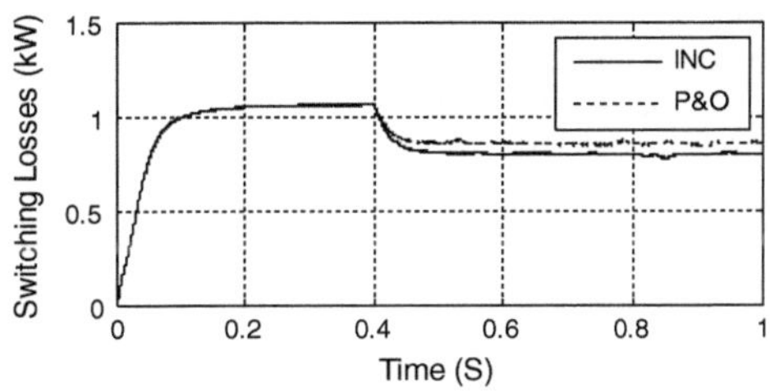


**Fig. 5.23** Comparison of switching losses of IGBT and diode

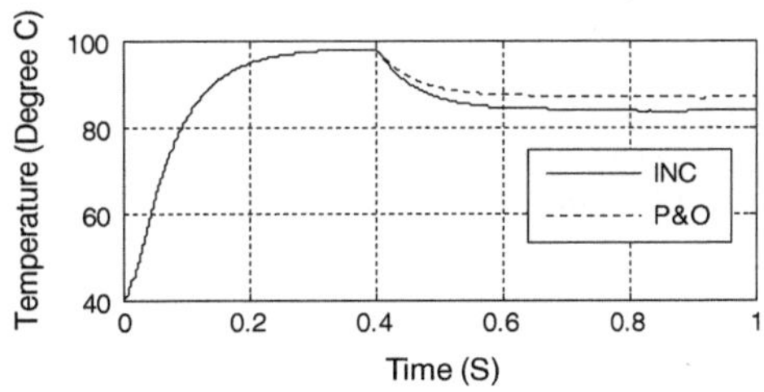


**Fig. 5.24** Comparison of case temperature of IGBT and diode

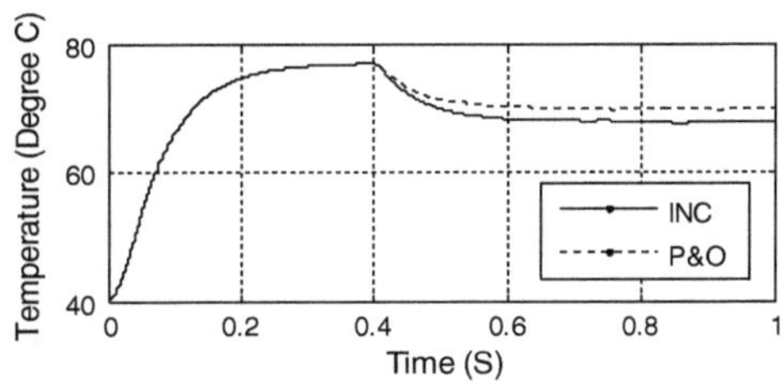


**Fig. 5.25** Comparison of sink temperature of IGBT and diode

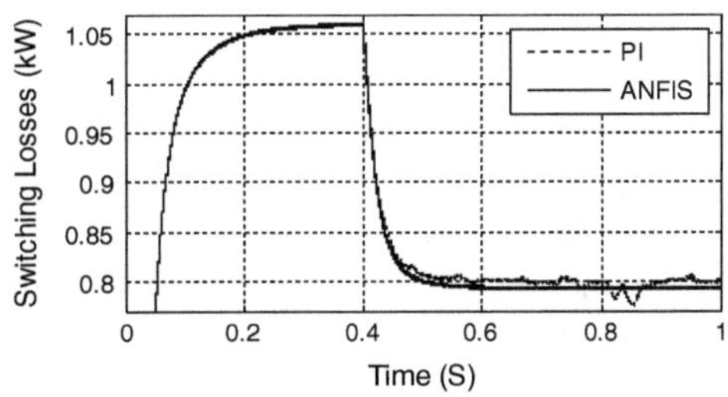


**Fig. 5.26** Comparison of switching losses using PI and ANFIS controller

MPPT is on (at 0.4 s as shown in Fig. 5.23) the switching losses starts decreasing. The switching losses are less using INC MPPT as compared to P&O MPPT.

The Figs. 5.24 and 5.25 clearly indicates that the case and sink temperature of switching devices are also less using INC MPPT as compared to P&O MPPT in the DC–DC converter. The switching losses of the IGBT module in DC–DC boost converter start decreasing after the application of MPPT algorithm (at 0.4 s) as shown in the Fig. 5.26. The Fig. 5.26 clearly indicates that the switching losses using ANFIS controller in INC Algorithm are less as compared to the PI controller.

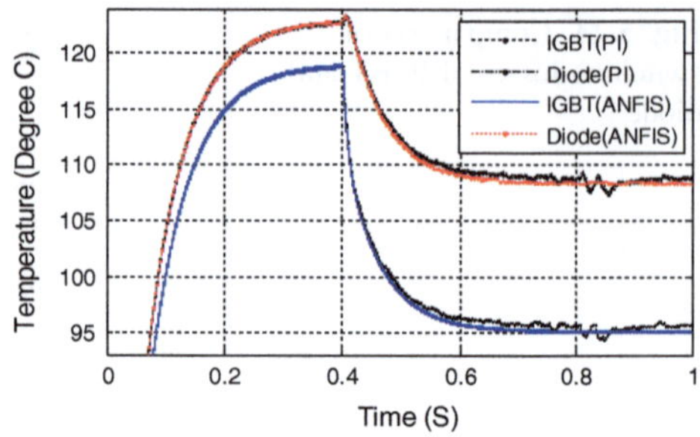


**Fig. 5.27** Comparison of junction temperature of IGBT and diode using PI and ANFIS controller

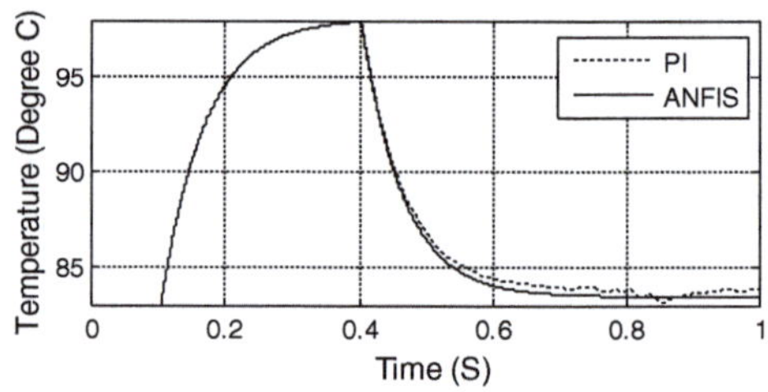


**Fig. 5.28** Comparison of case temperature for PI and ANFIS controller

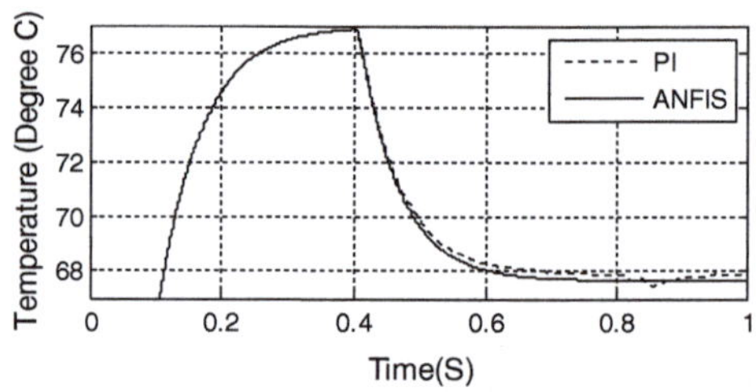


**Fig. 5.29** Comparison of sink temperature of PI and ANFIS controller

The junction temperature of the IGBT module (Including IGBT and FD) is very important. Figure 5.27 clearly indicates that the junction temperature if IGBT module start decreasing when MPPT is activated. The junction temperature of IGBT module is less using ANFIS controller as compared to PI controller. The case temperature is also less using ANFIS controller as compared to PI controller as shown in the Fig. 5.28. As the Fig. 5.29 clearly indicates that the sink temperature is also less using ANFIS controller as compared to PI controller.

## 5.10 Concluding Remarks

This chapter presents a performance analysis of grid connected SPV system for different MPPT algorithms. The SPV array output injected into grid can be maximized using MPPT control systems, which consist of a power conditioner to interface the PV output to grid and a control unit which derives the power conditioner such that it extracts the maximum power from a PV array. In this chapter, two different Maximum Power Point Tracking (MPPT) algorithm, Incremental Conductance (INC) and Perturb and Observe (P&O) MPPT algorithm are compared for DC–DC boost converter. Present work first gives the detailed mathematical model of grid connected three-phase SPV system. A parametric model of SPV cell is also presented. Second, thermal modeling and switching loss calculation of switching devices has been discussed and then the performance evaluations are be carried out for P&O and INC based MPPT algorithms for various operating conditions of the SPV array, in terms of energy injected to grid, switching losses, junction temperature and sink temperature, for switching in the DC–DC boost converter. Using the method of loss calculation, the power loss of IGBT and junction temperature can be estimated in the power conversion system. It can be used to improve the efficiency of the system and the ultimate thermal design. Also it can predict the working temperature of the IGBT and diode devices in order to avoid faults of the devices. The simulation result shows that the MPPT algorithms increase the SPV output energy injected into the grid. The switching losses and junction temperature of the switches are calculated and compared with two different MPPT algorithms. The performance of INC found to be better as compared to P&O. Later, in this present work, a nonlinear adaptive neuro-fuzzy inference system (ANFIS) is proposed to control the DC–DC boost converter instead of a conventional PI controller. The ANFIS have been trained with the input and output data of the conventional PI controller for different operating conditions. The training of ANFIS controllers has been done by combining the back-propagation gradient descent learning algorithm to choose the parameters related to membership functions and the least-squares estimation to determine the consequent parameters. Performance of the system is compared with two different controllers, i.e. (PI and ANFIS). The simulation result shows that the performance of the system in terms of switching losses, junction temperature, sink temperature and case temperature are better using ANFIS controller as compared to PI controller.

The vast potential of Computational Intelligence (CI) based techniques has yet to be explored for power electronic systems control applications. Hybrid CI techniques, particularly neuro-fuzzy techniques, have enormous potential for application. Similarly, there are numerous MPPT techniques are available in literature and practice, associated with many respective advantages and disadvantages. The area is vast, and the authors provided a discussion on the subjects which are the most relevant to the SPV system.

**Acknowledgments** This work was supported by the Department of Science and Technology, Government of India, under the Science and Engineering Research Board Fast Track Scheme for Young Scientists (SERC/ET-0123/2012).

## References

1. Esram T, Chapman PL (2007) Comparison of photovoltaic array maximum power point tracking techniques. IEEE Trans Energy Convers Summer Meet 22(2):439–449
2. Liu F, Duan S, Liu F, Liu B, Kang Y (2010) A variable step size INC MPPT for PV system. IEEE Trans Ind Electron 55:2622–2628
3. Data Sheet, IGBT Module U-Series 1200/600A, 1MBI600UB-120
4. Dell'Aquila RV, Balboni L (2010) A new approach: modelling simulation, development and implementation of a commercial grid-connected transformerless PV inverter. In: International symposium on power electronics, pp 1422–1429
5. Schoeman JJ, vanWyk JD (1982) A simplified maximal power controller for terrestrial photovoltaic panel arrays. In: 13th Annual IEEE power electronics specialists conference, pp 361–367
6. Buresch M (1983) Photovoltaic energy systems. McGraw Hill, New York
7. Hart GW, Branz HM, Cox CH (1984) Experimental tests of open loop maximum-power-point tracking techniques. Solar Cells 13:185–195
8. Patterson DJ (1990) Electrical system design for a solar powered vehicle. In: Proceedings of 21st annual IEEE power electronics conference, pp 618–622
9. Masoum MAS, Dehbonei H, Fuchs EF (2002) Theoretical and experimental analysis of photovoltaic systems with voltage and current based maximum power point tracking. IEEE Trans Energy Convers 17(4):514–522
10. Noh H-J, Lee D-Y, Hyun D-S (2002) An improved MPPT converter with current compensation method for small scaled PV-applications. In: Annual conference on industrial electronic society, pp 1113–1118
11. Kobayashi K, Matsuo H, Sekine Y (2004) A novel optimum operating point tracker of the solar cell power supply system. In: IEEE power electronics specialists conference, pp 2147–2151
12. Bekker B, Beukes HJ (2004) Finding an optimal PV panel maximum power point tracking method. In: 7th AFRICON conference in Africa, pp 1125–1129
13. Noguchi T, Togashi S, Nakamoto R (2000) Short-current pulse based adaptive maximum-power-point tracking for photovoltaic power generation system. In: IEEE international symposium on industrial electronics, pp 157–162
14. Mutoh N, Matuo T, Okada K, Sakai M (2002) Prediction-data-based maxi mum-power-point-tracking method for photovoltaic power generation systems. In: IEEE power electronics specialists conference, pp 1489–1494
15. Yuvarajan S, Xu S (2003) Photo-voltaic power converter with a simple maximum-power-point-tracker. In: International symposium on circuits and system, pp 399–402
16. Wasynczuk O (1983) Dynamic behaviour of a class of photovoltaic power systems. IEEE Trans Power App Syst 102(9):3031–3037
17. Hua C, Lin JR (1996) DSP-based controller application in battery storage of photovoltaic system. In: IEEE IECON 22nd international conference on industrial electronics, pp 1705–1710
18. Slonim MA, Rahovich LM (1996) Maximum power point regulator for 4 kW solar cell array connected through inverter to the AC grid. In: 31st intersociety energy conversion eng conf, pp 1669–1672
19. Al-Amoudi A, Zhang L (1998) Optimal control of a grid-connected PV system for maximum power point tracking and unity power factor. In: 7th international conference on power electronics and variable speed drives, pp 80–85
20. Kasa N, Iida T, Iwamoto H (2000) Maximum power point tracking with capacitor identifier for photovoltaic power system. In: 8th international conference on power electronics and variable speed drives, pp 130–135

21. Zhang L, Al-Amoudi A, Bai Y (2000) Real-time maximum power point tracking for grid-connected photovoltaic systems. In: 8th international conference on power electronics variable speed drives, pp 124–129
22. Hua C-C, Lin J-R (2001) Fully digital control of distributed photovoltaic power systems. In: IEEE international symposium on industrial electronics, pp 1–6
23. Chiang M-L, Hua C-C, Lin J-R (2002) Direct power control for distributed PV power system. In: Proceedings of power conversion conference, pp 311–315
24. Chomsuwan K, Prisuwanna P, Monyakul V (2002) Photovoltaic grid connected inverter using two-switch buck-boost converter. In: 29th IEEE photovoltaic specialist conference, pp 1527–1530
25. Hsiao Y-T, Chen C-H (2002) Maximum power tracking for photovoltaic power system. In: 37th IAS annual meeting of industry application conference, pp 1035–1040
26. Boehringer AF (1968) Self-adapting dc converter for solar spacecraft power supply. IEEE Trans Aerosp Electron Syst 4(1):102–111
27. Costogue EN, Lindena S (1976) Comparison of candidate solar array maximum power utilization approaches. In: Intersociety energy conversion engineering conference, pp 1449–1456
28. Harada J, Zhao G (1989) Controlled power-interface between solar cells and ac sources. In: IEEE telecommunication power conference, pp 22.1/1–22.1/7
29. Hussein KH, Mota I (1995) Maximum photovoltaic power tracking: An algorithm for rapidly changing atmospheric conditions. In: IEE proceedings on generation, transmission and distribution, pp 59–64
30. Brambilla A, Gambarara M, Garutti A, Ronchi F (1999) New approach to photovoltaic arrays maximum power point tracking. In: 30th annual IEEE power electronics specialist conference, pp 632–637
31. Irisawa K, Saito T, Takano I, Sawada Y (2000) Maximum power point tracking control of photovoltaic generation system under non-uniform isolation by means of monitoring cells. In: 28th IEEE photovoltaic specialist conference, pp 1707–1710
32. Kim T-Y, Ahn H-G, Park SK, Lee Y-K (2001) A novel maximum power point tracking control for photovoltaic power system under rapidly changing solar radiation. In: IEEE international symposium on industrial electronics, pp 1011–1014
33. Kuo Y-C, Liang T-J, Chen J-F (2001) Novel maximum-power-point tracking controller for photovoltaic energy conversion system. IEEE Trans Ind Electron 48(3):594–601
34. Yu GJ, Jung YS, Choi JY, Choy I, Song JH, Kim GS (2002) A novel two-mode MPPT control algorithm based on comparative study of existing algorithms. In: 29th IEEE photovoltaic specialist conference, pp 1531–1534
35. Kobayashi K, Takano I, Sawada Y (2003) A study on a two stage maximum power point tracking control of a photovoltaic system under partially shaded insolation conditions. In: IEEE power engineering society general meeting, pp 2612–2617
36. Wilamowski BM, Li X (2002) Fuzzy system based maximum power point tracking for PV system. In: 28th annual conference on IEEE industrial electronics, pp 3280–3284
37. Veerachary M, Senjyu T, Uezato K (2003) Neural-network-based maximum-power-point tracking of coupled-inductor interleaved-boost converter-supplied PV system using fuzzy controller. IEEE Trans Ind Electron 50(4):749–758
38. Khaehintung N, Pramotung K, Tuvirat B, Sirisuk P (2004) RISC microcontroller built-in fuzzy logic controller of maximum power point tracking for solar-powered light-flasher applications. In: 30th annual conference on IEEE industrial electronics society, pp 2673–2678
39. Ro K, Rahman S (1998) Two-loop controller for maximizing performance of a grid-connected photovoltaic-fuel cell hybrid power plant. IEEE Trans Energy Convers 13(3):276–281
40. Hussein A, Hirasawa K, Hu J, Murata J (2002) The dynamic performance of photovoltaic supplied dc motor fed from DC–DC converter and controlled by neural networks. In: International joint conference on neural network, pp 607–612

41. Sun X, Wu W, Li X, Zhao Q (2002) A research on photovoltaic energy controlling system with maximum power point tracking. In: Power conversion conference, pp 822–826
42. Zhang L, Bai Y, Al-Amoudi A (2002) GA-RBF neural network based maximum power point tracking for grid-connected photovoltaic system. In: International conference on power electronics machines and drives, pp 18–23
43. Xu D, Lu H, Hung Liu L, Azuma SS(2010). Power losses and junction temperature analysis of power semiconductor devices. IEEE Trans Ind Electron 38(5):1426–1431
44. Afgoul H, Krim F (2012) Intelligent energy management in a photovoltaic installation using neuro-fuzzy technique. In: IEEE ENERGYCON conference and exhibition, pp 20–25
45. Sugeno M (1993) Industrial applications of fuzzy control. Elsevier Science Publication, Amsterdam
46. Shing J, Jang R (1993) ANFIS: adaptive-network based fuzzy inference system. IEEE Trans Syst Man 23:665–685

# Chapter 6
# Optimal Siting and Sizing of Wind Turbines Based on Genetic Algorithm and Optimal Power Flow

**Geev Mokryani and Pierluigi Siano**

**Abstract** This chapter proposes a hybrid optimization method for optimal siting and sizing of wind turbines (WTs) that combines genetic algorithm (GA) and market-based optimal power flow (OPF). The method jointly minimizes total energy losses and maximizes social welfare (SW) considering different combinations of wind generation and load demand. The GA is used to choose the optimal size while the market-based OPF to determine the optimal number of WTs at each candidate bus. WTs are modeled with a PQ generator model, with a constant power factor. The stochastic nature of both load demand and wind power generation is modeled by hourly time series analysis. The interrelationships between demand and generation potential are preserved with their joint probability defining the number of coincident hours over a year. For each generation level, each WT is modeled with equivalent number of blocks in the WT's offer with the same price dependent on the WT's size. The method is conceived for distribution network operators (DNOs) to strategically allocate a number of WTs among different potential combinations. The effectiveness of the method is demonstrated with an 84-bus 11.4 kV radial distribution system.

**Keywords** Wind turbines · Social welfare maximization · Genetic algorithm · Distribution network operator

---

G. Mokryani (✉)
Department of Electrical and Electronic Engineering, Imperial College London, South Kensington, London SW7 2AZ, UK
e-mail: gmokryanni@gmail.com; g.mokryani@imperial.ac.uk

P. Siano
Department of Industrial Engineering, University of Salerno, 84084 Fisciano, SA, Italy
e-mail: psiano@unisa.it

J. Hossain and A. Mahmud (eds.), *Renewable Energy Integration*,
Green Energy and Technology, DOI: 10.1007/978-981-4585-27-9_6,

## 6.1 Introduction

### *6.1.1 Motivation and Approach*

Wind energy is one of the attractive forms of renewable energy sources (RES) for electricity production to achieve goals such as carbon-dioxide emission reduction, energy independence and enhanced infrastructure reliability. Hence, a lot of European countries are following policies to augment wind energy exploitation by means of incentives and financial options. The increasing number of wind turbines (WTs) connected to distribution networks creates numerous challenges to distribution network operators (DNOs) such as voltage variation, power losses, voltage stability and reliability [1]. The reason is the mismatch between the location of RES and the local network capability to allocate the new distributed generators (DGs) into the network.

The major role of a DNO is to supply loads at an acceptable voltage and loading level. A DNO has to develop a rational operating strategy taking into account dispatching DGs, interrupting loads, and purchasing power from the wholesale market while keeping system security. In some cases, DNOs play the role of retailers which buy power on the wholesale market at volatile prices and sell it again at fixed tariffs to small consumers. DNOs and retailers are separate market entities with different purposes, networks, and sizes [2]. In the approach presented here, the DNO is defined as the market operator of the DNO acquisition market, which does the price estimation and the optimization process for the hourly acquisition of active power [3].

Assuming that the DNOs purpose is to maximize their benefits, two different regulatory cases can be considered: (1) DG-owning DNO—permitted to possess DG and can exploit the financial benefits brought by considering new generation as an option for distribution system investments, (2) Unbundled DNO—prohibited from DG ownership but can maximize benefits based on a set of incentives [4, 5].

European Directive 2003/54/EC describes the technical and legal existing limitations among the different market actors of European electricity markets. Particularly, it establishes the unbundling regulations that DNOs have to be unbundled from generation interests, hence, prohibiting DNOs from DG ownership. It separates the electricity distribution from retail supply where distribution utilities are not responsible to sell power to customers.

The US approach for the ownership of DG is driven by the traditional structure of distribution networks in which they are responsible for supplying consumers throughout purchasing power from various sources in addition to owning and operating the wires. The financial profits of DG allocation to the utility from deferred generation and distribution investments are well recognized and utilities are permitted to site DG at strategic places on the grid to defer network upgrade costs and reduce peak-hour supply costs.

Both DG-owning and Unbundled DNOs are subject to remunerations for loss reductions and penalties for increases in comparison with a target level. DNOs should be encouraged to carry out the essential investment to reduce power losses. The ability of DG penetration to reduce losses could incentivize DNOs to constrain connections within their networks [4–6].

In this chapter, a novel method for optimal siting and sizing of WTs that combines the GA and the market-based OPF to jointly minimize the total energy losses and maximize the social welfare (SW) over a year is proposed. The GA is used to choose the optimal size of WTs while the market-based OPF to determine the optimal number of WTs. The method is conceived for by DG-owning DNOs to find the optimal numbers and sizes of WTs among different potential combinations. The DNO is defined as the market operator of the DNO acquisition market, which does the price estimation and the optimization process for the hourly acquisition of active power.

The uncertainty in wind power generation and load demand is modeled through hourly time-series model of load demand and wind generation. The interrelationships between load demand and wind generation potential are preserved with their joint probability defining the number of coincident hours over a year.

The proposed method can help DNOs to better allocate WTs by considering cost reduction and consumers' benefits as well as total energy losses reduction.

By following this approach it is expected that WTs will be allocated at buses where they are more advantageous, that is near higher loads or in parts of the networks where the loads have maximum value and the consumers' benefit is higher.

## 6.1.2 Literature Review and Contribution

A lot of previous works have been carried out to seek the optimal capacities and locations of DGs. In [7, 8], the authors proposed a genetic algorithm (GA)-based method to determine the optimal sizes and locations of multiple DGs in order to minimize the network losses considering network constraints. In [9], the authors proposed a hybrid optimization method to minimize the annual system power losses to find optimal locations and capacities of WTs. The method combines GA, gradient-based constrained nonlinear optimization algorithm and sequential Monte Carlo simulation. In [10], a Tabu search method to obtain the optimal sizes and locations of DGs has been proposed. In [11], the authors proposed a cost based model to allocate DGs in distribution networks in order to minimize DG investment and total operation costs of the network. The objective function is solved using an ant colony optimization (ACO) method. In [12], a method for optimal placement of WTs in distribution networks to minimize the annual energy losses has been proposed. The method is based on generating a probabilistic generation-load model that combines all possible operating circumstances of the WTs and load levels with their probabilities. The problem is formulated as mixed integer non-linear programming with an objective function for the minimization of annual energy losses. In [13], the authors proposed a stochastic optimization algorithm to minimize the power losses of the network by controlling the power factor of WTs. The stochastic models of WTs and load demand were used to take into consideration the stochastic variation. In [14], a novel method for optimal allocation of DGs in distribution systems to minimize the network losses and to guarantee the acceptable reliability level and voltage profile has been proposed. In [15, 16], the use of active management schemes such as the coordinated voltage regulation of on-load tap-changers and the power factor control of DGs, including WTs and diesel generators, integrated in the OPF for the optimization of objective function have been investigated.

To the best of our knowledge, no wind power investment method in distribution level from the point of view of DNOs in market environment by using hybrid market-based OPF and GA has been reported in the technical literature.

### *6.1.3 Chapter Organization*

The rest of the chapter is organized as follows. Section 6.2 explains the model features. Sections 6.3 and 6.4 describe GA implementation and DNO market acquisition formulation, respectively. Section 6.5 explains the 84-bus test system while Sect. 6.6 presents some numerical results. Discussions and conclusions are presented in Sect. 6.7.

## 6.2 Model Features

### *6.2.1 Modeling of Time-Varying Load Demand and Wind Power Generation*

Based on their joint probability of occurrence, defining the number of coincident hours over the year, wind availability and demand have been aggregated into a number of wind/demand scenarios. Actual data for both demand and wind production have been taken from [17]. The approach reduces hourly time-series data to a number of scenarios where, for each hour, load demand and wind generation are allocated to a series of bins. Defining the number of coincident hours over the target year allows preserving the interrelationships between load demand and wind generation potential with their joint probability [17].

In order to reduce the computational burden of a full time series analysis, wind availability and demand are aggregated into a manageable number of wind/

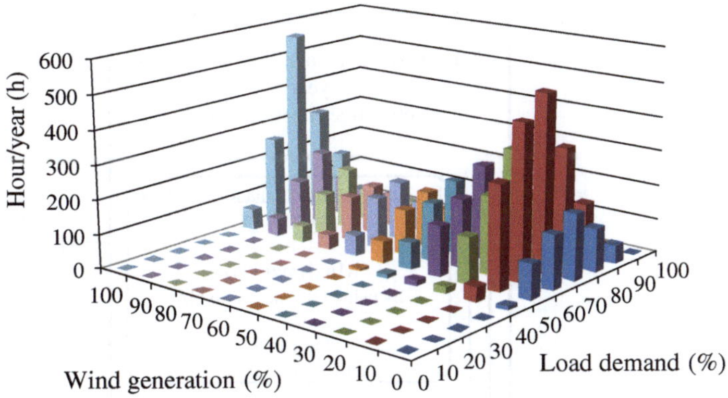


| 40 | 50 | 60 | 70 | 80 | 90 | 100 | % of generation capacity ↓ |
|---|---|---|---|---|---|---|---|
| 10 | 103 | 158 | 192 | 127 | 53 | 2 | 0 |
| 43 | 303 | 451 | 515 | 339 | 156 | 11 | 10 |
| 20 | 136 | 226 | 336 | 175 | 73 | 15 | 20 |
| 16 | 147 | 201 | 276 | 138 | 45 | 6 | 30 |
| 11 | 79 | 170 | 212 | 113 | 41 | 4 | 40 |
| 7 | 63 | 130 | 161 | 84 | 33 | 7 | 50 |
| 0 | 60 | 147 | 172 | 85 | 41 | 4 | 60 |
| 1 | 40 | 132 | 143 | 95 | 33 | 4 | 70 |
| 0 | 48 | 123 | 176 | 90 | 42 | 8 | 80 |
| 2 | 54 | 144 | 212 | 110 | 48 | 6 | 90 |
| 0 | 63 | 257 | 559 | 305 | 152 | 16 | 100 |

% of peak demand →

**Fig. 6.1** Coincident hours for demand/generation scenarios

demand scenarios on the basis of their joint occurrence probability. The duration of each scenario represents the number of coincident hours as shown in Fig. 6.1 (bottom). It splits the demand and generation into a series of bins. In order to show the procedure, ten ranges for demand (i.e. [0, 10 %], [10, 20 %],..., [90, 100 %]) and 11 ranges for wind generation (i.e. {0}, [0, 10 %], [10, 20 %],..., [90, 100 %]) are used. It is seen that with demand higher than 30 %, 74 non-zero scenarios are considered in the analysis. Furthermore, low load demand, i.e. 40 %, and high wind generation, i.e. 60–100 %, present few coincident hours.

The uncertainty in wind power generation and load demand are represented via scenarios. Each demand level is characterized by 11 wind generation levels, i.e. 0–100 %. There are seven load demand and eleven wind generation levels. Therefore, jointly considering the load demand and wind power generation levels results in 77 scenarios, i.e. seven load demand levels, with two blocks per level with different sizes and the same price, by eleven wind power generation levels with four blocks per level with the same size and the same price for all blocks.

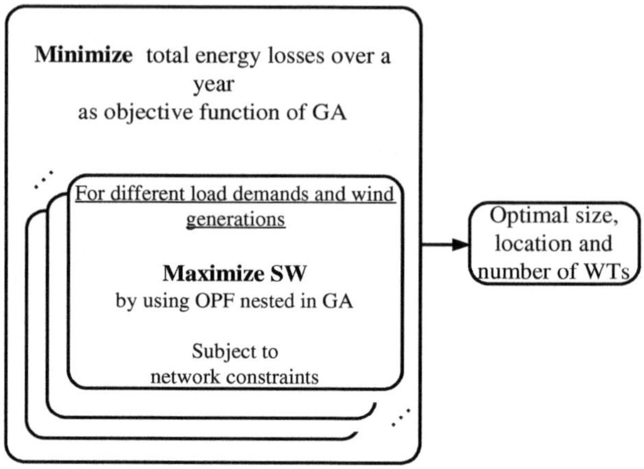


**Fig. 6.2** The structure of the proposed method

## *6.2.2 The Structure of the Proposed Method*

The structure of the proposed hybrid optimization method for optimal siting and sizing of WTs is shown in Fig. 6.2. The method jointly minimizes the total energy losses over a year and maximizes the SW for each scenario.

The WTs' sizes and locations are represented by the variable of the GA: a vector of integers, called chromosome, in the range [0, $N_{sizes}$] with a length equal to the number of candidate buses $N_c$ such that each element of the vector is associated to a candidate bus as shown in Fig. 6.3. In such a way, different vectors allow representing different investments in WTs both in terms of selected locations and sizes.

Different sizes of WTs identified with a label in the range [0, $N_{sizes}$] are considered on the basis of their rating power and power coefficients. Three different sizes of WTs of sizes 1.2, 2 and 3 MW, namely sizes A, B and C, respectively, are considered by the DNO. For example, WTs of size A are related to the first element of the vector while WTs of sizes B and C are related to the last two elements of the vector. According to this formulation, WTs of the same size can be allocated at a candidate bus.

For each chromosome, the annual energy losses are evaluated considering the different scenarios: the energy losses in coincidence with a given scenario derived from the SW maximization.

The GA is able to find the optimal locations and sizes of WTs while the market-based OPF, nested in the GA, to determine the optimal number of WTs of a given size to be selected at a chosen location. In particular, the optimal number is the maximum number of WTs identified by the market-based OPF among all the 77 scenarios.

By solving the abovementioned problems, the optimal locations, sizes and numbers of WTs to be allocated at candidate buses are determined.

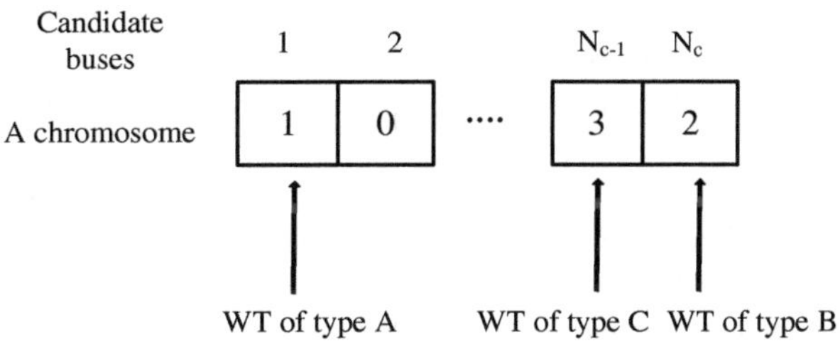


**Fig. 6.3** Schematic of the GA chromosome

### *6.2.3 Simulation Procedure*

The proposed hybrid optimization method to determine the optimal locations, sizes and numbers of WTs runs as follows.

(1) Define GA parameters and WTs types (size and speed-power curves) to be allocated by the DG-owning DNO.
(2) Set the candidate buses according to wind energy availability.
(3) Model uncertainties through hourly time series analysis as explained in Sect. 6.2.1.
(4) Calculate the power output of different sizes of WTs based on their speed-power curves and calculate WTs offer price as is explained in Sect. 6.5.1.
(5) For each chromosome, maximize the SW for the considered scenarios and evaluate the hourly energy losses.
(6) Evaluate the annual energy losses.
(7) If one of the stopping criteria is reached go to step (9) otherwise repeat steps (5)–(7) until one of the stopping criteria is reached.
(8) The products of the proposed method are the optimal locations, sizes and numbers of WTs.
(9) Print the solution.

## 6.3 Genetic Algorithm Implementation

The GA starts with an initial population whose elements are called chromosomes. Chromosomes encode candidate solutions and evolve toward better solutions. The evolution starts from a population of randomly generated chromosomes. During each iteration, called generation, the objective function for each chromosome in the population is assessed and, based on this assessment; a new population of candidate solution is formed. The new population generated in the next iteration is usually better than those in the current population. The GA uses three types of rules at each step namely selection rules, crossover rules and mutation rules to create the next generation from the current population and it continues until some stopping criteria is reached. The algorithm is usually terminated when either a maximum number of generations have been generated or a satisfactory fitness level has been reached for the population [18].

The GA generates the initial populations by defining a set of vector in the range [0, 3]. The number of chromosomes and iterations are set. Each chromosome has a size $N_c$, where $N_c$ is the number of candidate buses. A new set of improved individuals, according to their objective function, at each generation is created by choosing the individuals. After the selection of new population the genetic operators are applied to selected individuals. The operators are simply crossover and binary mutation, an elitism mechanism is also considered. The iteration procedure is repeated until one of the following stopping criteria is reached: (1) the maximum generation number exceeds 300, (2) there isn't improvement in the objective function for five consecutive generations, and (3) the cumulative variation in the objective function in five generations is less than 1e-6.

Sensitivity analyses have been carried out to consider different values for the GA parameters such as stopping criteria, population size and genetic operators. From these analyses, it can be seen that the values used here guarantee the convergence of the algorithm to a satisfactory solution.

The energy losses over the year to be minimized are represented as follows:

$$\text{Minimize} \quad E_{loss}(x) = \sum_{k=1}^{8760} E_{loss}^{k}(x) \tag{6.1}$$

where

$E_{loss}(x)$ is the energy losses over the study period, $E_{loss}^{k}(x)$ is the total system energy losses during $k$th hour and $x$ is the decision variable of the GA that is a vector of integers in the range [0, $N_{sizes}$] with a length equal to the number of candidate buses. For each chromosome, hourly energy losses are obtained considering the DNO acquisition market formulation as described in the following section for all the considered scenarios.

## 6.4 DNO Acquisition Market Formulation

Usually, DNO purchases energy from the wholesale market at a high voltage level and transfers and delivers the energy to final consumers. Nevertheless, due to the power industry restructuring and emerging new agents and producers such as WTs, conventional DNO functions are unbundled into technical and commercial tasks. A DNO energy acquisition market model, called the DNO acquisition market, with WTs and DL is presented here under a distribution market structure based on pool and bilateral contracts within DNO's control area at low voltage level. The DNO is defined as the market operator of the acquisition market, which determines the price estimation and the optimization process for the acquisition of active power. DLs and WTs send active power offers and bids to the DNO acquisition market in form of blocks for each hour [3].

The DNO behaves as a sort of aggregator of distributed energy resources that allows the distribution grid participating in the wholesale market, buying and selling active and reactive power to the bulk grid and optimizing the local (renewable) production capabilities. It takes into account the bids received by energy producers and consumers.

When buying active and reactive power from the grid, the DNO tries to maximize the benefit function of demand while minimizing the costs of energy. When selling active and reactive power to the bulk power grid, due to an excess of low price renewable generation, the DNO also tries to maximize revenues by exchanging power with the grid.

In other words, the complementary operations executed by the DNO acquisition market are:

(1) A one day-ahead schedule of DGs and loads according to the market prices, with every trade day including 24 h trading periods. The dispatch schedules are determined for every trading period [17].
(2) A real-time intraday optimization operation that every 15 min changes the scheduling to take into account the operation and economic requirements.

Under the assumed DNO acquisition market, the market clearing quantity and price are determined by maximizing the social welfare while keeping the distribution network's security. Its maximization implies not only the minimization of the costs related to energy production but also the maximization of the consumers' benefit function. The optimization problem is formulated as follows:

$$\text{Maximize } SW = \sum_j B_j(d_j) - \sum_i C_i(g_i) \tag{6.2}$$

where

$$B_j(d_j) = \frac{1}{2} m_d d_j^2 + b_j d_j \tag{6.3}$$

$$C_i(g_i) = \frac{1}{2} m_g g_i^2 + b_i g_i \tag{6.4}$$

$C_i(g_i)$ and $B_j(d_j)$ are the production cost and benefit of consumers, respectively, $p_i$ is the price at which producer $i$ is willing to supply in €/MWh:

$$p_i = b_i + m_g g_i, \text{ for } i = 1, 2, \cdots, I \tag{6.5}$$

where: $b_i$ is the intercept (reservation price $b_i > 0$) in €/MWh, $m_g$ is the slope ($m_g > 0$) in €/MW$^2$h, $g_i$ is the supply in MW, $p_j$ is the price at which consumer $j$ is willing to pay in €/MWh

$$p_j = b_j + m_d d_j, \text{ for } j = 1, 2, \cdots, J \tag{6.6}$$

where: $b_j$ is the intercept (reservation price $b_j > 0$) in €/MWh, $m_d$ is the slope ($m_d < 0$) in €/MW$^2$h, $d_j$ is the demand in MW.

The optimization variable of the market-based OPF include vector $\mathrm{L} = [V_i, \theta_i, P_g, P_d]$ where $V_i$ and $\theta_i$ are voltage magnitude and voltage angle at the buses, respectively. $P_g$ is the active power generated from WTs and $P_d$ is active power absorbed by DLs.

### 6.4.1 Constraints

(1) Active and reactive power constraints for the interconnection to the external network (slack bus):

$$P_b^{\min} \le P_b \le P_b^{\max}, \; Q_b^{\min} \le Q_b \le Q_b^{\max} \tag{6.7}$$

where $P_b$ and $Q_b$ are active and reactive power of the slack bus, respectively.
(2) Voltage level constraints at the buses

$$V_i^{\min} \le V_i \le V_i^{\max} \tag{6.8}$$

where $V_i^{\min}$ and $V_i^{\max}$ are the lower and upper bounds of the bus voltage, respectively.
(3) Thermal limits of the lines connecting the buses
The maximum thermal capacity $S_k^{\max}$ of network also bounds the apparent power transfer, $S_k$:

$$S - S_k^{\max} \le 0 \tag{6.9}$$

(4) WTs power constraint

$$0 \le P_g \le P_g^{\max} \tag{6.10}$$

It is assumed that WTs operate at constant power factor.

$$\cos\varphi = \frac{P_g}{\sqrt{P_g^2 + Q_g^2}} = \text{constant} \tag{6.11}$$

where $P_g$ and $Q_g$ are the generated active and reactive powers by WTs, respectively.
(5) DLs power constraints

$$P_d^{\min} \le P_d \le 0 \tag{6.12}$$

It is assumed that DLs operate at constant power factor.

$$\cos\varphi = \frac{P_d}{\sqrt{P_d^2 + Q_d^2}} = \text{constant} \tag{6.13}$$

where $P_d$ and $Q_d$ are active and reactive power of DLs, respectively.

### *6.4.2 Dispatchable Load Modeling*

One way to model price-sensitive or DLs is modeling them as negative generators with related negative costs. This is carried out by specifying a generator with a negative output ranging from a minimum injection equal to the negative of the largest possible load to a maximum injection of zero [19]. Here, it is assumed that DLs have a constant power factor. Furthermore, an additional equality constraint to enforce a constant power factor for any "negative generator" is utilized to model a DL. Note that with the definition of DLs as negative generators, if the negative cost corresponds to a benefit for consumption, minimizing the cost of generation is equivalent to maximizing SW [19, 20].

### *6.4.3 Constrained Cost Variable Formulation*

As the formulation of standard OPF is unable to solve the non-smooth piecewise linear cost functions that results from discrete offers and bids, when such cost functions are convex they can be modeled by means of CCV method [19–21]. The piecewise linear cost function $c(x)$ is substituted by a helper variable y and linear constraints that form a convex "basin" requiring the cost variable y to put into the function $c(x)$. Figure 6.4 shows a convex n-segment piecewise linear cost function.

$$c(x) = \begin{cases} m_1(x - x_1) + c_1, & x \le x_1 \\ m_2(x - x_2) + c_2, & x_1 < x \le x_2 \\ \vdots & \\ m_n(x - x_n) + c_n, & x_{n-1} < x \end{cases} \tag{6.14}$$

A convex n-segment piecewise linear cost function is defined by a sequence of points $(x_j, c_j), j = 0 \cdots n$ where $m_j$ denotes the slope of the $j$th segment,

$$m_j = \frac{c_j - c_{j-1}}{x_j - x_{j-1}}, \; j = 1, \cdots, n \tag{6.15}$$

and $x_0 < x_1 < \cdots < x_n$ and $m_1 \le m_2 \le \cdots < m_n$.

The "basin" corresponding to the cost function is formed by the following n constraints on the helper variable y.

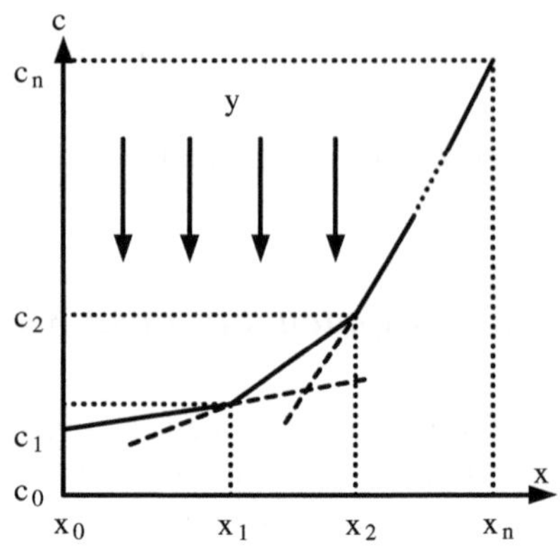


**Fig. 6.4** Constrained cost variable

$$y \geq m_j(x - x_j) + c_j,\ j = 1 \cdots n \tag{6.16}$$

The cost term added to the objective function instead of $c(x)$ is the variable y. The CCV method can convert any piecewise linear costs of active or reactive power into the appropriate helper variable and the related constraints.

By applying CCV method, every piecewise function in the objective is substituted with a helper variable. A number of inequality constraints, one for each piece of the piecewise function, are placed on that variable. CCV is a way that formulates a piecewise linear cost function on a new variable that is linearly constrained.

In this chapter, the offers of WTs and bids of DLs are taken and treated as marginal cost and marginal benefit functions, respectively, then they are converted into the equivalent total cost and total benefit functions by integrating the marginal cost and benefit functions and plugged into a matrix as piecewise linear costs. CCV approach conquers the difficulty of disruptive Lagrange derivatives by expanding the optimization into a higher dimensional space and counting on good constrained optimization methods to resolve the transformed smooth optimization problem.

### *6.4.4 Step-Controlled Primal Dual Interior Point Method*

The primal dual interior point method (PDIPM) and its numerous variations have become the algorithms of choice for solving OPFs in the last years [22–24].

Even if the PDIPM fits properly with classical OPFs that utilize smooth polynomial cost function, it is not able to solve the market-based OPFs with non-differentiable piecewise cost. When piecewise cost is considered, the gradient and Hessian variables change from iteration to iteration considerably. Also, the descending of Newton steps is not obtained. The SCPDIPM [20, 21] that is used in this chapter overcomes this difficulty by monitoring the accuracy of the quadratic

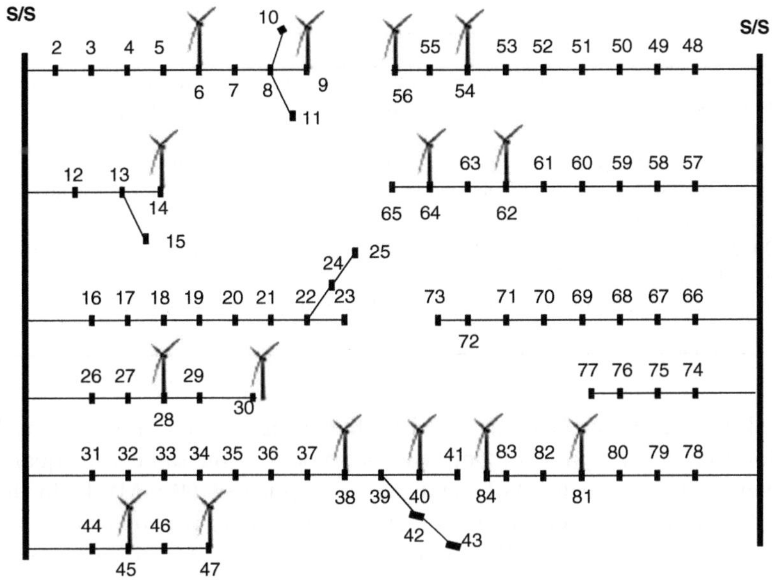


**Fig. 6.5** 84-bus radial distribution network with candidate locations for WTs

approximation of the Lagrangian during the OPF computation and reducing the Newton step if any unexpected change of derivative results in an inaccurate approximation. It is efficient applying such step control method when the normal PDIPM step is not able to improve the gradient condition. By adjusting steps, SCPDIPM is able to reduce both system cost and gradients.

## 6.5 Test System Description

In this section, the distribution system used to test the proposed method is described. The following analyses are based on an 84-bus 11.4-kV radial distribution system whose data given in [4]. The eleven feeders are supplied by two 20 MVA, 33/11.4 kV transformers. The one line diagram of the distribution system is shown in Fig. 6.5. The candidate buses in the test system are included in the set {6, 9, 14, 28, 30, 38, 40, 45, 47, 54, 56, 62, 64, 81, 84}. The WTs operate at power factor of 0.95 lagging. Voltage limits are taken to be ± 6 % of nominal value, i.e. $V_{max} = 1.06$ p.u. and $V_{min} = 0.94$ p.u., and the feeders' thermal limits are given in Table 6.1 and vary between 90 and 480 A. Dispatchable and fixed loads, with constant power factor equal to 0.95, are served by both the grid and the WTs. The total maximum fixed loads are 5.4 MW.

The different sizes of WTs 1.2, 2 and 3 MW are considered by the DNO. At each candidate bus it is assumed that maximum four WTs of each size can be

Table 6.1 Existing wires

| Wires | Amps |
|---|---|
| 1–2, 2–3, 3–4, 4–5, 5–6, 6–7, 8–11, 1–12, 12–13, 1–16, 18–19, 1–26, 26–27, 1–31, 31–32, 1–44, 44–45, 1–48, 48–49, 1–57, 57–58, 1–66, 66–67, 1–74, 74–75, 1–78, 78–79, 80–81 | 480 |
| 7–8, 13–15, 16–17, 17–18, 27–28, 32–33, 33–34, 34–35, 35–36, 36–37, 37–38, 38–39, 44–45, 49–50, 50–51, 51–52, 52–53, 58–59, 59–60, 60–61, 61–62, 67–68, 68–69, 70–71, 79–80, 80–81 | 330 |
| 19–20, 20–21, 21–22, 22–23, 22–24, 24–25, 28–29, 39–42, 42–43, 53–54, 54–55, 55–56, 62–63, 69–70, 71–72, 72–73, 75–76, 76–77 | 180 |
| 8–9, 8–10, 13–14, 29–30, 39–40, 40–41, 45–46, 46–47, 63–64, 64–65, 81–82, 82–83, 83–84 | 90 |

allocated; this requirement is regulated by the accessible land for building WTs. Thus, for each generation level there are four blocks of the same size equal to the rated power of the selected WTs and the same price of 60 €/MWh. In the following subsection, the method of estimating WTs' offers is explained.

The maximum number of WTs that can be allocated at a given bus is represented by an equivalent number of blocks in the WT's offer. The offers are assumed at price of 80 €/MWh for the bus connecting the distribution network to the transmission one.

With regards to the bids for DLs, there are two blocks per demand bid with different sizes as presented in Table 6.2 and the same price of 140 €/MWh for all blocks.

### 6.5.1 Calculation of WTs' Offers from the Point of View of DNO

In order to calculate WTs' offers, financial data, i.e. WTs' life time, installation cost, depreciation time, interest rate, are considered as summarized in Table 6.3 [25, 26]. The annual cost for WTs is calculated as follows [26]:

$$Ann\_Cost = \frac{r(1+r)^n}{(1+r)^n - 1} \times Inst\_Cost \tag{6.17}$$

where $r$ is the interest rate, $n$ is the depreciation period in years, *Inst_Cost* is the installation cost, and *Ann_Cost* is the annual cost for depreciation. The capacity factor is evaluated according to the wind generation data and the WTs capability curves. For example, for a 1.2 MW WT the capacity factor is about 40 %, i.e. 3,504 MWh/MW. Therefore, by dividing *Ann_Cost* by equivalent number of hours i.e. 3,504 h, the WTs' offers with no subsidy is about 56 €/MWh. Therefore, the 1.2 MW WTs' offers, without subsidies, are assumed as 60 €/MWh. The same approach can be applied considering WTs of different sizes and capacity factors. In

**Table 6.2** Bids of dispatchable loads

| Load no. | Bus no. | Block 1 (MW) | Block 2 (MW) | Load no. | Bus no. | Block 1 (MW) | Block 2 (MW) |
|---|---|---|---|---|---|---|---|
| 1 | 6 | 0.84 | 0.84 | 16 | 44 | 2.10 | 1.05 |
| 2 | 9 | 0.84 | 0.84 | 17 | 46 | 2.10 | 1.05 |
| 3 | 14 | 0.84 | 0.84 | 18 | 51 | 2.10 | 1.05 |
| 4 | 28 | 0.84 | 0.84 | 19 | 53 | 2.10 | 4.20 |
| 5 | 30 | 0.84 | 0.84 | 20 | 55 | 2.10 | 4.20 |
| 6 | 38 | 0.84 | 0.84 | 21 | 59 | 4.20 | 2.10 |
| 7 | 40 | 0.84 | 0.84 | 22 | 63 | 4.20 | 2.10 |
| 8 | 45 | 0.84 | 0.84 | 23 | 67 | 2.10 | 0.84 |
| 9 | 47 | 2.10 | 1.05 | 24 | 69 | 2.10 | 0.84 |
| 10 | 54 | 2.10 | 1.05 | 25 | 73 | 2.10 | 2.10 |
| 11 | 56 | 2.10 | 1.05 | 26 | 75 | 4.20 | 2.10 |
| 12 | 62 | 2.10 | 1.05 | 27 | 77 | 0.84 | 0.84 |
| 13 | 64 | 2.10 | 1.05 | 28 | 79 | 4.20 | 2.10 |
| 14 | 81 | 2.10 | 2.10 | 29 | 82 | 0.84 | 0.84 |
| 15 | 84 | 1.05 | 1.05 | | | | |

**Table 6.3** Financial data for estimating the offers for a 1.2 MW WT

| Life time (years) | 20 |
|---|---|
| Installation cost (€/kW) | 1,200 |
| Depreciation time (years) | 10 |
| Interest rate (%) | 10 |
| Number of equivalent hours (h) | 3,504 |
| Capacity factor (%) | 40 |
| Annual cost (€/kW-year) | 195.27 |

order to simplify the analysis, the same offers are considered for all the sizes of WTs.

## 6.6 Case Study and Simulation Results

The proposed method is applied to the abovementioned distribution network. According to a sensitivity analysis, the number of generations and the population size are chosen as, 300 and 20, respectively. The method has been implemented in MATLAB® incorporating some features of MATPOWER suite [19] and MATLAB® toolbox for GA [27] on a laptop with core i7, 1.6 GHz processor and 4 GB of RAM.

The minimum energy losses over the year are about 7,532 MWh. The optimal sizes and numbers of WTs at each candidate bus found by the proposed method are given in Table 6.4. It is evident from Table 6.4 that buses 54, 62, and 81 have the

Table 6.4 The optimal numbers, sizes and capacities of WTs obtained by the proposed method

| Bus no. | Size | Number | Capacity (MW) |
|---|---|---|---|
| 6 | – | – | – |
| 9 | C | 1 | 3 |
| 14 | A | 2 | 2.4 |
| 28 | – | – | – |
| 30 | C | 1 | 3 |
| 38 | A | 4 | 4.8 |
| 40 | – | – | – |
| 45 | A | 4 | 4.8 |
| 47 | – | – | – |
| 54 | B | 4 | 8 |
| 56 | B | 2 | 4 |
| 62 | B | 4 | 8 |
| 64 | – | – | – |
| 81 | B | 4 | 8 |
| 84 | – | – | – |
| Total capacity | | | 46 |

Table 6.5 Comparison of the results with ACO

| Method | Total energy losses (MWh) |
|---|---|
| ACO | 7,651 |
| GA | 7,532 |

largest installed capacity (i.e. 8 MW) while bus 14 has the lowest one (i.e. 2.4 MW). The installed capacity, in fact, is limited by voltage and thermal limits as well as by the bids' values at each bus. For instance, the installed capacity at bus 14 is limited to 2.4 MW (two WTs of size A) and this is mainly due to the lowest value of both thermal limit of the line connecting the buses 13–14 (i.e. 90 A) and the bids' values of DLs if compared to those at the other lines and buses.

The installed capacity at buses 38 and 45 is 4.8 MW (four WTs of size A). These buses have the same bid values and the higher thermal limits (i.e. 330 A) of the lines 37–38 and 44–45 connecting the WTs if compared to previous case.

At buses 62 and 81, the bids of DL are higher and the thermal limits of the lines 61–62 and 80–81 are the same if compared to the previous case; consequently, the highest capacity is installed at these buses (i.e. four WTs of size B).

In order to evaluate and compare the obtained results, ACO is used with the same population size for the GA, the same number of ants, i.e. 20 and the same number of iterations, i.e. 300. It can be observed from Table 6.5 that the total energy losses obtained by GA are lower than that obtained by ACO.

As regards with the SW, it increases proportionally to both load demand and wind generation as shown in Fig. 6.6. It is worth mentioning that, in all scenarios, the SW is higher if compared to that without WTs in the network.

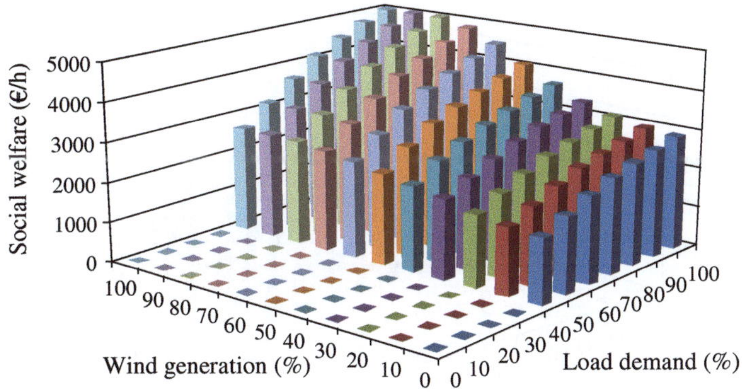


Fig. 6.6 Social welfare

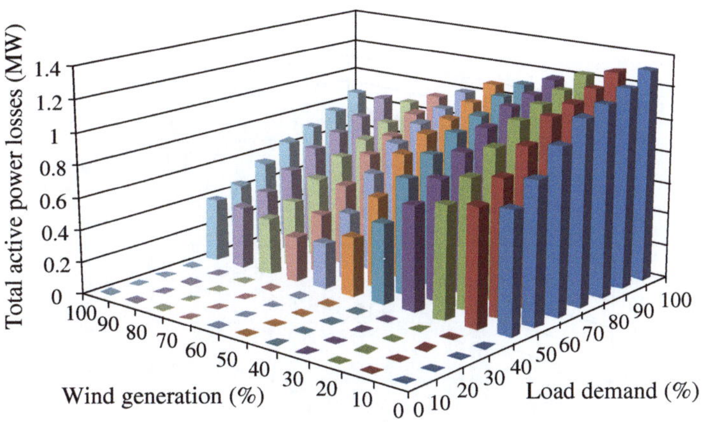


Fig. 6.7 Total active power losses

In the case of minimum load (i.e. 40 %) and maximum wind generation level (i.e. 100 %), the SW is equal to about 2,500 €/h and in the case of maximum load and minimum wind generation level the SW is equal to about 3,000 €/h. Instead, in the case of maximum wind generation level and maximum load demand this value is equal to around 5,000 €/h.

It is seen from Fig. 6.7 that by increasing the generation, active power losses are decreased. In all scenarios, the total active power losses are lower if compared to the case with no WTs in the network.

In the case of maximum wind power generation and minimum load demand the total active power losses are reduced by about 50 % if compared to the case with no WTs in the network.

It is evident that total active power losses have inverse relation with the wind generation and direct relation with load demand.

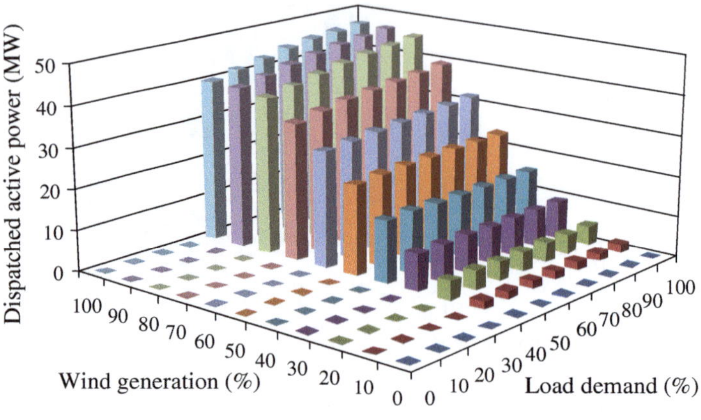


**Fig. 6.8** Dispatched active power

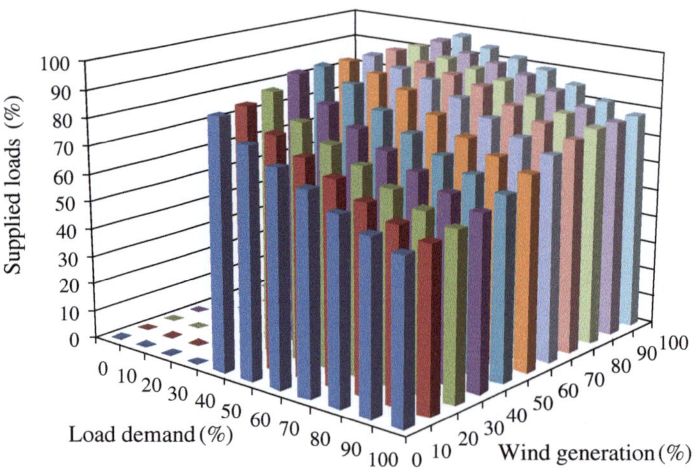


**Fig. 6.9** Percentage of supplied loads

Figure 6.8 shows the total dispatched active power by WTs in different scenarios of wind generations and load demands. It is evident that the dispatched active power has the direct relation with both load demand and wind generation.

The supplied loads, shown in Fig. 6.9, evidences its direct relation with wind generation and its inverse relation with load demand due to the network constraints that limit load increase when constraints are binding.

The cases considering different wind energy potential at candidate buses can be addressed by considering different capacity factors for each location and calculating the WTs' offers as described in Sect. 6.5.

## 6.7 Conclusion

In this chapter, a hybrid optimization method that combines the GA and the market-based OPF to optimally siting and sizing WTs from the point of view of the DG-owning DNO is proposed. The method jointly minimizes the annual energy losses and maximizes the SW considering different combination of wind generations and load demands to determine the optimal locations, sizes and numbers of WTs to be allocated at candidate buses. The GA is used to select the optimal locations and sizes among different sizes of WTs while the market-based OPF to determine the optimal number of WTs. The DNO acts as the market operator of the DNO acquisition market that estimates the market clearing price and the optimization process for the active power hourly acquisition. The stochastic nature of both load and wind is modeled by hourly time series analysis.

With the proposed method not only WTs are optimally allocated but also the total power losses is decreased and the SW, dispatched active power and supplied loads are increased if compared to the case with no WTs in the network.

The proposed method is consistent with the topology of the distribution system, thus taking into account the demand willingness to buy energy at different buses and can be used to assist DNOs to evaluate the performance of the network and to plan the WTs integration into distribution networks. Simulation results confirmed the capability and effectiveness of the proposed method in optimally siting and sizing of WTs in distribution networks.

## References

1. Banosa R, Manzano-Agugliaro F, Montoya FG, Gila C, Alcayde A, Gómez J (2011) Optimization methods applied to renewable and sustainable energy: a review. Renew Sustain Energ Rev 15(4):1753–1766
2. Kirschen D, Strbac G (2004) Fundamentals of power system economics. Wiley, New York
3. Palma-Behnke R, Cerda JLA, Vargas L, Jofre A (2005) A distribution company energy acquisition market model with the integration of distribution generation and load curtailment options. IEEE Trans Power Syst 20(4):1718–1727
4. Piccolo A, Siano P (2009) Evaluating the impact of network investment deferral on distributed generation expansion. IEEE Trans Power Syst 24(3):1559–1567
5. Siano P, Ochoa LF, Harrison GP, Piccolo A (2009) Assessing the strategic benefits of distributed generation ownership for DNOs. IET Gener Transm Distrib 3(3):225–236
6. Harrison GP, Piccolo A, Siano P, Wallace AR (2007) Exploring the tradeoffs between incentives for distributed generation developers and DNOs. IEEE Trans Power Syst 22(2):821–828
7. Harrison GP, Piccolo A, Siano P, Wallace AR (2008) Hybrid GA and OPF evaluation of network capacity for distributed generation connections. Elect Power Syst Res 78(3):392–398
8. Harrison GP, Piccolo A, Siano P, Wallace AR (2007) Distributed generation capacity evaluation using combined genetic algorithm and OPF. Int J Emerg Electr Power Syst 8(2):1–13

9. Chen P, Siano P, Chen Z, Bak-Jensen B (2010) Optimal allocation of power-electronic interfaced wind turbines using a genetic algorithm-Monte Carlo hybrid optimization method, wind power systems. Springer, Berlin, pp 1–23
10. Nara K, Hayashi Y, Ikeda Y, Ashizawa K (2007) Optimal allocation of distributed generation and reactive sources considering tap positions of voltage regulators as control variables. Eur Trans Electr Power 17(3):219–239
11. Falaghi H, Haghifam MR (2007) ACO based algorithm for distributed generation sources allocation and sizing in distribution systems. In: IEEE Power Technologies, pp 555–560
12. Atwa YM, El-Saadany EF (2011) Probabilistic approach for optimal allocation of wind-based distributed generation in distribution systems. IET Renew Power Gener 5(1):79–88
13. Chen P, Siano P, Bak-Jensen B, Chen Z (2010) Stochastic optimization of wind turbine power factor using stochastic model of wind power. IEEE Trans. Sustain Energ 1(1):19–29
14. Borges CLT, Falcao DM (2006) Optimal distributed generation allocation for reliability, losses, and voltage improvement. Electr Power Energ Syst 28(6):413–420
15. Siano P, Chen P, Chen Z, Piccolo A (2010) Evaluating maximum wind energy exploitation in active distribution networks. IET Gener Transm Distrib 4(5):598–608
16. Cecati C, Citro C, Piccolo A, Siano P (2011) Smart operation of wind turbines and diesel generators according to economic criteria. IEEE Trans Ind Electron 58(10):4514–4525
17. Cecati C, Citro C, Siano P (2011) Combined operations of renewable energy systems and responsive demand in a smart grid. IEEE Trans Sustain Energ 2(4):468–476
18. Srinivas M, Patnaik LM (1994) Adaptive probabilities of crossover and mutation in genetic algorithms. IEEE Trans Syst Man Cybernet 24(4):656–667
19. Zimmerman RD (2011) Manual MATPOWER http://www.pserc.cornell.edu/matpower/manual.pdf
20. Zimmerman RD, Murillo-Sánchez CE, Thomas RJ (2011) MATPOWER: steady-state operations, planning, and analysis tools for power systems research and education. IEEE Trans Power Syst 26(1):12–19
21. Wang H, Murillo-Sánchez CE, Zimmerman RD, Thomas RJ (2007) On computational issues of market-based optimal power flow. IEEE Trans Power Syst 22(3):1185–1193
22. Castronuovo ED, Campagnolo JM, Salgado R (2001) On the application of high performance computation techniques to nonlinear interior point methods. IEEE Trans Power Syst 16(3):325–331
23. Jabr RA, Coonick AH, Cory BJ (2001) A primal-dual interior point method for optimal power flow dispatching. IEEE Trans Power Syst 17(3):654–662
24. Qiu W, Flueck AJ, Tu F (2005) A new parallel algorithm for security constrained optimal power flow with a nonlinear interior point method. IEEE PES general meeting, pp 2422–2428
25. MICROGRIDS-Large scale integration of micro-generation to low voltage grids (2002) EU Contract ENK5-CT-2002-00610, Technical Annex. Available http://microgrids.power.ece.ntua.gr
26. Tsikalakis AG, Hatziargyriou ND (2008) Centralized control for optimizing microgrids operation. IEEE Trans Energ Convers 23(1):241–248
27. MathWorks (2004) Genetic algorithms and direct search toolbox: user guide

# Chapter 7
# Power Flow Analysis and Reactive Power Compensation of Grid Connected Wind Energy Conversion Systems

**J. Ravishankar**

**Abstract** The power flow analysis is the basic tool for analyzing the steady state operation of any power system. It also provides the necessary initial conditions to investigate the dynamic performance of the system. This chapter discusses the power flow analysis of grid connected wind energy conversion systems (WECS). The power flow analysis with WECS is quite complicated, unlike the analysis with conventional sources, because: 1. The power injected into the grid by WECS depends on the instantaneous wind speed, which varies unpredictably. 2. Most WECS use induction generators. Therefore, the operating slip of the machine has to be determined. The machine operates at a slip for which the mechanical power developed by the turbine is equal to the electrical power developed by the induction generator. This chapter outlines two methods (1) sequential method of power flow and (2) simultaneous method of power flow analysis for grid connected WECS. Both the methods of power flow are tested on a sample system and the results are presented. This chapter also looks into the effect of various types of reactive power compensation, namely, shunt, series and series-shunt compensation, on the steady state performance of WECS equipped with squirrel-cage induction generators. A cost effective method to strengthen the given network between the point of common coupling (PCC) and rest of the grid is proposed. The effect of compensation in improving the penetration level of wind energy into the grid is also analyzed.

**Keywords** Wind energy · Induction generators · Point of common coupling · Power flow analysis · Reactive power compensation

J. Ravishankar (✉)
School of Electrical Engineering and Telecommunications, The University of New South Wales, Sydney, NSW 2052, Australia
e-mail: jayashri.ravishankar@unsw.edu.au

J. Hossain and A. Mahmud (eds.), *Renewable Energy Integration*,
Green Energy and Technology, DOI: 10.1007/978-981-4585-27-9_7,

## 7.1 Introduction

One of the most critical issues for wind energy exploitation in developing countries like India has been the transmission capacity of the grid in the areas where the wind farms exist. Wind farms are concentrated in the rural areas where the existing transmission grids are very weak. In addition, the wind farms were developed during a comparatively short period of time in a few areas, and the reinforcement of the transmission systems in these areas has lagged behind the fast development of wind energy [1].

One of the problems with wind energy generation is the dependence of the injected power on the wind speed. The wind speed cannot be predicted, but the probability of a particular wind speed occurring can be estimated. This can be done by assuming a Weibull or a Rayleigh probability distribution. Once the wind speed is known, the power injected into the grid can be calculated by means of the wind turbine power curve. Therefore, assessment of steady state performance of electrical networks with WECS is not as simple as with conventional generation and can be planned from a probabilistic point of view only.

Most WECS are equipped with induction generators. Early systems used squirrel-cage generators as fixed-speed types and the current systems use doubly-fed induction generators (DFIG) that are capable of producing a variable-speed operation. For the dynamic analysis of a system, correct initialization is required otherwise time will be wasted as the system tries to find a steady-state operating point as the initial condition [2]. Initializing most components' variables is a straightforward, direct process. However, initializing induction machine variables requires an indirect, iterative process [3]. Therefore, the power flow analysis in systems with induction machines needs additional iterative procedure.

Several attempts have been made in the past to investigate the steady state behaviour of induction machines in a wind farm [4–9]. All these attempts involve a sequential approach to calculate the state variables of the wind generators. This sequential iterative approach is rather attractive because it is straightforward to implement in existing power flow programs, but caution has to be exercised because it will yield no quadratic convergence [10], and an additional set of nonlinear algebraic equations have to be solved to obtain the values of the wind generator's state variables [11, 12].

In this Chapter, the power flow of a radial system with WECS is simulated, using two methods (a) sequential and (b) simultaneous method of power flow. The WECS is modeled with squirrel-cage induction generators. Simulation carried out on a 9-bus radial system shows that the power flow results of both the methods are comparable. Additionally, the simultaneous method of power flow analysis exhibits better convergence characteristics.

From the power flow results, it is observed that the grid connected induction generator imports heavy reactive power from the grid, resulting in poor power factor and reducing the voltage at the WECS terminals. Therefore, reactive power support is needed at WECS terminals for voltage regulation and improvement of low voltage ride-through capabilities. As the wind speed continuously changes, the voltage at the PCC fluctuates. A possible way to improve this situation is by incorporating reactive power compensation. Simulation results show that the combination of series and shunt compensation results in a very good voltage profile at PCC and a significant reduction in transmission line current due to the reduction of reactive power flow through the line. This also allows more turbines to be connected at the PCC.

## 7.2 Steady State Model of Induction Generator

For power flow analysis, the induction generator can be represented by the well-known equivalent circuit shown in Fig. 7.1.

In Fig. 7.1, $R_s$—stator resistance, $R_r$—rotor resistance, $X_s$—stator leakage reactance, $X_r$—rotor leakage reactance, $X_m$—magnetizing reactance, $\bar{I}_1$—stator current, $\bar{I}_2$—rotor current, $\bar{V}$—terminal voltage, s—slip given by $(\omega_s\text{-}\omega_r)/\omega_s$.

From Fig. 7.1, the current $I_1$ can be written as,

$$\bar{I}_1 = \frac{\bar{V}}{(R_s + R_e) + j(X_s + X_e)} \tag{7.1}$$

where,

$$R_e + jX_e = \frac{jX_m\left(\frac{R_r}{s} + jX_r\right)}{\frac{R_r}{s} + j(X_m + X_r)} \tag{7.2}$$

The per-unit active power transferred from the rotor to the stator through the air gap, called air gap power, is readily calculated from the equivalent circuit as,

$$P_g = I_2^2 \frac{R_r}{s} \tag{7.3}$$

The electrical power developed in the rotor is,

$$P_e = I_2^2 \frac{R_r}{s}(1 - s) \tag{7.4}$$

where the slip is negative. The electrical torque developed is then given by,

$$\tau_e(V, s) = \frac{V^2 X_m^2 \frac{R_r}{s}}{\left[\left(R_x + \frac{R_r}{s}\right)^2 + (X_x + X_r)^2\right]\left[R_s^2 + (X_s + X_m)^2\right]} \tag{7.5}$$

where

$$R_x + jX_x = \frac{jX_m(R_s + jX_r)}{R_s + j(X_s + X_m)} \tag{7.6}$$

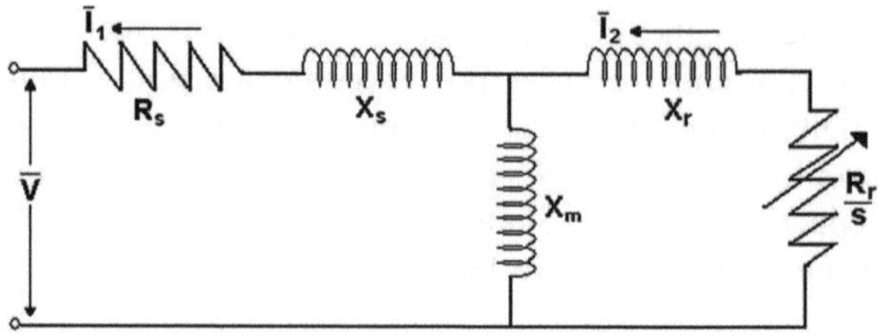


**Fig. 7.1** Steady state equivalent circuit of induction generator

## 7.3 Turbine Aggregation

It is possible to aggregate the machines in the wind farm, consisting of N parallel turbines, by a single machine equivalent with a re-scaled power capacity. The following are valid:

- The wind turbines are identical, including the parameters of the generator used.
- Wind speeds at the wind farm are uniform.
- Each wind turbine runs at the same operating condition at all times. Thus the voltage, current and power of each wind turbine are identical.
- With the aggregation procedure, the equivalent wind turbine of the entire wind farm is a scale up of a single wind turbine, i.e. the base power becomes N times the base power of a single wind turbine in the farm.
- Similarly, the equivalent generator impedance becomes 1/N times the impedance of the generator in the individual turbine.

In this Chapter, the single machine equivalent is applied and found to reduce the computational time.

## 7.4 Power Flow Analysis

In steady state, the induction generator is modeled based on the following two facts [13]:

(i) The machine can be simulated in steady state as impedance if its parameters and slip are known.

(ii) The slip of the machine can be calculated if its power coefficient curve and wind speed are known.

When a wind farm with asynchronous generation is included in the power flow analysis, the PQ and RX [14, 15] buses are the most commonly used. DFIG machines can be modeled as PV buses, however, handling this is not always easy especially when multiple and different types are combined [16]. When the conventional PQ bus model is used, the real and reactive powers have constant values, although these values can be modified in order to represent loads depending either on the voltage or on the frequency [17].

### 7.4.1 Sequential Method of Power Flow Analysis

In the sequential method, for a given wind speed and a given rotor speed the power extracted from the WECS is calculated from power coefficient curve. To the same rotor speed, the power developed by the induction generator is calculated from the results of power flow analysis. These two powers are then compared for convergence. Thus, in this method power flow analysis is carried out for every value of calculated slip. The analysis is continued until the operating slip is obtained.

The algorithmic steps for the sequential solution method for a given wind speed, are presented below:

1. Begin with a slip $s = s_{rated}$. With this value of s, calculate the impedance Z of the induction generator.
2. With these values, model the wind farm as impedance by including the admittance of the generators in the admittance matrix.
3. Run the power flow and obtain the solution. With these results, calculate power developed by the generator by using Eq. (7.4).
4. With the value of s, assumed in the step 1, calculate the tip-speed ratio, $C_p$ and power extracted from the wind using Eqs. (7.7), (7.8) and (7.10) respectively. The tip-speed ratio is given by,

$$\lambda = \frac{\omega R}{v} \tag{7.7}$$

The power coefficient is computed using,

$$C_p(\lambda, \theta) = C_1(C_2\frac{1}{\Lambda} - C_3\theta - C_4\theta^x - C_5)e^{(-C_6\frac{1}{\Lambda})} \tag{7.8}$$

where

$$\frac{1}{\Lambda} = \frac{1}{\lambda + 0.008\theta} - \frac{0.035}{1 - \theta^3} \tag{7.9}$$

and $C_1$ to $C_6$, x are constants. The power extracted from the turbine is then,

$$P_t = C_P \times \frac{1}{2}\rho A v^3 \tag{7.10}$$

where, $\rho$ is the density of dry air, $v$ is the speed of the wind in m/s and A is the swept area of the blades in $m^2$

5. Compute $\Delta P_m = P_t - P_e$. If $|\Delta P_m| >$ Specified tolerance value, $\varepsilon$, then update the slip by using, $s^{new} = s + \Delta s$ and go to Step 2. Else stop the process.

   Here, $\Delta s = -J^{-1} * \Delta P_m$, where the Jacobian J is given by,

$$J = \frac{\partial(P_t(s) - P_m(s))}{\partial s} \tag{7.11}$$

### 7.4.2 Simultaneous Method of Power Flow Analysis

This method determines simultaneously the state variables corresponding to nodal voltage magnitudes and angles of the network and slip of induction generators [18]. In this method, the WECS is modeled as a variable PQ bus. Assuming adequate initial conditions, the method retains Newton's quadratic convergence. In this method the N-R power flow algorithm is reformulated to include the mismatch equation $\Delta P_m$. The unified power flow formulation is,

$$\begin{bmatrix} [\Delta P] \\ [\Delta Q] \\ [\Delta P_m] \end{bmatrix} = \begin{bmatrix} \left[\frac{\partial P}{\partial \theta}\right] & \left[|V|\frac{\partial P}{\partial |V|}\right] & \left[\frac{\partial P}{\partial s}\right] \\ \left[\frac{\partial Q}{\partial \theta}\right] & \left[|V|\frac{\partial Q}{\partial |V|}\right] & \left[\frac{\partial Q}{\partial s}\right] \\ \left[-\frac{\partial \Delta P_m}{\partial \theta}\right] & \left[-|V|\frac{\partial \Delta P_m}{\partial |V|}\right] & \left[-\frac{\partial \Delta P_m}{\partial s}\right] \end{bmatrix} \times \begin{bmatrix} [\Delta \theta] \\ [\Delta |V|/|V|] \\ [\Delta s] \end{bmatrix} \tag{7.12}$$

where $\left[\frac{\partial \Delta P_m}{\partial s}\right]$ is a diagonal matrix whose order is equal to the number of wind farms in the network. Their elements are given by,

$$\frac{\partial \Delta P_m}{\partial s} = \frac{1}{2}\rho a v^3 [C_2 - EC_1] C_1 e^{C_6 \Lambda} \frac{\omega r/v}{D} + \frac{A}{sC^2}(B - C) \tag{7.13}$$

where the values of A to E are given by,

$$A = \frac{V^2 X_m^2 R_r/s}{R_s^2 + (X_s + X_m)^2} \tag{7.14}$$

$$B = \frac{2R_r}{s}(R_x + \frac{R_r}{s}) \tag{7.15}$$

$$C = (R_x + \frac{R_r}{s})^2 + (X_x + X_r)^2 \tag{7.16}$$

$$D = (\lambda + 0.08\theta)^2 \tag{7.17}$$

$$E = C_2 \Lambda - C_3 \theta - C_5 \tag{7.18}$$

Here, $\Delta P_m$ is the difference between the power extracted from the turbine and electrical power developed in the machine and $\Delta s$ is the vector of incremental changes in induction generator's slip. It is to be noted that when the induction machine slip varies, its generated electrical power varies. This in-turn impacts the mismatch vector $\Delta P_m$. Thus, $\Delta s$ has a significant influence on the power flow result.

In the above unified power flow Eq. (7.12) the dimension of the Jacobian for the power system having $N_g$ generator buses, $N_l$ load buses and $N_w$ wind farm buses with induction generators is $(N_g + 2N_l + 3N_w - 1) \times (N_g + 2N_l + 3N_w - 1)$.

### *7.4.3 Convergence Characteristics of Simultaneous and Sequential Methods of Power Flow*

Both the methods of power flow are tested on a 9-bus radial system given in Fig. 7.2 [19]. The data for the same is provided in Appendix I.

Simulations are carried out for wind speeds ranging from 5 m/s (cut-in speed) to 25 m/s (cut-out speed) in steps of 2 m/s and with one turbine connected at bus 9. It should be noted that the ideal location of the WECS is identified to be at bus 9, by running power flow simulations with WECS at buses 2, 7, 8 and 9. Voltage magnitudes of all the buses and power system losses of each case were compared and it was noted that WECS interconnected at bus 9, produced the best voltage magnitude and minimum losses. The variation of voltage at the WECS terminals (bus 9), real power generated and reactive power consumed by the WECS using the simultaneous and sequential methods of power flow are presented in Table 7.1. For the system considered, the Jacobian matrix for the simultaneous method is a $(17 \times 17)$ matrix. The elements of the Jacobian are given in Appendix II.

The analysis reveals that the results obtained in both methods are comparable. It is also observed that the number of iterations with sequential method is 20 (5 times power flow $\times$ 4 iterations in every power flow), whereas with simultaneous method it is only 6. Thus the convergence with simultaneous method is better.

In Table 7.1, P is the power generated at WECS terminals (measured at bus 9), Q is the reactive power consumed by WECS, V and $\delta$ represent the voltage magnitude and phase angles of bus 9.

## 7.5 Grid Strengthening

The number of wind turbines on-line determines the loading of the transmission system. Thus, it affects the voltage at the WECS terminals. To illustrate this, the simulation is initially run for one wind turbine connected at bus 9. The wind turbines considered are uncompensated. Repeated runs are performed by increasing the number of turbines connected at bus 9. In the simulation, the impedance of induction generators connected at bus 9 is calculated as Z/N, where N is the number of turbines. The plots of voltage at WECS terminals, real power generated and reactive power consumed by WECS, for wind speeds ranging from 5 m/s to 25 m/s, obtained through simulation are given in Figs. 7.3, 7.4 and 7.5 respectively. It is clear that as the number of turbines increases, the reactive power consumed increases and as a result the voltage drops. These are reflected in Figs. 7.5 and 7.3.

As more turbines are connected, the generated power increases as illustrated in Fig. 7.4. But when the number of wind turbines connected is more than 2, the aerodynamic power of the wind turbine over-powers the power developed by induction generator and thus the wind turbine goes into a runaway condition. This is

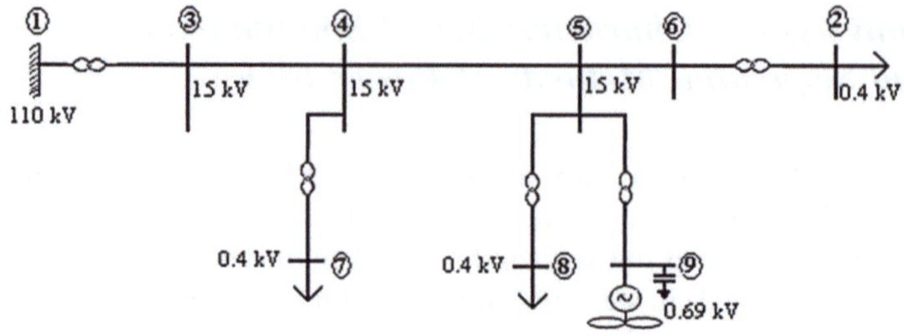


**Fig. 7.2** Single-line diagram of 9-bus radial system

**Table 7.1** Power flow results of 9-bus radial system for bus 9

| Wind Speed (m/s) | Sequential method | | | | Simultaneous method | | | |
|---|---|---|---|---|---|---|---|---|
| | V (PU) | $\delta$ (deg) | P (kW) | Q (kVAR) | V (PU) | $\delta$ (deg) | P (kW) | Q (kVAR) |
| 5 | 0.9787 | 0.9278 | 82.9141 | 214.6557 | 0.9780 | 0.9047 | 82.8858 | 213.6035 |
| 7 | 0.9843 | 1.9068 | 211.1735 | 221.9829 | 0.9838 | 1.8937 | 211.1126 | 223.4978 |
| 9 | 0.9895 | 3.4596 | 367.4732 | 241.6829 | 0.9890 | 3.1213 | 367.5418 | 242.8460 |
| 11 | 0.9918 | 4.6453 | 517.8517 | 263.7065 | 0.9922 | 4.3267 | 517.4311 | 264.0342 |
| 13 | 0.9927 | 5.3267 | 636.3004 | 289.5034 | 0.9933 | 5.3135 | 636.7911 | 291.5365 |
| 15 | 0.9927 | 5.9654 | 715.0093 | 309.2507 | 0.9934 | 5.9736 | 714.7080 | 310.7337 |
| 17 | 0.9928 | 6.2827 | 750.0036 | 325.7328 | 0.9932 | 6.2816 | 750.0001 | 325.8205 |
| 19 | 0.9928 | 6.2826 | 749.2659 | 318.9891 | 0.9932 | 6.2749 | 749.6892 | 318.5931 |
| 21 | 0.9928 | 6.2839 | 720.8318 | 310.4878 | 0.9933 | 6.2748 | 720.6771 | 310.3805 |
| 23 | 0.9928 | 5.7648 | 672.8567 | 301.7757 | 0.9934 | 5.6124 | 672.2771 | 302.4401 |
| 25 | 0.9925 | 5.2987 | 612.9106 | 289.0685 | 0.9932 | 5.1093 | 612.3689 | 289.6878 |

**Fig. 7.3** Effect of increasing the connected wind turbines on voltage at WECS terminals

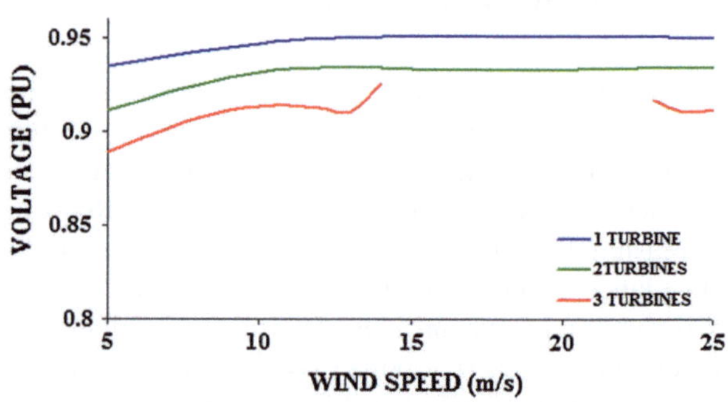


reflected by the non-convergence of the power flow in the operating wind speed region considered (shown by red line in Figs. 7.3, 7.4 and 7.5).

The discontinuity of the curve (red line) can be explained as follows. The induction generator operates at a slip where the aerodynamic power equals the electrical power developed. When the number of turbines connected at the PCC increases to 3, the aerodynamic power increases 3 fold. Correspondingly, with 3

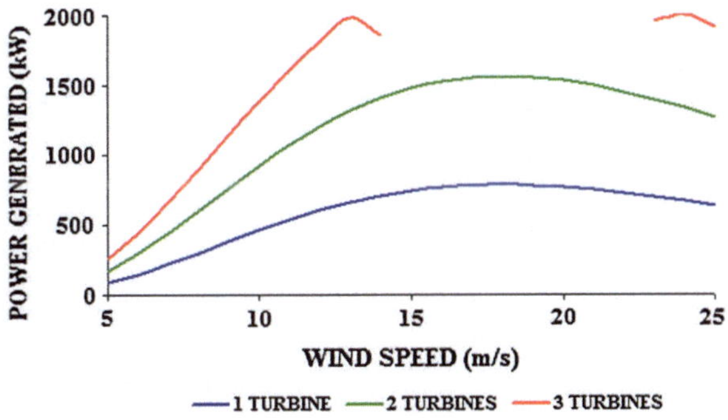


**Fig. 7.4** Effect of increasing the connected wind turbines on real power generated by WECS

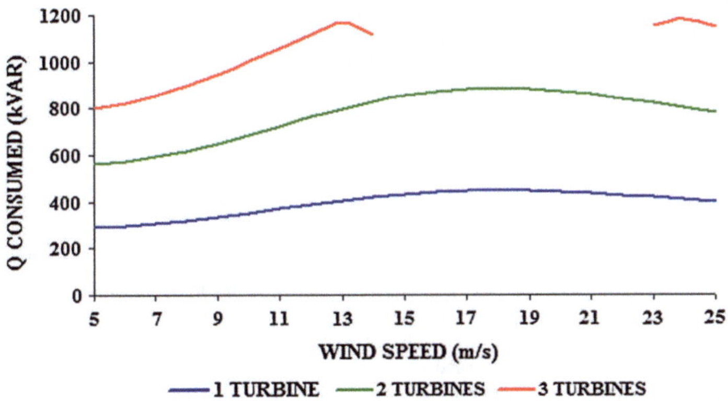


**Fig. 7.5** Effect of increasing the connected wind turbines on reactive power consumed by WECS

induction generators in parallel, the electrical power developed should also increase. At the same time, due to the nature of induction machines, the Q consumed by WECS also increases. This decreases the voltage at the PCC, which in-turn reduces the power generated. This results in a condition where the aerodynamic power is much greater than the electrical power. Thus, it is clear that the considered 9-bus radial system cannot accommodate more than two turbines.

To meet the demand of increased power production from wind, more wind turbines are required to be connected to the wind farms. Therefore, the grid needs to be strengthened to accommodate more turbines. Figures 7.6, 7.7 and 7.8 respectively give the voltage at WECS terminals, real power generated and reactive power consumed by the WECS after strengthening of the transmission line between the PCC (bus 5 in Fig. 7.2) and bus 3 in Fig. 7.2. It is observed that when a double circuit line is used, 5 uncompensated wind turbines can be connected. The power flow converges for all wind speeds in the range, i.e., from cut-in to cut-out speeds. As the effective line impedance between the PCC and the grid is now reduced by half, the grid becomes stiffer. As seen in Fig. 7.8, obviously the

Fig. 7.6 Grid strengthening—Voltage at WECS terminals

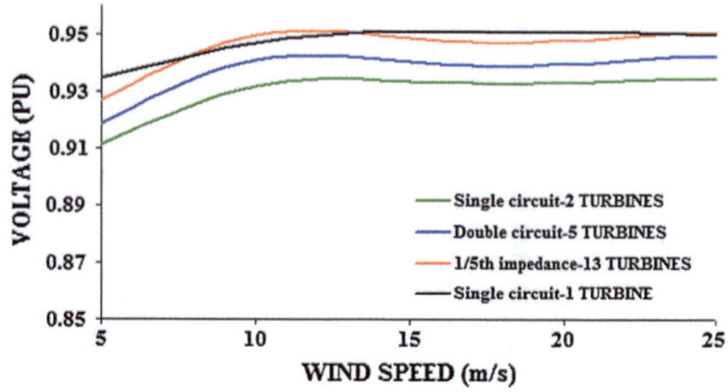


Fig. 7.7 Grid strengthening—Real power generated by WECS

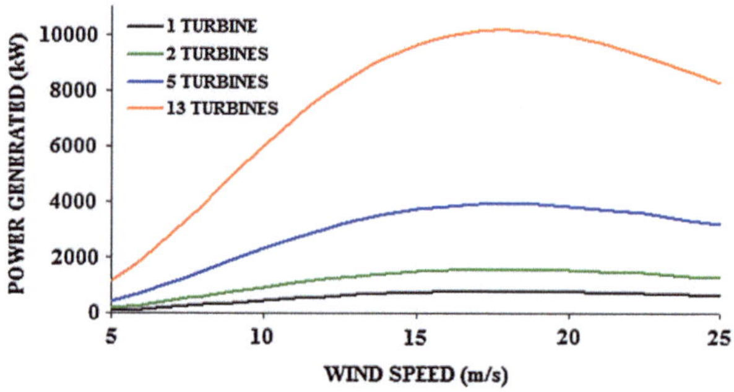


Fig. 7.8 Grid strengthening—Reactive power consumed by WECS

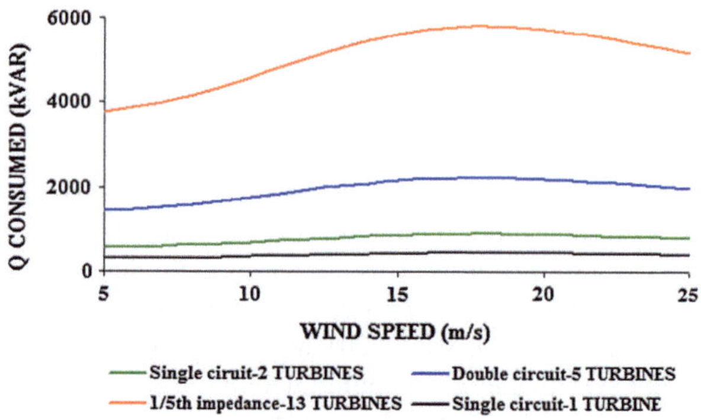


reactive power drawn from the grid increases, but at the same time the voltage at WECS terminals also increases as seen in Fig. 7.6. This increase in voltage results from grid strengthening. When the grid is further strengthened by reducing the impedance of the transmission line 5 times, 13 uncompensated wind turbines can be connected at bus 9. It is seen from Figs. 7.6 and 7.8, that although the reactive power consumption with 13 turbines is quite high, the voltage increases to the original value. This proves that the transmission capacity of the line is improved.

Grid strengthening may not be advisable beyond a certain level due to the following reasons:

(i) Strengthening the grid by reducing the impedance of transmission lines, leads to the use of many lines in parallel. This is not at all economical.
(ii) In this analysis, uncompensated WECS is assumed. In general, fixed capacitors are connected in parallel to the induction generators for reactive power compensation. Therefore, there is a danger of overcompensation if this method of grid strengthening is adopted.

Thus, the option of using different types of fixed reactive power compensation need to be studied.

## 7.6 Reactive Power Compensation

The reactive power compensation for WECS is traditionally done with capacitor banks, which is an economic and relatively simple solution. In order to avoid problems with over voltages, especially in island conditions, no-load compensation is most commonly used. No-load compensation means, that the compensation is designed to counterbalance the consumption of reactive power in the no-load situation, when the WECS operates at cut-in speed.

### *7.6.1 Shunt Compensation*

Shunt (parallel) compensation is a common practice in WECS to improve the power factor of each turbine. Some WECSs use more than one capacitor bank at their terminals to compensate for reactive power at different wind speeds [20]. In case of a fixed shunt capacitor, the reactive power output of the capacitor is proportional to the square of the voltage across it. The reactive power required by the induction generator varies with operating slip. Thus with fixed shunt capacitor, the voltage varies with the slip of the induction generator and the number of turbines on-line.

The power flow analysis is carried out for different KVAR ratings of shunt capacitor compensation and with the grid strengthened using double circuit lines between PCC and the grid. A total of 5 wind turbines are assumed to be connected at bus 9. Simulation results are shown in Figs. 7.9, 7.10, 7.11 and 7.12. The uncompensated system is used as the base line. Figure 7.9 shows that the terminal voltage drops to 92 % when the system is uncompensated. With compensation, the voltage at WECS terminals is seen to be within the limit of ±5 % of rated value. Figure 7.10 shows that the reactive power demand reduces significantly. Figure 7.11 shows a reduction in line current with compensation. This proves that the advantage of an improved power factor is the reduction of total current, which,

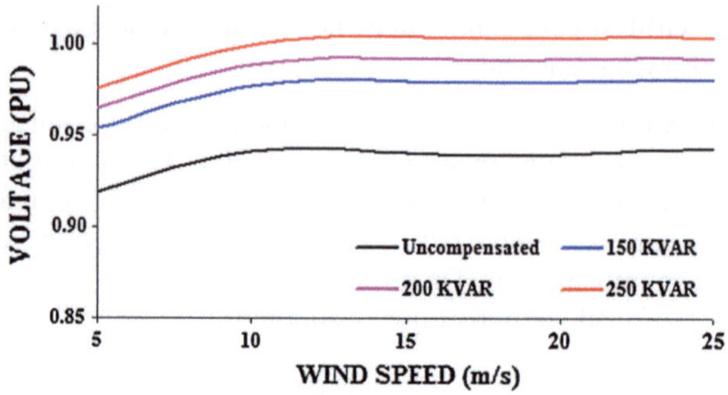


**Fig. 7.9** Voltage at WECS terminals with various values of shunt compensation

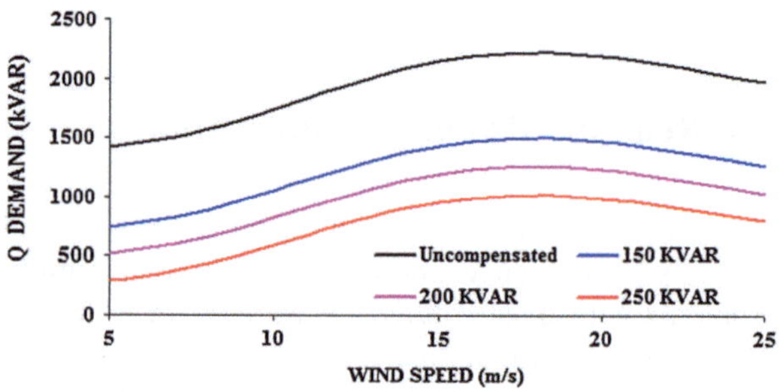


**Fig. 7.10** Reactive power demand with various values of shunt compensation

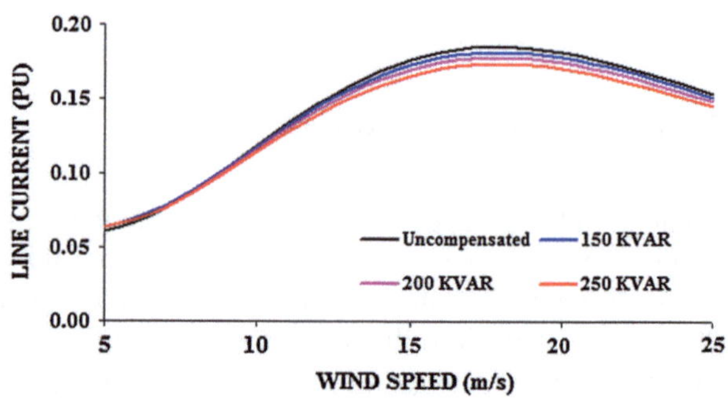


**Fig. 7.11** Line current with various values of shunt compensation

in turn, reduces transmission loss and improves voltage regulation. This is also reflected in the PV curve of Fig. 7.12.

In order to study the effect of compensation in improving the number of turbines connected at bus 9, the power flow analysis is carried out for cut-in speed and for different ratings of shunt capacitor compensation. The results are shown in Figs. 7.13 and 7.14.

Figure 7.13 shows that with a compensation of 250 kVAR, an additional 12 turbines can be connected for the same transmission line (assuming that the thermal limit is not reached). Without shunt compensation, only 5 turbines can be connected.

Figure 7.14 shows that the slope of PQ curve reduces with compensation, reflecting the improvement in power factor.

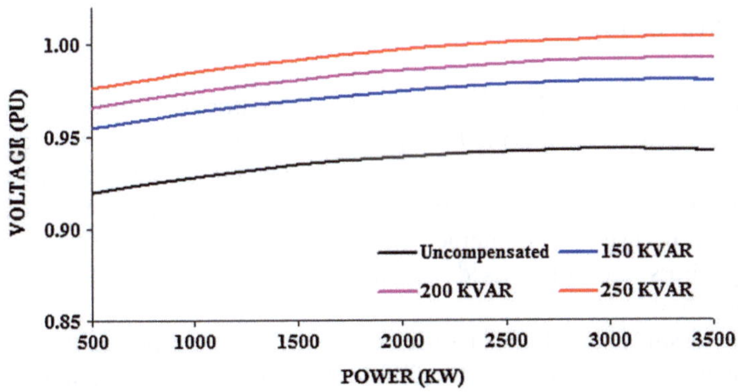


**Fig. 7.12** Power generated versus voltage at WECS terminals with various values of shunt compensation

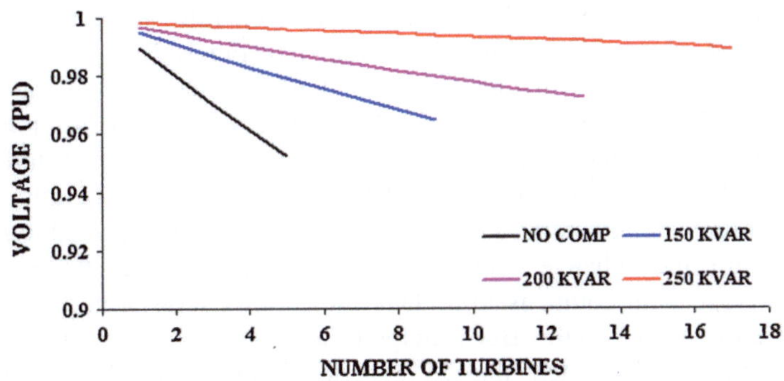


**Fig. 7.13** Effect of increasing the number of turbines with shunt compensation—Voltage at WECS terminals

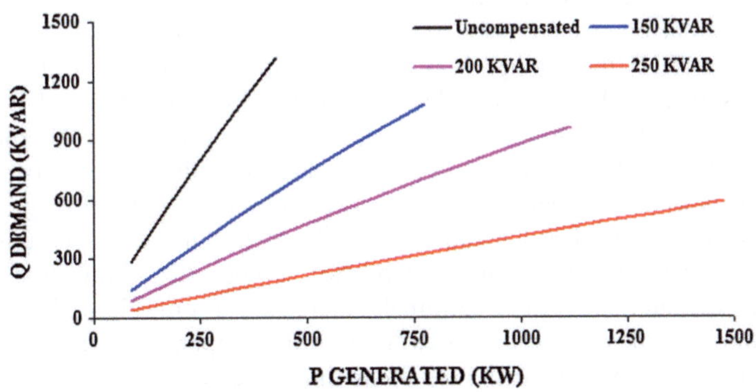


**Fig. 7.14** Effect of increasing the number of turbines with shunt compensation—P Vs Q of WECS

### 7.6.2 Series Compensation

For the system shown in Fig. 7.2, the series capacitor is connected in the transmission line between the infinite bus and the PCC i.e., in the line between buses 4 and 5. The voltage across the series capacitor has a 180° phase shift with respect to the voltage drop across the line reactance. Thus, the voltage across the series capacitor will be used to counteract the voltage drop across the line impedance. Series capacitors are often used to improve the power transfer capability of transmission lines.

In order to study the effect of series compensation, the power flow analysis is carried out for cut-in speed and for different values of series capacitor reactance measured as a percentage of transmission line reactance. Series capacitor reactance equal to 25, 50 and 75 % of transmission line reactance are selected, because 100 % compensation for line reactance is not practically feasible. The maximum compensation is governed by sub-synchronous resonance and by series capacitor short-circuit overcurrent protection needs and it is generally about 80 % [21]. The results obtained are shown in Figs. 7.15 and 7.16. Figure 7.15 shows that with series capacitor reactance equal to that of 75 % of line reactance, 22 turbines can be connected with the same double circuit lines. With shunt compensation, only 17 turbines can be connected. Even with 50 % series compensation, 18 turbines can be installed. Thus series compensation becomes a cost effective alternative to transmission line strengthening. Figure 7.16 shows that the slope of the PQ curve does not significantly change with series compensation. In fact, there is a slight increase in slope with compensation. This proves that series compensation does not improve the power factor. But it helps to accommodate more turbines. From Figs. 7.15 and 7.16, it is clear that 25 % series compensation has no significant effect.

To see the effect of series compensation on the steady state improvement, in the entire wind speed range, simulation is carried out with 10 turbines on-line and for 50 and 75 % series compensation. Although, with series compensation more than 20 turbines can be connected, only 10 turbines is chosen for simulation, because as seen from Fig. 7.15, the voltage is reasonably within limits for 10 turbines. The results are shown in Figs. 7.17, 7.18, 7.19 and 7.20. It is observed from these Figures that with 10 turbines on-line, complete convergence over the entire wind speed range, from cut-in to cut-out speed, is achieved with series compensation. It is seen that without compensation, the power flow does not converge over the entire speed range considered. From Figs. 7.17 and 7.18 it is observed that there is power and voltage increase with compensation. Figure 7.19 shows that the reactive power demand at WECS terminals increases. From Fig. 7.20, it can be observed that the line current increases with series compensation in comparison to the uncompensated system. It is now clear that series compensation reduces the voltage drop across the transmission line, thus improving the voltage at WECS terminals and improving the power transferred. The power factor of the wind farm is not affected by series compensation.

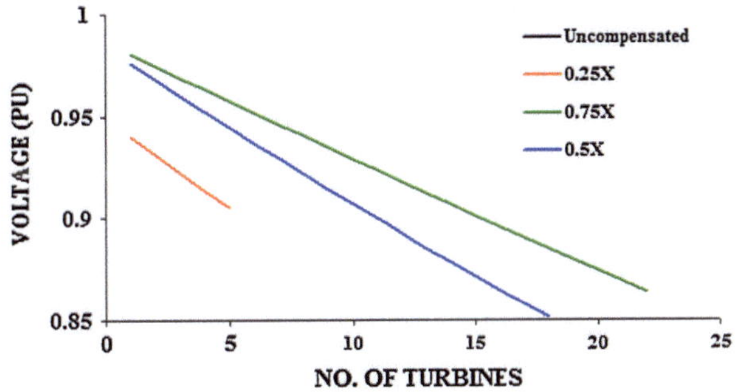


**Fig. 7.15** Effect of increasing the number of turbines with series compensation—Voltage at WECS terminals

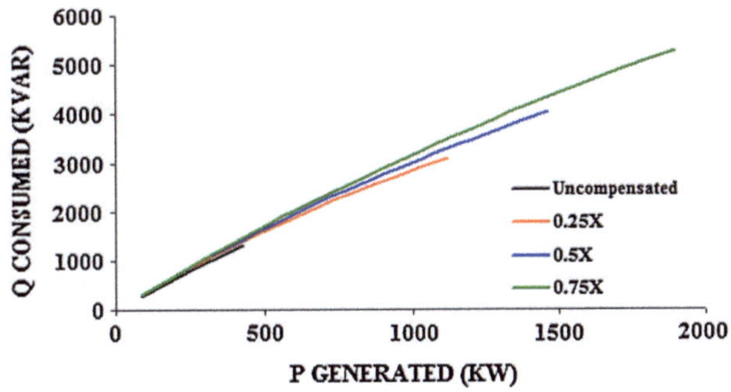


**Fig. 7.16** Effect of increasing the number of turbines with series compensation—P Vs Q of WECS

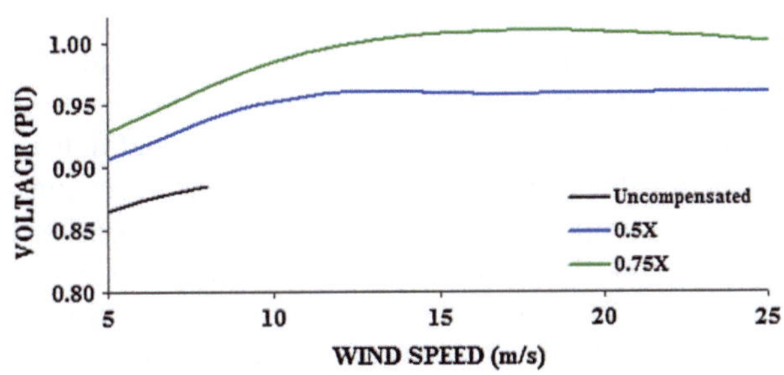


**Fig. 7.17** Voltage at WECS terminals with various values of series compensation

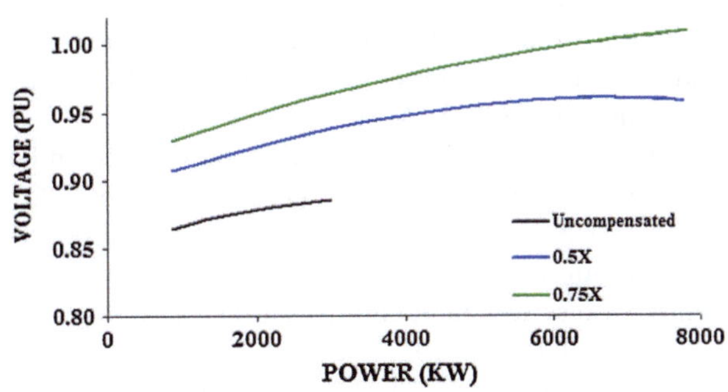


**Fig. 7.18** Real power generated Vs voltage at WECS terminals with various values of series compensation

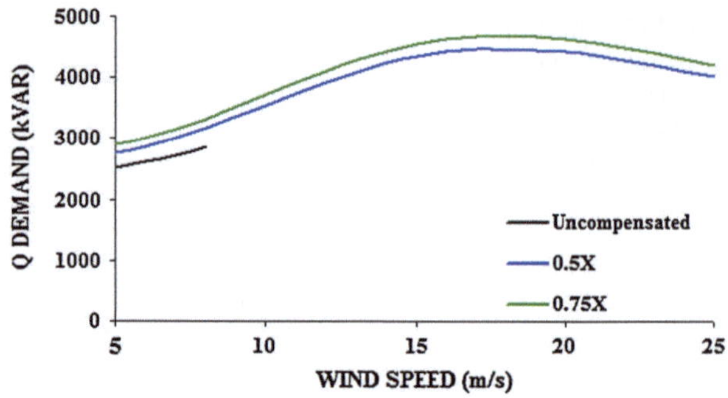


**Fig. 7.19** Reactive power demand with various values of series compensation

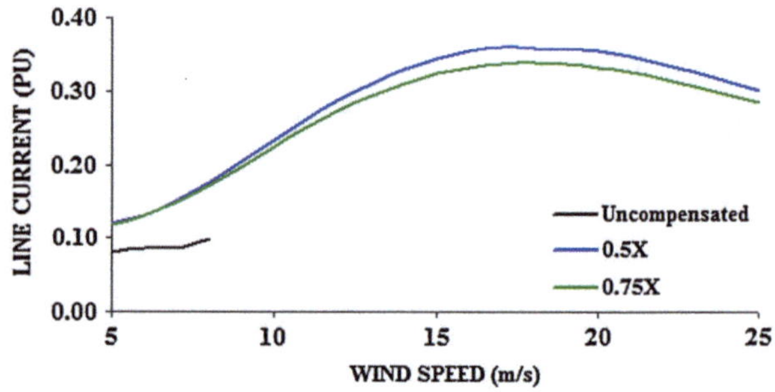


**Fig. 7.20** Line current with various values of series compensation

### *7.6.3 Series-Shunt Compensation*

It is clear from the above analysis, that the advantage of both shunt and series compensation of an AC capacitor can be made use of. In shunt compensation, the capacitor is used to compensate the individual induction generators. In series compensation, the capacitor is used to compensate the line impedance. The overall result will be an improvement in the voltage and lower losses on the transmission line.

To know the effect of the combined series and shunt compensations, simulations are carried out for two different combinations of series-shunt compensation. The same 10 wind turbines are connected at bus 9. Figure 7.21 illustrates the real power generated by WECS versus the voltage at the WECS terminals with various combinations of series-shunt compensations. It is seen that the voltage profile is high compared to individual series or shunt compensation. In fact, there is over compensation with series capacitor equal to 75 % of line reactance and shunt compensation of 200 kVAR.

Figure 7.22 shows that reactive power demand decreases significantly when compared to those with series compensation alone. As shown in Figs. 7.20 and 7.23, a comparison between the line current for series compensation and series-shunt compensation, it is observed that the transmission line current is reduced significantly with series-shunt compensation.

Comparing Figs. 7.12, 7.18 and 7.21, which show the variation of voltage with power for different types of compensation, it is evident that series-shunt

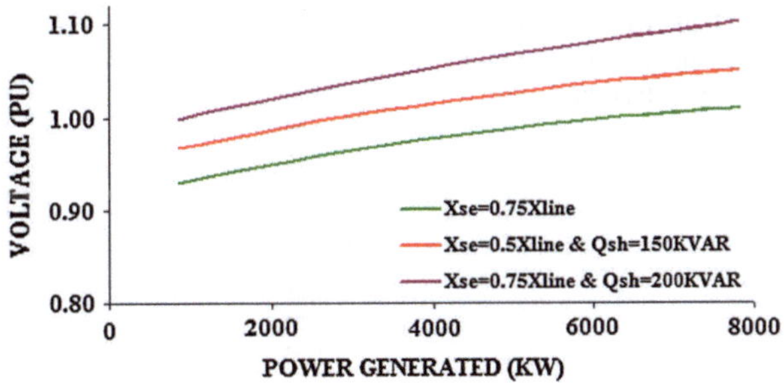


**Fig. 7.21** Real power generated Vs voltage with series-shunt compensation

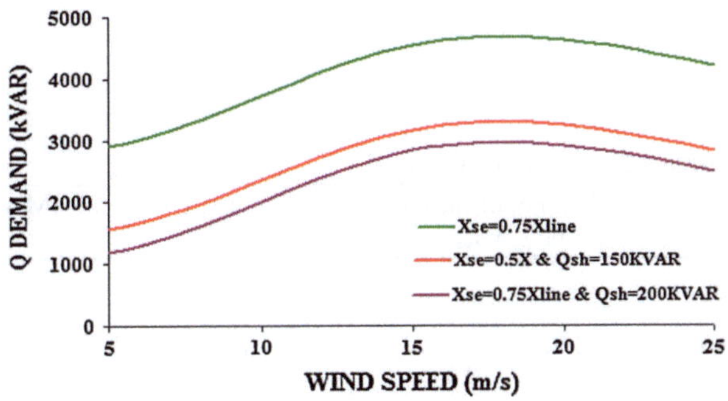


**Fig. 7.22** Reactive power demand with series-shunt compensation

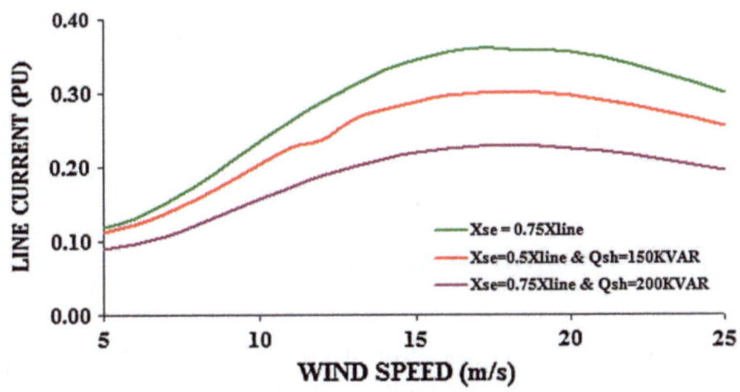


**Fig. 7.23** Line current with series-shunt compensation

compensation provides a very good voltage profile at WECS terminals and increases the wind power penetration, as more turbines can be connected.

From Fig. 7.24, it is observed that series-shunt compensation with 50 % series compensation and 150 kVAR shunt compensation gives better voltage profile than that obtained by strengthening the grid with 5 parallel lines. Also, for such capacitor combination, the reactive power demand and transmission line current are reduced significantly as seen from Figs. 7.22 and 7.23. Thus it can be concluded that series-shunt compensation is an economical alternative to grid strengthening.

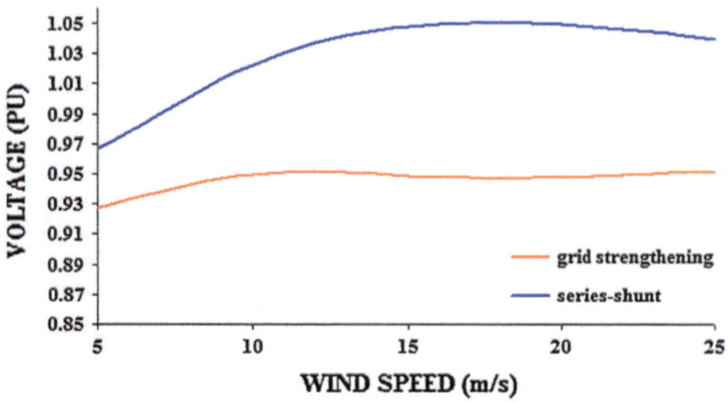


**Fig. 7.24** Comparison of series-shunt compensation with grid strengthening

**Table 7.2** Summary of results

| | Max. no. of turbines that can be connected | V (PU) | P (MW) | Q (MVAR) | I (PU) |
|---|---|---|---|---|---|
| A1 | 2 | 0.9329 | 1.56 | 0.88 | 0.0735 |
| A2 | 5 | 0.9389 | 3.89 | 2.21 | 0.1833 |
| A3 | 13 | 0.9472 | 10.12 | 5.79 | 0.4325 |
| B1 | 9 | 0.9788 | 3.89 | 1.49 | 0.1807 |
| B2 | 13 | 0.9912 | 389 | 1.25 | 0.1771 |
| B3 | 17 | 1.0031 | 3.89 | 1.01 | 0.1729 |
| C1 | 18 | 0.9580 | 7.77 | 4.43 | 0.3604 |
| C2 | 22 | 1.0090 | 7.77 | 4.68 | 0.3443 |
| D1 | 13 | 1.0014 | 7.77 | 2.27 | 0.1782 |
| D2 | 20 | 1.0619 | 7.77 | 1.91 | 0.1253 |

## 7.7 Discussion of Results

The convergence characteristics of the newly developed simultaneous method of power flow is compared with the existing sequential method of power flow analysis and it is found that the simultaneous method results in less number of iterations.

For the particular WECS modelled, the effect of increasing the number of turbines with and without compensation is summarized in Table 7.2. From Table 7.2 it is seen that when the transmission line between the PCC and the grid is strengthened (with 5 parallel lines), 13 turbines can be connected and thus more power can be fed into the grid. The voltage profile at the WECS terminals is also found to improve, although the reactive power drawn from the grid is very large. Yet such strengthening is not cost effective, being the equivalent of having 5 lines in parallel. Therefore, the effect of different types of compensation is studied.

Table 7.2 shows that with shunt compensation, there is improvement in voltage profile and decrease in the reactive power demand. The line current reduces thereby reducing transmission losses. The net effect is an improvement in power factor. With series compensation, there is improvement in voltage profile, although there is increase in the reactive power demand. The line current drawn is comparatively large, yet more turbines can be connected on-line. The net effect is an improvement in real power penetration. With series-shunt compensation there is a considerable improvement in voltage profile. At the same time, the reactive power demand and the line current drawn are both reduced compared to series compensation alone and are comparable with shunt compensation. A combination of series compensation equal to 50 % of line reactance and shunt compensation of 150 kVAR, shows better performance as compared to either series or shunt compensation alone, for the 9-bus radial system considered. Thus with the correct choice of capacitor sizes, a combination of series and shunt compensation can be used to improve the overall system performance.

In Table 7.2, V—voltage at WECS terminals at rated speed, P—Real power generated at rated speed, Q—Reactive power demand at rated speed, I—Line current at rated speed, A—Uncompensated system (1—without line strengthening, 2—with double circuit line, 3—with 5 lines in parallel), B—Shunt Compensation (1–150, 2–200, 3–250 kVAR), V, P, Q and I values with 5 turbines on-line, C—Series Compensation (1–0.5 $X_{line}$, 2–0.75 $X_{line}$)—V, P, Q and I values with 10 turbines on-line, D—Series-shunt Compensation (1–$X_{se} = 0.5\ X_{line}$ and $Q_{sh} = 150$ kVAR, 2–$X_{se} = 0.75\ X_{line}$ and $Q_{sh} = 200$ kVAR)—V, P, Q and I values with 10 turbines on-line.

To summarize, the following are recommended:

1. To strengthen the grid to evacuate the maximum power output from the wind farms, corresponding to installed rated capacity. This is to ensure harvesting of the wind power in the windy seasons.
2. To go in for appropriate value of series-shunt capacitor compensation with grid connected induction generators.

## Appendix I

Data for 9-Bus Radial System

**1. Transmission Line Data** (All Lines)

| | |
|---|---|
| Resistance | 0.24 Ω/km |
| Reactance | 0.48 Ω/km |
| Susceptance | 2.80 μS/km |
| Length | 20.0 km |

**Table A.1 Transformer data**

| Parameter | Load transformer data (all) | Step up transformer data (at the wind bus) | Feeding transformer data |
|---|---|---|---|
| Rated apparent power | 0.63 MVA | 1.0 MVA | 25 MVA |
| Rated voltage of MV side | 15 kV | 15 kV | 110 kV |
| Rated voltage of LV side | 0.4 kV | 0.69 kV | 15 kV |
| Nominal short-circuit voltage | 6 % | 6 % | 11 % |
| Copper loss at rated power | 6 kW | 13.58 kW | 110 kW |

### 2. Load Data (All Loads)

0.150 + j0.147 MVA

### 3. Coefficients of $C_P$—$\lambda$ Curve

| | |
|---|---|
| $C_1$ | 0.5 |
| $C_2$ | 67.56 |
| $C_3$ | 0.4 |
| $C_4$ | 0 |
| $C_5$ | 1.517 |
| $C_6$ | 16.286 |
| Gear box ratio | 67.5 |

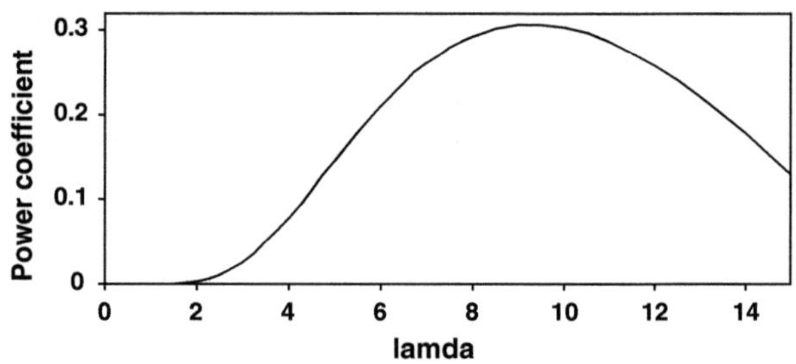


**Fig. A.1** $C_P$—$\lambda$ curve for the considered wind turbine

### Asynchronous Generator Data (Δ-Connection)

| | |
|---|---|
| Stator resistance | 0.0034 Ω |
| Rotor resistance | 0.003 Ω |
| Stator leakage reactance | 0.055 Ω |
| Rotor leakage reactance | 0.042 Ω |
| Magnetizing reactance | 1.6 Ω |

# Appendix II

Jacobian Matrix for the 9-bus System

For the system considered the Jacobian is a (17 × 17) matrix as below. The numbers in the matrix represent the row and column of the elements. The analytical expressions for these elements are given in the subsequent sections.

$$\begin{bmatrix} (1,1) & \cdots & (1,8) & (1,9) & \cdots & (1,16) & (1,17) \\ \vdots & \ldots J_1 \ldots & \vdots & \vdots & \ldots J_2 \ldots & \vdots & \ldots J_5 \ldots \\ (8,1) & \cdots & (8,8) & (8,9) & \cdots & (8,16) & (8,17) \\ & & & & & & \\ (9,1) & & (9,8) & (9,9) & & (9,16) & (9,17) \\ \vdots & \ldots J_3 \ldots & \vdots & \vdots & \ldots J_4 \ldots & \vdots & \ldots J_6 \ldots \\ (16,1) & & (16,8) & (16,9) & & (16,16) & (16,17) \\ & & & & & & \\ (17,1) & \ldots J_7 \ldots & (17,8) & (17,9) & \ldots J_8 \ldots & (17,16) & (17,17) J_9 \end{bmatrix}$$

## 1 Elements of $J_1$

$$\frac{\partial P_i}{\partial \theta_i} = imag\left[\sum_{j \neq i} V_i^* V_j Y_{ij}\right]$$

$$\frac{\partial P_i}{\partial \theta_j} = -imag\left[\sum V_i^* V_j Y_{ij}\right]$$

## 2 Elements of $J_2$

$$\frac{\partial P_i}{\partial |V_i|}|V_i| = real\left[\sum_{j \neq i} V_i^* V_j Y_{ij}\right] + 2|V_i|^2 real(Y_{ij})$$

$$\frac{\partial P_i}{\partial |V_j|}|V_j| = real\left[\sum V_i^* V_j Y_{ij}\right]$$

Diagonal element for WECS bus (bus 9),

$$J(8,16) = J(8,16) - \frac{2|V_9|^2 r_e}{(r_e^2 + x_e^2)}$$

## 3 Elements of $J_3$

$$\frac{\partial Q_i}{\partial \theta_i} = real\left[\sum_{j \neq i} V_i^* V_j Y_{ij}\right]$$

$$\frac{\partial Q_i}{\partial \theta_j} = -real\left[\sum V_i^* V_j Y_{ij}\right]$$

### 4 Elements of $J_4$

$$\frac{\partial Q_i}{\partial |V_i|}|V_i| = -imag\left[\sum_{j \neq i} V_i^* V_j Y_{ij}\right] - 2|V_i|^2 real(Y_{ij})$$

$$\frac{\partial Q_i}{\partial |V_j|}|V_j| = -imag\left[\sum V_i^* V_j Y_{ij}\right]$$

Diagonal element for WECS bus (bus 9),

$$J(16,16) = J(16,16) - \frac{2|V_9|^2 x_e}{(r_e^2 + x_e^2)}$$

$$\text{where}, Z_e = Z_S + Z_m||Z_r$$

### 5 Elements of $J_5$

$$J(8,17) = \frac{\partial}{\partial s}\left[\frac{V^2 r_e}{(r_e^2 + x_e^2)}\right] = \frac{|V_9|^2}{(r_e^2 + x_e^2)^2}\left[\left(x_e^2 - r_e^2\right)\frac{\partial r_e}{\partial s} - 2r_e x_e \frac{\partial x_e}{\partial s}\right]$$

### 6 Elements of $J_6$

$$J(16,17) = \frac{\partial}{\partial s}\left[\frac{-V^2 r_e}{(r_e^2 + x_e^2)}\right] = \frac{-|V_9|^2}{(r_e^2 + x_e^2)^2}\left[\left(r_e^2 - x_e^2\right)\frac{\partial x_e}{\partial s} - 2r_e x_e \frac{\partial r_e}{\partial s}\right]$$

### 7 Elements of $J_7$ : 0

### 8 Elements of $J_8$

$$J(17,16) = \frac{2|V_9|x_m^2 R_r/s}{\left[\left(R_1 + \frac{R_r}{s}\right)^2 + (x_1 + x_r)^2\right]\left[R_s^2 + (x_s + x_m)^2\right]}$$

$$\text{where}, Z_1 = Z_s||Z_m$$

### 9 Elements of $J_9$

$$J(17,17) = \frac{\partial \Delta P_m}{\partial s}$$

These are given by Eqs. (7.13) to (7.18).

## References


1. Liangzhong Y, Phill C, Laurent S, Xiao-Ping Z (2005) Congestion management of transmission systems using FACTS. In: IEEE/PES transmission and distribution conference and exhibition, Asia and Pacific, Dalian, China, pp 1–5
2. Slootweg J, Polinder H, Kling W (2001) Initialization of wind turbine models in power system dynamics simulations. In: IEEE Porto power tech proceedings, p 7

3. Molzahn DK, Lesieutre BC (2013) Initializing dynamic power system simulations using eigenvalue formulations of the induction machine and power flow models. IEEE Trans Circuits Syst 60(3):690–702
4. Hatziargyriou D, Karakatsanis TS, Papadopoulos M (1993) Probablistic load flow in distribution systems containing wind power generation. IEEE Trans Power Syst 8(1):159–165
5. Persaud S, Fox B, Flynn D (2000) Impact of remotely connected wind turbines on steady state operation of radial distribution networks. IEE Proc Gener Transm Distrib 147(3):157–163
6. Meliopoulos APS (2001) Distributed energy sources: needs for analysis and design tools. In: IEEE power engineering society summer meeting, pp 548–550
7. Boulaxis NG, Papathanassiou SA, Papadopoulos MP (2002) Wind turbine effect on the voltage profile of distribution networks. Renewable Energy 25:401–415
8. Muljadi E, Wan Y, Butterfield CP, Parsons B (2002) Study of a wind farm power system. NREL report
9. Divya KC, Nagendra Rao PS (2005) Models for wind turbine generating systems and their application in load flow studies. Electr Power Syst Res 76:844–856
10. Fuerte-Esquivel CR, Acha E (1997) A Newton-type algorithm for the control of power flow in electrical power networks. IEEE Trans Power Syst 12:1474–1480
11. Liu Y, Wang W, Xul L, Ni P, Wang L (2008) Research on power flow algorithm for power system including wind farm. In: IEEE International Conference on Electrical Machines and Systems ICEMS, pp 2551–3255
12. Feijóo AE (2009). On PQ models for asynchronous wind turbines. IEEE Trans Power Syst 24(4):1890–1891
13. Feijoo AE, Cidras J (2000) Modeling of wind farms in the load flow analysis. IEEE Trans Power Syst 15(1):110–115
14. Pecas JA, Maciel FP, Cidras J (1991) Simulation of MV distribution networks with asynchronous local generation sources. In: Proceedings of IEEE Melecom
15. Cidras J, Martinez JA, Pecas JA, Maciel FP (1992) Modelling of nonlinear nodal admittances in load flow analysis. In: Proceedings IFAC
16. Salem E, Mohammed B, Joydeep M (2012) Power flow analysis of distribution systems with embedded induction generators. In: 44th North American power symposium, Sept 9–11, Illinois
17. EI-Sadek MZ, Dessauky MM, Mahmaud GA, Rashed WI (1997) Load representation for steady-state voltage stability studies. Electr Power Syst Res 43:187–195
18. Jayashri R, Kumudini Devi RP (2006) Analysis of the impact of interconnecting wind turbine generators to the utility grid. Wind Eng 30:303–315
19. Lubosny Z (2003) Wind turbine operation in electric power systems advanced modelling. Springer, Berlin
20. Chompoo-inwai C, Lee Wei-Jen, Fuangfoo Pradit, Williams Mitch, Liao James R (2005) System impact study for the interconnection of wind generation and utility system. IEEE Trans Ind Appl 41(1):163–168
21. Laughton MA, Warne DJ (2003) Electrical engineer's reference book, 16th edn. Elsevier, Oxford

# Chapter 8
# Contribution of Variable-Speed Wind Generators to Frequency Regulation and Oscillation Damping in the United States Eastern Interconnection

**Yong Liu, J. R. Gracia, T. J. King and Yilu Liu**

**Abstract** The United States Eastern Interconnection (EI) has been going through some profound changes due to the increasing penetration of wind power in this bulk grid, including the worsening of the frequency response and inter-area oscillation. However, the fast response speed of electronics devices makes it possible that the kinetic energy stored and/or wind power reserve (if it exists) in variable-speed wind generators could be injected into the power grid in a fast manner. This portion of fast-injected active power could contribute to frequency regulation or oscillation damping of the EI significantly if implemented with appropriate control schemes. In this chapter, a user-defined wind electrical control model with fast active power controllers is built in PSS®E. Then, based on the 16,000-bus EI system dynamic model, the potential contributions of variable-speed wind generators to the EI frequency regulation and oscillation damping are evaluated respectively. Simulation results have demonstrated that current and future penetrations of wind generation are promising in providing frequency regulation and oscillation damping in the EI.

**Keywords** Eastern interconnection · Fast active power control · Frequency regulation · Inter-area oscillation damping · Variable-speed wind generation

---

Y. Liu (✉) · Y. Liu
Department of Electrical Engineering and Computer Science, The University of Tennessee, Knoxville, 1520 Middle Drive, Knoxville, TN, USA
e-mail: yliu66@utk.edu

Y. Liu
e-mail: liu@utk.edu

J. R. Gracia · T. J. King · Y. Liu
Energy and Transportation Science Division, Oak Ridge National Laboratory, One Bethel Valley Road, Oak Ridge, TN, USA
e-mail: graciajr@ornl.gov

T. J. King
e-mail: kingtjjr@ornl.gov

J. Hossain and A. Mahmud (eds.), *Renewable Energy Integration*,
Green Energy and Technology, DOI: 10.1007/978-981-4585-27-9_8,

## 8.1 Introduction

Today, most major power grids around the world are experiencing an increase in wind in their generation mix. This is causing concerns over stability of the grid and other difficulties in system operations. Though some subjects such as voltage and reactive power control, power quality requirements, and fault ride-through capability, have already been thoroughly examined and corresponding technical solutions are commercially available, some other rising issues still remain to be addressed. For example, conventional synchronous generators now play the pivotal role for system frequency support. Up to now, due to the absence of synchronization torque for variable-speed wind generators, higher penetrations of wind power may gradually lead to the decrease of overall system inertia as well as the reduction of system frequency regulation capabilities [1–5].

Another serious issue is the inter-area oscillation. Many studies have repeatedly shown that replacing conventional generation with double-fed induction generator (DFIG)-based wind farms may have a negative effect on the interconnected system oscillation damping [6–9]. As for the EI, the major portion of wind generation now and in the near future is located in northwestern area that is far from the load centers in the northeast, thus probably introducing new, difficultly damped inter-area modes as well as influence existing ones.

However, thanks to the development of modern power electronics technology, the answer to these problems may lie in wind generation itself. Though the steady-state wind energy delivered to the grid depends solely on the mechanical energy transferred from wind turbines, the electrical power of variable-speed wind generators can be effectively and fast controlled by modern electronics converters. As a result, variable-speed wind generations have considerable capabilities to engage in frequency regulation and oscillation damping, especially when the wind penetration is high. In this chapter, fast active power control techniques of variable-speed wind generators are discussed and applications of those controls to the EI frequency regulation and oscillation damping are demonstrated. This chapter is structured as follows: In Sect. 8.2, fast active power control techniques of variable-speed wind generators that could help with frequency regulation and oscillation damping are summarized; in Sect. 8.3, the wind generator and power system model used in this chapter's simulation is introduced; and in Sects. 8.4 and 8.5, the contributions of variable-speed wind generators to frequency regulation and oscillation damping are studied respectively.

## 8.2 Fast Active Power Control Techniques of Variable-Speed Wind Generators

Variable-speed wind generators mainly refer to DFIGs and permanent-magnet synchronous generators (PMSGs). A PMSG contains a multi-pole magnetic rotor and a back-to-back AC/DC/AC converter attached to the stator. The converter transmits energy from the stator to the grid and, consequently, decouples the generator fully from the grid. For a DFIG, the rotor supplies the AC excitation current and is connected to the power grid through the converter while its stator is connected directly to the power grid. During normal conditions, power electronics converters enable the variable-speed wind generators to capture wind energy over a wide range of wind speeds to their maximum extent (maximum power point tracking, MPPT), improve power quality and regulate both active and reactive power.

During transient conditions, additional controllers could be installed on the converters or pitch controllers of variable-speed wind generators to control active power in a fast manner for frequency regulation and oscillation damping [10–17]. In some studies, variable-speed wind generators could transiently support system frequency by implementing the kinetic energy-based active power control techniques such as "hidden" inertia emulation, fast power reserve emulation, inertia droop control, etc., to enable wind generators to have an "artificial" inertial response that behaves in a manner similar to the conventional generation [17]. This type of control utilizes the kinetic energy stored in wind turbine blades to inject more active power into the power grid in seconds. The active power output of wind generators are only increased temporally, thus the normal operation point remains unchanged. In fact, GE already has a commercial inertia droop product known as WindINTERIA [18]. Besides inertial response, the function of power system stabilizer (PSS) in conventional generations could also be emulated [19–26]. In this case, variable-speed generators could also contribute to the system stability significantly. It should be noted that the sustained loss of kinetic energy would cause the rotational speed of a wind turbine to decrease and the turbine to stall if the rotational speed falls too low. Consequently, limits need to be imposed to avoid removing too much kinetic energy.

As mentioned above, wind generation normally works on the MPPT curve in order to harvest wind energy to the maximum extent, since wind energy is both environment-friendly and low-cost. However, if wind turbines do not operate over the MPPT curve and save some power as reserve by rotational speed control (increasing the rotational speed from the MPPT value, usually referred to as over-speeding) or pitch control, certain control techniques could be implemented to enable wind generators to emulate the frequency droop characteristics ("governor response") of conventional synchronous generators. Because the time constant of pitch control is relatively large and frequent adjustments of pitch blades would also dramatically reduce their lifetime, pitch control is not considered in this chapter. Conversely, rotational speed control is provided by the power electronics converters, thus making possible the utilization of power reserve saved by over-speeding to balance the instantaneous power consumption and provide support to the long-term frequency regulation [10]. Furthermore, though no research has covered this topic yet, wind generation could also act to automatic generation control (AGC) regulation order from the operator as effectively as conventional generation if reserves exist.

In sum, by implementing fast active power control techniques, variable-speed wind generators could emulate both the inertial and governor response of conventional generations to provide frequency support. Furthermore, PSS function could also be implemented on variable-speed wind generators in order to contribute to system small-signal stability.

## 8.3 Power System and Variable-Speed Wind Generator Model

This chapter demonstrates the potential contribution of variable-speed wind generators to frequency regulation and oscillation damping in the EI. PSS®E has the capability to handle both large-scale power system dynamic simulation and wind generator modeling, thus is employed as the simulation tool in this chapter. The variable-speed wind generator model used and the buildup of the EI dynamic simulation scenario will be introduced in Sect. 8.3.

### *8.3.1 Wind Generator Model*

Since DFIG is the prevailing wind generation type nowadays, variable-speed wind generator is modeled as DFIG in this study. WT3 wind model has been developed in PSS®E to simulate the performance of a DFIG-based wind generator with basic active power control [27]. There are four basic components within DFIG model (usually referred to as WT3 model, whose structure shown in Fig. 8.1), namely:

- WT3G: generator/converter model
- WT3E: electrical control model
- WT3T: mechanical control (wind turbine) model
- WT3P: pitch control model.

Typical parameters of a GE 1.5 MW wind generator from PSS®E are employed in the simulation. And in order to add fast active power control functions to current DFIG model in PSS®E, a user-defined electrical control model is developed by the author (control structure shown by Fig. 8.2). Note that since pitch control and reactive power control are not considered in this study, WT3P part is ignored and constant reactive power control mode is selected in all the simulations.

### *8.3.2 Power System Model*

The EI is one of the two major power grids in North America. It reaches from central Canada eastward to the Atlantic coast (excluding Quebec), south to Florida,

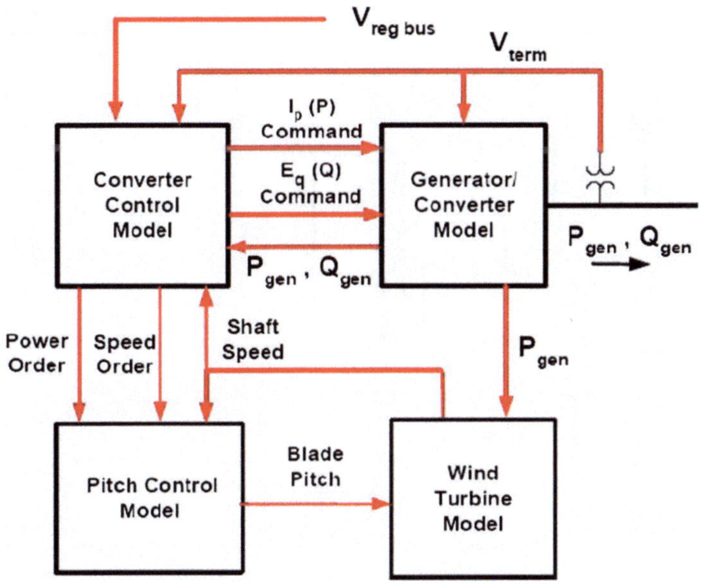


**Fig. 8.1** Structure of DFIG wind model in PSS®E [27]

**Fig. 8.2** User-defined wind electrical control model structure (active power control part)

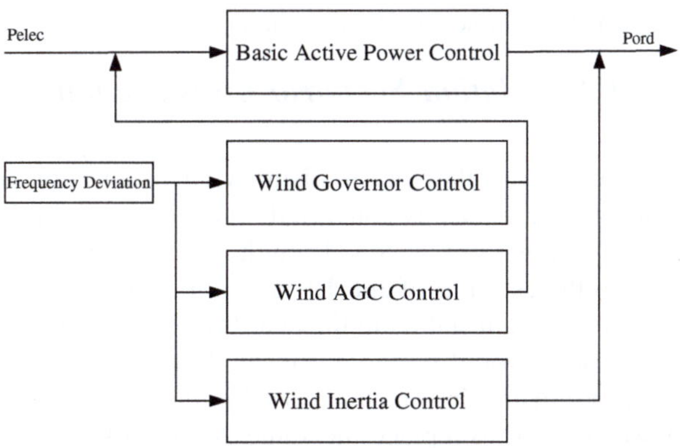


and back west to the foot of Rocky Mountain (excluding most of Texas). Its 16,000-bus dynamic model is selected as the base case in this study's simulation. The total generation capacity of this model is around 590 GW, which includes roughly 3,000 generators (This model is protected by Non-disclosure Agreement with Tennessee Valley Authority (TVA), thus no detailed information is allowed to be released). However, no wind generation is modeled in the original model; therefore a simulation scenario with realistic wind penetration needs to be developed by the author.

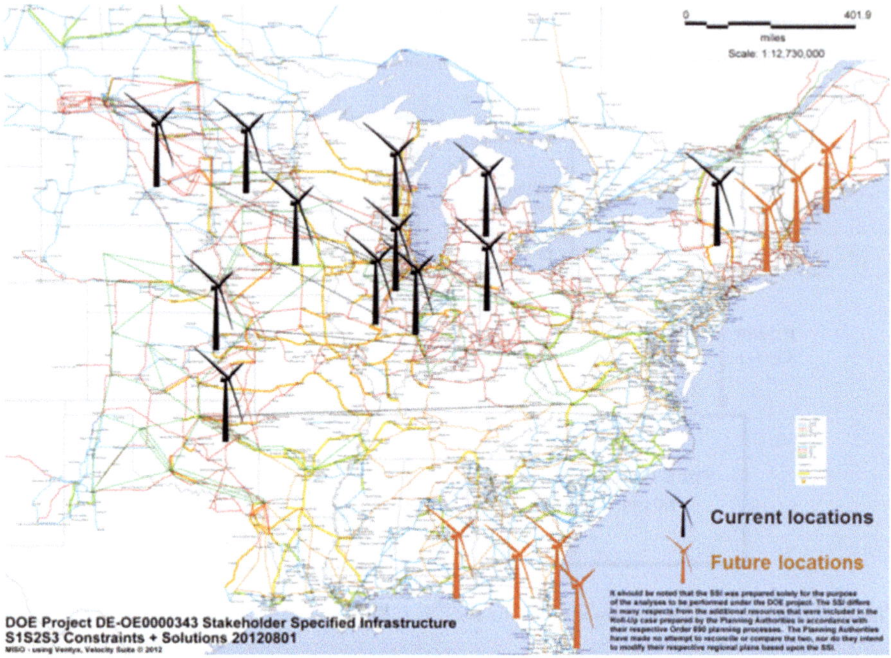


**Fig. 8.3** Major wind farm locations in the EI

### *8.3.3 Simulation Scenario Construction*

To simulate a relatively realistic wind penetration in the EI system, 5 % of EI generation capacity is converted to wind power in this study. Their locations are roughly shown in Fig. 8.3. Though most of current wind generators are located in the Northwest area of the EI, a significant number of off-shore wind farms are expected to be installed in the near future, as shown in Fig. 8.3. Additionally, wind generations would be largely located on the edges of the EI, which makes them more promising in inter-area oscillation damping. Last but not least, since wind power is used to replace the same amount of conventional generations in the EI, congestion is very unlikely to occur, thus not considered in the scope of this chapter.

## 8.4 Contribution of Variable-Speed Wind Generators to the EI Frequency Regulation

In order to enable variable-speed wind generators engage in frequency regulation, several additional controllers are employed in the user-defined PSS®E DFIG electrical control model, including Wind Inertia Control, Wind Governor Control and Wind AGC Control, which are introduced respectively in Sect. 8.5.

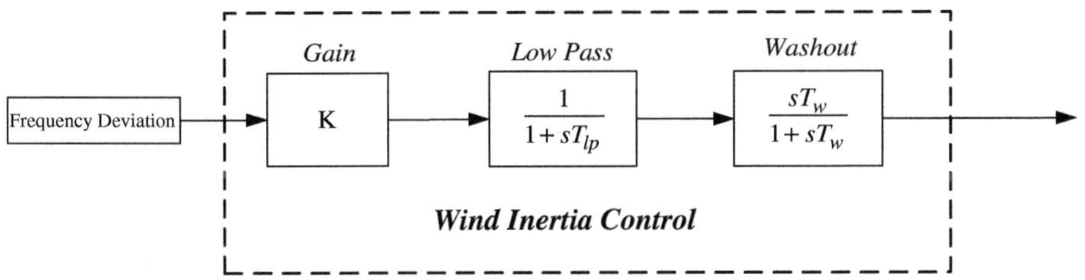


**Fig. 8.4** Wind inertia control structure

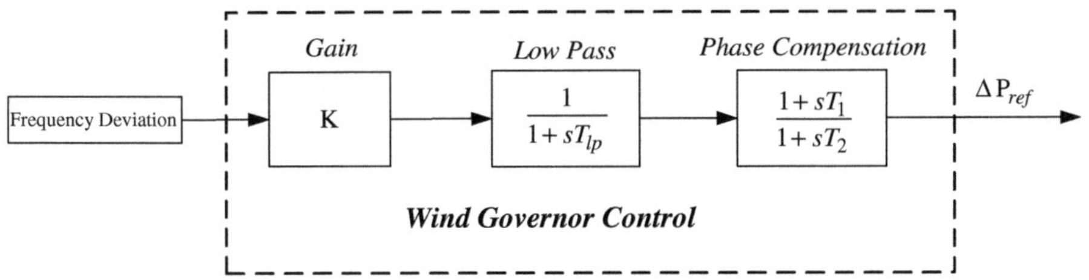


**Fig. 8.5** Wind governor control structure

### *8.4.1 Wind Inertia Control*

Wind Inertia Control here is of the same philosophy as GE WindINERTIA® technology, whose objective is to make wind generation provide the "artificial" inertial response like conventional generators. Droop control is utilized to produce active power output change that is proportional to the frequency deviation, which is given by

$$\Delta f = f_{meas} - f_{ref}$$

where $f_{meas}$ is the measured system frequency and $f_{ref}$ is the reference frequency. The structure of Wind Inertial Droop control is shown in Fig. 8.4.

### *8.4.2 Wind Governor Control*

As discussed in Sect. 8.1, wind turbines could decrease rotational speed to release the power reserve in seconds if they have been operating in the over-speeding zone. Therefore, governor response could be emulated on wind generators utilizing the wind power reserve. Again, droop control is employed and the corresponding Wind Governor Control structure is shown in Fig. 8.5.

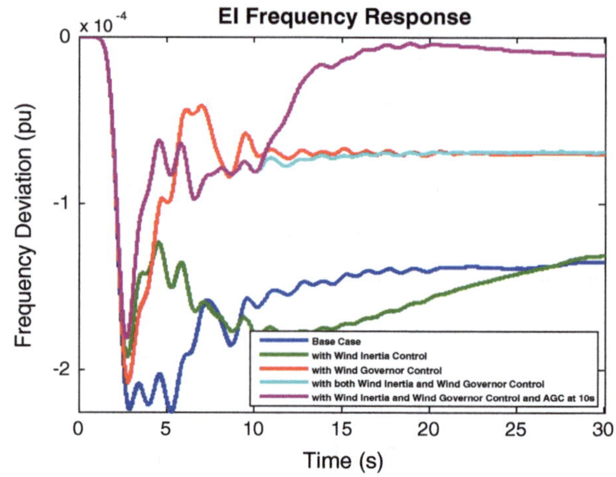


**Fig. 8.6** Frequency responses of the EI after a generation trip event with different controls

## 8.4.3 Wind AGC Control

AGC plays a key role in maintaining the balance of power grid generation and consumption. Though no research has covered this topic yet, variable-speed wind generators could act to the AGC regulation order from the operator as effeciently as conventional generation. Actually, thanks to the fast response speed of power electronics converters, the active power output of variable-speed wind generators could ramp up/down quickly, which makes wind generators even better candidates for AGC regulation. The only disadvantage is the waste of wind energy during normal conditions due to the reserve.

## 8.4.4 Case Study: Generation Trip

To demonstrate the contribution of variable-speed wind generators to the EI frequency regulation, a 1000 MW generation trip is simulated in Sect. 8.4.4. The frequency responses with different controls following the event are given in Fig. 8.6 and the active power outputs of a typical wind generator are shown in Fig. 8.7. (Note that the parameters of wind controllers in this chapter's case studies are all manully tuned by the author.)

In Fig. 8.7, if only with Wind Inertia Control (shown by the green line), the wind generator increases its active power temporally in the several seconds following the generation trip disturbance by use of the kinetic energy stored in the wind blade. However, as discussed previously, the active power increase could not be sustained and after the temporal active power surge, the wind generator active power output would go through a "dip", during which the kinetic energy of turbine blades are "recharged". As for the turbine speed, it would be observed to go

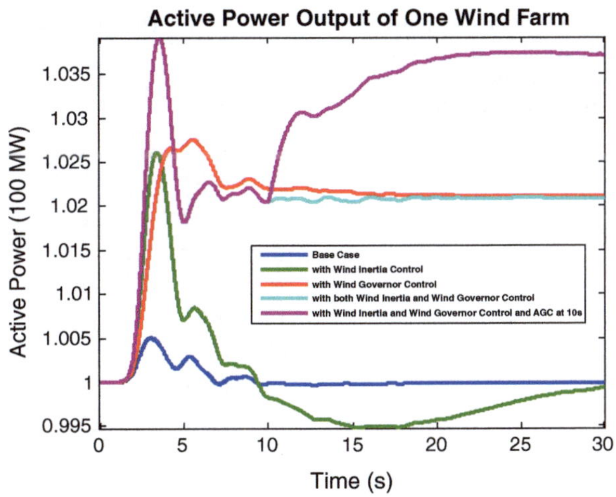


**Fig. 8.7** Active power outputs of a typical wind generator after a generation trip event with different controls

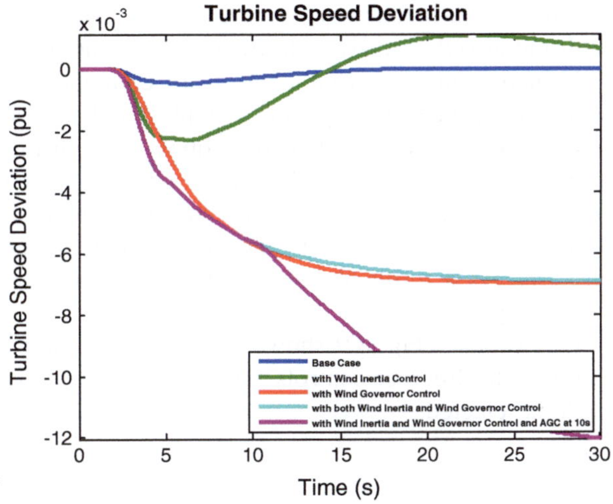


**Fig. 8.8** Turbine speed deviations of a typical wind generator after a generation trip event with different controls

through the phase of "deceleration" to release some amount of kinetic energy and then the phase of "acceleration" to "recharge", both of which are illustrated clearly in Fig. 8.8. Apparently, while the Wind Inertia Control function does not contribute to long-term frequency recovery but it does help reduce the frequency drop nadir, which is shown by green line in Fig. 8.6.

If only deployed in Wind Governor Control mode, the wind generator would behave just like the conventional generators but in a much faster manner. As shown by the red line in Fig. 8.7, the wind generator active power output would ramp up to a higher level because of the existence of the "governor" function and the wind power reserve would be put into use by moving the operation point away from over-

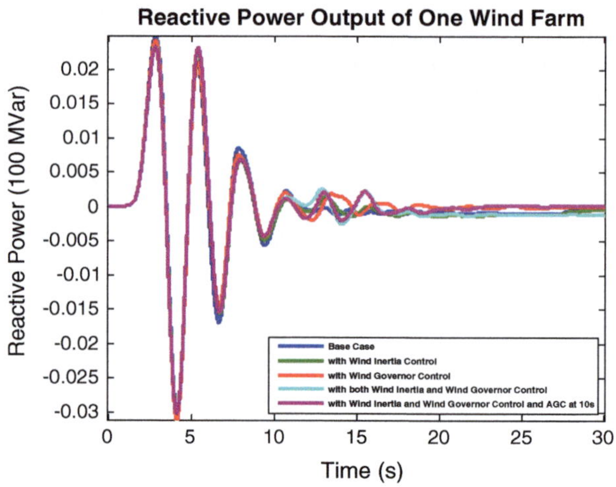


**Fig. 8.9** Reactive power outputs of a typical wind generator after a generation trip event with different controls

speeding zone. In this way, Wind Governor Control could help reduce not only the nadir but also the steady-state frequency deviation (red line in Fig. 8.6).

If inertia and governor control are combined together, the frequency nadir would be reduced to a greater extent and the steady-state frequency deviation would also decrease (shown by turquoise line in Fig. 8.6). However, in order to further reduce steady-state frequency deviation, AGC control should be considered. In the last scenario, all the wind generators received AGC orders from operator at 10-th seconds to increase their active power outputs (purple line in Fig. 8.7). As a result, the steady-state frequency deviation is eliminated almost completely.

Furthermore, Fig. 8.9 shows that reactive power of variable-speed wind generators with different controls are similar since a constant reactive power control mode is selected.

### *8.4.5 Case Study-Load Shedding*

Note that variable-speed wind generators could also contribute to the frequency response improvement in load shedding cases. In those cases, instead of decreasing wind turbine speed and release the reserve, wind turbines accelerate to store more energy in the blades and save more reserve so the frequency excursion would be reduced. To demonstrate this, a load shedding event of 724 MW is simulated. Similar to generation trip case, system frequency response (Fig. 8.10), active power (Fig. 8.11), turbine speed (Fig. 8.12) and reactive power (Fig. 8.13) are given below, respectively.

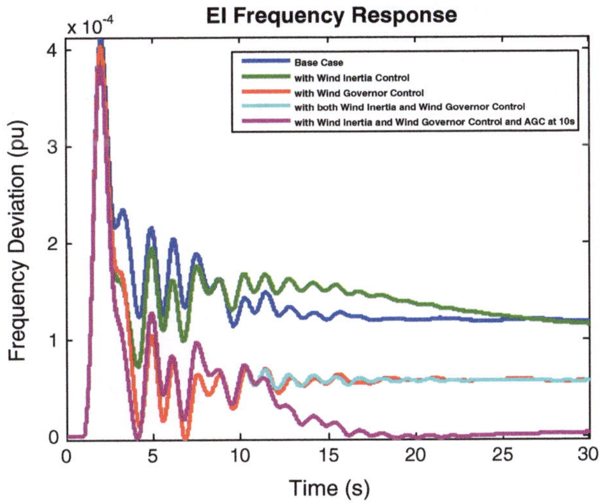


**Fig. 8.10** Frequency responses of the EI after a load shedding event with different controls

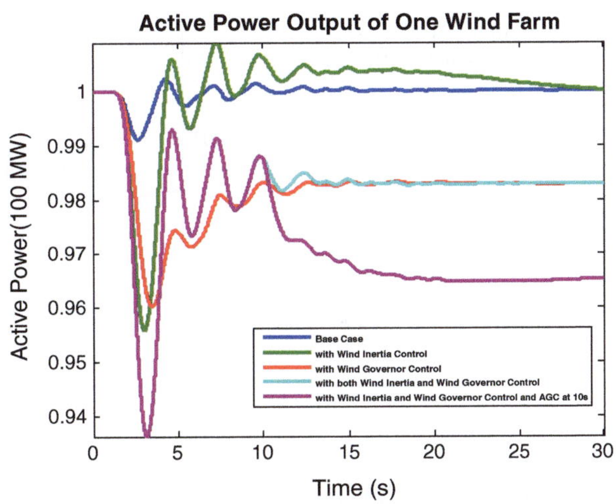


**Fig. 8.11** Active power outputs of a typical wind generator after a load shedding event with different controls

### 8.4.6 *Movie Display of Wind Generator Control for Frequency Regulation*

To demonstrate the overall effect of variable-speed wind generator control on the EI frequency regulation, a movie is made based on the EI simulation data of a generation trip event, one snapshot of which is given below in Fig. 8.14. Each red dot on the right maps in the movie stands for one sample bus in the EI and contour drawing is used to give the whole picture of the EI frequency deviation. On the upper side of the snapshot, all three wind controls including Wind Inertia Control, Wind Governor Control and Wind AGC control are employed in the simulation

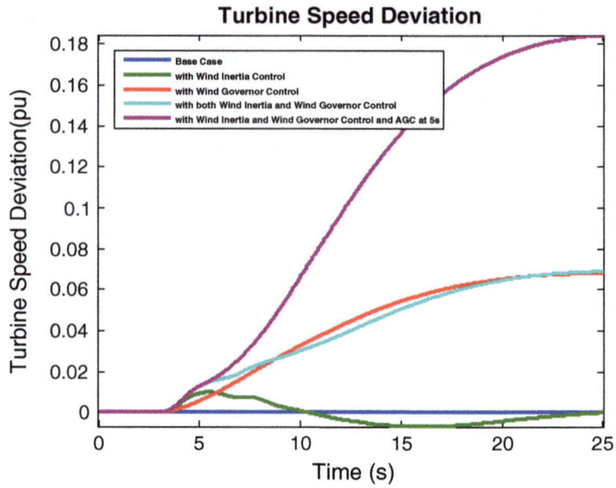


**Fig. 8.12** Turbine speed deviations of a typical wind generator after a load shedding event with different controls

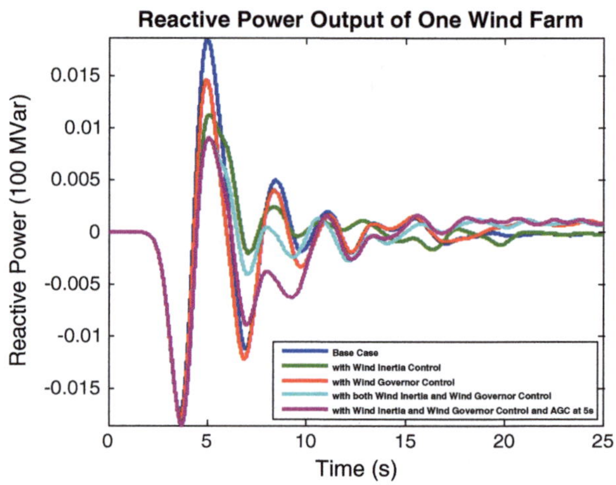


**Fig. 8.13** Reactive power outputs of a typical wind generator after a load shedding event with different controls

while on the lower side no control at all. Through comparison, it's obvious that the EI frequency response after generation loss could be improved dramatically because of the wind generator control.

## *8.4.7 Discussion*

Though the penetration of wind generation is still pretty low (3.4 % of all generated electrical energy in U.S. in 2012 and 5 % of total capacity in this scenario), variable-speed wind generators have already been proven to have the potential to effectively contribute to the EI frequency regulation. If wind generation penetration reaches 20 % of U.S. electricity by 2030 as expected by the U.S. Department

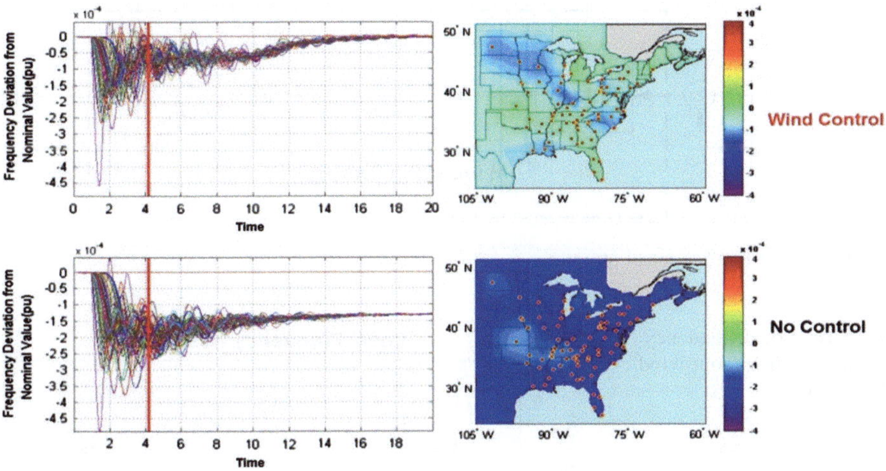


**Fig. 8.14** Movie display of the EI wind generator control for frequency regulation

of Energy's 2008 report, wind generation would be the game-changer of the frequency regulation service market. Therefore the corresponding control strategies should be carefully studied.

## 8.5 Contribution of Variable-Speed Wind Generators to the EI Oscillation Damping

Inter-area oscillation is another serious issue increasing wind generation brings about. However, the wind generation's potential in oscillation damping has been noticed by many researchers [19–26]. In Sect. 8.5, the potential of variable-speed wind generators' contribution to the EI oscillation damping is investigated based on the EI 16,000-bus dynamic model.

### *8.5.1 Wind PSS Control*

For conventional generators, a PSS adds damping to the generator rotor oscillations by producing a component of electrical torque in phase with the rotor speed deviations. A control structure similar to conventional generation PSS could be also employed in variable-speed wind generators for oscillation damping (control strucutre shown in Fig. 8.15), which is referred to as Wind PSS in this chapter. The signal washout block serves as a high-pass filter, with time constant $T_w$ high enough to allow signals associated with oscillations to pass unchanged. Either local or wide-area control signals can be used as the input to Wind PSS controller.

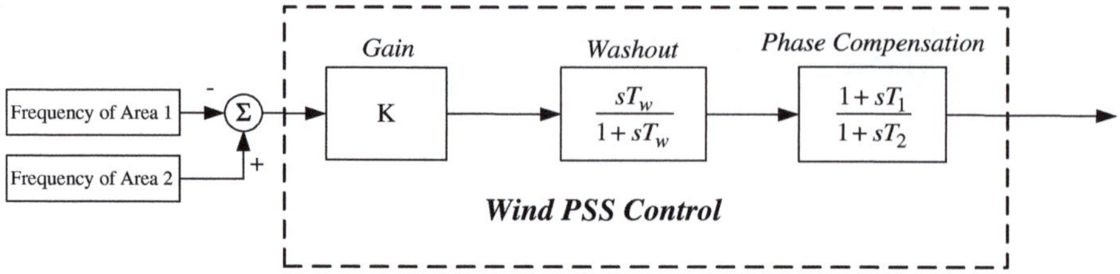


**Fig. 8.15** Wind PSS control structure

**Fig. 8.16** Local frequency response with/without wind PSS control

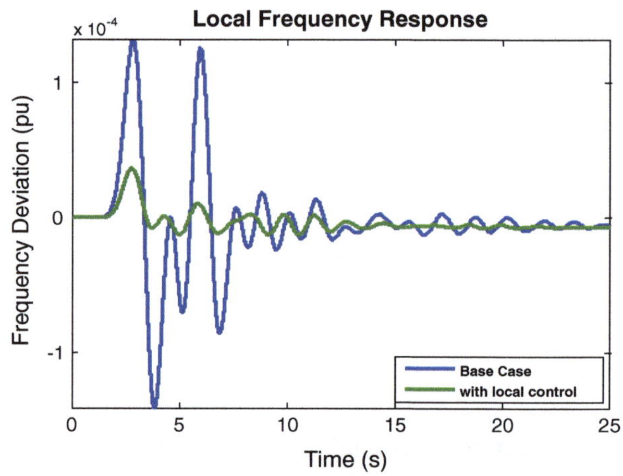


## *8.5.2 Oscillation Damping Using Local Feedback Signals*

Fed by a local frequency signal, Wind PSS could efficiently damp local frequency fluctuations or oscillations. A case study of wind generators in damping local frequency oscillation triggered by a line trip disturbance is given below. From Fig. 8.16, the local frequency oscillation is well damped by Wind PSS controllers with local feedback signals. The corresponding active power fluctuation in Fig. 8.17 demonstrates that the variable-speed wind generator could provide damping electrical torque effectively if with Wind PSS Control.

Just like conventional PSS, Wind PSS could also play a significant role in inter-area oscillation damping. In this case, a generation trip is simulated to trigger the inter-area oscillation between the Northwestern and Southern parts (NW-South) of the EI, as monitored in the real grid [28]. Fed by local frequency signal, the inter-area oscillation damping effect of Wind PSS could also be significant, as illustrated in Fig. 8.18.

**Fig. 8.17** Active power output with/without wind PSS control

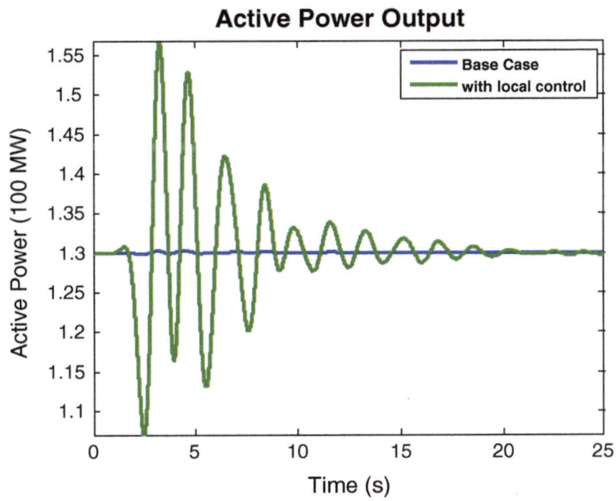


### *8.5.3 Inter-Area Oscillation Damping Using Wide-Area Feedback Signals*

A lot of studies have revealed that introducing wide-area feedback signals to PSSs could improve the inter-area oscillation damping effect to a great extent. In this case study, if introducing the frequency difference between NW and South as the input signal of Wind PSS controller, the inter-area oscillation between NW-South could be damped much more effectively than just using local frequency feedback signal, which is shown in Fig. 8.19. It should be noted that the originally weak inter-area oscillation between Northwest and Northeast (NW-NE) would be negatively influenced (as indicated by Fig. 8.20), which implies the necessity of a coordinated wide-area control strategy.

### *8.5.4 Coordinated Inter-Area Oscillation Damping Using Wide-Area Feedback Signals*

To further improve the overall oscillation damping effect, a simple coordinated strategy is developed in this chapter. According to the measurement experience [28], the inter-area oscillations in the EI mainly involve three areas: Northwest (NW), Northeast (NE) and South. The proposed coordinated wide-area Wind PSS controller would first decide which area the wind generator belongs to, for example NW, and then compare the frequency differences between the area it belongs to and the other two areas, such as NW-South and NW-NE, and then

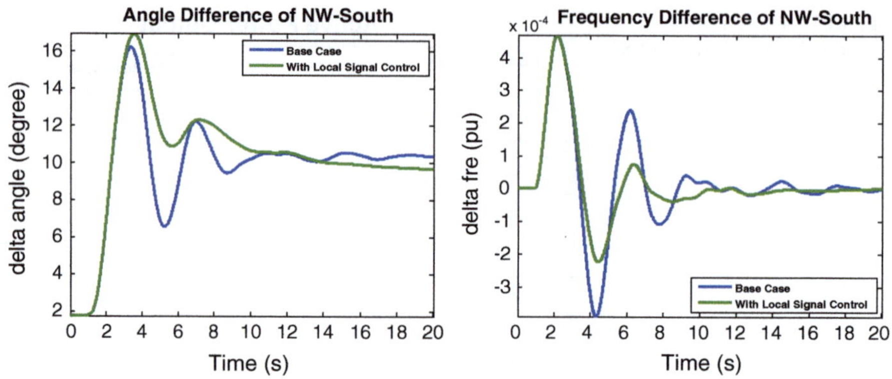


**Fig. 8.18** NW-South inter-area oscillation damping using local feedback signals

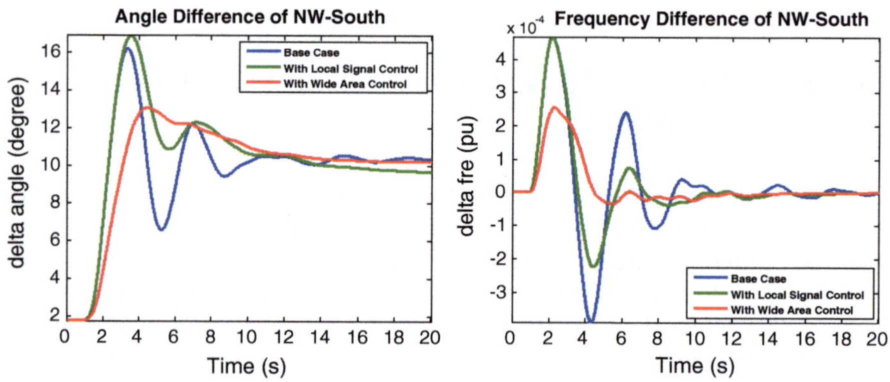


**Fig. 8.19** NW-South inter-area oscillation damping using wide-area feedback signals

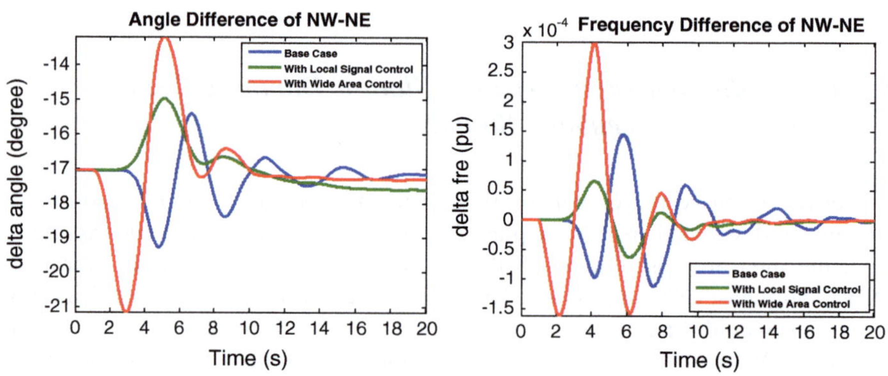


**Fig. 8.20** NW-NE inter-area oscillation

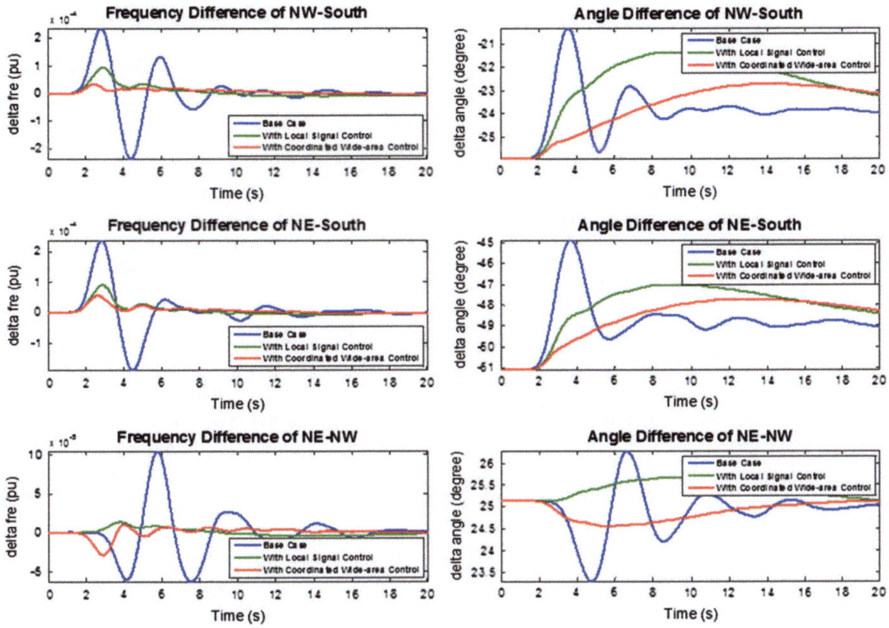


**Fig. 8.21** Inter-area oscillation damping using coordinated wide-area control

always choose the larger oscillation to damp. In this way, a coordinated oscillation damping effect could be achieved.

As a case study, a generation trip of 814 MW is simulated in the EI Florida region to incite the inter-area oscillations among the three main areas. The oscillation damping results using both local feedback signals and coordinated wide-area feedback signals are presented in Fig. 8.21. Through comparison, it's apparent that, with coordinated wide-area feedback signals, the inter-area oscillation could be damped much more effectively than just using local signals. Typical wind generators' active power outputs and local frequency responses are also given in Fig. 8.22.

### *8.5.5 Movie Display of Wide-Area Wind Generator Control for Inter-Area Oscillation Damping*

To demonstrate the overall effect of wide-area wind generator control on the EI inter-area oscillation damping, a movie based on the EI simulation data is created, one snapshot of which is given below in Fig. 8.23. In this movie, each bus's frequency deviation from the whole system average (instead of the nominal value) is plotted, therefore the inter-area oscillations could be demonstrated clearly.

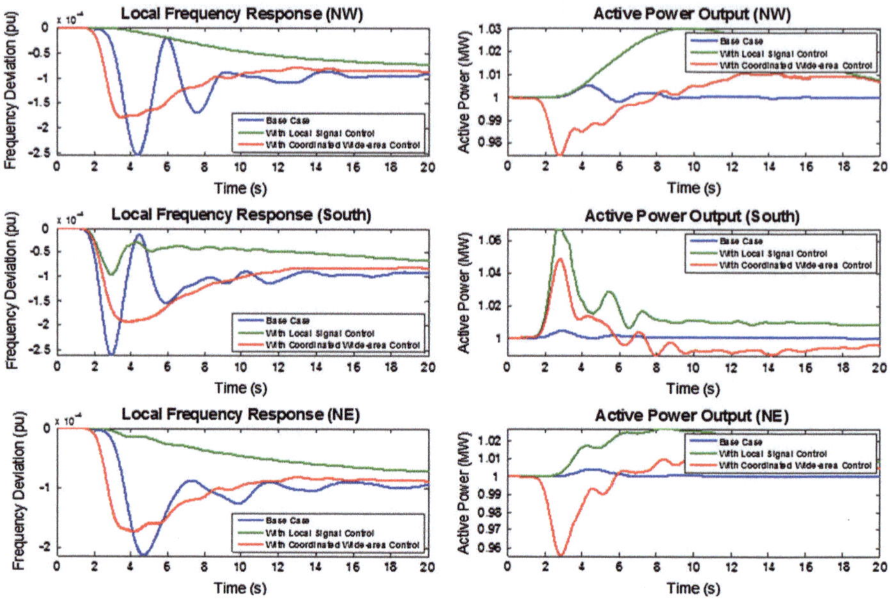


**Fig. 8.22** Typical wind generator behaviors with inter-area oscillation damping controls

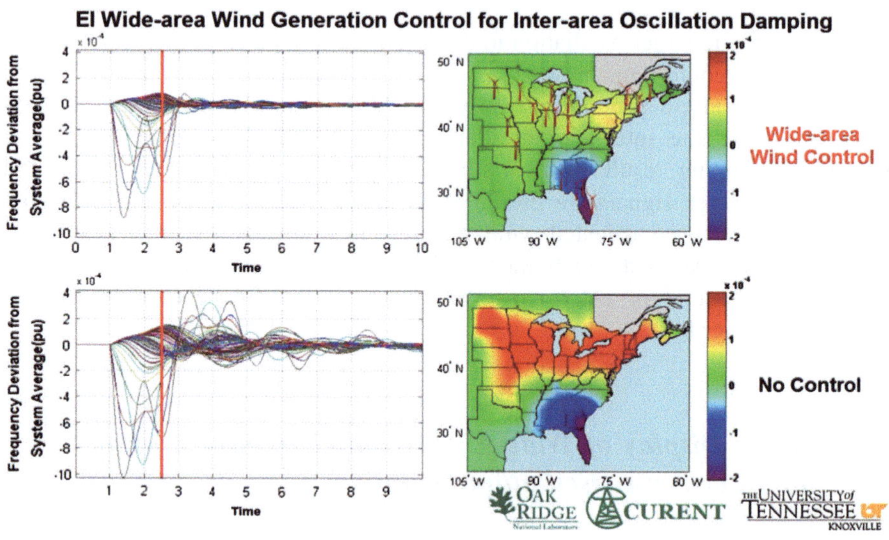


**Fig. 8.23** Movie display of the EI wide-area wind generator control for inter-area oscillation damping

The wind turbine icons indicate the locations of Wind PSS controllers. Through comparison, the EI inter-area oscillations are damped much more effectively with the wide-area wind generator control than without.

## 8.6 Discussion

Though promising as shown in Sect. 8.6, only very simple controller design is employed in this study. The controller design definitely needs to be improved and the parameters need to be tuned.

## 8.7 Conclusion

Based on the 16,000-bus EI model and the user-defined wind generator electrical control model, a relatively realistic scenario of the EI with a significant penetration of wind generators was utilized to evaluate the potential contributions of variable-speed wind generators to the EI frequency regulation and oscillation damping. Simulation results demonstrate that, at current and future penetration levels, wind generators represent a promising tool for frequency regulation and oscillation damping in the EI.

## References

1. Miller NW, Shao M, Venkataraman S, Loutan C, Rothleder M (2012) Frequency response of California and WECC under high wind and solar conditions. Power and energy society general meeting, pp 1–8
2. Ruttledge L, Miller NW, O'Sullivan J, Flynn D (2012) Frequency response of power systems with variable speed wind turbines. IEEE Trans Sustain Energ 3(4):683–691
3. Haan JES, Frunt J, Kling WL (2011) Grid frequency response of different sized wind turbines. Universities' power engineering conference (UPEC), pp 1–6
4. Mackin P, Daschmans R, Williams B, Haney B, Hunt R, Ellis J, Eto JH (2013) Dynamic simulation study of the frequency response of the Western interconnection with increased wind generation. International conference on system sciences (HICSS), pp 2222–2229
5. Villena-Lapaz J, Vigueras-Rodriguez A, Gomez-Lazaro E, Molina-Garcia A, Fuentes-Moreno JA (2012) Evaluation of frequency response of variable speed wind farms for reducing stability problems in weak grids. Power electronics and machines in wind applications (PEMWA), pp 1–5
6. Slootweg JG, Kling WL (2003) The impact of large scale wind power generation on power system oscillations. Electr Power Syst Res 67:9–20
7. Thakur D, Mithulananthan N (2009) Influence of constant speed wind turbine generator on power system oscillation. Electr Power Compon Syst 37:478–494
8. Modi N, Saha TK, Mithulananthan N (2011) Effect of wind farms with doubly fed induction generators on small-signal stability—a case study on Australian equivalent system. IEEE PES innovative smart grid technologies Asia (ISGT), pp 1–7
9. Vowles DJ, Samarasinghe C, Gibbard MJ, Ancell G (2008) Effect of wind generation on small-signal stability: a New Zealand example. IEEE Power Energy Society General Meeting, pp 1–8
10. Janssens NA, Lambin G, Bragard N (2007) Active power control strategies of DFIG wind turbines. Power Tech, pp 516–521

11. Mauricio JM, Marano A, Gomez-Exposito A, Martinez Ramos JL (2009) Frequency regulation contribution through variable-speed wind energy conversion systems. IEEE Trans Power Syst 24(1):173–180
12. Conroy JF, Watson R (2008) Frequency response capability of full converter wind turbine generators in comparison to conventional generation. IEEE Trans Power Syst 23(2):649–656
13. Ramtharan G, Ekanayake JB, Jenkins N (2007) Frequency support from doubly fed induction generator wind turbines. IET Renew Power Gener 1(1):3–9
14. Lalor G, Mullane A, O'Malley M (2005) Frequency control and wind turbine technologies. IEEE Trans Power Syst 20(4):1905–1913
15. Almeida RG, Castronuovo ED, Peas Lopes JA (2006) Optimum generation control in wind parks when carrying out system operator requests. IEEE Trans Power Syst 21(2):718–725
16. Almeida RG, Peas Lopes JA (2007) Participation of doubly fed induction wind generators in system frequency regulation. IEEE Trans Power Syst 22(3):944–950
17. Sun Y-Z, Zhang Z-S, Li G-J, Lin J (2010) Review on frequency control of power systems with wind power penetration. International conference on power system technology (POWERCON), pp 1–8
18. WindINERTIA® Control Fact Sheet (2009) GE Energy. (http://site.geenery.com/prod_serv/products/renewable_energy/en/downloads/GEA17210.pdf)
19. Tsourakis G, Nomikos BM, Vournas CD (2009) Contribution of doubly fed wind generators to oscillation damping. IEEE Trans Energy Convers 24(3):783–791
20. Ledesma P, Gallardo C (2007) Contribution of variable-speed wind farms to damping of power system oscillations. Power Tech, pp 190–194
21. Zhixin M, Lingling F, Osborn D, Yuvarajan S (2008) Control of DFIG based wind generation to improve inter-area oscillation damping. Power and energy society general meeting-conversion and delivery of Electrical Energy in the 21st Century, pp 1–7
22. Liu X, McSwiggan D, Littler TB, Kennedy J (2010) Measurement-based method for wind farm power system oscillations monitoring. IET Renew Power Gener 4(2):198–209
23. Lingling F, Haiping Y, Zhixin M (2011) On active/reactive power modulation of DFIG-based wind generation for inter-area oscillation damping. IEEE Trans Energy Convers 26(2):513–521
24. Knuppel T, Nielsen JN, Jensen KH, Dixon A, Ostergaard J (2011) Power oscillation damping capabilities of wind power plant with full converter wind turbines considering its distributed and modular characteristics. IET conference on renewable power generation, pp 1–6
25. Knuppel T, Nielsen JN, Jensen KH, Dixon A, Otergaard J (2011) Power oscillation damping controller for wind power plant utilizing wind turbine inertia as energy storage. Power and energy society general meeting, pp 1–8
26. Tsourakis G, Nomikos BM, Vournas CD (2009) Contribution of doubly-fed wind generators to oscillation damping. IEEE Trans Energy Convers 24(3):783–791
27. PSSE version 32 user manual, 2012
28. Zhiyong Y, Tao X, Yingchen Z, Lang C, Markham, PN, Gardner RM, Yilu L (2010) Inter-area oscillation analysis using wide area voltage angle measurements from FNET. Power and energy society general meeting, pp 1–7

# Chapter 9
# Power Management of Low and Medium Voltage Networks with High Density of Renewable Generation

**M. A. Barik, H. R. Pota and J. Ravishankar**

**Abstract** This chapter presents a review of existing control techniques for load-sharing in low and medium voltage networks. The advantages and major drawbacks of each method are described here. An overall comparison is made to find out the best suitable method for the distribution systems of the future. Finally, the limitations of existing methods and future directions for this research are indicated.

**Keywords** Microgrid · Load-sharing control · LV and MV networks · Renewable energy sources

## 9.1 Introduction

The conventional power system uses fossil fuel to generate electrical power which affects the environment [1]. As a result, there is interest in integrating renewable energy sources (RESs) in them [2–4]. RESs are environmental friendly, but some technical challenges must be addressed to integrate RESs with the grid. A major difference from the conventional generation is that RESs are connected to the grid

M. A. Barik (✉) · H. R. Pota
School of Engineering and Information Technology, The University of New South Wales, PO Box 7916Canberra BC, ACT 2610, Australia
e-mail: md.barik@student.adfa.edu.au

H. R. Pota
e-mail: h.pota@adfa.edu.au

J. Ravishankar
School of Electrical Engineering and Telecommunications,
The University of New South Wales, Sydney 2052, Australia
e-mail: jayashri.ravishankar@unsw.edu.au

J. Hossain and A. Mahmud (eds.), *Renewable Energy Integration*,
Green Energy and Technology, DOI: 10.1007/978-981-4585-27-9_9,

via power electronic interface and have low mechanical inertia [1, 5, 6]. Also, RESs are small and distributed throughout the network as distributed generators (DGs) [1]. The integration of DGs into a distribution system reduces the system's power loss, improves its voltage support and increases its efficiency and reliability [7]. On the other hand, automatic load-sharing is an issue with increasing penetration of RESs, because they are inertia-less DGs and connected to the mesh distribution network via inverters [8–11].

Droop control method is a popular way for active and reactive power sharing in a power system. This method has been primarily designed for high voltage (HV) transmission lines and high-inertia based generators. In HV networks, line impedances are inductive, whereas they are resistive in low voltage (LV) and medium voltage (MV) networks. Also, RESs are zero or low-inertia generators. For these reasons, conventional droop-based control does not work well [3, 12]. Thus, it is necessary to develop advanced control techniques for load-sharing in LV and MV networks.

Research studies aimed at improving the load-sharing of LV and MV networks have been conducted. These studies are usually based on modifications of the conventional droop control method. Some studies use communication-based control techniques. In this chapter, an overview of load-sharing in LV and MV networks with high densities of RESs is provided. Firstly, the basic principle and limitations of the conventional droop control method are explained. Then, the results from existing research into mitigating its drawbacks are presented and compared. After that, the load-sharing of LV and MV networks using a communication link is discussed and compared. Afterwards, a droop control-based load-sharing control method with communication is presented and its advantages and disadvantages are highlighted. Some limitations of existing research on the load-sharing of LV and MV networks are determined. Finally, this chapter suggests future directions for overcoming the limitations.

## 9.2 Load-Sharing Control Techniques for LV and MV Networks

The main objective of load-sharing is to distribute the active and reactive loads among the available DGs while maintaining voltage regulation and accommodating various types of loads [13]. In LV and MV networks, load-sharing becomes challenging due to the following factors.

- Most DGs have some local loads,
- Most DGs are non-dispatchable RESs, and
- The stability and reliability of the system gain importance, apart from the cost.

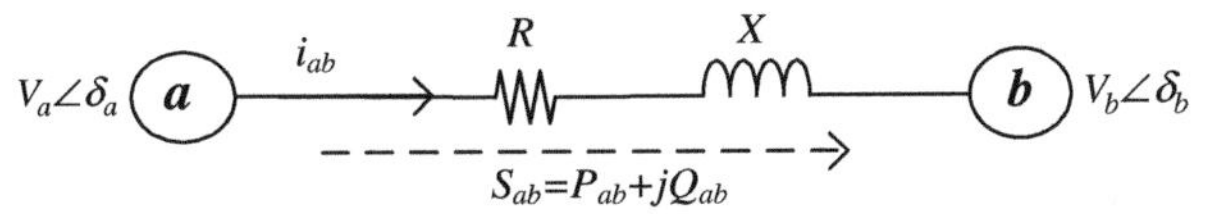


**Fig. 9.1** Power flow via transmission line from nodes *a* to *b* [12]

Three different methods of load sharing are commonly employed: droop-based control, communication-based control, and droop-based control with communication link, as described in the following subsections.

## *9.2.1 Droop-Based Control Techniques*

### 9.2.1.1 Conventional Droop Control Method

A widely used method for load-sharing is the droop control method. A rotating generator's frequency and real power are closely related. Its load torque increases with increasing load, causing its speed to decrease. Thus, its frequency changes with a change in its real load while, its terminal voltage changes with a similar change in its reactive load [14]. The droop control method is based on the concept of controlling the generator's frequency and voltage separately, in order to achieve real and reactive load sharing [5, 15–31]. For an inverter-based generator without a rotating part, better load-sharing can be achieved by the angle droop than the frequency droop method [3, 32]. To explain the droop control method for load-sharing, a complex power flow ($S_{ab}$) from node *a* to node *b* via a transmission line, as shown in Fig. 9.1, is considered. The equation of the complex power can be written as,

$$S_{ab} = P_{ab} + jQ_{ab} = v_a i_{ab}^* = v_a \left(\frac{v_a - v_b}{z}\right)^* = Y\left(V_a^2 e^{j\theta} - V_a V_b e^{j(\theta+\delta_{ab})}\right) \qquad (9.1)$$

where $P_{ab}$, $Q_{ab}$, and $i_{ab}$ are the real power, reactive power, and current flow from node *a* to node *b*, respectively, $z = Z\angle\theta = R + jX$ is the corresponding line impedance, $v_a = V_a\angle\delta_a$ and $v_b = V_b\angle\delta_b$ are the voltages of nodes *a* and *b*, respectively, $Y = \frac{1}{Z}$ is the admittance of the transmission line, and $\delta_{ab} = \delta_a - \delta_b$ is the bus angle difference between nodes *a* and *b*. From (9.1), real and reactive power flow can be expressed as:

$$P_{ab} = Y(V_a^2 \cos\theta - V_a V_b \cos(\theta + \delta_{ab})) \qquad (9.2)$$

$$Q_{ab} = Y(V_a^2 \sin\theta - V_a V_b \sin(\theta + \delta_{ab})) \qquad (9.3)$$

In Eqs. (9.2) and (9.3), it can be seen that the power flows and node voltages are dependent on each other because the line parameters are constant for a given line. Then, these equations can be rewritten for small changes in power flows as:

$$\Delta P_{ab} = Y[(2V_a \cos\theta - V_b \cos(\theta + \delta_{ab}))\Delta V_a + V_a V_b \sin(\theta + \delta_{ab})\Delta\delta_a] \qquad (9.4)$$

$$\Delta Q_{ab} = Y[(2V_a \sin\theta - V_b \sin(\theta + \delta_{ab}))\Delta V_a - V_a V_b \cos(\theta + \delta_{ab})\Delta\delta_a] \quad (9.5)$$

For HV transmission lines, line reactances are very high compared to line resistances, that is, $\theta \approx 90°$, $\cos\theta \approx 0$ and $\sin\theta \approx 1$. Thus, Eqs. (9.4) and (9.5) can be rewritten as:

$$\Delta P_{ab} = Y[V_b \sin\delta_{ab}\Delta V_a + V_a V_b \cos\delta_{ab}\Delta\delta_a] \quad (9.6)$$

$$\Delta Q_{ab} = Y[(2V_a - V_b \cos\delta_{ab})\Delta V_a + V_a V_b \sin\delta_{ab}\Delta\delta_a] \quad (9.7)$$

Again, as the $\delta_{ab}$ is very small, cos $\delta_{ab} \gg$ sin $\delta_{ab}$, Eqs. (9.6) and (9.7) can be modified as:

$$\Delta P_{ab} \approx YV_a V_b \Delta\delta_a \text{ and } \Delta Q_{ab} \approx Y(2V_a - V_b)\Delta V_a \quad (9.8)$$

or

$$\Delta P_{ab} \propto \Delta\delta_a \text{ and } \Delta Q_{ab} \propto \Delta V_a \quad (9.9)$$

Equation (9.9) shows that the real power flow change depends on the bus angle and the reactive power flow change depends on the bus voltage. Thus, load-sharing control can be achieved by independently controlling voltage magnitude($V$) and bus angle($\delta$), as presented in Eqs. (9.10) and (9.11). The characteristics and block diagram of this control scheme are shown in Figs. 9.2 and 9.3, respectively.

$$\delta - \delta_r = -k_p(P - P_r) \quad (9.10)$$

$$V - V_r = -k_q(Q - Q_r) \quad (9.11)$$

where $P$ and $Q$ are the real and reactive powers injection to the grid, $V_r$, $\delta_r$, $P_r$, and $Q_r$ are the reference values for the bus voltage, bus angle, real power, and reactive power, respectively, and $k_p$ and $k_q$ are the constants.

#### 9.2.1.2 Limitations of the Conventional Droop Control Method

The conventional droop control method was designed for HV transmission systems in which generators are rotational and transmission lines are inductive. Again, this method was derived from the power flow equation considering X $\gg$ R. Its performance is very good and very easy to implement [33]. In LV and MV networks, this method becomes ineffective for load-sharing due to the following major drawbacks.

- Most generators are connected to the grid via inverters which are inertia-less and have highly resistive line impedances [3].
- In a highly resistive network, coupling between the *P-f* and *Q–V* droops are unavoidable [34].
- For a large load variation, which is common in LV and MV networks, the transient current is very high [35].

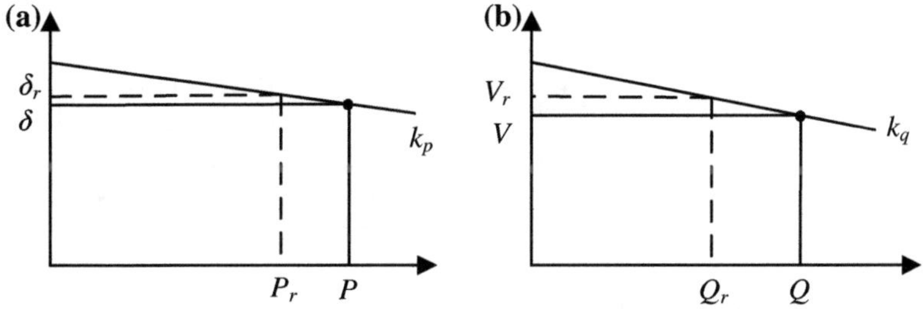


**Fig. 9.2** Characteristics of the conventional droop control scheme: **a** $P$-$\delta$ droop, **b** $Q$–$V$ droop

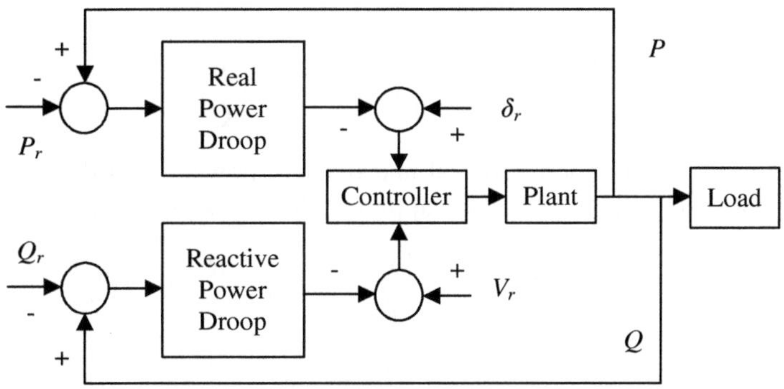


**Fig. 9.3** Block diagram of the conventional droop control scheme

- Due to the unequal line impedances, the accuracy of reactive load-sharing decreases [3].
- The majority of DGs have local loads which also decrease the accuracy of reactive load-sharing [3].
- The load-sharing depends on the inverter's output impedance which degrades the performance [34].
- The load-sharing accuracy is low and voltage regulation is poor [36].
- In a weak system, a high gain of the angle droop is required for the proper sharing of load which has negative effects on the stability of the system [37].

#### 9.2.1.3 Modified Droop Control Methods

To overcome the limitations of the conventional droop control method, several research studies have been conducted. Each of the studies has both benefits and drawbacks, as described in the following subsections.

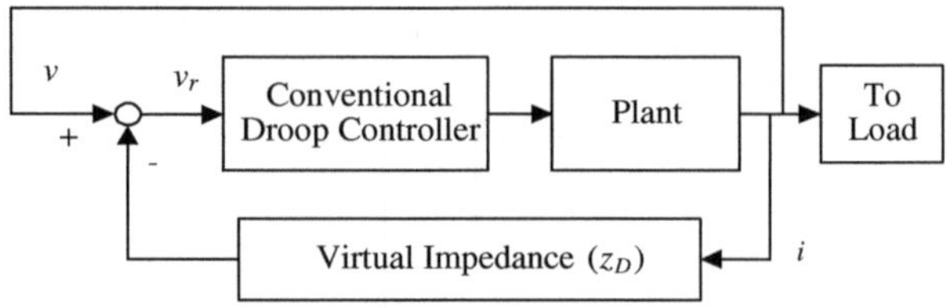


**Fig. 9.4** Block diagram of a virtual impedance control scheme [43]

**Frame transformation method**

The frame transformation technique considers both resistive and inductive line impedances for load-sharing [33, 38, 39]. In this method, an orthogonal linear rotational transformation matrix (*T*) is used to modify the real and reactive powers as:

$$\begin{bmatrix} P' \\ Q' \end{bmatrix} = T\begin{bmatrix} P \\ Q \end{bmatrix} = \begin{bmatrix} \frac{X}{Z} & \frac{-R}{Z} \\ \frac{R}{Z} & \frac{X}{Z} \end{bmatrix}\begin{bmatrix} P \\ Q \end{bmatrix} \tag{9.12}$$

where $P'$ and $Q'$ are the modified real and reactive powers, and the conventional droop control method is modified by applying this transformation as:

$$\delta - \delta^0 = -k_p\left(P' - P'^0\right) = -k_p\left[\frac{X}{Z}\left(P - P^0\right) - \frac{R}{Z}\left(Q - Q^0\right)\right] \tag{9.13}$$

$$V - V^0 = -k_q\left(Q' - Q'^0\right) = -k_q\left[\frac{R}{Z}\left(P - P^0\right) + \frac{X}{Z}\left(Q - Q^0\right)\right] \tag{9.14}$$

The advantage and major limitation of this method are [40]:

- It improves the real load-sharing accuracy and stability of the system, but
- The reactive load-sharing error exists.

**Virtual output impedance method**

The coupling between real and reactive load-sharing degrades load-sharing accuracy. An inductor is connected in series with the inverter to improve the sharing accuracy [41], but it is heavy, bulky and costly. This method considers that a virtual impedance ($z_D$) is connected to the output of the inverter to reduce the imbalance of the line impedance [3, 23, 34, 42–44]. The relationship between the virtual impedance and reference voltage ($v_r$) is presented in Eq. (9.15) and the control scheme is shown in Fig. 9.4.

$$v_r = v - z_D i \tag{9.15}$$

where $i$ and $v$ are the output current and voltage, respectively.

In this control scheme, the output current is fed via a virtual impedance to improve the voltage regulation which also reduces the coupling between real and reactive load-sharing. This increases the reactive load-sharing error because of increasing droops in the impedance voltage [3]. To improve the accuracy of reactive load-sharing, a method based on an additional control signal is proposed

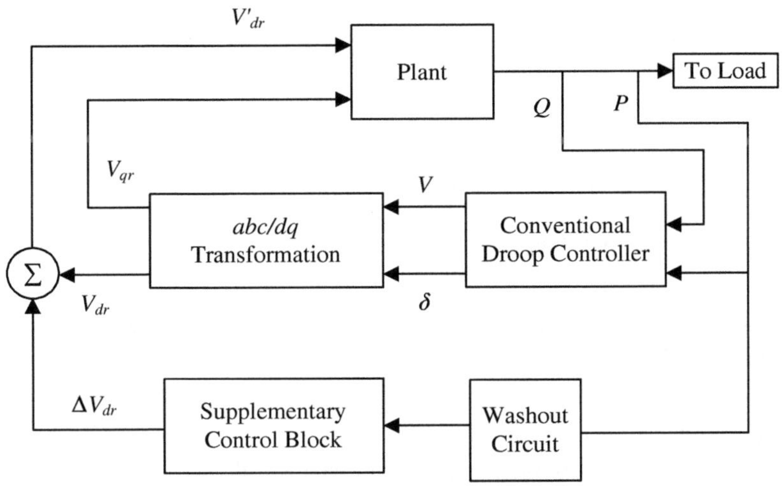


**Fig. 9.5** Block diagram of a supplementary droop control scheme, where $V_{qr}$ and $V'_{dr}$ are the *q*-axis voltage reference and the modified *d*-axis reference voltage, respectively [37]

in [5]. This method is complex and has a possibility of creating line current distortions. Another approach, which adds $\frac{\Delta V}{Q}$ slopes into voltage droop control, increases the reactive load-sharing accuracy, and reduces the effects of local loads [3]. Another technique is obtained by enforcing the resistive output impedance [34]. This technique offers the advantages of automatic harmonic sharing and improves the dynamic response of the paralleled system. Advantages and limitation of virtual impedance methods are [40]:

- It improves voltage regulation and load-sharing accuracy, and
- It reduces the imbalance of the line impedance, but
- It may not work properly in some situations due to the plug-and-play features of DGs and loads.

**Supplementary control loop method**

In a weak system, the high gain of the angle droop controller is required for the proper sharing of load causing a stability problem in the system. A supplementary control scheme is presented in [37]. This scheme considers closed-loop stability over a range of operating conditions. Also, its reduces the effect of the high gain by using a supplementary loop with conventional droop control as shown in Fig. 9.5. In this method, the output real power from the inverter passes through a high-pass washout circuit with a 0.05 s time constant to capture the oscillatory behaviour and eliminate the DC component. The supplementary control block generates a supplementary control signal *($\Delta V_{dr}$)* which modulates the output from the droop controller to modify the *d*-axis reference voltage *($V_{dr}$)*. This method is based on the local measurement and modulation of the *d*-axis voltage reference of each inverter. Finally, the supplementary loop increases the operating range of a weak system to

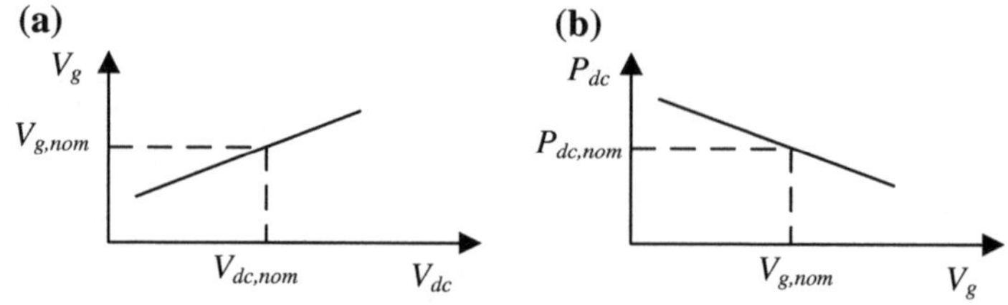


**Fig. 9.6** **a** *Vg/Vdc* droop, **b** *P/Vg* droop [46]

ensure satisfactory load-sharing. Its advantages are increase in the operating range and reduction of the load-sharing error. However, this method considers only the stability issue.

**Voltage-based droop control method**

In an inertia-based generator, the power imbalance affects the frequency of the generator which can be managed by controlling this frequency. In an inverter-based generator, the power balance is achieved by controlling the DC-link voltage. Also, the power flow in a distribution network depends on the voltage magnitude due to the network's highly resistive transmission lines. Based on these, a voltage-based droop control scheme with two parts, a $V_g/V_{dc}$ droop and a $P/V_g$ droop, is presented in [1, 45, 46] and shown in Fig. 9.6. In the $V_g/V_{dc}$ droop, the RMS value of the terminal voltage ($V_g$) changes according to the change in the DC-link voltage ($V_{dc}$) as:

$$V_g = V_{g,nom} + m(V_{dc} - V_{dc,nom}) \tag{9.16}$$

where $V_{g,nom}$ and $V_{dc,nom}$ are the nominal values of the RMS voltage and DC-link voltage, respectively and $m$ is a constant. Again, voltage variations can occur due to the control action taken by the $V_g/V_{dc}$ droop. In every system, there is a certain tolerance level for voltage variation. If it is exceeded, the output power from the DGs is changed by the $P/V_g$ droop. This controls the DC-link power ($P_{dc}$) which depends on the constant power bandwidth of the DGs, as shown in Fig. 9.7. The advantages and disadvantages of this method are:

- It improves sharing accuracy and voltage regulation,
- It is more beneficial for delaying RESs' power changes than a dispatchable unit which encourages the integration of more renewable energy, and
- It opposes hard curtailment which reduces ON–OFF oscillations, but
- It does not address the reactive power-sharing accuracy.

#### 9.2.1.4 Comparison of Modified Droop Control Methods

All the modified droop control methods for the proper sharing of load in a distribution system improve sharing performance. In the frame transformation

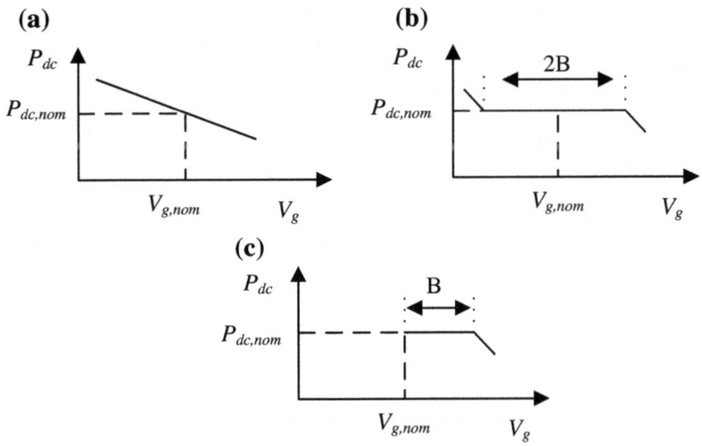


**Fig. 9.7** Constant power bands: **a** dispatchable unit (no bandwidth limit), **b** less dispatchable unit (bandwidth = 2B), and **c** non-dispatchable unit (bandwidth = B), where $P_{dc,\ nom}$ is the nominal value of the DC-link power [1]

method, load-sharing is improved by considering both resistive and inductive line impedances. A reactive load-sharing error exists in this method due to the coupling between real and reactive load-sharing. The virtual output impedance method reduces the effect of unbalanced line impedance and increases load-sharing accuracy. Sometimes this method may not work properly due to the plug-and-play features of DGs and loads. On the other hand, stability can be improved by adding a supplementary control loop to the supervisory control loop. Finally, voltage-based droop control method considers all the generator types, such as dispatchable, non-dispatchable and inertia-less DGs, and considers resistive transmission lines while encouraging the integration of more RESs. This method is more suitable for load-sharing in a distribution system because it considers the system's maximum constraints. Also, it highlights the sharing accuracy of the real power, but it does not address that of the reactive power very well.

## *9.2.2 Communication-Based Control Techniques*

### 9.2.2.1 Centralised Control Schemes

In these control techniques, load-sharing is centrally controlled by coordinating all DGs and loads, with a central control unit measuring the load demand of the system. A communication link sends reference signals to the local controllers which are responsible for controlling the generating unit to meet the reference value. There are two centralised control techniques, the central limit and master, which are described in the following subsections.

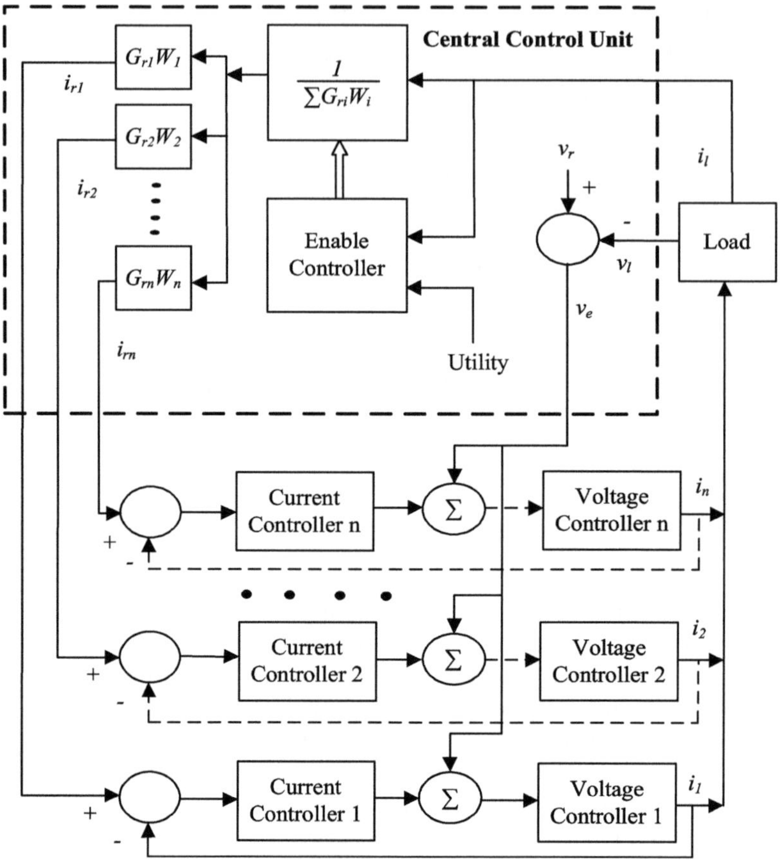


**Fig. 9.8** Block diagram of a central limit control scheme [47]

**Central limit control scheme**

The central control unit measures the real and reactive load values and load voltages, and calculates the reference currents ($i_r$) for each generator and the voltage error term ($v_e$). The central limit control scheme is presented in Fig. 9.8. The reference current for a generator can be calculated by dividing the total load current ($i_l$) by the weighting factors ($W$) of the generators, i.e., the summation of all the reference currents is equal to the load current, where the weighting factors depend on the ratings of the generators ($G_r$). The voltage error term can be calculated by comparing the load voltage ($v_l$) with the reference voltage ($v_r$). Local controllers control the output currents and terminal voltages by considering the reference current and voltage error [47–49]. A phase-lock loop (*PLL*) is used to synchronise the central control unit and local controllers. This method has some superior characteristics and also some limitations, as described below [35, 48–50].

- Its control algorithm is very simple.
- Current-sharing is forced during all times, including transients.
- Accurate load-sharing and voltage regulation are achieved in the steady state as well as during transients.
- The communication links and supervisory control centre required are expensive.
- It is difficult to apply in a large and highly distributed system especially during system expansion.
- It is difficult to achieve a fast response for power distribution control due to the relatively slow response of the PLL.
- Neglecting the line impedances in the control strategies is a significant disadvantage.
- If the sum of the weighting factors differs from one due to reasons like the shutdown of a unit or a programming fault, the load current will not be supplied properly.

**Master control scheme**

This technique is almost the same as the central limit control technique [48], where all the local control units control both the voltage and current. In master control scheme, master unit is responsible for only the voltage regulation. The load current is divided among the other units according to their weighting factors. An advantage of this method is that the master unit can provide transient current at the time of a wrong weighting factor because it does not have a current controller.

In this control scheme, the master unit can be selected as a fixed, arbitrarily chosen or maximum crest current unit and in the grid-connected mode, the main grid can be used. Another option in this method is that the master unit can take the responsibility of a central control unit and operate as a voltage source inverter. In this case, master unit measures the load demand of the system, controls the grid voltage, calculates the reference current for each generator and sends it to the relevant generator. This technique has some advantages over the central limit control approach as well as some disadvantages, as given below [35].

- In the case of failure of a unit, the system will still be operational because the master unit will supply the transient current.
- Its load-sharing performance is good because the master unit controls the grid voltage while other units are responsible for controlling only the output current.
- At a time, one signal has to be distributed to each local controller.
- It can operate without a central control unit.
- As the instantaneous voltages and currents are distributed throughout the system, a high bandwidth is required for communication.
- The system will not work if the master unit fails because all the other units depend on it.
- A high transient current can cause a dangerous situation because the master unit does not have any current controller.
- As a relatively higher bandwidth is required in the transient than steady-state condition, the system can fail if it has a low bandwidth.

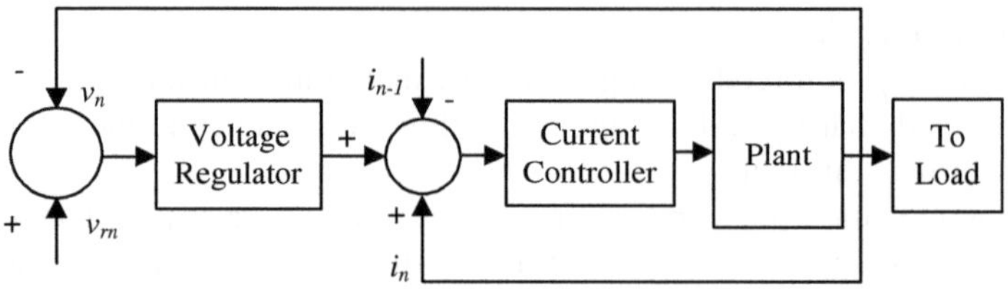


**Fig. 9.9** Block diagram of 3C scheme [13]

#### 9.2.2.2 Circular Chain Control (3C) Scheme

The 3C method is presented in [13]. In this method, each module tracks the output current from the previous module, with the first modules tracking those from the last as reference values to share equal current, as depicted in Fig. 9.9. The voltage control loop is used to ensure voltage regulation. This method requires less communication than other techniques because each module communicates with only the previous one and its dynamic response is very fast. Its advantage and major drawback are:

- It requires less communication, but
- All units must be successively connected.

#### 9.2.2.3 Distributed Control Through Frequency Partition

The distributed control technique is based on frequency partitioning between the central and local controllers [50, 51]. The central controller is responsible for controlling the low-frequency term and the local controllers the high-frequency term. Information on the modules' voltage references, current references and average feedback voltages is shared among the modules. This scheme presents a control algorithm which combines a low-pass filter ($H_{LF}$) with a matched high-pass filter ($1-H_{LF}$). The filters are used for perfect sharing of the control spectrum between two controllers, as depicted in Fig. 9.10. This control scheme uses a limited bandwidth of the communication signal. Its advantages and disadvantages are [35, 50]:

- Transient load-sharing is improved,
- The system will continue to run if a module breaks down,
- A limited bandwidth communication link is used to maintain load-sharing between the units,
- The local controller rejects the harmonic component, and
- The power quality of the system is improved under linear, nonlinear, balanced and unbalanced loads, but
- Interconnections between the DGs are required, and
- Higher performance can be achieved with high bandwidth which is costly.

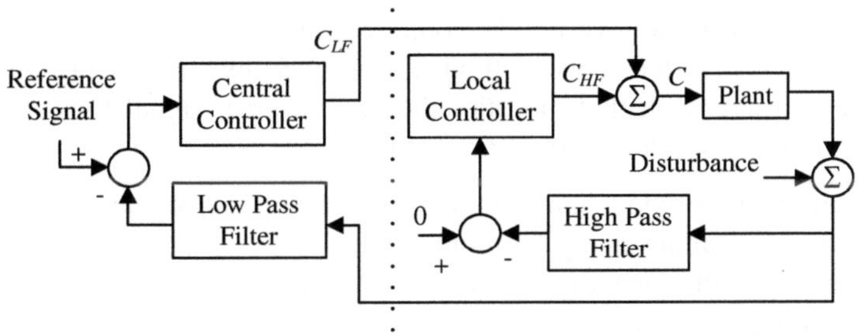


**Fig. 9.10** Block diagram of a distributed control scheme, where $C$ is the control signal [51]

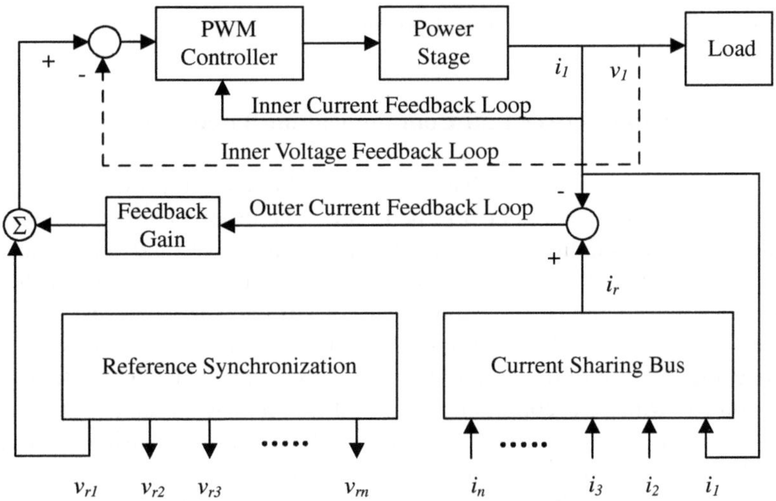


**Fig. 9.11** Block diagram of an instantaneous average current-sharing control scheme [53]

#### 9.2.2.4 Instantaneous Average Current-Sharing Scheme

For proper load-sharing control, the RMS values of the output currents are shared among the modules [26] and the average active and reactive loads shared [52]. These methods have good performances, but their current-sharing responses are very slow [53]. To overcome this issue, an instantaneous average current-sharing scheme based on sharing the instantaneous average current values among the inverters was designed [53–56]. In this method, a current-sharing bus measures deviations of the individual output currents and generates a current reference value for all DGs, as shown in Fig. 9.11. The voltage reference of each generator is different, but synchronisation is required to make the voltage phase angles of all inverters the same to ensure equal load-sharing. Each inverter has three control loops, inner current, outer current and inner voltage. The inner voltage and current loops control the inverter to provide good sharing in both steady-state and transient conditions for each single inverter. The outer current loop ensures equal sharing of the load by each inverter. This method is improved by introducing an adaptive gain-scheduling approach that modifies the current error signal [54]. The advantage and limitations of this method are [35]:

- It performs well for both current-sharing and voltage regulation, even if the output currents contain many harmonics, but
- The necessary interconnections between the inverters limit the flexibility of the system and degrade redundancy,
- The highest current control deteriorates the current distribution and output voltage regulation,
- The non-identical component characteristics and input voltage variations of the paralleled inverters might also deteriorate system performance, and
- This method is designed for equal power-sharing only.

#### 9.2.2.5 Comparison Between all Communication-Based Methods

Various types of communication-based control techniques proposed for sharing the load in a distribution system perform better than the droop control method. These methods need a communication link, which is expensive, and interconnections among all modules, which reduces the reliability of the system. Of all the communication-based methods, the centralised control ones are the most simple and accurate for sharing the load among DGs in LV and MV networks. They are expensive and difficult to implement in a complex system. The 3C method offers a simple control with less communication and has a first dynamic response, but requires successive connections of all DGs which may not visible in practical cases. The instantaneous average current-sharing control scheme controls the output current from each generator by measuring the deviations of individual output currents to ensure equal sharing of the load currents. This method only produces good results for equal current-sharing. Finally, distributed control through the frequency partition method performs well in terms of load-sharing under both steady-state and transient conditions. Also, this method is effective for linear, non-linear, balanced, and unbalanced loads. Its performance depends on the communication bandwidth. A high bandwidth would achieve better performance. As this is costly, this method considers a limited bandwidth communication link for communicating with all DGs.

### *9.2.3 Droop Control Method with Communication*

The main limitation of the droop control method is the coupling between real and reactive load-sharing which decreases the sharing accuracy. The high cost of a communication link is the major drawback of the communication-based control

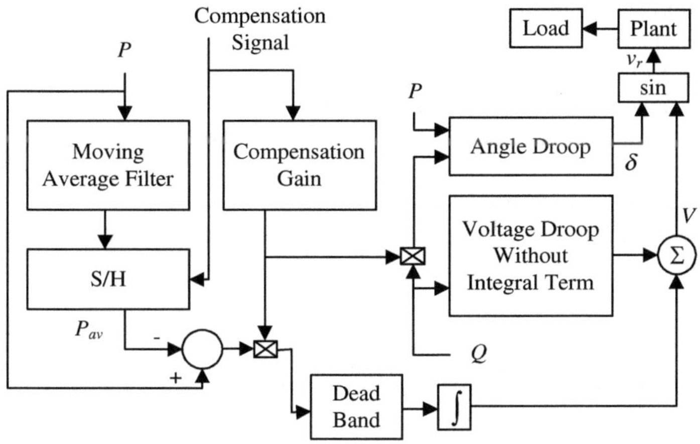


**Fig. 9.12** Block diagram of a reactive power compensation control scheme [40]

method. Considering these constraints, a method is proposed in [40] which achieves good load-sharing at a relatively low communication link cost. This is a reactive power compensation scheme based on the droop control method. This scheme uses a low bandwidth synchronisation flag signal to obtain the reactive-sharing error from the central controller.

A reactive power compensation control scheme for load-sharing in LV and MV networks is shown in Fig. 9.12. Here, load-sharing is undertaken by coordinating the central and local controllers. The central controller measures the reactive load-sharing error and transmits it to the local controllers via a uni-directional low-bandwidth communication link. Initially, each local controller uses the conventional droop control method for load-sharing and measures the average power ($P_{av}$). Each local controller stores the average power until it receives an error signal from the central controller. Then, a modified droop control method is used to compensate the reactive load-sharing as:

$$\delta - \delta^0 = -(k_p P + k_q Q) \tag{9.17}$$

$$V - V^0 = -k_q Q + \int k_i (P - P_{av}) \tag{9.18}$$

where $P_{av}$ is the last value of the average active power saved before the error signal is received. In Eq. (9.17), angle droop control is achieved by coupling the real and reactive powers in the case of a reactive power error, where $k_q Q$ is used as an unequal offset and is present only for a compensating reactive power-sharing error. The integral term in Eq. (9.18) is used to maintain the active power at $P_{av}$ during the time of reactive power compensation. A dead band is used before the integral block to limit the impact of load variations during that period. The advantages and disadvantages of this method are:

- A low bandwidth uni-directional flag signal is used to obtain the reactive power error signal from the central controller,
- Local load effects, unequal voltage drops in virtual and physical impedances, and variations in droop slopes are considered, and
- Its reactive load-sharing accuracy is very good like its frequency droop in the steady state, but
- Measuring the average power is not straightforward, and
- Its transient performance is not very good, particularly for a sudden large load variation.

## 9.3 Conclusions and Future Directions

An appropriate technique for controlling generating units in a distribution system to ensure the proper flows of real and reactive powers while maintaining the stability of the system is required. Designing a control scheme for load-sharing in a distribution system is a big challenge because its transmission lines are resistive, line impedances are unequal, most of its DGs are inverter-interfaced with some local loads, and it is based on inertia-less RESs which have significant effects on load-sharing.

A brief discussion of existing research on load-sharing of LV and MV networks is presented in this chapter. Existing research shows that the load-sharing of the distribution system can be enhanced by modifying the conventional droop control method, using communication link, or using modified droop control method with communication link. In all approaches, load sharing performance is improved. Among them, communication-based methods achieve better performance for load-sharing as well as better voltage regulation. These methods have not achieved popularity due to high cost of the communication link and the complexity of implementation. Most of the present research focuses on modifying the droop control method for load sharing. Existing modified droop control methods give a significant better performance on load-sharing, but reactive power-sharing accuracy, of the modified droop control methods, needs a significant improvement for practical applications. Finally, droop-based control method with communication link improves reactive power sharing accuracy, but advanced technique is required to ensure the fast transient response.

Overall, it can be concluded that the major limitation of existing research is that its results in terms of reactive power-sharing accuracy in the transient condition are not up to the mark, but could be improved by adding some extra features using the reactive power compensation method, such as:

(1) a better tuning algorithm for selecting the controller gain to reduce compensation time,
(2) a scheme for predicting loads and generation capability in advance, and
(3) a robust controller which has a better transient response capability.

## Appendix-I: List of Symbols

| Symbols | Variable names |
|---|---|
| $s_{ab}$ | Power flow from node *a* to *b* |
| $P_{ab}$ | Real power flow from node *a* to *b* |
| $Q_{ab}$ | Reactive power flow from node *a* to *b* |
| $i_{ab}$ | Current flow from node *a* to *b* |
| $R$ | Line resistance |
| $X$ | Line reactance |
| $z$ | Line impedance |
| $Z$ | Magnitude of the line impedance |
| $\theta$ | Phase angle of the line impedance |
| $Y$ | Line admittance |
| $z_D$ | Virtual impedance |
| $v$ | Terminal voltage/output voltage of a generator |
| $v_a$ | Voltage of node *a* |
| $v_b$ | Voltage of node *b* |
| $v_r$ | Reference voltage of a generator |
| $v_l$ | Load voltage |
| $v_n$ | Terminal voltage of the *n*th generator |
| $v_e$ | Voltage error term |
| $V$ | Terminal voltage magnitude |
| $V_a$ | Voltage magnitude of node *a* |
| $V_b$ | Voltage magnitude of node *b* |
| $V_r$ | Reference value for *V* |
| $V_g$ | RMS value of terminal voltage |
| $V_{dc}$ | DC-link voltage |
| $V_{dr}$ | Direct axis voltage reference |
| $V_{qr}$ | Quadratic axis voltage reference |
| $V_{g,\ nom}$ | Nominal value of RMS voltage |
| $V_{dc,\ nom}$ | Nominal value of DC-link voltage |
| $\Delta V_{dr}$ | Supplementary control signal for modifying direct axis voltage reference |
| $V'_{dr}$ | Modified direct axis voltage reference |
| $\delta$ | Bus angle |
| $\delta_a$ | Bus angle of node *a* |
| $\delta_b$ | Bus angle of node *b* |
| $\delta_r$ | Reference value for $\delta$ |
| $\delta_{ab}$ | Bus angle difference between nodes *a* and *b* |
| $i$ | Output current of a generator |
| $i_r$ | Reference current |
| $i_l$ | Total load current |
| $i_n$ | Output current of the *n*th generator |
| $P$ | Real power injection to the grid |
| $Q$ | Reactive power injection to the grid |
| $P_r$ | Reference value for *P* |
| $Q_r$ | Reference value for *Q* |

(continued)

(continued)

| Symbols | Variable names |
|---|---|
| $P'$ | Modified real power injection to the grid |
| $Q'$ | Modified reactive power injection to the grid |
| $P_{dc}$ | DC-link power |
| $P_{av}$ | Average power |
| $P_{dc,\ nom}$ | Nominal value of DC-link power |
| $G_r$ | Rating of generator |
| $W$ | Weighting factor of the generators |
| $C$ | Control signal |
| $C_{LF}$ | Control signal from central controller |
| $C_{HF}$ | Control signal from local controller |
| $m$, $k_p$ and $k_q$ | Constants |

## References

1. Vandoorn TL et al (2013) Voltage-based droop control of renewables to avoid on–off oscillations caused by overvoltages. IEEE Trans Power Delivery 2:845–854
2. Maisonneuve N, Gross G (2011) A production simulation tool for systems with integrated wind energy resources. IEEE Trans Power Syst 26(4):2285–2292
3. Li YW, Kao C-N (2009) An accurate power control strategy for power-electronics-interfaced distributed generation units operating in a low-voltage multibus microgrid. IEEE Trans Power Electron 24(12):2977–2988
4. Mohamed Y, El-Saadany EF (2008) Adaptive decentralized droop controller to preserve power sharing stability of paralleled inverters in distributed generation microgrids. IEEE Trans Power Electron 23(6):2806–2816
5. Tuladhar A et al (2000) Control of parallel inverters in distributed AC power systems with consideration of line impedance effect. IEEE Trans Ind Appl 36(1):131–138
6. Barik MA, Pota HR (2012) Complementary effect of wind and solar energy sources in a microgrid. In: IEEE PES innovative smart grid technologies, pp 1–6
7. Vandoorn TL et al (2013) Voltage-based control of a smart transformer in a microgrid. IEEE Trans Industr Electron 60(4):1291–1305
8. Walling RA et al (2008) Summary of distributed resources impact on power delivery systems. IEEE Trans Power Delivery 23(3):1636–1644
9. Foote CET et al (2008) A power-quality management algorithm for low-voltage grids with distributed resources. IEEE Trans Power Delivery 23(2):1055–1062
10. Li Y, Vilathgamuwa DM, Loh PC (2005) Microgrid power quality enhancement using a three-phase four-wire grid-interfacing compensator. IEEE Trans Ind Appl 41(6):1707–1719
11. Dondi P et al (2002) Network integration of distributed power generation. J Power Sources 106(12):1–9
12. Barik MA, Pota HR, Ravishankar J (2013) An automatic load sharing approach for a DFIG based wind generator in a microgrid. In: The 8th IEEE conference on industrial electronics and applications (ICIEA), pp 589–594
13. Wu T-F, Chen Y-K, Huang Y-H (2000) 3C strategy for inverters in parallel operation achieving an equal current distribution. IEEE Trans Industr Electron 47(2):273–281
14. Bollman AM (2009) An Experimental Study of frequency droop control In a Low-inertia microfrid. The Graduate College of the University of Illinois

15. Diaz G et al (2010) Scheduling of droop coefficients for frequency and voltage regulation in isolated microgrids. IEEE Trans Power Syst 25(1):489–496
16. Vasquez JC et al (2009) Adaptive droop control applied to voltage-source inverters operating in grid-connected and islanded modes. IEEE Trans Industr Electron 56(10):4088–4096
17. Nikkhajoei H, Lasseter RH (2009) Distributed generation interface to the CERTS microgrid. IEEE Trans Power Delivery 24(3):1598–1608
18. Sao CK, Lehn PW (2008) Control and power management of converter fed microgrids. IEEE Trans Power Syst 23(3):1088–1098
19. Barklund E et al (2008) Energy management in autonomous microgrid using stability-constrained droop control of inverters. IEEE Trans Power Electron 23(5):2346–2352
20. Sao CK, Lehn PW (2005) Autonomous load sharing of voltage source converters. IEEE Trans Power Delivery 20(2):1009–1016
21. Katiraei F, Iravani MR, Lehn PW (2005) Micro-grid autonomous operation during and subsequent to islanding process. IEEE Trans Power Delivery 20(1):248–257
22. Chung I-Y et al (2005) Operating strategy and control scheme of premium power supply interconnected with electric power systems. IEEE Trans Power Delivery 20(3):2281–2288
23. Guerrero JM et al (2004) A wireless controller to enhance dynamic performance of parallel inverters in distributed generation systems. IEEE Trans Power Electron 19(5):1205–1213
24. Borup U, Blaabjerg F, Enjeti PN (2001) Sharing of nonlinear load in parallel-connected three-phase converters. IEEE Trans Ind Appl 37(6):1817–1823
25. Chandorkar MC, Divan DM, Adapa R (1993) Control of parallel connected inverters in standalone AC supply systems. IEEE Trans Ind Appl 29(1):136–143
26. Kawabata T, Higashino S (1988) Parallel operation of voltage source inverters. IEEE Trans Ind Appl 24(2):281–287
27. Tuladhar A et al (1997) Parallel operation of single phase inverter modules with no control interconnections. In: IEEE twelfth annual applied power electronics conference and exposition, pp 94–100
28. Meng Y et al (2000) Research on voltage source inverters with wireless parallel operation. In: The IPEMC third international power electronics and motion control conference, pp 808–812
29. Byun YB et al (2000) Parallel operation of three-phase UPS inverters by wireless load sharing control. In: INTELEC. Twenty-second international telecommunications energy conference, pp 526–532
30. Barsali S et al (2002) Control techniques of dispersed generators to improve the continuity of electricity supply. In: IEEE power engineering society winter meeting, pp 789–794
31. Katiraei F, Iravani MR (2006) Power management strategies for a microgrid with multiple distributed generation units. IEEE Trans Power Syst 21(4):1821–1831
32. Pogaku N, Prodanovic M, Green TC (2007) Modeling, analysis and testing of autonomous operation of an inverter-based microgrid. IEEE Trans Power Electron 22(2):613–625
33. De Brabandere K et al (2007) A voltage and frequency droop control method for parallel inverters. IEEE Trans Power Electron 22(4):1107–1115
34. Guerrero JM et al (2007) Decentralized control for parallel operation of distributed generation inverters using resistive output impedance. IEEE Trans Industr Electron 54(2):994–1004
35. Vandoorn TL et al (2013) Review of primary control strategies for islanded microgrids with power-electronic interfaces. Renew Sustain Energy Rev 19:613–628
36. Coelho EAA, Cortizo PC, Garcia PFD (2002) Small-signal stability for parallel-connected inverters in stand-alone AC supply systems. IEEE Trans Ind Appl 38(2):533–542
37. Majumder R et al (2010) Improvement of stability and load sharing in an autonomous microgrid using supplementary droop control loop. IEEE Trans Power Syst 25(2):796–808
38. Mehrizi-Sani A, Iravani R (2010) Potential-function based control of a microgrid in islanded and grid-connected modes. IEEE Trans Power Syst 25(4):1883–1891
39. Barik MA, Pota HR, Ravishankar J (2013) A decentralized coordinated controller for load sharing in a microgrid with renewable generation. In: IEEE power and energy society general meeting pp 1–5

40. He J, Li YW (2012) An enhanced microgrid load demand sharing strategy. IEEE Trans Power Electron 27(9):3984–3995
41. Hua C-C, Liao K-A, Lin J-R (2002) Parallel operation of inverters for distributed photovoltaic power supply system. In: IEEE 33rd annual power electronics specialists conference, pp 1979–1983
42. Lee CT, Chu CC, Cheng PT (2013) A new droop control method for the autonomous operation of distributed energy resource interface converters. IEEE Trans Power Electron 28(4):1980–1993
43. Guerrero JM et al (2005) Output impedance design of parallel-connected UPS inverters with wireless load-sharing control. IEEE Trans Industr Electron 52(4):1126–1135
44. Chiang SJ, Yen CY, Chang KT (2001) A multimodule parallelable series-connected PWM voltage regulator. IEEE Trans Industr Electron 48(3):506–516
45. Vandoorn TL et al (2011) Active load control in islanded microgrids based on the grid voltage. IEEE Trans Smart Grid 2(1):139–151
46. Vandoorn TL et al (2011) A control strategy for islanded microgrids with DC-link voltage control. IEEE Trans Power Delivery 26(2):703–713
47. Siri K, Lee CQ, Wu TF (1992) Current distribution control for parallel connected converters. II. IEEE Trans Aerosp Electron Syst 28(3):841–851
48. Chen J-F, Chu C-L (1995) Combination voltage-controlled and current-controlled PWM inverters for UPS parallel operation. IEEE Trans Power Electron 10(5):547–558
49. Banda J, Siri K (1995) Improved central-limit control for parallel-operation of DC–DC power converters. In: 26th Annual IEEE power electronics specialists conference, pp 1104–1110
50. Milan P (2004) Power quality and control aspects of parallel connected inverters in distributed generation. Imperial College University of London
51. Prodanovic M, Green TC (2006) High-quality power generation through distributed control of a power park microgrid. IEEE Trans Industr Electron 53(5):1471–1482
52. Shanxu D et al (1999) Parallel operation control technique of voltage source inverters in UPS. In: Proceedings of the IEEE international conference on power electronics and drive systems, pp 883–887
53. Sun X, Lee YS, Xu D (2003) Modeling, analysis, and implementation of parallel multi-inverter systems with instantaneous average-current-sharing scheme. IEEE Trans Power Electron 18(3):844–856
54. Roslan AM et al (2011) Improved instantaneous average current-sharing control scheme for parallel-connected inverter considering line impedance impact in microgrid networks. IEEE Trans Power Electron 26(3):702–716
55. Chen Y-K et al (2003) ACSS for paralleled multi-inverter systems with DSP-based robust controls. IEEE Trans Aerosp Electron Syst 39(3):1002–1015
56. Xing Y et al (2002) Novel control for redundant parallel UPSs with instantaneous current sharing. In: Proceedings of the power conversion conference, pp 959–963

# Chapter 10
# Integration of Green Energy into Power Distribution Systems: Study of Impacts and Development of Control Methodology

**N. K. Roy and H. R. Pota**

**Abstract** Distributed generation (DG) is gaining popularity as it has a positive environmental impact and the capability to reduce high transmission costs and power losses. Although the integration of renewable energy-based DG will help reduce greenhouse gas emissions, it will rely heavily on new ways of managing system complexity. As traditional distribution networks were not designed to accommodate power generation facilities, various technical issues arise in the integration of distributed energy resources (DERs) into grids. This chapter presents an analysis of the major obstacles to the integration of green energy into power distribution systems (PDSs). Static and dynamic analyses are carried out with solar photovoltaic (PV) generators connected to different test systems to gain a clear understanding of the effect of PVs in PDSs. The results are compared with the existing utility standards to determine the critical issues in the integration of PVs into PDSs. A novel $H_\infty$ based control methodology is proposed to ensure grid code-compatible performances of PV generators. During the controller design, special attention is given to the dynamics of the load compositions of distribution systems. It is found that the proposed controller enhances the voltage stability of distribution systems under varying operating conditions.

**Keywords** Composite loads · Distributed generation · D-STATCOM · Voltage stability · Photovoltaic generator · Robust control and uncertainty

N. K. Roy (✉) · H. R. Pota
School of Engineering and Information Technology, The University of New South Wales, PO Box 7916, Canberra, ACT 2610, Australia
e-mail: n.roy@student.unsw.edu.au

H. R. Pota
e-mail: h.pota@adfa.edu.au

J. Hossain and A. Mahmud (eds.), *Renewable Energy Integration*, Green Energy and Technology, DOI: 10.1007/978-981-4585-27-9_10,

## 10.1 Introduction

Power supply reliability, efficiency and sustainability are the major concerns in the development of future energy systems. Many countries of the world have set their targets for power generation from green energy to reduce greenhouse gas emissions, as shown in Table 10.1 [1]. To reach the goal of green energy, distributed generation (DG) is a promising option and, therefore, is receiving a great deal of interest.

Photovoltaic (PV) and wind energy-based systems are the most important green energy resources. Research on the integration of solar PV in distribution systems is still in its infancy. At present, the main concerns about high penetration levels of solar PV generation in a distribution system are the effects of intermittency on the system protection when multiple sources are connected to a radial feeder or network [2, 3]. Close attention has been paid to modeling generators and their associated controls, and distribution system equipment. A model of a PV array is proposed in [4] which uses theoretical and empirical equations, together with data provided by the manufacturer, solar radiation, cell temperature and other variables, to predict the current–voltage curve. To study interactions of PV generators within the power system, a model of PV generator developed based on experimental results [5] suggests that the maximum power point tracking (MPPT) part of the control system of a PV generator dominates the dynamic behavior of the system. A mathematical model suitable for stability analysis that includes the nonlinear behavior of grid-connected PV modules is presented in [6]. Simulations which include the entire power converter are performed in [6] to support the mathematical analysis and it is concluded that the system is more susceptible to instability under high loading levels, i.e., when operating close to its maximum power point.

Grid integration of solar PV systems is gaining more interest than traditional stand-alone systems because of the following benefits:

- under favorable conditions, a grid-connected PV system supplies the excess power, beyond the consumption required by the connected load, to the utility grid;
- it is comparatively easy to install as it does not require a battery system because the grid is used as a backup;
- no storage losses are incurred; and
- it has potential cost advantages.

It is expected that grid-connected PV systems in medium-voltage networks will be commercially accepted in the near future [7]. Therefore, it is necessary to accurately predict the dynamic performance of three-phase grid-connected PV systems under different operating conditions in order to make a sound decision on the ancillary services that need to be provided to utilize their maximum benefits without violating grid constraints. Thus, the spread and growth of solar and other distributed renewable energy has led to significant modeling and engineering analyses of distribution systems. Although DG has several potential benefits, the

**Table 10.1** Renewable energy targets in different countries

| Country | Target (%) | Year |
|---|---|---|
| Australia | 20 | 2020 |
| Austria | 34 | 2020 |
| Belgium | 13 | 2020 |
| China | 15 | 2020 |
| Denmark | 30 | 2025 |
| Finland | 38 | 2020 |
| France | 23 | 2020 |
| Germany | 18 | 2020 |
| Netherlands | 14 | 2020 |
| New Zealand | 90 | 2025 |
| Spain | 20 | 2020 |
| Sweden | 49 | 2020 |
| UK | 15 | 2020 |
| US | 25 | 2025 |

connection of it in the existing distribution network will increase the fault level of the system. The impact of DG on the local voltage level can be significant during a contingency. Typical contingencies on a distribution network can occur in the form of single or multiple outages, such as unplanned losses of generators or distribution feeders. Several internal and external causes are responsible for equipment outages [8]. The internal causes arise from phenomena, such as insulation breakdown, over-temperature relay action or simply incorrect operation of relays. The external causes result from some environmental effects, such as lightning, high winds and icy conditions or non-weather related events, such as a vehicle or aircraft coming into contact with equipment or even human or animal direct contact. These contingencies can result in partial or full power outage in a distribution network unless an appropriate control action is taken.

A higher PV penetration level could possibly cause instability problems when a large percentage of the system load is supplied by PVs. Therefore, it is becoming more important to understand the behavior of a DG-integrated system under disturbances with practical distribution network loads since variations in loads physically close to generators are a large fraction of the generation. A practical system load is a combination of various types of loads and it is referred to as a composite load. The accurate modeling of loads is a difficult task due to several factors, such as [9]:

- large number of diverse load components;
- ownership and location of load devices in customer facilities not directly accessible to the electric utility;
- changing load composition with time of day and week, seasons, weather and through time;
- lack of precise information on the composition of the load; and
- uncertainties regarding the characteristics of many load components, particularly for large voltage or frequency variations.

**Table 10.2** Interconnection system response to abnormal voltages

| Voltage range (pu) | Clearing time (s) |
|---|---|
| $V < 0.5$ | 0.16 |
| $0.5 \leq V < 0.88$ | 2.00 |
| $1.1 < V < 1.2$ | 1.00 |
| $V \geq 1.2$ | 0.16 |

Load modeling is qualitatively different from generator modeling in many aspects. Generally, only the aggregate behavior of load is required for power system stability studies rather than a whole collection of individual component behaviors [10, 11]. Thus, one cannot escape the necessity to analyze the impact of bus load compositions in the distribution system behavior.

It is known that the majority (more than 60 %) of power system loads are induction motors, the impact of which must be taken into account during network analysis [12]. A higher proportion of induction motors in the composite loads could cause voltage stability problem which could disconnect DG units from the network as per the current utility practice [13, 14] which demands that a system voltage should recover to an acceptable level after a disturbance within the time indicated in Table 10.2. The unnecessary disconnection of generators reduces the expected benefits of DG and should be avoided because of the increasing importance of DG. In this context, it is particularly important to supply reactive power to the dynamic load to maintain system stability, thereby keeping DG units connected.

Although inverter-connected PV systems have their own reactive power capability, they are not allowed to operate in voltage control mode to avoid controller interactions [13, 14]. Moreover, in order to contribute more real power into a system, small-scale PV units are operated at unity power factor (pf) [15]. It is not advised to use PV inverters with a variable pf because, at high penetration levels, this may increase the number of balanced conditions of load demands and generations and, subsequently, increase the probability of islanding which is a safety hazard [7]. If DG units are not allowed to regulate voltage, additional sources of reactive power need to be installed at critical locations to supply the reactive power of a system.

It is well-known that a static synchronous compensator (STATCOM) has excellent performance in terms of its response speed and capabilities to reduce system power loss and harmonics, improve voltage level and stability, and decrease occupation area [16, 17]. The internal controls of a Distribution STATCOM (D-STATCOM) play a very important role in maintaining the system voltage. The use of suitable control methods in a D-STATCOM may offer a better performance along with making possible tracking of the desirable references more efficiently. Conventional controllers for D-STATCOMs are mainly PI controllers [18, 19], the tuning of which is a complex task for a nonlinear system with switching devices. To avoid the limitations of PI controllers, a linear quadratic regulator (LQR) method is used in [20] to design a STATCOM controller which has a superior performance. Compared to the LQR method [20], which uses states as feedback, linear quadratic Gaussian (LQG) controller is more realistic as it can be designed using only measurable outputs and state variables estimated from them [21]. However, linear controllers designed based on the given operating point are not suitable in the event of large variations in the system model. To improve the performance of an LQG controller, model mismatches or uncertainties can be bounded by an $H_\infty$ norm which provides robust closed-loop stability as well as optimal performance [22]. As distribution networks have different types of loads, a controller designed without regard to a tight bound on variations in load compositions will not lead to a satisfactory performance. However, this important issue has not been considered in the existing literature. The superior dynamic performance of the D-STATCOM compared to many other compensating devices encourages further refining its control scheme to achieve robust performance without enhancing its reactive power capacity.

This chapter makes three contributions: (a) an investigation of the static voltage stability of distribution networks through Q–V analysis for different contingencies; (b) an examination of the impact of different load compositions on the dynamic behavior of distribution networks; and (c) a novel D-STATCOM controller design which is robust to variations in load compositions in a practical distribution system.

The organization of this chapter is as follows. Section 10.2 presents the static voltage stability analysis of the system. The mathematical model of the system for dynamic simulation is described in Sect. 10.3. Section 10.4 demonstrates the impact of various load compositions on the distribution system. The controller design is given in Sect. 10.5 and the performance of the designed controller is evaluated for different operating conditions in Sect. 10.6. Concluding remarks are given in Sect. 10.7.

## 10.2 Static Voltage Stability Analysis

A 16-bus 3-feeder distribution test system [23], as shown in Fig. 10.1, is used in this study. Two PV generators, one at bus 2 and one at bus 3, are connected to the distribution system. The total load on the system is 28.7 MW, 17.3 MVAr. The test system data with distributions of system loads in different nodes is given in Appendix-I. The PV system used in this analysis is operated at unity pf to comply with the grid code requirement [13].

According to the Q–V method [12], the bus voltage magnitude (V) increases as the reactive power (Q) injection at the same bus is increased. When the voltage of any bus decreases with an increase in the Q for that bus, the system is said to be unstable. The reactive power margin (Q-margin) is measured as the distance between the lowest MVAr point of the Q–V curve and the voltage axis, as shown in Fig. 10.2 [24]. Based on the Q–V analysis, the impacts of high PV penetration and the behavior of the system under contingencies are described below.

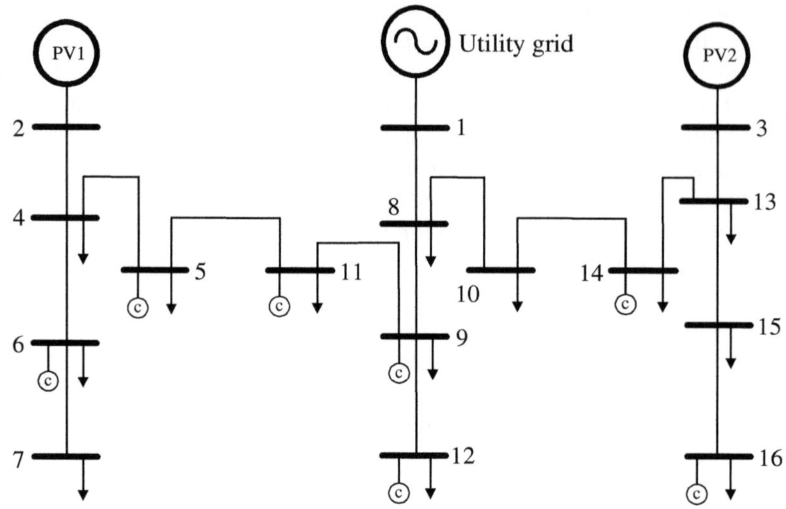


**Fig. 10.1** Single *line* diagram of 16-bus distribution test system

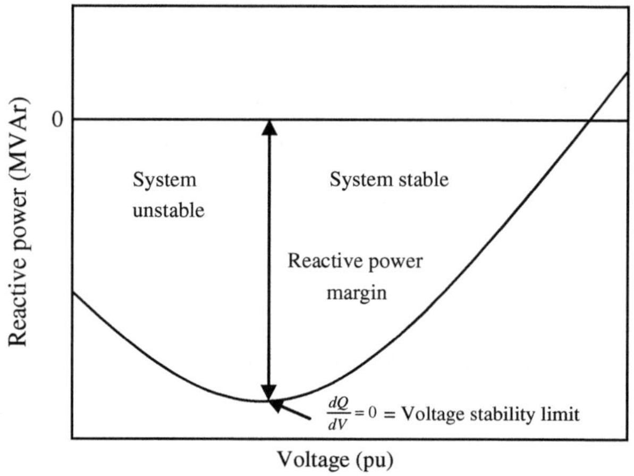


**Fig. 10.2** Typical reactive power-voltage (Q–V) *curve*

### *10.2.1 Impact of High PV Penetration*

The penetration level of PV (%PV penetration) is given by

$$\%\ \text{PV penetration} = \frac{P_{\text{PV}}}{P_{\text{load}}} \times 100 \qquad (10.1)$$

where $P_{\text{PV}}$ is the total power delivered by the PV generator and $P_{\text{load}}$ is the peak load demand.

**Table 10.3** Q-margins (MVAr) for different PV penetration levels

| Bus no. | Penetration level | | | | |
|---|---|---|---|---|---|
| | 20 % | 40 % | 60 % | 80 % | 100 % |
| 2 | 29.80 | 31.07 | 32.19 | 33.18 | 34.06 |
| 3 | 37.78 | 39.17 | 40.42 | 41.55 | 42.57 |
| 4 | 35.59 | 36.93 | 38.04 | 38.99 | 39.80 |
| 5 | 46.21 | 47.69 | 48.89 | 49.88 | 50.69 |
| 6 | 26.43 | 27.56 | 28.57 | 29.46 | 30.28 |
| 7 | 24.77 | 25.85 | 26.82 | 27.69 | 28.49 |
| 8 | 167.60 | 169.96 | 171.62 | 172.74 | 173.42 |
| 9 | 82.43 | 84.25 | 85.65 | 86.73 | 87.55 |
| 10 | 82.07 | 83.97 | 85.55 | 86.85 | 87.92 |
| 11 | 52.46 | 53.99 | 55.22 | 56.21 | 57.02 |
| 12 | 55.12 | 56.52 | 57.73 | 58.83 | 59.79 |
| 13 | 48.45 | 49.89 | 51.13 | 52.21 | 53.14 |
| 14 | 69.37 | 71.12 | 72.58 | 73.80 | 74.82 |
| 15 | 37.95 | 39.21 | 40.34 | 41.37 | 42.32 |
| 16 | 34.97 | 36.15 | 37.23 | 38.22 | 39.13 |

The Q-margins of the system for increasing PV penetration levels are shown in Table 10.3 which signify that PV generators have positive impacts on the system's Q-margin. In this scenario, the generations at both the PV buses are increased by the same amount. As the penetration level of PV increases, static voltage stability of the system also increases.

### *10.2.2 Behavior of System Under Contingencies*

To get an idea of the system's vulnerability, Q–V analysis is performed when PV units are tripped as a result of disturbances in the system. Before application of the contingency, it is assumed that PV1 and PV2 are supporting the total system load equally. Figure 10.3 shows the Q-margins of the load buses when PV generators are tripped from the system. It can be seen that the tripping of multiple PV generators reduces the system stability compared to the tripping of a single PV generator.

It is expected that the penetration level of grid-connected distributed energy resources (DERs) will increase substantially over the next few decades. With the increased penetration of DG, tripping of massive generation due to local disturbances can further risk the stability of the whole system.

The most common operational problems in distribution systems are the overloads due to high demand in peak hours and inadequate voltage levels at system buses. Load change exceeding the normal boundary can overload the lines and transformers which is a safety hazard. To investigate the effect of overloading, it is considered that initially, PV1 and PV2 are equally supporting the nominal system load and then loading of the system is increased keeping the penetration level of

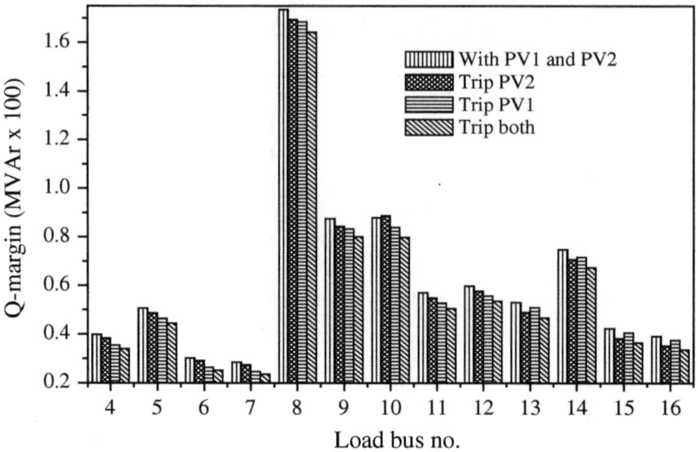


**Fig. 10.3** Q-margins of load buses under contingencies

**Table 10.4** Q-margins of load buses before and after contingency

| Load bus no. | Q-margin (MVAr) | | | |
|---|---|---|---|---|
| | No contingency | 20 % overload | 30 % overload | Outage of line 2–4 and 30 % overload |
| 4 | 39.80 | 37.41 | 36.22 | 31.56 |
| 5 | 50.69 | 47.85 | 46.44 | 41.65 |
| 6 | 30.28 | 28.40 | 27.45 | 23.19 |
| 7 | 28.49 | 26.71 | 25.80 | 21.70 |
| 8 | 173.42 | 168.49 | 166.05 | 160.71 |
| 9 | 87.55 | 83.76 | 81.89 | 77.13 |
| 10 | 87.92 | 84.68 | 83.01 | 78.78 |
| 11 | 57.02 | 53.99 | 52.48 | 47.73 |
| 12 | 59.79 | 56.93 | 55.50 | 51.11 |
| 13 | 53.14 | 51.04 | 49.97 | 47.54 |
| 14 | 74.82 | 72.00 | 70.56 | 66.93 |
| 15 | 42.32 | 40.60 | 39.73 | 37.90 |
| 16 | 39.13 | 37.53 | 36.72 | 35.05 |

PV generators same. The effect of overloading of the system is tabulated in Table 10.4 from which it can be investigated that a sudden overloading reduces the voltage stability of the system. The impact of multiple contingencies is more severe than a single contingency in a distribution network. Sudden outage of a line and growth in demand significantly reduces the system's stability margin, as depicted in Table 10.4.

In this study, contingency analysis is carried out using a steady-state or power flow model of the distribution system. Dynamic stability assessment is described in the next section.

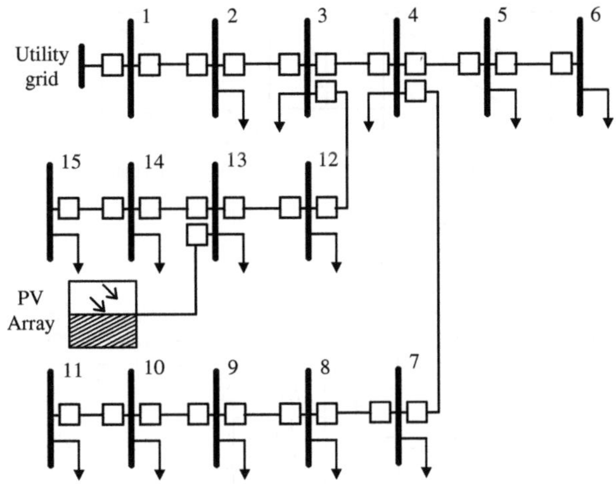


**Fig. 10.4** Single *line* diagram of Kumamoto 15-bus distribution system

## 10.3 System Model for Dynamic Analysis

To investigate the dynamic behavior of system, a Japanese practical distribution system shown in Fig. 10.4 is considered [25]. A PV generator is connected at bus 13 and bus 1 is connected to the utility grid. The total load on the system is 6.301 MW, 0.446 MVAr. The PV generator has the capability to support 50 % of the system load and the rest is supplied by the grid. The test system data with loads is given in Appendix-I.

### *10.3.1 Solar PV*

A PV generator consists of arrays of solar PV modules. PV modules produce DC power which is converted into AC power by an inverter. It is essential to create aggregate equivalent generators instead of modeling each individual inverter. Small-scale PV generators are typically aggregated at the distribution interface buses. Figure 10.5 shows a PV system connected to the grid through a DC–DC converter and a DC-AC inverter. The equivalent circuit of a PV system is shown in Fig. 10.6. The dynamics of the PV generator can be described by the following equations [6, 26].

$$\frac{di_{\mathrm{pv}}}{dt} = \frac{1}{\vartheta L_{\mathrm{pv}}} \ln\left(\frac{I_L - i_{\mathrm{pv}} - \frac{v_{\mathrm{pv}} + R_s i_{\mathrm{pv}}}{R_{\mathrm{sh}}}}{I_s} + 1\right) - \frac{1}{L_{\mathrm{pv}}}(v_{\mathrm{pv}} + R_s i_{\mathrm{pv}}) \tag{10.2}$$

$$\frac{dv_{\mathrm{pv}}}{dt} = \frac{1}{C_{\mathrm{pv}}}[i_{\mathrm{pv}} - Ni_{\mathrm{dc}}] \tag{10.3}$$

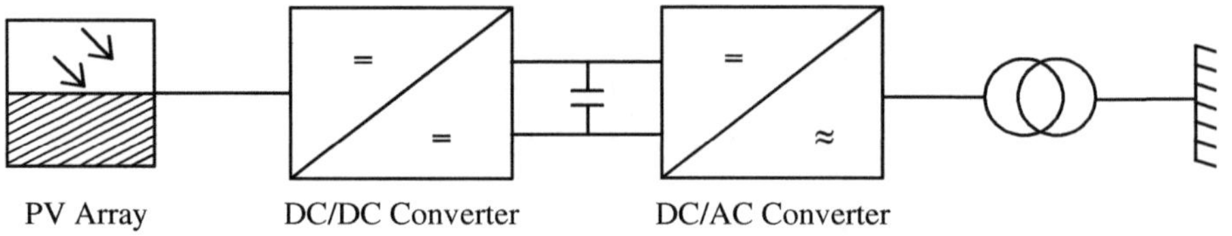


**Fig. 10.5** Grid-connected PV generator

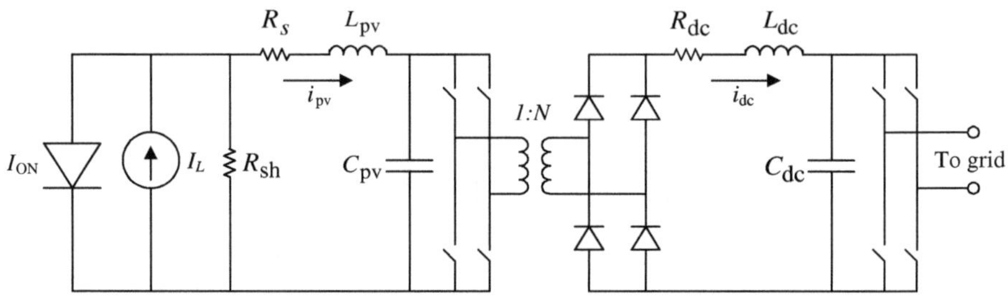


**Fig. 10.6** Equivalent model of grid-connected PV generator

$$\frac{di_{\mathrm{dc}}}{dt} = \frac{1}{L_{\mathrm{dc}}}[Nv_{\mathrm{pv}} - R_{\mathrm{dc}}i_{\mathrm{dc}} - v_{\mathrm{dc}}] \tag{10.4}$$

$$\frac{dv_{\mathrm{dc}}}{dt} = \frac{1}{v_{\mathrm{dc}}C_{\mathrm{dc}}}[P_{\mathrm{pv}} - P] \tag{10.5}$$

where $\vartheta = \frac{q}{nskT}$, $k = 1.3807 \times 10^{-23}$ JK$^{-1}$ is the Boltzmann's constant, $q = 1.6022 \times 10^{-19}$ C is the charge of electron, $T = 298$ K is the cell temperature and $ns$ is the number of series cells in the PV array, $I_L$ is the light-generated current, $I_{ON}$ is the dark diode characteristics of photocells, $L_{\mathrm{pv}}$ and $C_{\mathrm{pv}}$ is the wiring inductance and capacitance of the PV cells, respectively, $I_s = 9 \times 10^{-11}$ A is the saturation current, $R_s$ and $R_{\mathrm{sh}}$ are the series and shunt resistance of the array, respectively, $i_{\mathrm{pv}}$ is the current flowing through the array, $v_{\mathrm{pv}}$ is the output voltage of the array, $N$ is the turns-ratio of the step-up transformer, $R_{\mathrm{dc}}$ is the resistance, $L_{\mathrm{dc}}$ is the reactance, $C_{\mathrm{dc}}$ is the capacitance, $G$ is the solar irradiance, $i_{\mathrm{dc}}$ is the current flowing through and $v_{\mathrm{dc}}$ is the output voltage of the DC-link, $P_{\mathrm{pv}}$ is the power of the PV array which is a function of $v_{\mathrm{dc}}$, $G$ and $T$, and $P = 3/2(v_d i_d + v_q i_q)$ is the output power, where $v_d$ and $v_q$ are the direct- and quadrature-axis components of the output voltage, respectively, and $i_d$ and $i_q$ are the direct- and quadrature-axis components of the current, respectively.

The converter control of a solar PV system is shown in Fig. 10.7. The real and reactive power generated and absorbed by the voltage source converter (VSC) can be controlled by controlling its firing angle, $\alpha_{\mathrm{pv}}$ and modulation index, $m_{\mathrm{pv}}$. In this study, the reactive power, $Q_{\mathrm{ref}}$ is set to zero to operate the PV generator at unity pf and $Q = 3/2(v_q i_d$ - $v_d i_q)$. A shunt AC filter is not considered in the PV model since its effect is more relevant for an electromagnetic transient analysis [27].

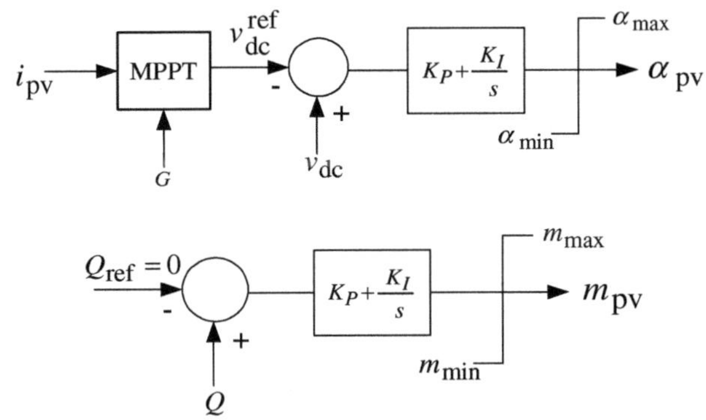


**Fig. 10.7** Converter control of PV generator

## *10.3.2 D-STATCOM*

A D-STATCOM is a shunt-connected flexible alternating current transmission system (FACTS) device that regulates the voltage of an AC bus. The regulation is done by a VSC connected to a DC capacitor, as shown in Fig. 10.8. The dynamics of this voltage source is governed by the charging and discharging of a large (nonideal) capacitor. The capacitor voltage can be adjusted by controlling the phase angle difference between line voltage, $v_s$ and VSC voltage, $E$ ($E = kv_{\text{dcs}}\angle\alpha$). If the phase angle of the line voltage is taken as a reference, that of the VSC voltage is the same as the firing angle, $\alpha$, of the VSC. Thus, if $\alpha$ is slightly advanced, the DC voltage, $v_{\text{dcs}}$, decreases and reactive power flows into the D-STATCOM. Conversely, if $\alpha$ is slightly delayed, the DC voltage increases and the D-STATCOM supplies reactive power to the bus. By controlling $\alpha$, the reactive power can be supplied from or absorbed by the D-STATCOM and, thus, the voltage regulation can be achieved.

The dynamics of a D-STATCOM can be described by the following equation:

$$\dot{v}_{dcs}(t) = -\frac{P_s}{C_s v_{\text{dcs}}} - \frac{v_{\text{dcs}}}{R_{cs} C_s}, \tag{10.6}$$

where $v_{\text{dcs}}$ is the capacitor voltage, $C_s$ is the DC capacitance, $R_{cs}$ is the internal resistance of the capacitor, $P_s$ is the power supplied by the system to the D-STATCOM to charge the capacitor which is a nonlinear function of ($\alpha$, $k$, $E$, $v_{dc}$, $v_d$ and $v_q$) [28].

The terminal voltage of the D-STATCOM is measured using a transducer with first-order dynamics,

$$\dot{v}_{sm} = -\frac{v_{sm}}{T_m} + K_m v_s, \tag{10.7}$$

where $v_{sm}$ is the sensor output and $v_s$ is the voltage at the connection point of the D-STATCOM, $K_m$ is a constant and $T_m$ is the time constant of the voltage transducer. In this chapter, the constant associated with the inverter, $k$ is fixed and

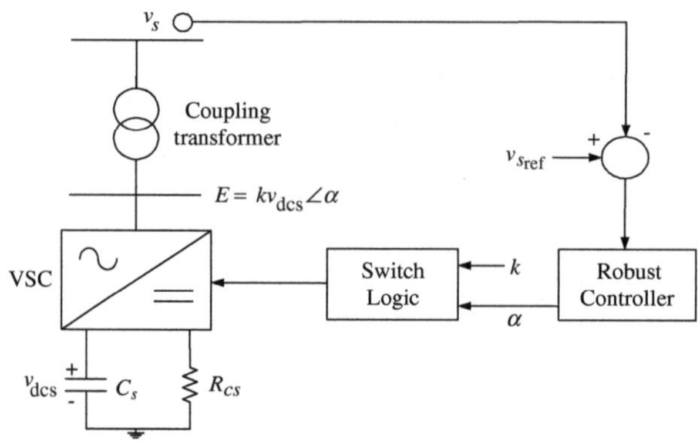


**Fig. 10.8** D-STATCOM equivalent circuit with its control

the firing angle, $\alpha$ is used as the control variable. In the load flow, the D-STATCOM is modeled as a generator bus (PV bus) with $P_G = 0$.

### 10.3.3 Load

The following exponential forms are used to represent a static load.

$$P(V) = P_0\left(\frac{V}{V_0}\right)^a, \tag{10.8}$$

$$Q(V) = Q_0\left(\frac{V}{V_0}\right)^b, \tag{10.9}$$

where $P$ and $Q$ are the active and reactive components of the load, respectively and $V$ is the bus voltage magnitude. The subscript 0 identifies the values of the respective variables at the initial operating condition. The parameters of this model are the exponents $a$ and $b$. With these exponents equal to 0, 1 or 2, the model represents the constant power, constant current or constant impedance characteristics of load components, respectively.

For composite loads, the load components for the nominal system are aggregated assuming that a load delivery point consists of 30 % static loads (space heating, cooking, water heater, etc.), 10 % fluorescent lamps [24] and 60 % induction motors [24]. In this study, the active components of static loads are represented by constant current models and reactive components by constant impedance models, as recommended in [9] for dynamic simulations.

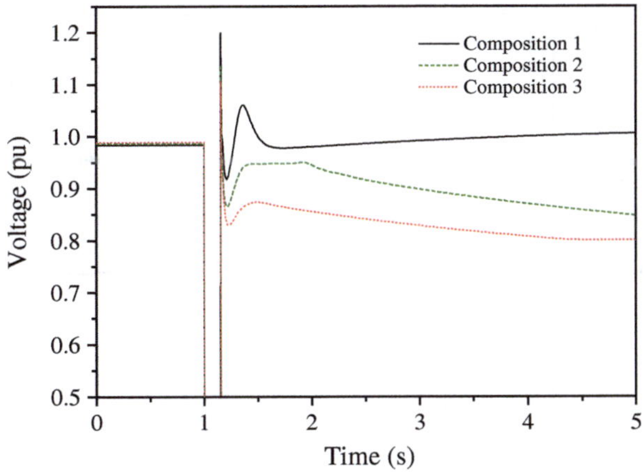


**Fig. 10.9** Voltage at PCC (bus 13) for various load compositions

**Table 10.5** Different load compositions in composite load model

| No. | Static (%) | Fluorescent lighting (%) | Induction motor (%) |
|---|---|---|---|
| Composition 1 | 45 | 15 | 40 |
| Composition 2 | 30 | 10 | 60 |
| Composition 3 | 15 | 5 | 80 |

## 10.4 Impact of Different Load Compositions

In order to investigate the dynamic behavior of the system with the composite load model, the system is tested with a sudden three-phase short-circuit fault at bus 2 for a fault duration of 150 ms. Figure 10.9 shows the voltage at the point of common coupling (PCC) for different load compositions in Table 10.5. The figure illustrates that the composition of load has a great impact on the system stability. From Fig. 10.9, it can be seen that the system has transient over-voltage in all cases and it fails to return to pre-fault condition as the proportion of induction motors in the load composition increases. The voltage profiles at different load buses with the load composition 3 under the same disturbance are shown in Fig. 10.10. Real and reactive powers drawn by the loads in those buses are shown in Figs. 10.11 and 10.12, respectively. From Figs. 10.11 and 10.12, it can be observed that the real powers are less sensitive to the disturbance, however, dynamic loads cause voltage instability due to their high reactive power consumption after a sudden disturbance. To overcome the instability problem, the system requires dynamic compensating devices to maintain the grid code requirements in Table 10.2.

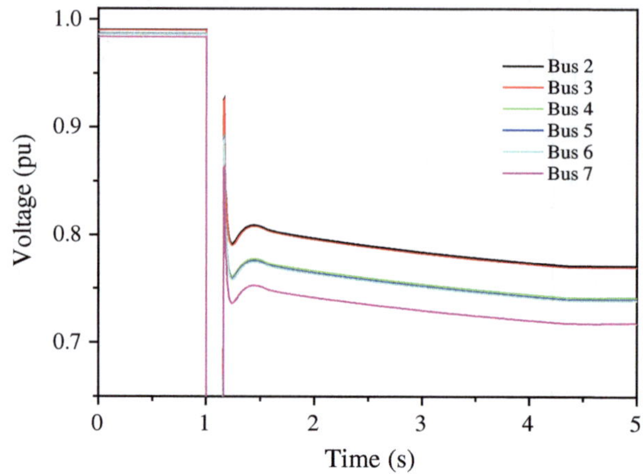


**Fig. 10.10** Voltages at different load buses of system without D-STATCOM

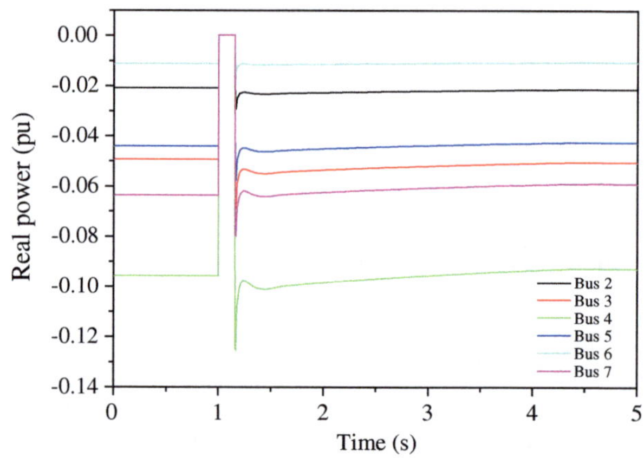


**Fig. 10.11** Real power drawn by loads at different buses without D-STATCOM

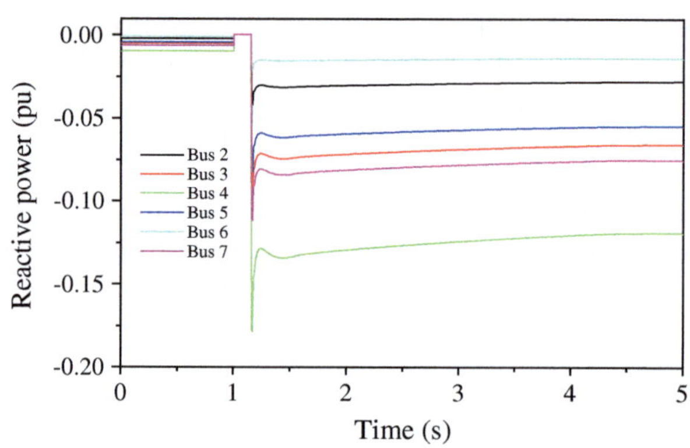


**Fig. 10.12** Reactive power consumed by loads at different buses without D-STATCOM

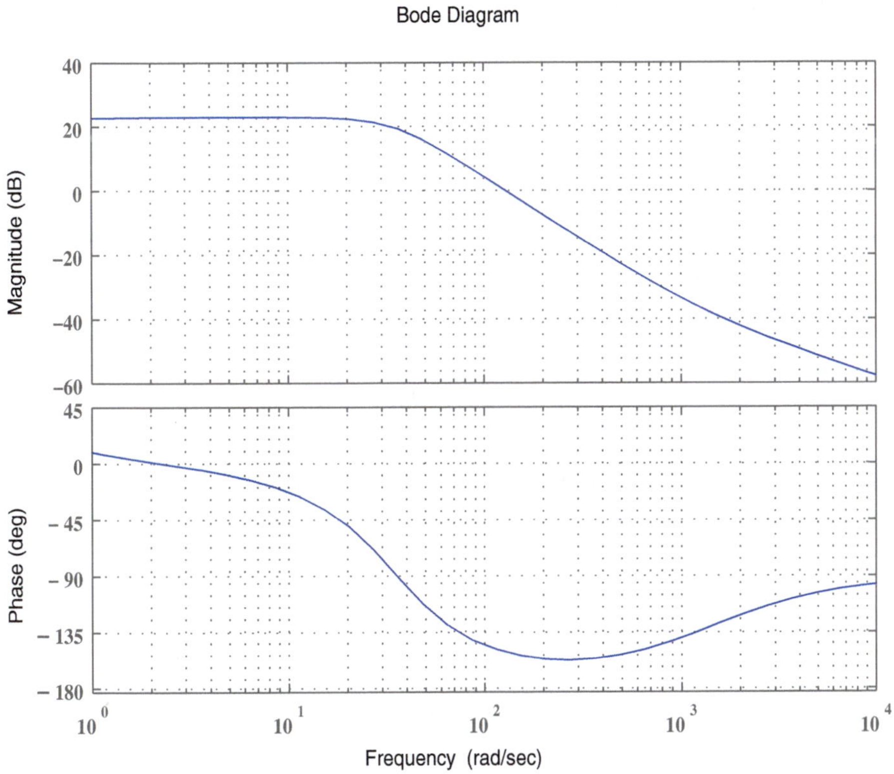


**Fig. 10.13** Bode plot of open-loop system with converter control—transfer function from input (firing angle) to terminal voltage

## 10.5 Controller Design for D-STATCOM

The problem considered here is the design of a robust controller which works for various load compositions in the composite load. Modal analysis is performed on the grid-connected PV system to obtain an idea of the dominant mode which needs to be controlled [12]. The dominant mode of the nominal test system is monotonic with an eigenvalue at −0.00059. The normalized participation factors indicate that system voltage states ($v_{pv} = 1.0$, $v_{dc} = 0.96$, $v_{dcs} = 0.57$) have significant contributions in this mode. This marginally stable mode causes unstable operation during large disturbances in the system. The open-loop frequency response of the system with converter controller is shown in Fig. 10.13 in which it can be observed that the system has an inadequate margin (phase margin 28.55 deg) in order to withstand disturbances.

### 10.5.1 LQG Controller

The system to be controlled can be written as

$$\begin{aligned} \dot{x}(t) &= Ax(t) + Bu(t), \\ y(t) &= Cx(t), \end{aligned} \tag{10.10}$$

where $A$ is the system matrix, $B$ is the control or input matrix, $C$ is the output matrix, $x = \left[i_{\text{pv}}, v_{\text{pv}}, i_{\text{dc}}, v_{\text{dc}}, v_{\text{dcs}}, v_{sm}\right]^T$, input, $u = \alpha$ and output, $y = v_{sm}$.

The standard LQR cost function can be defined as

$$J = \lim_{T\to\infty} \frac{1}{T} E \int_0^T \left(x^T Q_r x + u^T R_r u\right) dt, \tag{10.11}$$

where $Q_r$ is a positive semi-definite state weighting matrix and $R_r$ is a positive definite control weighting matrix, that is, $Q_r = Q_r^T \geq 0$, $R_r = R_r^T > 0$ and $E$ is the expectation operator.

The controller for the nominal system that consists of a LQR and a Kalman-Bucy filter (KBF) is given by [22]

$$\dot{x}_c = A_c x_c + B_c(\Delta V_{\text{ref}} - V_{sm}), \tag{10.12}$$

$$y_c = C_c x_c, \tag{10.13}$$

where

$$A_c = A - BR_r^{-1}B^T P_r - P_f C^T R_f^{-1} C, \tag{10.14}$$

$$B_c = P_f C^T R_f^{-1}, C_c = R_r^{-1} B^T P_r. \tag{10.15}$$

The matrices $P_r$ and $P_f$ are symmetric positive-definite solutions of the control and filter algebraic Riccati equations given by equations (10.16) and (10.17), respectively.

$$P_r A + A^T P_r - P_r B R_r^{-1} B^T P_r + Q_r = 0, \tag{10.16}$$

$$P_f A^T + A P_f - P_f C^T R_f^{-1} C P_f + Q_f = 0. \tag{10.17}$$

For the given test system, the weighting matrices for the standard LQG controller are chosen as $Q_r = \text{diag}(1, 1, 1, 10, 15, 5)$ and $R_r = 1$ to calculate the optimal gain. As IEEE Std. 1547 [13, 14] does not allow to regulate the voltage at the PCC actively, in this chapter, the location of the D-STATCOM is determined based on the weak bus oriented D-STATCOM placement planning to increase the system's Q-margin [29]. As, according to [29], the most suitable location for the D-STATCOM in this study is bus 15, which is the most critical bus in this

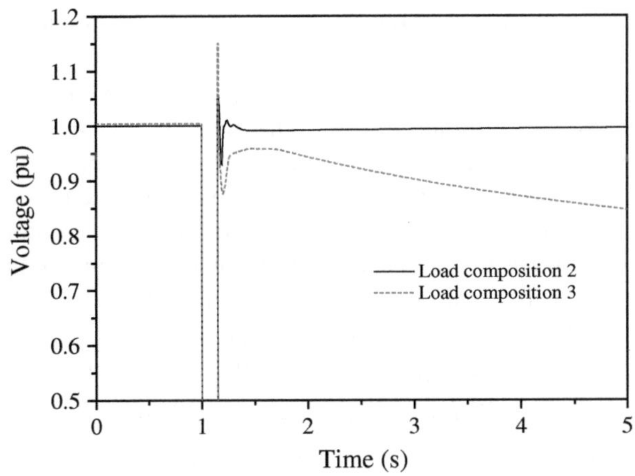


**Fig. 10.14** Performances of standard LQG controller for different load compositions

network, to evaluate the performance of the designed controller, simulations are carried out on the nonlinear model of the system by connecting the D-STATCOM at bus 15.

To demonstrate the performance of the standard LQG controller [30], the voltage profile of the system with the controller is shown in Fig. 10.14 for a sudden three-phase short-circuit fault of 150 ms at bus 2. From Fig. 10.14, it can be seen that the standard LQG controller performs well to recover the voltage within the utility's time frame for the nominal system load composition. As the standard LQG controller is designed for the nominal system (with load composition 2), it can ensure the performance for the plant for which it is designed. However, it cannot provide robust performance when the load composition changes (load composition 3), as depicted in Fig. 10.14. The results of stability simulations indicate that the performance of the controller can be drastically affected by the variations in load compositions which motivates the design of a robust controller for the D-STATCOM.

### *10.5.2 Norm-Bounded LQG Controller*

To facilitate the control design, the uncertain system can be represented as

$$\begin{aligned} \dot{x}(t) &= A(p)x(t) + B(p)u(t), \\ y(t) &= C(p)x(t), \end{aligned} \tag{10.18}$$

where $p \in \mathbb{R}^k$ denotes the vector of uncertain parameters; $A(p), B(p), C(p)$ are appropriately dimensioned matrices assumed to be affine functions of the parameter vector, $p$. The parameters $p_i, i = 1, 2, \ldots, k$, which are components of $p$, are assumed to lie in a convex polytopic region $\zeta$ in the parameter space, bounded by vertices $p^j, j = 1, 2, \ldots, l$.

During the design procedure of the controller, the system with load composition 2 is considered as the nominal system. The first step in the design of robust controller using the proposed method is to linearize the system in the region given by the nominal load, $p_0$ and changes in the system model due to variations in load compositions, $\Delta p$. From the simulations of the variations in load compositions, we obtain $p$, which is made up of the end-points of the region of interest. $A(p)$ is calculated for all possible combinations of these quantities. The corner points for $A(P)$ are given by $\bar{P}_{ls} = P_{ls0} + 0.15\,\text{pu}$, $\underline{P}_{ls} = P_{ls0} - 0.15\,\text{pu}$; $\bar{Q}_{ls} = Q_{ls0} + 0.15\,\text{pu}$, $\underline{Q}_{ls} = Q_{ls0} - 0.15\,\text{pu}$; $\bar{P}_{lf} = P_{lf0} + 0.05\,\text{pu}$, $\underline{P}_{lf} = P_{lf0} - 0.05\,\text{pu}$; $\bar{Q}_{lf} = Q_{lf0} + 0.05\,\text{pu}$, $\underline{Q}_{lf} = Q_{lf0} - 0.05\,\text{pu}$; $\bar{P}_{lm} = P_{lm0} + 0.20\,\text{pu}$, $\underline{P}_{lm} = P_{lm0} - 0.20\,\text{pu}$; $\bar{Q}_{lm} = Q_{lm0} + 0.20\,\text{pu}$, $\underline{Q}_{lm} = Q_{lm0} - 0.20\,\text{pu}$, where $P_{ls}$ and $Q_{ls}$ are the real and reactive power of static loads, respectively, $P_{lf}$ and $Q_{lf}$ are the real and reactive power of fluorescent lighting loads, respectively, and $P_{lm}$ and $Q_{lm}$ are the real and reactive power of induction motors, respectively, in the aggregated model.

In this study, the power supplied by the system, $P_s$, to the D-STATCOM to charge the capacitor is a nonlinear function of the control input, $\alpha$. Considering 10 % uncertainty in the input, corner points for $B(P)$ are given by $\bar{P}_s = P_{s0} + 0.1\text{pu}$, $\underline{P}_s = P_{s0} - 0.1\text{pu}$.

The output matrix $C$ is defined as $C = [0, 0, 0, 0, 0, 1]$. Considering 5 % uncertainty due to the measurement error in the output, the corner points for $C(p)$ are given by $\bar{V}_s = V_{s0} + 0.05\text{pu}$, $\underline{V}_s = V_{s0} - 0.05\,\text{pu}$.

The problem is to obtain a controller which minimizes the performance function (10.11) under the constraint (10.18). The approach considered here is to find the smallest upper bound on the $H_\infty$ norm of the uncertain system and then an optimal controller is designed with this bound.

Suppose that there is a positive-definite symmetric matrix $P$ and a scalar $\sigma > 0$ such that [22]

$$Z(p^j, P) := \begin{bmatrix} A(p^j)^T P + PA(P^j) + C(p^j)^T C(p^j) & PB(p^j) \\ \mathrm{B}(p^j)^T P & -\sigma I \end{bmatrix} \leq 0, \tag{10.19}$$

for $j = 1, 2, \ldots, l$.

To find the smallest upper bound on $G(p^j, s) = C(p^j)[sI - A(p^j)]^{-1} B(p^j)$, it is necessary to minimize $\sigma$.

If the inequality (10.19) is satisfied for all $p^j$ then,

$$\| G(p^j, s) \|_\infty \leq \gamma := \sqrt{\sigma}, \forall p \in \zeta$$

For the nominal system $(A, B, C)$, $\| G(s) \|_\infty \leq \gamma$, iff there is a symmetric positive-definite matrix $P$ such that the following inequality is satisfied.

$$\begin{bmatrix} A^T P + PA + C^T C & PB \\ B^T P & -\gamma^2 I \end{bmatrix} \leq 0. \tag{10.20}$$

The problem is to find an LQG controller whose infinity norm is bounded by a given number $\mu$ [22],

$$\| \mathrm{G_c(s)} \|_\infty \leq \mu, \text{such that } 0<\mu<1/\gamma,$$

where $G_c(s) = C_c(sI - A_c)^{-1}B_c$.

Without loss of generality (Appendix-II), to reduce computational complexity, an equivalent condition (10.25) is derived which is in the form of inequality (10.20) with $P = I$ and the resulting transformed system takes the form of equation (10.24).

Suppose that [22] the LQG performance function weights are such that $Q_r > 0$ and

$$R_r \geq I, \tag{10.21}$$

$$Q_r > \mu^{-2} P_r \hat{C}^T R_f^{-2} \hat{C} P_r - \hat{C}^T R_f^{-1} \hat{C} P_r - P_r \hat{C}^T R_f^{-1} \hat{C}, \tag{10.22}$$

$$Q_f = -(\hat{A} + \hat{A}^T) + \hat{C}^T R_f^{-1} \hat{C}. \tag{10.23}$$

Then, the controller is norm bounded by $\mu$ $(0<\mu<\frac{1}{\gamma})$, if (10.21)–(10.23) are satisfied.

### 10.5.3 Design Steps

The design steps of the LQG optimal norm-bounded controller for the nominal system are given below:

Step 1: Obtain the region of interest by varying load compositions in the composite load.

Step 2: Find the smallest upper bound $\gamma$ on the norm of the uncertain system so that the inequality (10.19) is satisfied.

Step 3: Obtain a solution $P$ to the linear matrix inequality (10.20) for the nominal system and transform the system matrices using the similarity transformation, $z = \Gamma x$.

Step 4: Choose the LQR weighting matrices $R_r \geq I$ and $Q_r > 0$ based on the performance requirements and choose the scalar $\mu$, $0<\mu<\frac{1}{\gamma}$.

Step 5: Solve equation (10.16) for $P_r$ and choose $R_f$ such that (10.22) is satisfied.

Using the above steps, the required controller is given by equations (10.14) and (10.15) with $P_f = I$.

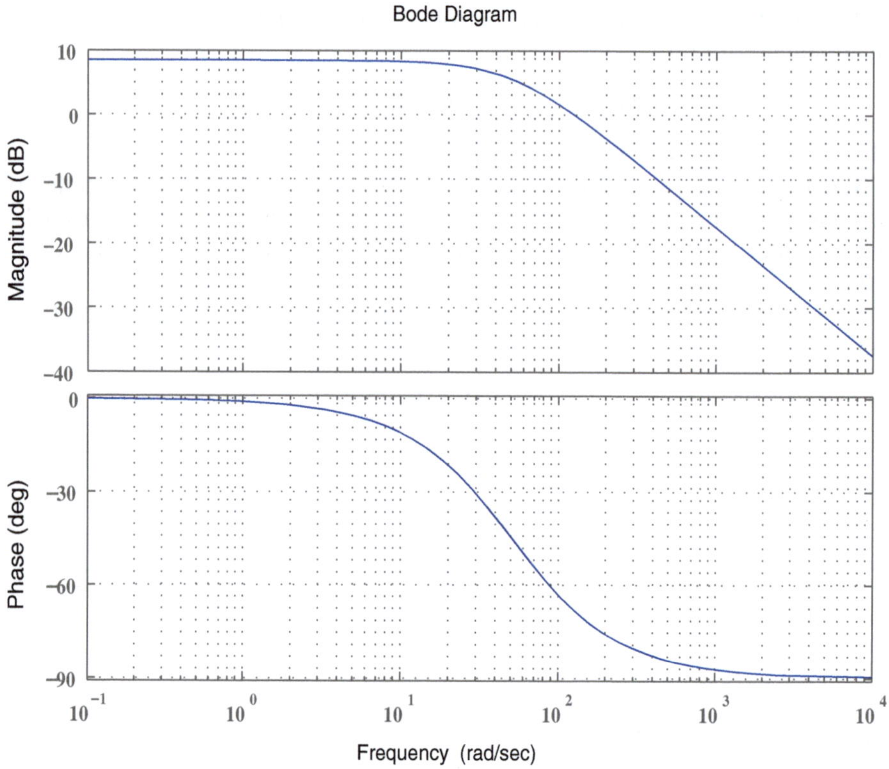


**Fig. 10.15** Bode plot of robust D-STATCOM controller—firing angle versus terminal voltage

## 10.6 Performance Evaluation

In order to improve the system's performance under varying operating conditions, a robust LQG controller is designed according to the Steps 1–5 described in Subsection 10.5.3. For the proposed approach, the smallest $H_\infty$ norm-bound is found to be $\gamma = 1.15$. A norm-bounded controller is designed to have an $H_\infty$ norm less than $1/\gamma$. The weighting matrices are chosen as

$$Q_r = \text{diag}(1, 1, 1, 12, 20, 10), R_r = 1.5, R_f = 1.$$

$Q_r$ and $R_r$ are chosen so as to give good regulator performance and $R_f$ is chosen so as to satisfy (10.22). There is a considerable freedom in choosing the regulator weighting matrices $Q_r$ and $R_r$, but limited freedom in choosing the filter weighting matrices $Q_f$ and $R_f$. This is to be expected because the objective is to limit the overall controller gain which consists of the product of the LQR and KBF gains. A suitable balance between the LQR and KBF gains can be obtained by adjusting $Q_r$ and $R_r$ in such a way as to permit a smaller $R_f$ [22].

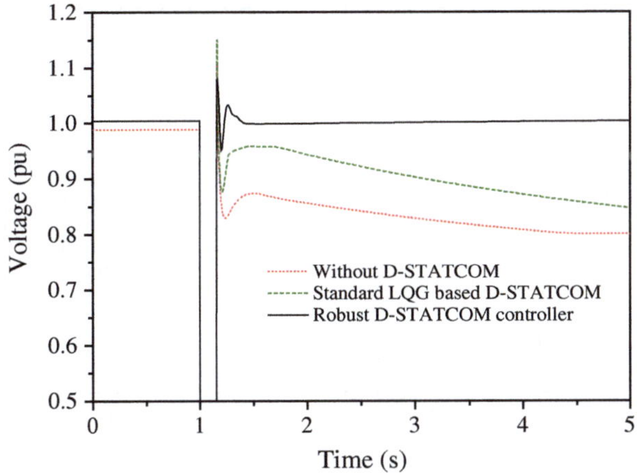


**Fig. 10.16** Voltage at PCC (bus 13) for load composition 3 (80 % induction motor in composite load model)

The frequency response of the proposed controller in Fig. 10.15 signifies that it provides adequate bandwidth and margins to stabilize the system under different operating conditions. The closed-loop eigenvalue of the dominant mode is found at −4.512. To examine the robustness of the proposed controller, the following contingencies are applied:

- a symmetrical three-phase short-circuit fault close to the substation; and
- an asymmetrical fault close to the DG unit.

### *10.6.1 Contingency I: Three-Phase Short-Circuit Fault Close to Substation*

In the studies of cases involving a short-circuit fault, the load characteristics during a fault might deserve far more attention as bus voltages are depressed and eventually affect the stability of the system. In the demonstrated test system, main source of reactive power support to the load is the substation located at bus 1. To test the controller performance under a severe disturbance, the system is operated with the load composition 3 and a three-phase short-circuit fault is applied near the substation at bus 2 for 150 ms. From Fig. 10.16, it can be observed that although the standard LQG controller fails to operate for a change in the load composition, the proposed $H_\infty$ norm-bounded LQG controller provides the grid code-compatible performance due to the incorporation of load uncertainties in the design process. It can be also observed from Fig. 10.16 that the system with the robust D-STATCOM causes less transient over-voltage than the case without a D-STATCOM. Voltages at the different load buses under the same disturbance

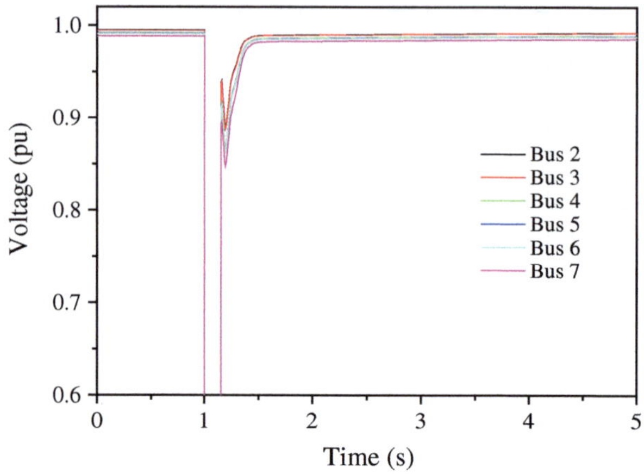


**Fig. 10.17** Voltages at different load buses of system with robust D-STATCOM

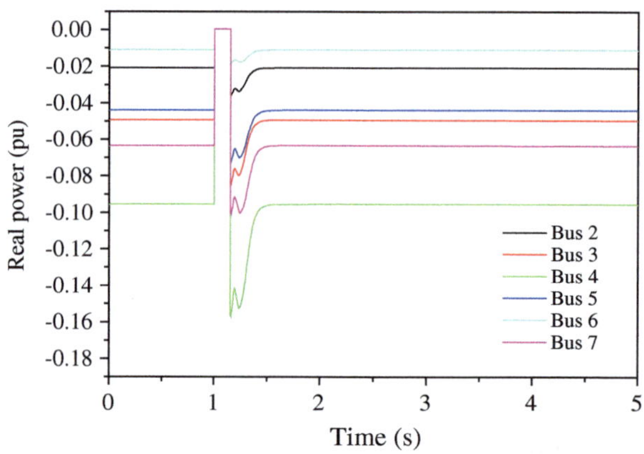


**Fig. 10.18** Real power drawn by loads at different buses with robust D-STATCOM

with the proposed control scheme can be seen from Fig. 10.17. Real and reactive power drawn by the load before, during and after the disturbance are given in Figs. 10.18 and 10.19, respectively. In all cases, the system returns to pre-fault condition within a short-time. The outputs from the D-STATCOM for the standard LQG and robust LQG controller are shown in Fig. 10.20 in which it can be observed that the robust LQG controller supports reactive power to the system instantaneously following a disturbance, thereby improving voltage performance. The results obtained for the fault condition reveal superior dynamic performance and fast fault recovery of the designed D-STATCOM.

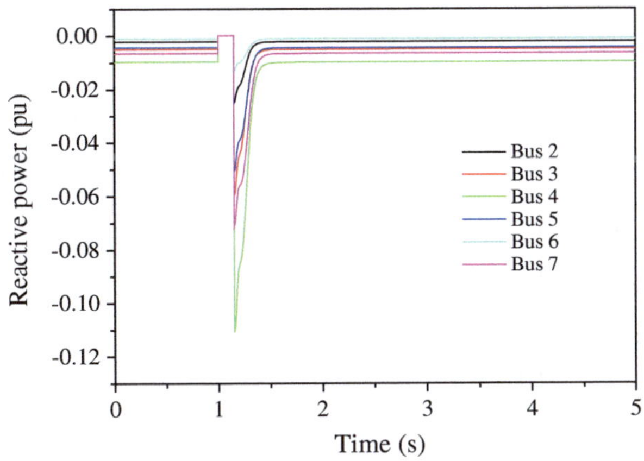


**Fig. 10.19** Reactive power consumed by loads at different buses with robust D-STATCOM

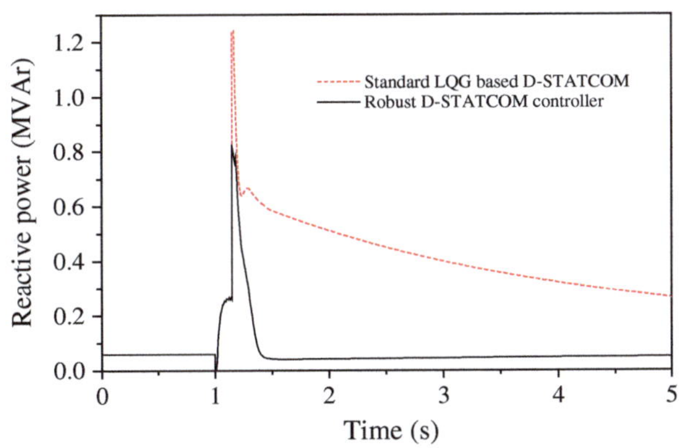


**Fig. 10.20** Reactive power output of D-STATCOM (sudden three-phase fault at bus 2)

### *10.6.2 Contingency II: Asymmetrical Fault Close to DG Unit*

To observe the effect of an unbalanced fault, a single-line-to-ground fault is applied in the middle of the line connecting the buses 12 and 13. In this case, the system is operated with 70 % induction motor loads, 10 % fluorescent lamps and 20 % static loads in the composite load model. The fault is applied at 1 s and cleared after 150 ms. The voltage profile of the system at the PCC is shown in Fig. 10.21. It can be investigated form Fig. 10.21 that although the controller is designed for the nominal system, it performs well for a variation in the system model and the original system is restored after the clearance of the fault. Reactive power consumed by the load at bus 12 is shown in Fig. 10.22. The robust D-STATCOM

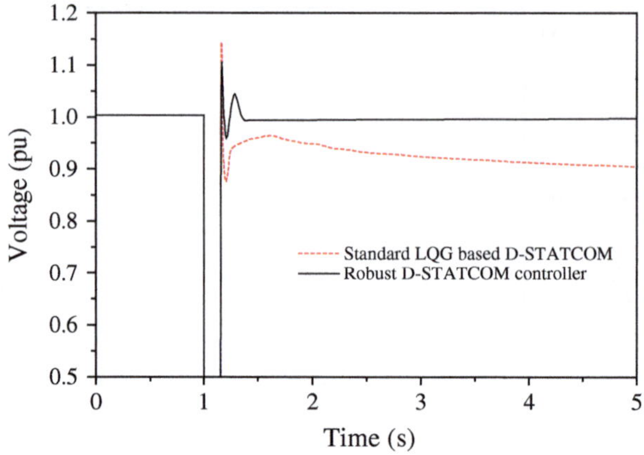


**Fig. 10.21** Voltage at PCC (bus 13) for asymmetrical fault (70 % induction motor in composite load model)

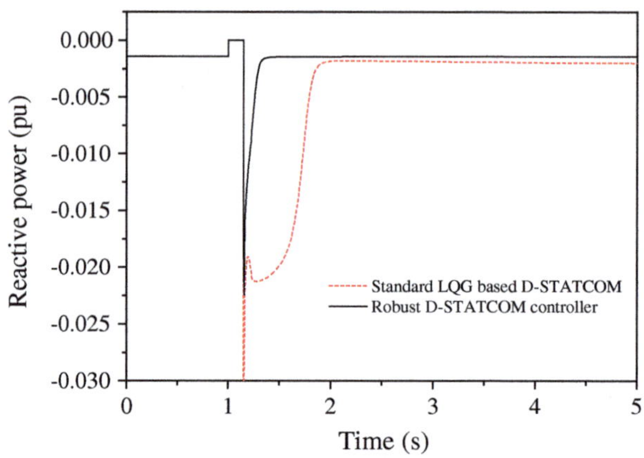


**Fig. 10.22** Reactive power consumed by load at bus 12

injects reactive power to the load following a disturbance and the system voltage starts to recover. The applied control effort $\alpha$ by the designed controller to stabilize the system can be observed in Fig. 10.23.

The above investigations demonstrate that the proposed controller is robust against changes in operating conditions and stabilizes the system under large disturbances to ensure DG units remain connected to the network.

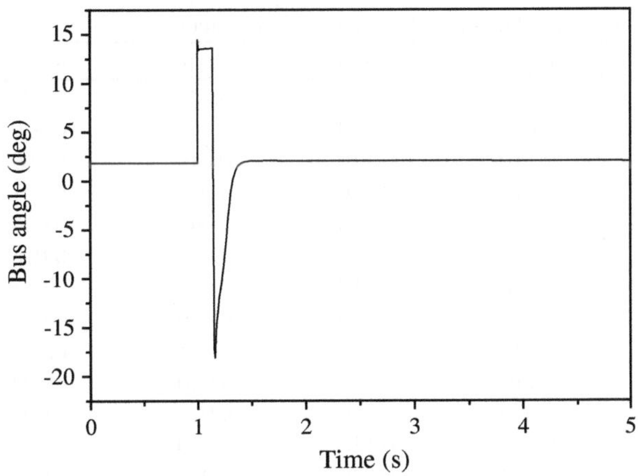


**Fig. 10.23** Control effort ($\alpha$) with designed controller for single-line-to-ground fault in middle of line 12−13

**Table 10.6** PV system data

| Number of PV cells per module | 54 |
|---|---|
| Number of parallel modules in each row | 150 |
| Module current rating | 2.8735 A |
| Module voltage rating | 43.5 V |
| Voltage temperature coefficient | −0.1 V/K |
| Current temperature coefficient | 0.003 A/K |
| Diode ideality factor | 1.3 |
| VSC switching frequency | 3060 Hz |
| On-state resistance of VSC valves | 1 mΩ |
| Solar irradiance, G | 1.0 kW/m$^2$ |
| Series resistance, $R_s$ | 0.0819 Ω |
| Shunt resistance, $R_{sh}$ | 72 kΩ |
| PV inductance, $L_{pv}$ | 1.0 μH |
| PV capacitance, $C_{pv}$ | 10 mF |
| DC-link resistance, $R_{dc}$ | 0.1 Ω |
| DC-link inductance, $L_{dc}$ | 1.0 mH |
| DC-link capacitance, $C_{dc}$ | 400 μF |
| Transformer ratio, N | 15 |

**Table 10.7** Line and load data of 16-bus distribution system

| SE | RE | $R$ (pu) | $X$ (pu) | $P_l$ (pu) | $Q_l$ (pu) | $C$ (pu) |
|---|---|---|---|---|---|---|
| 1 | 8 | 0.110 | 0.110 | 0.040 | 0.027 | 0.000 |
| 8 | 9 | 0.080 | 0.110 | 0.050 | 0.030 | 0.012 |
| 8 | 10 | 0.110 | 0.110 | 0.010 | 0.009 | 0.000 |
| 9 | 11 | 0.110 | 0.110 | 0.006 | 0.001 | 0.006 |
| 9 | 12 | 0.080 | 0.110 | 0.045 | 0.020 | 0.037 |
| 2 | 4 | 0.075 | 0.110 | 0.020 | 0.016 | 0.000 |
| 4 | 5 | 0.080 | 0.110 | 0.030 | 0.015 | 0.011 |
| 4 | 6 | 0.090 | 0.180 | 0.020 | 0.008 | 0.012 |
| 6 | 7 | 0.040 | 0.040 | 0.015 | 0.012 | 0.000 |
| 3 | 13 | 0.110 | 0.110 | 0.010 | 0.009 | 0.000 |
| 13 | 14 | 0.090 | 0.120 | 0.010 | 0.007 | 0.018 |
| 13 | 15 | 0.080 | 0.110 | 0.010 | 0.009 | 0.000 |
| 15 | 16 | 0.040 | 0.040 | 0.021 | 0.010 | 0.018 |
| 5 | 11 | 0.040 | 0.040 | 0.000 | 0.000 | 0.000 |
| 10 | 14 | 0.040 | 0.040 | 0.000 | 0.000 | 0.000 |

(Base power is 100 MVA and base voltage is 23 kV, SE and RE are the sending and receiving end nodes, respectively)

**Table 10.8** Line and load data of 15-bus Kumamoto distribution system

| SE | RE | $R$ (pu) | $X$ (pu) | $B$ (pu) | $P_l$ (pu) | $Q_l$ (pu) |
|---|---|---|---|---|---|---|
| 1 | 2 | 0.003145 | 0.075207 | 0.00000 | 0.02080 | 0.0021 |
| 2 | 3 | 0.000330 | 0.001849 | 0.00150 | 0.04950 | 0.0051 |
| 3 | 4 | 0.006667 | 0.030808 | 0.03525 | 0.09580 | 0.0098 |
| 4 | 5 | 0.005785 | 0.014949 | 0.00250 | 0.04420 | 0.0045 |
| 5 | 6 | 0.014141 | 0.036547 | 0.00000 | 0.01130 | 0.0012 |
| 4 | 7 | 0.008001 | 0.036961 | 0.03120 | 0.06380 | 0.0066 |
| 7 | 8 | 0.008999 | 0.041575 | 0.00000 | 0.03230 | 0.0033 |
| 8 | 9 | 0.007000 | 0.032346 | 0.00150 | 0.02130 | 0.0022 |
| 9 | 10 | 0.003666 | 0.016940 | 0.00350 | 0.02800 | 0.0029 |
| 10 | 11 | 0.008999 | 0.041575 | 0.00200 | 0.21700 | 0.0022 |
| 3 | 12 | 0.027502 | 0.127043 | 0.00000 | 0.01320 | 0.0014 |
| 12 | 13 | 0.031497 | 0.081405 | 0.00000 | 0.00290 | 0.0003 |
| 13 | 14 | 0.039653 | 0.102984 | 0.00000 | 0.01610 | 0.0016 |
| 14 | 15 | 0.016070 | 0.004153 | 0.00000 | 0.01390 | 0.0014 |

(Base power is 10 MVA and base voltage is 6.6 kV, SE and RE are the sending and receiving end nodes, respectively.)

## 10.7 Conclusions

In this chapter, the impacts of solar PV generators and composite loads on distribution systems are analyzed and a novel control methodology is proposed to ensure grid code-compatible performances of PV generators connected to distribution networks. Static analyses show that the loss of multiple solar PV generators, overloading and line outages from a distribution network can be a risk for the system stability. Dynamic analyses illustrate that the proportion of induction motor load in the composite load model has a significant impact on the stability of a distribution network. It is investigated that a high percentage of induction motors in the composite load model affects the dynamic voltage stability of the system. A novel controller for a D-STATCOM is proposed to solve this problem. The proposed control scheme determines the smallest upper bound on the $H_\infty$ norm of the uncertain system. The effectiveness of the proposed controller is verified through time-domain simulation studies conducted on a widely used test system. The results show that the distributed PV system with the proposed robust D-STATCOM controller maintains its stability despite major variations in load compositions. As, in a practical system, there are always continual variations in the load compositions itself, the proposed control method provides a good reflection of the reality for supporting DG integration.

## Appendix-I

The data of the PV system is given in Table 10.6. D-STATCOM parameters: 1.0 MVA, $C_s$= 300 μF, $R_{cs}$= 0.01 pu. The data of 16-bus and 15-bus distribution test systems are given in Table 10.7 and Table 10.8, respectively.

## Appendix-II

Suppose that $\Gamma \in \mathbb{R}^{n\times n}$ is a square root of $P$, i. e., $P = \Gamma^T \Gamma$. Using the coordinate transformation ($\xi = \Gamma x$), the system becomes

$$\dot{\xi} = \hat{A}\xi + \hat{B}u, \qquad y = \hat{C}\xi, \tag{10.24}$$

where $\hat{A} = \Gamma A \Gamma^{-1}, \hat{B} = \Gamma B, \hat{C} = C\Gamma^{-1}$.

Pre-multiplying (10.20) by $diag[\Gamma^{-T}, I]$ and post-multiplying it by its transpose, we have [22]

$$\begin{bmatrix} \hat{A}^T + \hat{A} + \hat{C}^T\hat{C} & \hat{B} \\ \hat{B}^T & -\gamma^2 I \end{bmatrix} := -\begin{bmatrix} Q_A & Q_{12} \\ Q_{12}^T & R \end{bmatrix} \leq 0. \tag{10.25}$$

Therefore, without loss of generality, it is assumed that the system is in this form [i.e., it satisfies the inequality (10.20) with $P = I$].

## References

1. Renewable Energy Policy Network for 21st Century, Renewables 2010 global status report. http://www.ren21.net. Accessed 22 June 2013
2. Sengupta M, Keller J (2012) PV ramping in a distributed generation environment: a study using solar measurements. In: IEEE Photovolt Spec Conf pp 586–589
3. Schauder C (2011) Impact of FERC 661-A and IEEE 1547 on photovoltaic inverter design. In: IEEE PES General Meeting pp 1–6
4. Molina MG, Mercado PE (2008) Modeling and control of grid-connected photovoltaic energy conversion system used as a dispersed generator. In: IEEE/PES Transmission and Distribution Conference and Exposition: Latin America, pp 1–8
5. Tan YT, Kirschen DS, Jenkins N (2004) A model of PV generation suitable for stability analysis. IEEE Trans Energy Convers 19(4):748–755
6. Rodriguez C, Amaratunga GAJ (2004) Dynamic stability of grid-connected photovoltaic systems.In: IEEE Power Engineering Society General Meeting pp 2193–2199
7. Eltawil MA, Zhao Z (2010) Grid-connected photovoltaic power systems: technical and potential problems—a review. Renew Sustain Energy Rev 14(1):112–129
8. Atwa YM, El-Saadany EF (2009) Reliability evaluation for distribution system with renewable distributed generation during islanded mode of operation. IEEE Trans Power Systems 24(2):572–581
9. IEEE Task Force (1993) Load representation for dynamic performance analysis. IEEE Trans on Power Systems 8(2):472–482
10. Concordia C, Ihara S (1982) Load representation in power system stability studies. IEEE Trans Power Apparatus Syst PAS-101(4): 969–977
11. Roy NK, Hossain MJ, Pota HR (2011) Effects of load modeling in power distribution system with distributed wind generation.In: 21st Australasian Universities Power Engineering Conference pp 1–6
12. Kundur P (1994) Power system stability and control. McGraw-Hill, New York
13. IEEE Std. 1547 (2003) IEEE standard for interconnecting distributed resources with electric power systems
14. IEEE Std 1547.2 (2008) IEEE application guide for IEEE Std 1547, IEEE standard for interconnecting distributed resources with electric power systems
15. Mahmud MA, Pota HR, Hossain MJ, Roy NK (2014) Robust partial feedback linearizing stabilization scheme for three-phase grid-connected photovoltaic systems. IEEE J Photovolt 4(1): 423–431
16. IEEE Power and Energy Series—Song YH, Johns AT(eds) (1999) Flexible AC transmission systems (FACTS). The Institution of Electrical Engineers, London
17. Hingorani NG, Gyugyi L (2000) Understanding FACTS-concepts and technology of flexible AC transmission systems. IEEE Press, New York
18. Yunus AMS, Masoum MAS, Abu-Siada A (2011) Effect of STATCOM on the low-voltage ride-through capability of Type-D wind turbine generator. In: Innovative Smart Grid Technologies Asia pp 1–5
19. Yang K, Cheng X, Wang Y et al (2012) PCC voltage stabilization by D-STATCOM with direct grid voltage control strategy. In: Innovative Smart Grid Technologies Asia pp 1–5
20. Rao P, Crow ML, Yang Z (2000) STATCOM control for power system voltage control applications. IEEE Trans Power Delivery 15(4):1311–1317
21. Seo JC, Kim TH, Park JK , Moon SI (1996) An LQG based PSS design for controlling the SSR in power systems with series-compensated lines. IEEE Trans Energy Conversion 11(2):423–428
22. Joshi SM, Kelkar AG (2002) Design of norm-bounded and sector-bounded LQG controllers for uncertain systems. J Optim Theory Appl 113(2):269–282
23. Civanlar S, Grainger JJ, Yin H , Lee SSH (1988) Distribution feeder reconfiguration for loss reduction. IEEE Trans Power Delivery 3(3):1217–1223

24. Taylor CW (1994) Power system voltage stability. McGraw-Hill, New York
25. Li S, Tomsovic K, Hiyama T (2000) Load following functions using distributed energy resources. In: IEEE Power Engineering Society Summer Meeting pp 1756–1761
26. Hossain MJ, Saha TK, Mithulananthan N, Pota HR (2012) Robust control strategy for PV system integration in distribution systems. Appl Energy 99:355–362
27. Milano F (2010) Power system modeling and scripting. Springer-Verlag, London
28. Hossain MJ, Pota HR, Ramos RA (2011) Robust STATCOM control for the stabilisation of fixed-speed wind turbines during low voltages. Renewable Energy 36(11):2897–2905
29. Roy NK, Pota HR, Hossain MJ (2013) Reactive power management of distribution networks with wind generation for improving voltage stability. Renewable Energy 58:85–94
30. Anderson BDO, Moore JB (1990) Optimal control: linear quadratic methods. Prentice-Hall, New Jersey

# Chapter 11
# Integrating Smart PHEVs in Future Smart Grid

**F. R. Islam and H. R. Pota**

**Abstract** In a smart power network, PHEVs can act as either loads or distributed sources of energy. The two terms most commonly used to describe the interconnection of a power network and electric vehicle are 'Grid-to-Vehicle (G2V)' and 'Vehicle-to-Grid (V2G)'. When electric vehicles are connected into the grid to recharge their batteries or supply energy to it, they act as loads known as the G2V or V2G modes of operation respectively. This chapter reviews the impact of implementing the G2V mode, and the benefits and drawbacks of, and strategies for, the V2G interfacing of individual vehicles with a PHEV park. The performance of a power system network can be improved using V2G technology, which offers reactive power support, power regulation, load balancing, and harmonics filtering, which in turn, improve its quality, efficiency, reliability and stability. To implement V2G technology, a power network might require significant changes in its structure, components and controls, the issues for which include battery life, the need for concentrated communication between vehicles and the grid, the effects on distribution accessories, infrastructure changes, and social, political, cultural and technical concerns. As storage is essential for a power system, distributed electric vehicles can be an economical storage solution if it has a good plan for buying and selling its energy. Bidirectional power flow technologies of V2G systems need to be addressed and the economic benefits of V2G technologies depend on vehicle aggregation and G2V/V2G strategies. In the future, it is expected that their benefits will receive greater attention from grid operators and vehicle owners.

**Keywords** Smart Grid · Renewable energy · PHEV · V2G · G2V

F. R. Islam (✉) · H. R. Pota
School of Engineering and Information Technology, The University of New South Wales, PO Box 7916 Canberra BC, ACT 2610, Australia
e-mail: fr.islam81@yahoo.com.au

H. R. Pota
e-mail: h.pota@adfa.edu.au

J. Hossain and A. Mahmud (eds.), *Renewable Energy Integration*,
Green Energy and Technology, DOI: 10.1007/978-981-4585-27-9_11,

**Table 11.1** Charging power levels

| Power level | Description | Power level |
|---|---|---|
| Level 1 | Opportunity charger (any available outlet) | 1.4 kW (12 A)<br>1.9 kW (20 A) |
| Level 2 | Primary dedicated charger | 4 kW (17 A)<br>19.2 kW (80 A) |
| Level 3 | Commercial fast charger | Up to 100 kW (12 A) |

## 11.1 Introduction

Due to environmental and climate issues, along with the rising cost of petroleum, energy security and limited reserves of fossil fuels [1–3], PHEV technology has become of increasing interest. However, it is in an early stage of development and faces a few problems before it can be adopted worldwide, such as technical limitations, sociocultural obstacles and the fact that PHEVs currently cost more than conventional vehicles [4]. According to the EPRI, penetration of PHEVs into the USA's vehicle market will be 35 % by the year 2020 [5]. To attain a stable and versatile interfacing between a grid and PHEVs, standards and codes for system requirements are developed by various organizations, such as the automotive sector, the IEEE, the Society of Automotive Engineers (SAE) and the EPRI. In this dissertation, PHEVs are chosen for analysis as they have a few advantages over hybrid electric vehicle (HEV) and internal combustion engine (ICE) vehicles as they can act in the discharge mode as V2G devices and in the charging mode as G2V devices [6]. This chapter reviews V2G/G2V technologies on grids and customer requirements, cost analysis, challenges and policies for V2G interfaces of both individual PHEVs and vehicle fleets. To assess the impacts and utilization of PHEVs in utility distribution or transmission networks, their controls and usage prototypes need to be evaluated. The SAE has defined three levels of charger for PHEVs [7], as summarized in Table 11.1. A PHEV behaves as a load when it needs to recharge its battery in the G2V mode and as a generator when a utility grid takes power from its battery in the V2G mode of operation. Its recharging and discharging characteristics depend on a few factors, such as its geographical location, the number of PHEVs in that particular area, its charging levels (charging current and voltage), battery state and capacity, and the connection type used (unidirectional or bidirectional) [8, 9].

## 11.2 Impact of G2V on Grid

First-generation mass-market PHEVs, such as the Chevrolet Volt and Nissan Leaf [10, 11], connect to the grid for only battery charging, which is the most basic configuration. G2V includes conventional and fast battery charging systems, and the latter can stress a grid distribution network because its power is high, as a typical PHEV requires more than double an average household's load [25]. Charging practices in different locations also have an effect on the amount of power taken from an electric grid by a fleet of PHEVs; for example, charging at work in congested urban centres can lead to undesirable peak load [12], which could require significant investments in expensive peak generation. Injected harmonics and a low power factor can be serious problems if the charger does not employ a state-of-the-art conversion for charging PHEVs at night, which has minimal impact on the power grid given suitable choices for intelligent controls [5, 13–17]. The increasing exploitation of PHEVs is still a topical area of research. One of the foremost recent studies of smart-grid development with PHEVs is [18] that recognized the complexity of studying the impact of PHEVs on a smart grid, with the results depending on many factors (power level, timing, duration of PHEV connection to the grid) and possibly affecting several variables (capacity needs, emissions generated). As mentioned above, as a charging PHEV may present a load to an electrical grid twice the order of magnitude of that of a typical home, connecting it may create power quality problems, such as momentary voltage drops. An interesting point about the simulations in [19], which assumed no control over the charging of vehicles, is that they showed voltage drops between 5 and 10.3 % depending on the time of day and season. The simulation results showed how the voltage supplied to a house changed without and with PHEV charging where, for the latter, the drop in it increased from 1.7 to 4.3 % while, for the former, much more random behaviour was exhibited with average voltage drops of around 4 % (although this eventually reached a value close to 1.7 % once the PHEV was charged). These results point to a need to improve the quality of electrical energy delivery by utilizing smart technology to coordinate the charging of PHEVs. The study in [19] was based on simulations using residential power consumption profiles. The shorter period of AC power consumption became a switched load, with levels similar to what would be expected to be observed in PHEVs with Level 2 charging profiles, that is, the notion that the grid had not experienced high-magnitude loads with a random switching profile was not true. This switching high-power AC consumption profile was masked by its aggregated effect on the grid. One other important aspect when studying the effect of PHEVs on an electric grid is the grid's stability for which damping components will need to be introduced into future designs for controlling PHEV charging.

## 11.3 V2G Technology

V2G describes a system in which PHEVs communicate with the power grid to sell demand response services by delivering electricity into the grid or throttling their charging rate. PHEVs can serve as stored and distributed energy resources as well as reserves for unexpected outages when they have proper on-board power electronics, smart connections to the grid and inter-active charger hardware control [14, 20–23]. A bidirectional charging system is essential to support energy injection into the grid [24–27], as a unidirectional charger, although simple and easy to use in terms of control, can be used only for a G2V system.

A smart charging system and proper management can shift loads and avoid peaks while a proper controller can minimize the impact of PHEVs on the utility grid [16, 17, 28]. Smart metering, communication and control systems play important roles in the direct coordination of the V2G and G2V modes of operation. The real-time, nonlinear pricing of a utility bill is one of the important factors for obtaining higher returns from grid-connected PHEVs [29]. In the V2G mode, interconnection between the grid and vehicle is essential and an individual vehicle or even a fleet of PHEVs can take part as spinning reserves for the grid. A group of cars in a park is more convenient to manage as a load for the grid [30] and more helpful as it can work as a distributed energy resource when necessary [31]. The potential benefits and economic issues of V2G technology are of great concern for researchers nowadays [28–47]. Another current issue is the use of RESs in a power network as, due to their sporadic natures, they need storage devices for which a PHEV's battery can be a solution as it offers the opportunity to store wind and solar energy at times of excess generation and provide possible backup when necessary [6, 32–34]. The implementation of V2G technology has been explored in a number of ways through different research, such as for reactive power support [28], active power regulation, load balancing by valley filling [35, 36, 48], and peak load shaving [37, 38]. These systems can enable such ancillary services as frequency control and spinning reserves [6, 30, 39–42], improve grid efficiency, stability, reliability [43] and generation dispatch [44], and reduce utility operating costs and, potentially, even generate revenue [29]. In addition, PHEVs owners benefit when electricity is cheaper than fuel for equivalent distances. Researchers have estimated that potential net returns from V2G methods range from $90 to 4,000 per year per vehicle based on the power capacity of electrical connections, market value, PHEV penetration and the energy capacity of the PHEV battery [6, 19, 20, 45–47]. Besides the intrinsic benefits of PHEVs, emissions have been reduced [17, 49, 50], and it has been reported that V2G strategies have the potential to displace the equivalent of 6.5 million barrels of oil per day in the USA [37]. Peterson et al. estimated the annual net social welfare benefits from the grid to be $300–400 [51]. The design of a power system filter is one more option for using V2G technology, which in this dissertation, is described so that its implementation makes the grid smarter.

The expected increase in the number of electric vehicles produced could have a significant impact on the potential for utility-related energy storage as these vehicles can provide some of its benefits. Specifically, it may be cost-effective to charge electric vehicles when energy prices are low and then dispatch the power from them to support the grid, especially during grid emergencies. Using electric vehicles as distributed storage is an important complement to the expected increase in intermittently RESs, such as solar and wind power outputs, which are sometimes produced when the energy demand and price are low and can change rapidly [52]. If every suburb in Australia installed just one vehicle to a grid recharge point, Australia's V2G coverage would look like that in Fig. 11.1 [53].

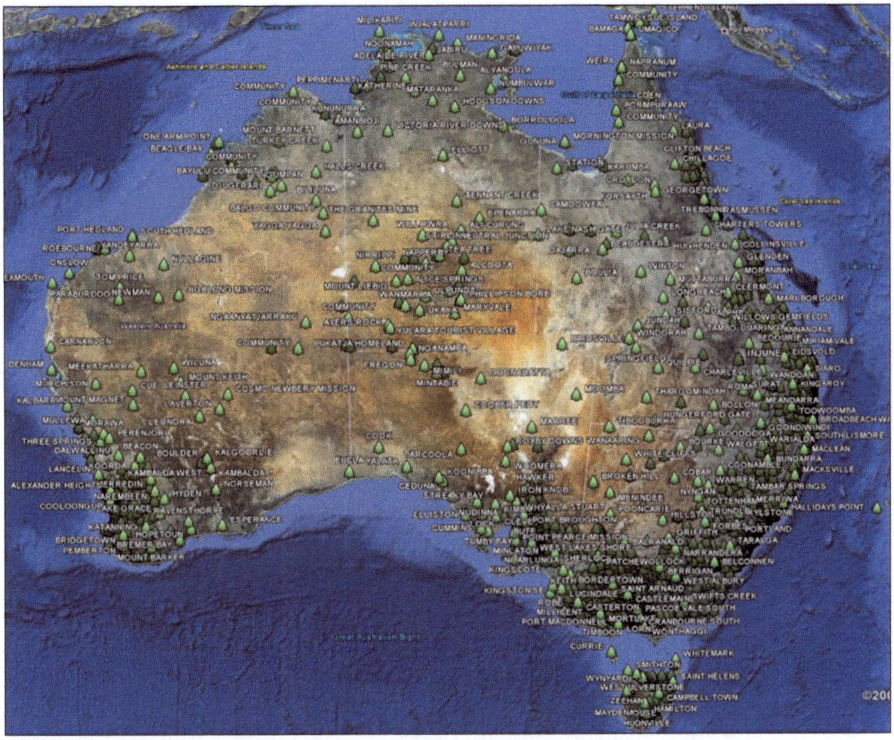

**Fig. 11.1** Potential V2G coverage in Australia

## 11.4 A Simple Structure of V2G System

Pang et al. summarized the requirements for a simple V2G structure with energy resources and an electrical utility as being an independent system operator and aggregator, a charging infrastructure and locations, a bidirectional electrical energy flow and communication between each PHEV and the aggregator, on-board and off-board intelligent metering and control, and the PHEV's battery charger and management [54]. In short, a power connection with a grid, suitable metering and control with effective communication can build a V2G system [55]. Figure 11.2 [56] shows a simple V2G system structure and Fig. 11.3 the power flows within the charger. In general, although communications must be bidirectional to report a battery's status and receive control commands [57, 58], achieving intelligent metering and control that are aware of a battery's capacity and state-of-charge (SOC) is challenging [6, 59–61]. Both on-board and off-board smart meters have been proposed to support V2G methods [28, 48, 62], smart metering can make PHEVs into controllable loads to help combine them with renewable energy [63], GPS locators and on-board meters are useful [43, 62] while sensors and smart meters on charging stations can monitor and exchange in- formation with the

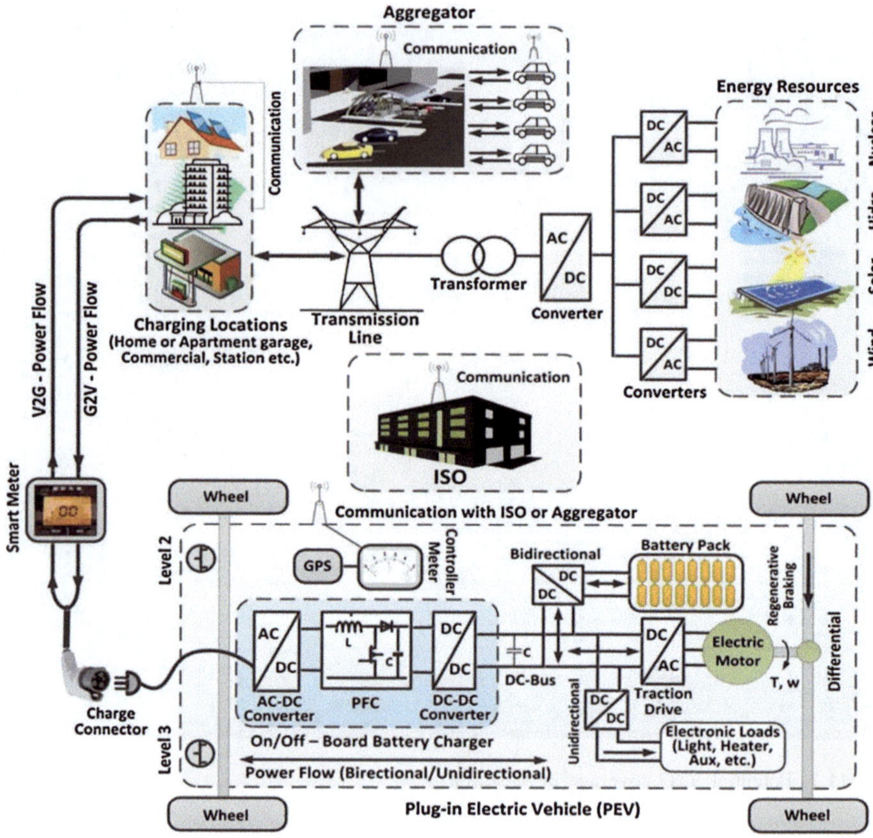


**Fig. 11.2** The components and power flow of a V2G system

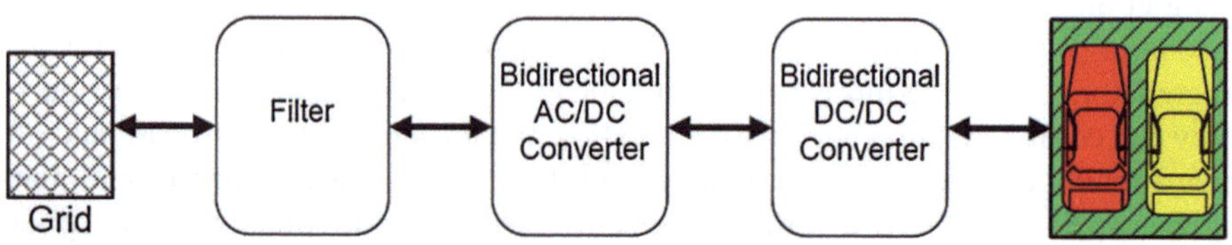


**Fig. 11.3** General unidirectional and bidirectional power flow topology

relevant control centre through a field area network [55]. Also, control and communication are essential for services such as dynamic adjustments that track intermittent resources and alter charging rates to track power prices, frequency or power regulation, and spinning reserves [6, 64–67], for which a variety of protocols have been discussed, including Bluetooth, Home-Plug, Z-Wave and ZigBee, [68–72]. In US, the IEEE and SAE provide the necessary communications requirements and specification [73–75] while the National Electric Infrastructure

Working Council (IWC) has defined a communications standard to enable PHEVs to communicate with chargers [76, 77]. PHEV chargers without state-of-the-art power electronics can produce deleterious harmonic effects on a distribution system [78]. The IEEE-519 [79], IEEE-1547 [80], SAE-2894 [81] and International Electrotechnical Commission's IEC-1000-3-6 [79, 81] standards limit the allowable harmonic and DC current injections into the grid with which PHEV chargers are usually designed to full fill. Sophisticated active power converter technology has been developed to reduce harmonic currents and provide a high power factor [25, 82–85], while shock hazard risk reduction for PHEV charging is addressed in the standard for personnel protection systems for PHEV supply circuits [86].

## 11.5 PHEV as Source of Stored Energy for Distribution Networks

A single PHEV's battery storage capacity is small relative to that of the grid. However, the better coordination and reliability of a smart grid can be achieved by aggregating PHEVs as storage devices [23, 30, 87]. An aggregator can be a communication or controller device, or an algorithm that plays an effective role between PHEV owners, the electricity market and distribution and transmission system operators [88–90]. Both aggregated vehicles and the grid need to be properly controlled to maintain the stability of the grid [91]. Figure 11.2 shows an aggregator in a V2G system. One of its ma jor roles is to manage PHEVs to operate in the V2G mode whenever the grid needs power [92]. Each PHEV can be contracted for this service in a cost-effective way by an aggregator that understands its battery's SOC condition [43, 93, 94]. In an aggregated smart grid environment, vehicles can engage and disengage while performing ancillary services of the grid [64] and maintaining the maximum and minimum contract limits. Considering each vehicle as an individual decision maker and the aggregator as the coordinator, Wu et al. proposed a method of smart pricing and optimal frequency regulation [95]. Another optimal frequency regulation controller for a V2G aggregator was designed by Han et al. [96] while the western Danish power system was used for the long-term aggregation of PHEVs in [97]. In the industrial networks MOBIE [98] and Better Place [99], the aggregation concept was successfully implemented, and it was found that control and communication with individual vehicles was much difficult than with the aggregator [67].

## 11.6 Benefits of V2G System

PHEVs can support V2G mode of operation because on average, in the USA they travel on the road for only 4–5 % of the day while sitting in home garages or parks for the rest of the time [23, 62, 65]. Several services, such as voltage and frequency regulation [23, 30, 39, 40], spinning reserves, reactive power support, peak shaving, valley filling (charging when demand is low), load following and energy balance [28, 37, 48] could be provided by PHEVs. These services are sometimes essential for power system while using V2G system overall costs could be reduced and, thereby, prices to customers, and selling energy to the grid could improve load factors and reduce emissions [5], and possibly replacing large-scale energy storages.

### 11.6.1 Renewable Energy Supporting

The power quality of intermittent source of energy wind and solar can be improved using PHEVs as storage and filter devices [23, 33, 34, 97, 100–103]. The combination of PHEVs and renewable energy sources can make the grid more stable and reliable. The unpredictable nature of wind speed makes the wind energy sources strongly intermittent and leading to imbalances [33, 104]. Solar radiations are available during the day while the peak energy demand occurs in the evening which refers the excess solar energy generation at the time when exciting grid does not need it [66].

A number of studies have been done to combine PHEVs with renewable energy sources for different purposes such as using as battery energy storage system (BESS) and reactive power support system. To overcome the fluctuation of wind power Kepton and Tomic [23] investigated the possibility of using V2G technology while Guille and Gross [30] proposed a structure using model predictive control (MPC) to analyze the positive effect of PHEVs on wind generator. To improve the power quality of a renewable energy based power network, Ota et al. [105] design a control scheme of PHEVs as distributed spinning reserve. Wand et al. [106] have provided a combination of demand response and wind power integration while Goransson et al. [107] elicited different strategies for integrating PHEVs into a wind-thermal power system.

A higher level penetration of renewable energy sources make the grid unstable, PHEVs can improve the situation by charging and discharging their battery during the period of excess generation and the period of peak load demand respectively. It can help the generation and load scheduling by consuming and supply energy whenever necessary [66]. Thus, V2G increases the flexibility of the grid to better utilize intermittent renewable sources.

### 11.6.2 Environmental Benefits

PHEVs have emissions benefits over conventional vehicles, even when considering power generation emissions. $CO_2$ emissions would fall significantly if PHEVs replaced conventional ICE vehicles [99]. In V2G mode of operation,

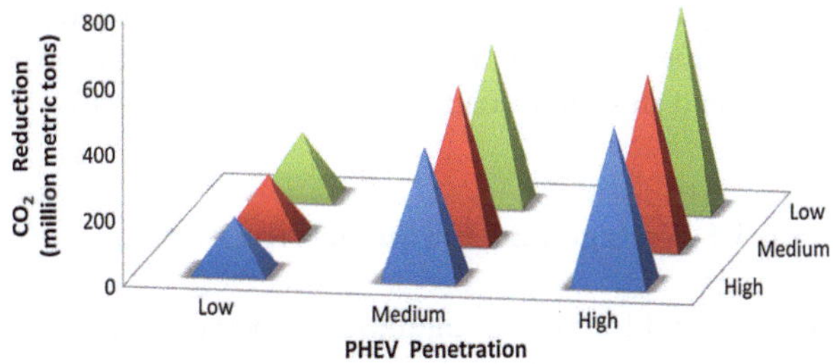


**Fig. 11.4** Annual greenhouse emission reduction from PHEVs

PHEVs could offer more environmental benefits and reduce greenhouse-gas (GHG) emissions [16, 29]. $CO_2$ emissions are estimated to drop from about 6.2–4 tons per year from a single vehicle [33, 57] while GHG emissions linked to driving depend on the type of fuel used for electricity generation. When the electricity is produced from fossil fuels, the environmental benefits of PHEVs are reduced, for renewable energy sources the GHG emissions almost 0 g/km while for coal-based plants it increases up to 155 g/km [108], even then their emissions may be 7–21 % lower than those of HEVs [17, 109] and 25 % fewer GHG emissions than ICE vehicles [110]. The estimated reductions for PHEVs range from 15 to 65 % in another USA-based study that examined low-carbon electricity sources [5, 17]. Long-term GHG reductions depend on reducing a grid's carbon intensity [111, 112], and using PHEVs more than 33 % emission can be reduced in future smart grid [113]. However automotive and oil companies allege that EVs would have a net negative effect on the environment because of lead discharges from battery manufacturing facilities and battery disposal [4, 114].

EPRI predicted the GHG impact of PHEVs over the years from 2010 to 2050 [115], as shown in Fig. 11.4, in which three scenarios represent levels of both $CO_2$ and total GHG emissions intensity and another three scenarios represent penetration of PHEVs. Nine different outcomes are possible from these two sets of scenarios, which determine the potential long term impacts.

From the analysis, it is found that each of the nine scenario combinations reduced annual GHG emissions significantly while reaching a maximum reduction of 612 million metric tons in 2050 (High PHEV fleet penetration, Low $CO_2$ intensity case) and reductions from 2010 to 2050 can range from 3.4 to 10.3 billion metric tons.

### *11.6.3 Auxiliary Services*

To maintaining stability, reliability, supply and load balancing and overall power quality, power system sometime needs auxiliary services from external and internal network devices. PHEVs with a bidirectional charger can provide higher quality ancillary services, such services are voltage and frequency regulation, load levelling and peak demand management. A few of them are described here in the light of literature. An aggregator can be the main part of the system by creating a larger and desirable load for the utility [30, 116].

#### 11.6.3.1 Voltage and Frequency Regulation

Voltage and frequency regulation in power system is always essential for the better quality power supply to the end user, V2G technology can provide this service and it could be one of the best service from PHEVs due to their high market value and minimal stress on a vehicle power storage system [62, 117]. An expensive process of cycling large generator in the network [95] is used to regulate the frequency in present grid system to balance supply and demand for active power [118] and the reactive power demand is balanced by voltage regulation [118]. The charging and discharging of PHEVs can be an alternative way of frequency regulation [23].

A proper logic of charging and discharging of PHEVs can be implanted in the battery charger with a voltage control to compensate reactive power, which will select the current phase angle to operate in inductive or capacitive mode of charging [118]. With an appropriate voltage control a PHEV can able to decide when it will charge or discharge it's battery. As for example, when the grid voltage becomes too low, vehicle charging can stop and, when it becomes too high, charging can start [66]. Although penetration of large number of PHEV for charging batteries from the grid could be a reason of line over loading and voltage instability at a low voltage network [119], it can regulate the reactive power within the local network by V2G operation [19].

The Union for the Coordination of Transmission of Electricity (UCTE) defined three types of control for the frequency stability in the distribution network: primary, secondary and tertiary frequency control [120].

There are two regulation in power system: regulation up and down and separate prices are given for regulation down and regulation up capacity, depending on bids submitted during an auction. If a vehicle providing regulation submits a bid below the market clearing price, it is contracted for its available capacity. Over the course of the contracted hour, the vehicle will charge or discharge some percentage of its contracted capacity. When the vehicle charges for regulation up, the owner will be charged for the energy consumed, and when it discharges for regulation down, the owner will be reimbursed for the energy provided. For secondary and tertiary frequency control, activation is also based on bids. When demand for regulation up arises, the lowest bid is activated first. Because delivering regulation down means charging at a lower price, this can be profitable for PEVs [45]. In [121], primary control is expected to have the highest value for V2G.

V2G research group at the University of Delaware compared the potential profit of V2G with existing grid regulation system and found that a PHEV with 10–15 kW power regulation capacity can earn \$3,777–4,000 per year [23, 62] and Brooks's calculation on California City's PHEVs shows the amount up to \$5,038 for V2G application [122].

#### 11.6.3.2 Load Shifting

By discharging during daily peaks and charging during low demand V2G can level the energy load. Local and global smart-charging control strategies could reduce the peak load [123]. Based on variation method an electricity pricing algorithm has been proposed for load levelling and identified an electricity price curve by Takagi et al. [36] that could realize an ideal bottom charge while PHEV owners could minimize their electricity bills. Sana showed that even 4 million PHEVs charging load could be accommodated with the existing grid of Californian [124] and for New York City it is observed that, up to 10 % of peak capacity could be safely contributed by PHEVs at penetration levels of around 50 %, which represented an economic benefit of $110 million per year [125]. Smart charger reduce peak load and shift energy demand [35, 126] while a little financial incentive for increased PHEV penetration when V2G is used for peak load reduction [51, 116].

## 11.7 Challenges to V2G Concept

V2G technology in a power distribution system may impact on its performance through overloading transformers and feeders, and in some cases this would reduce efficiency, produce voltage deviations and increase harmonics [127, 128]. The US Department of Energy reported [129] specific challenges and opportunities in terms of communication needs. Security issues are another challenge at public charging facilities [130]. Battery degradation, investment costs, energy losses, resistance of the automotive and oil sectors are also impediments and barriers to V2G systems.

Rapid charging and discharging of PHEV's battery for V2G concept may reduce the life of its battery. The rate of energy withdraws and cycling frequency determines the amount of battery degradation. Equivalent series resistance (ESR) and state of charge (SOC) are two major parameters proper controlling of which is a good way of slowing degradation [131–133].

According to Andersson [45], the investment cost of a battery is $300/kWh and a lifetime of 3,000 cycles at 80 % depth of charge (DOC), the degradation cost is $130/MWh. For a 16 kWh battery Peterson et al. [134] calculated the maximum net annual degradation cost of battery for V2G services, which is only $10–120.

Implementation of V2G technology in the present distribution network is likely to have a huge impact on equipment [135, 136]. Depending on the number and capacity of PHEV a distribution network could overload distribution transformers, increase voltage deviations, harmonic distortions and peak demand [137–142].

According to Dyke et al. [143], PHEVs penetration need a significant investment in electrical networks within the United Kingdom, while Fernandez et al. [144] presented the impacts of investments in distribution networks and incremental energy losses for different levels of PHEV penetration.

## 11.8 Scope of Research

From the above literature, there are several issues, which have not yet been taken into consideration by researchers. In this dissertation, some are discussed, with the main focus being on the following.

- Consideration of PHEV battery dynamics for load calculation and a cussed in the literature.
- Introduction of a novel ancillary service of PHEVs through designing a filter for a power system.
- Designs of virtual FACTS devices using PHEVs, which a few researchers have addressed.
- A complete power quality solution for a benchmark distribution network using V2G technology.

## 11.9 Chapter Summary

In this chapter, the impact of G2V and V2G technologies on power system and the benefits and challenges with the requirements and strategies for the interconnection between PHEVs and power system were reviewed. With the help of a bidirectional charger PHEVs can act like energy storage devices and serve the network whenever necessary. Unidirectional charger was the logical first step of the PHEV while the addition of on board bidirectional charger makes PHEV a smart part for the future smart grid with the opportunity of charging from any outlet of the grid and supports the network by injecting energy back to the grid. The economic benefits, $CO_2$ emissions, cost and the impact on the distribution system depend upon the cooperation between PHEV owner, aggregators and efficient strategy for grid operators.

Efficiency, stability, reliability and generation dispatch of a grid can be improved by using V2G operations. This mode of operation can offer active power regulation, reactive power support, power sources, current harmonic filtering, peak shaving and load balancing by valley filling for the grid. To improve the reliability of intermittent renewable energy sources PHEVs can provide possible backup as energy storage and as load at the time of excess generation. Several auxiliary supports can be provided by PHEVs to power system, such as voltage control, spinning reserves, reduce grid operating cost and generate revenue. Based on the power market value, the number of PHEVs and their battery energy capacities V2G mode of operation have the potential of net return between $90 and 4,000 per year per vehicle.

The V2G operation includes the cost of battery degradation, the need for smart communication between the vehicles and the grid, effects on the distribution system, the requirement for infrastructure changes, and political, social, technical and cultural issues. A number of proposed V2G technique have been discussed in this chapter, and it is shown that with a few disadvantage V2G technology is more economical and feasible from both owner and grid operator point of view. Political and environmental benefits can be ensured by the development of PHEVs. For the better interfacing of the PHEV and grid, the PHEV battery must have an extended life cycle with pre-determined standard of V2G and G2V connections.

## References

1. Ehsani M (2005) Modern electric, hybrid electric, and fuel cell vehicles. FL CRC Press, Boca Raton
2. Larminie J, Lowry J (2003) Electric vehicle technology explained. Wiley, New York
3. Wirasingha S, Emadi A (2011) Pihef: plug-in hybrid electric factor. IEEE Trans Veh Technol 60(3):1279–1284
4. Sovacool BK, Hirsh RF (2009) Beyond batteries: an examination of the benefits and barriers to plug-in hybrid electric vehicles (PHEVs) and a vehicle-to-grid (V2G) transition. Energy Policy 37(3):1095–1103. doi:10.1016/j.enpol.2008.10.005
5. Duvall M, Knipping E (2007) Environmental assessment of plug-in hybrid electric vehicles. Final report pp 156, Nationwide Greenhouse Gas Emissions, EPRI and NRDC
6. Kempton W, Tomi′c J (2005) Vehicle-to-grid power implementation: from stabilizing the grid to supporting large-scale renewable energy. J Power Sources 144(1):280–294
7. SAE electric vehicle and plug-in hybrid electric vehicle conductive charge coupler, SAE standard j1772 (January 2010)
8. Lassila J, Haakana J, Tikka V, Partanen J (2012) Methodology to analyze the economic effects of electric cars as energy storages. IEEE Trans Smart Grid 3(1):506–516
9. Li G, Zhang X-P (2012) Modeling of plug-in hybrid electric vehicle charging demand in probabilistic power flow calculations. IEEE Trans Smart Grid 3(1):492–499
10. GM-Volt: Latest Chevy volt battery pack and generator details and clarifications 2011
11. Nissan zero emission website, leaf specs (2010). www.nissan-zeroemission.com/EN/LEAF/specs.html
12. Heydt GT (1983) The impact of electric vehicle deployment on load management strategies. IEEE Power Eng Rev PER 3(5):41–42. doi:10.1109/MPER.1983.5519161
13. Meliopoulos S, Meisel J, Cokkinides G, Overbye T (2009) Power system level impacts of plug-in hybrid vehicles. Technical report, PSERC document, pp 09–12
14. Lund H, Kempton W (2008) Integration of renewable energy into the transport and electricity sectors through V2G. Energy Policy 36(9):3578–3587. http://www.sciencedirect.com/science/article/pii/S0301421508002838
15. Stephan CH, Sullivan J (2008) Environmental and energy implications of plug-in hybrid-electric vehicles. Environ Sci Technol 42(4):1185–1190
16. Sioshansi R, Denholm P (2009) Emissions impacts and benefits of plug-in hybrid electric vehicles and vehicle-to-grid services. Environ Sci Technol 43(4):1199–1204. http://pubs.acs.org/doi/pdf/10.1021/es802324j
17. Samaras C, Meisterling K (2008) Life cycle assessment of greenhouse gas emissions from plug-in hybrid vehicles: implications for policy. Environ Sci Technol 42(9):3170–3176
18. Hadley SW (2006) Impact of plug-in hybrid vehicles on the electric grid. Technical report, ORNL, Oct 2006
19. Clement-Nyns K, Haesen E, Driesen J (2010) The impact of charging plug-in hybrid electric vehicles on a residential distribution grid. IEEE Trans Power Syst 25(1):371–380
20. Larsen E, Chandrashekhara D, Ostergard J (2008) Electric vehicles for improved operation of power systems with high wind power penetration In: IEEE energy 2030 conference, pp 1–6

21. Locment F, Sechilariu M, Forgez C (2010) Electric vehicle charging system with PVgrid-connected configuration. In: IEEE vehicle power and propulsion conference (VPPC), pp 1–6
22. Gurkaynak Y, Khaligh A (2009) Control and power management of a grid connected residential photovoltaic system with plug-in hybrid electric vehicle (phev) load. In: Twenty-fourth annual IEEE applied power Electronics conference and exposition, pp 2086–2091
23. Kempton W, Tomi J (2005) Vehicle-to-grid power fundamentals: calculating capacity and net revenue. J Power Sources 144(1):268–279. doi:10.1016/j.jpowsour.2004.12.025
24. Singh B, Singh B, Chandra A, Al-Haddad K, Pandey A, Kothari D (2003) A review of single-phase improved power quality AC–DC converters. IEEE Trans Ind Electron 50(5):962–981
25. Singh B, Singh B, Chandra A, Al-Haddad K, Pandey A, Kothari D (2004) A review of three-phase improved power quality AC–DC converters. IEEE Trans Ind Electron 51(3):641–660
26. Zhou X, Lukic S, Bhattacharya S, Huang A (2009) Design and control of grid-connected converter in bi-directional battery charger for plug-in hybrid electric vehicle application. In: IEEE vehicle power and propulsion conference, pp 1716–1721
27. Zhou X, Wang G, Lukic S, Bhattacharya S, Huang A (2009) Multi-function bi-directional battery charger for plug-in hybrid electric vehicle application. In: IEEE energy conversion congress and exposition, pp 3930–3936
28. De Breucker S, Jacqmaer P, De Brabandere K, Driesen J, Belmans R (2006) Grid power quality improvements using grid-coupled hybrid electric vehicles. In: The 3rd IET international conference on power electronics, machines and drives, pp 505–509
29. Saber AY, Venayagamoorthy GK (2010) Intelligent unit commitment with vehicle-to-grid a cost emission optimization. J Power Sources 195(3):898–911. http://www.sciencedirect.com/science/article/pii/S037877530901341X
30. Guille C, Gross G (2009) A conceptual framework for the vehicle-to-grid V2G implementation. Energy Policy 37(11):4379–4390
31. Mitra P, Venayagamoorthy G, Corzine K (2011) Smartpark as a virtual STATCOM. IEEE Trans Smart Grid 2(3):445–455
32. Marano V, Rizzoni G (2008) Energy and economic evaluation of PHEVs and their interaction with renewable energy sources and the power grid. In: IEEE international conference on vehicular electronics and safety (ICVES), pp 84–89
33. Short W, Denholm P (2006) Preliminary assessment of plug-in hybrid electric vehicles on wind energy markets. Technical report, National Renewable Energy Lab NREL/TP-620-39729, April 2006
34. Ramos JMLA, Olmos L, Perez-Arriaga IJ (2008) Modeling medium term hydroelectric system operation with large-scale penetration of intermittent generation. In: XIV Latin and Iberian conference on operations research, 2008
35. Koyanagi F, Uriu Y (1998) A strategy of load leveling by charging and discharging time control of electric vehicles. IEEE Trans Power Syst 13(3):1179–1184
36. Takagi M, Iwafune Y, Yamaji K, Yamamoto H, Okano K, Hi-watari R, Ikeya T (2012) Electricity pricing for PHEV bottom charge in daily load curve based on variation method. In: IEEE innovative smart grid technologies (ISGT), pp 1–6, 2012
37. Kintner-Meyer KPSM, Pratt RG Impacts assessment of plug-in hybrid vehicles on electric utilities and regional US power grids part 1: technical analysis pacific northwest national laboratory. Technical report, Pacific Northwest National Lab PNNL-SA-61669
38. Sortomme E, El-Sharkawi M (2012) Optimal scheduling of vehicle-to-grid energy and ancillary services. IEEE Trans Smart Grid 3(1):351–359
39. Kempton W, Tomic J, Letendre ABS, Lipman T (2001) Vehicle-to-grid power: battery, hybrid, and fuel cell vehicles as resources for distributed electric power in California. Technical report, California Air Resources Board and California Environmental Protection Agency, CEPA, Los Angeles. Research Report, UCD-ITS-RR-01-03

40. Wirasingha S, Schofield N, Emadi A (2008) Plug-in hybrid electric vehicle developments in the US: trends, barriers, and economic feasibility. In: IEEE vehicle power and propulsion conference, pp 1–8, 2008
41. Dallinger D, Krampe D, Wietschel M (2011) Vehicle-to-grid regulation reserves based on a dynamic simulation of mobility behavior. IEEE Trans Smart Grid 2(2):302–313. doi:10.1109/TSG.2011.2131692
42. Keane E, Flynn D (2012) Potential for electric vehicles to provide power system reserve. In: IEEE innovative smart grid technologies (ISGT), pp 1–7, 2012
43. Srivastava AK, Annabathina B, Kamalasadan S (2010) The challenges and policy options for integrating plug-in hybrid electric vehicle into the electric grid. Electr J 23(3):83–91
44. Denholm P, Short W (2006) An evaluation of utility system impacts and benefits of optimally dispatched PHEVs. Technical report, National Renewable Energy Lab NREL/TP-620-40293 (October 2006)
45. Andersson S-L, Elofsson A, Galus M, Gransson L, Karlsson S, Johnsson F, Andersson G (2010) Plug-in hybrid electric vehicles as regulating power providers: case studies of Sweden and Germany. Energy Policy 38(6):2751–2762
46. Camus C, Esteves J, Farias T (2009) Electric vehicles and electricity sector regulatory framework: The Portuguese example. In: EVS24, 2009
47. Sioshansi R, Denkolm P (2010) The value of plug-in hybrid electric vehicles as grid resources. Energy Journal 31(3):1–23
48. Shireen W, Patel S (2010) Plug-in hybrid electric vehicles in the smart grid environment. In: IEEE transmission and distribution conference and exposition, pp. 1–4, 2010. doi:10.1109/TDC.2010.5484254
49. Grahn M, Azar C, Williander MI, Anderson JE, Mueller SA, Wallington TJ (2009) Fuel and vehicle technology choices for passenger vehicles in achieving stringent $CO_2$ targets: connections between transportation and other energy sectors. Environ Sci Technol 43(9):3365–3371
50. Goransson SKL, Johnsson F (2009) Plug-in hybrid electric vehicles as a mean to reduce $CO_2$ emissions from electricity production. In: Proceeding electric vehicle symposiums 24, Stavanger, Norway, 2009
51. Peterson SB, Whitacre J, Apt J (2010) The economics of using plug-in hybrid electric vehicle battery packs for grid storage. J Power Sources 195(8):2377–2384
52. Zahedi A (2012) Electric vehicle as distributed energy storage resource for future smart grid. In: 22nd Australasian universities power engineering conference (AUPEC), pp 1–4, 2012
53. Introduction to vehicle to grid (V2G). http://www.v2g.com.au/
54. Pang C, Dutta P, Kim S, Kezunovic M, Damnjanovic I (2010) PHEVs as dynamically configurable dispersed energy storage for V2B uses in the smart grid. In: 7th mediterranean conference and exhibition on power generation, transmission, distribution and energy conversion (Med Power 2010), pp 1–6, 2010. doi:10.1049/cp.2010.0903
55. Su W, Eichi H, Zeng W, Chow M-Y (2012) A survey on the electrification of transportation in a smart grid environment. IEEE Trans Ind Inf 8(1):1–10
56. Yilmaz M, Krein P (2012) Review of benefits and challenges of vehicle-to-grid technology. In: IEEE energy conversion congress and exposition (ECCE), pp. 3082–3089, 2012. doi:10.1109/ECCE.2012.6342356
57. DiPeso J (2008) Cars to grid: an electrifying idea. Environ Qual Manage 18(2):89–94
58. Turker H, Bacha S, Chatroux D, Hably A (2012) Aging rate of low voltage transformer for a high penetration of plug-in hybrid electric vehicles (PHEVs). In: IEEE innovative smart grid technologies (ISGT), pp 1–8, 2012
59. Saber A, Venayagamoorthy G (2011) Plug-in vehicles and renewable energy sources for cost and emission reductions. IEEE Trans Ind Electron 58(4):1229–1238
60. School of Public and Environmental Affairs at Indiana University (2011) Plug-in electric vehicles: a practical plan for progress, the report of an expert panel. Technical report, School of Public and Environmental Affairs at Indiana University, Feb 2011

61. Electrification of the transportation system, Technical report, MIT energy initiative symposium, MITei, USA, April 2010
62. Kempton JTW (2007) Using fleets of electric-drive vehicles for grid support. J Power Sources 168(2):459–468. http://www.sciencedirect.com/science/article/pii/S0378775307005575
63. Rua D, Issicaba D, Soares F, Almeida P, Rei R, Lopes J (2010) Advanced metering infrastructure functionalities for electric mobility. In: Innovative smart grid technologies conference Europe (ISGT Europe), IEEE PES, pp 1–7, 2010
64. Quinn C, Zimmerle D, Bradley TH (2010) The effect of communication architecture on the availability, reliability, and economics of plug-in hybrid electric vehicle-to-grid ancillary services. J Power Sources 195(5):1500–1509
65. Kempton W, Letendre S (1997) Electric vehicles as a new power source for electric utilities. Transp Res D 2(3):157–175
66. Clement-Nyns K, Haesen E, Driesen J (2011) The impact of vehicle-to-grid on the distribution grid. Electr Power Syst Res 81(1):185–192. http://www.sciencedirect.com/science/article/pii/S0378779610002063
67. Markel T, Kuss M, Denholm P (2009) Communication and control of electric drive vehicles supporting renewable. In: IEEE vehicle power and propulsion conference, pp 27–34, 2009
68. Su W, Zeng W, Chow M-Y (2012) A digital testbed for a PHEV/PEV enabled parking lot in a smart grid environment. In: Innovative smart grid technologies (ISGT), IEEE PES, pp 1–7, 2012
69. Chaudhry H, Bohn T (2012) Security concerns of a plug-in vehicle. In: IEEE innovative smart grid technologies (ISGT), pp 1–6, 2012
70. (2011) Technical report, Zigbee smart energy overview. http://www.zigbee.org/Standards/ZigBeeSmartEnergy/Overview.aspx
71. Kulshrestha P, Swaminathan K, Chow M-Y, Lukic S (2009) Evaluation of zigbee communication platform for controlling the charging of phevs at a municipal parking deck. In: IEEE vehicle power and propulsion conference, pp 1211–1214, 2009
72. Lee HALSKMK, Newman RE, Yonge L (2003) Homeplug 1.0 powerline communication lansprotocol description and performance results. Int J Commun Syst 16:447–473
73. Scholer R, Maitra A, Ornelas E, Bourton M et al (2010) "Communication between Plug-in Vehicles and the Utility Grid" SAE Technical Paper 2010-01-0837, doi:10.4271/2010-01-0837.
74. Communication between plug-in vehicles and the utility grid (2011)
75. Communication between plug-in vehicles and off-board dc chargers, SAE Stand J2847/2 (2011)
76. Communication between plug-in vehicles and the utility grid for reverse power flow, SAE Stand J2847/3 (2011)
77. Gungor V, Sahin D, Kocak T, Ergut S, Buccella C, Cecati C, Hancke G (2011) Smart grid technologies: communication technologies and standards. IEEE Trans Ind Inf 7(4):529–539
78. Ferreira J, Monteiro V, Afonso J, Silva A (2011) Smart electric vehicle charging system. In: IEEE intelligent vehicles symposium (IV), pp 758–763, 2011
79. Bai S, Lukic S (2011) Design considerations for DC charging station for plug-in vehicles. In: IEEE vehicle power and propulsion conference (VPPC), pp 1–6, 2011
80. Xu W (2000) Comparisons and comments on harmonic standards IEC 1000-3-6 and IEEE stdandard 519. In: Proceedings of ninth international conference on harmonics and quality of power, vol 1, pp 260–263, 2000
81. Geske PKM, Winkler T, Heideck G (2010) Controlled battery charger for electric vehicles. PIERS Online 6(6):532536
82. Power quality requirements for plug-in vehicle chargers-part 1: requirements, SAE Stand J2894/2 (2011)
83. Chan C, Chau K (1997) An overview of power electronics in electric vehicles. IEEE Trans Ind Electron 44(1):3–13

84. Aguilar C, Canales F, Arau J, Sebastian J, Uceda J (1995) An integrated battery charger/discharger with power factor correction. In: 26th annual IEEE power electronics specialists conference, vol 2, pp 714–719, 1995
85. Lee Y-J, Khaligh A, Emadi A (2009) Advanced integrated bidirectional ac/dc and dc/dc converter for plug-in hybrid electric vehicles. IEEE Trans Veh Technol 58(8):3970–3980
86. Gomez J, Morcos M (2003) Impact of EV battery chargers on the power quality of distribution systems. IEEE Trans Power Delivery 18(3):975–981
87. Aggeler D, Canales F, Zelaya-De La Parra H, Coccia A, Butcher N, Apeldoorn O (2010) Ultra-fast DC-charge infrastructures for ev-mobility and future smart grids. In: IEEE innovative smart grid technologies conference Europe (ISGT Europe), pp 1–8, 2010
88. Aabrandt A, Andersen P, Pedersen A, You S, Poulsen B, O'Connell N, Ostergaard J (2012) Prediction and optimization methods for electric vehicle charging schedules in the edison project. In: IEEE innovative smart grid technologies (ISGT), pp 1–7, 2012
89. Tuttle D, Baldick R (2012) The evolution of plug-in electric vehicle-grid interactions. IEEE Trans Smart Grid 3(1):500–505
90. Bessa R, Matos M, Soares F, Lopes J (2012) Optimized bidding of a ev aggregation agent in the electricity market. IEEE Trans Smart Grid 3(1):443–452
91. Bessa RJ, Matos MA Economic and technical management of an aggregation agent for electric vehicles: a literature survey. European Trans Electr Power. http://onlinelibrary.wiley.com/doi/10.1002/etep.565/abstract
92. Singh M, Kumar P, Kar I (2012) Implementation of vehicle to grid infrastructure using fuzzy logic controller. IEEE Trans Smart Grid 3(1):565–577
93. Wu D, Chau K, Liu C, Gao S, Li F (2012) Transient stability analysis of SMES for smart grid with vehicle-to-grid operation. IEEE Trans Appl Supercond 22(3):5701105
94. Kristoffersen KCTK, Meibom P (2011) Optimal charging of electric drive vehicles in a market environment. Appl Energy 88:1940–1948
95. Wu D, Aliprantis D, Ying L (2012) Load scheduling and dispatch for aggregators of plug-in electric vehicles. IEEE Trans Smart Grid 3(1):368–376
96. Wu C, Mohsenian-Rad H, Huang J (2012) Vehicle-to-aggregator interaction game. IEEE Trans Smart Grid 3(1):434–442
97. Han S, Han S, Sezaki K (2010) Development of an optimal vehicle-to-grid aggregator for frequency regulation. IEEE Trans Smart Grid 1(1):65–72
98. Pillai J, Bak-Jensen B (2011) Integration of vehicle-to-grid in the western Danish power system. IEEE Trans Sustain Energy 2(1):12–19
99. MOBIE (2011) http://www.mobie.pt
100. Jourabchi M (2008) Impact of plug-in hybrid vehicles on northwest power system: a preliminary assessment. Northwest Power and Conservation Council, Columbia
101 Schneider K, Gerkensmeyer C, Kintner-Meyer M, Fletcher R (2008) "Impact assessment of plug-in hybrid vehicles on pacific northwest distribution systems," Power and Energy Society General Meeting - Conversion and Delivery of Electrical Energy in the 21st Century, 2008 IEEE 1(6):20–24 July 2008
102. Islam FR, Pota HR (2011) Design a PV-AF system using V2G technology to improve power quality. In: 37th annual conference on IEEE industrial electronics society, IECON, pp 861–866, 2011
103. Islam FR, Pota HR, Ali MS (2011) V2G technology for designing active filter system to improve wind power quality. In: 21st Australasian universities power engineering conference (AUPEC), pp 1–6, 2011
104. Islam FR, Pota HR (2011) V2G technology to improve wind power quality and stability. In: Australian control conference (AUCC), pp 452–457, 2011
105. Islam FR, Pota HR (2014) Virtual active filters for HVDC networks using V2G technology. Int J Electr Power Energy Syst 54:399–407, ISSN 0142-0615. http://dx.doi.org/10.1016/j.ijepes.2013.07.028 (Jan 2014)

106. Ray PK, Mohanty SR, Kishor N (2011) Disturbance detection in grid-connected distributed generation system using wavelet and s-transform. Electr Power Syst Res 81(3):805–819. http://www.sciencedirect.com/science/article/pii/S037877961000283X
107. Ota Y, Taniguchi H, Nakajima T (2012) Autonomous distributed V2G satisfying scheduled charging. IEEE Trans Smart Grid 3(1):559–564
108. Wang J, Liu C, Ton D, Zhou Y, Kim J, Vyas A (2011) Impact of plug-in hybrid electric vehicles on power systems with demand response and wind power. Energy Policy 39(7):4016–4021. http://www.sciencedirect.com/science/article/pii/S0301421511000528
109. Goransson L, Karlsson S, Johnsson F (2010) Integration of plug-in hybrid electric vehicles in a regional wind-thermal power system. Energy Policy 38(10):5482–5492. http://www.sciencedirect.com/science/article/pii/S0301421510002740
110. Peterson SB, Whitacre JF, Apt J (2011) Net air emissions from electric vehicles: the effect of carbon price and charging strategies. Environ Sci Technol 45(5):1792–1797, arXiv: http://pubs.acs.org/doi/abs/10.1021/es102464y
111. Jaramillo P, Samaras C, Wakeley H, Meisterling K (2009) Greenhouse gas implications of using coal for transportation: life cycle assessment of coal-to-liquids, plug-in hybrids, and hydrogen pathways. Energy Policy 37(7):2689–2695. http://www.sciencedirect.com/science/article/pii/S0301421509001451
112. Jaramillo P, Samaras C (2007) Comparing life cycle ghg emissions from coal-to-liquids and plug-in hybrids. Technical reports, CEIC working
113. Axsen J, Kurani KS (2010) Anticipating plug-in hybrid vehicle energy im- pacts in California: constructing consumer-informed recharge profiles. Trans Res Part D: Trans Environ 15(5):212–219
114. Axsen J, Kurani KS, McCarthy R, Yang C (2011) Plug-in hybrid vehicle impacts in California: integrating consumer-informed recharge profiles with an electricity-dispatch model. Energy Policy 39(3):1617–1629. http://www.sciencedirect.com/science/article/pii/S0301421510009389
115. Sioshansi R, Fagiani R, Marano V (2010) Cost and emissions impacts of plug-in hybrid vehicles on the Ohio power system. Energy Policy 38(11):6703–6712. http://www.sciencedirect.com/science/article/pii/S0301421510005045
116. Stewart TA (2001) E-check: a dirty word in Ohio's clean air debate. Technical report, Capital University Law Review, vol 29, p 338341
117. EPRI, Environmental assessment of plug-in hybrid electric vehicles, nationwide greenhouse gas emissions. Technical report, vol.1 (July 2007)
118. White CD, Zhang KM (2011) Using vehicle-to-grid technology for frequency regulation and peak-load reduction. J Power Sources 196(8):3972–3980. http://www.sciencedirect.com/science/article/pii/S0378775310019142
119. De Los Rios A, Goentzel J, Nordstrom K, Siegert C (2012) Economic analysis of vehicle-to-grid (V2G)-enabled fleets participating in the regulation service market. In: IEEE innovative smart grid technologies (ISGT), pp 1–8, 2012
120. Wu C, Mohsenian-Rad H, Huang J, Jatskevich J (2012) PEV based combined frequency and voltage regulation for smart grid. In: IEEE innovative smart grid technologies (ISGT), pp 1–6, 2012
121. Lopes J, Soares F, Almeida P (2009) Identifying management procedures to deal with connection of electric vehicles in the grid. In: IEEE Bucharest power technology, pp 1–8, 2009
122. Operation handbook, Union for the coordination of transmission of electricity (UCTE) (2008)
123. Oudalov A, Chartouni D, Ohler C, Linhofer G (2006) Value analysis of battery energy storage applications in power systems. In: IEEE power systems conference and exposition, PSCE'06, pp 2206–2211, 2006
124. Brooks AN (2002) Vehicle-to-grid demonstration project: grid regulation ancillary service with a battery electric vehicle. Technical report, AC Propulsion Inc., San Dimas, CA

125. Mets K, Verschueren T, Haerick W, Develder C, De Turck F (2010) Optimizing smart energy control strategies for plug-in hybrid electric vehicle charging. In: IEEE/IFIP network operations and management symposium workshops (NOMS Wksps), pp 293–299, 2010
126. Sana L (2005) Driving the solution, the plug-in hybrid vehicle. EPRI J Fall:8–17
127. Chakraborty S, Shukla S, Thorp J (2012) A detailed analysis of the effective-load-carrying-capacity behavior of plug-in electric vehicles in the power grid. In: Innovative smart grid technologies (ISGT), 2012 IEEE PES, pp 1–8, 2012
128. IEEE Standard for Interconnecting Distributed Resources With Electric Power Systems, IEEE Std 1547-2003
129. Sortomme E, Hindi M, MacPherson S, Venkata S (2011) Coordinated charging of plug-in hybrid electric vehicles to minimize distribution system losses. IEEE Trans Smart Grid 2(1):198–205
130. Bojrup M, Karlsson P, Alakula M, Simonsson B (1998) A dual purpose battery charger for electric vehicles. In: 29th annual IEEE power electronics specialists conference, vol 1, pp 565–570, 1998
131. Communication requirements for smart grid technologies (2010) Technical report, US Department of Energy, Washington, 2010
132. Khurana H, Hadley M, Lu N, Frincke D (2010) Smart-grid security issues. IEEE Secur Priv 8(1):81–85
133. Dogger JD, Roossien B, Nieuwenhout F (2011) Characterization of li-ion batteries for intelligent management of distributed grid-connected storage. IEEE Trans Energy Convers 26(1):256–263
134. Han S, Han SH, Sezaki K (2012) Economic assessment on V2G frequency regulation regarding the battery degradation. In: IEEE innovative smart grid technologies (ISGT), pp 1–6, 2012
135. S Han, S Han, Sezaki K (2012) "Economic assessment on V2G frequency regulation regarding the battery degradation," Innovative Smart Grid Technologies (ISGT), 2012 IEEE PES pp. 16–20
136. Quinn C, Zimmerle D, Bradley T (2012) An evaluation of state-of-charge limitations and actuation signal energy content on plug-in hybrid electric vehicle, vehicle-to-grid reliability, and economics. IEEE Trans Smart Grid 3(1):483–491
137. Peterson SB, Apt J, Whitacre J Lithium-ion battery cell degradation resulting from realistic vehicle and vehicle-to-grid utilization. J Power Sources 195(8):2385–2392. http://www.sciencedirect.com/science/article/pii/S0378775309017443
138. Rutherford M, Yousefzadeh V (2011) The impact of electric vehicle battery charging on distribution transformers. In: Twenty-Sixth annual IEEE applied power electronics conference and exposition (APEC), pp 396–400, 2011
139. Desbiens C (2012) Electric vehicle model for estimating distribution transformer load for normal and cold-load pickup conditions. In: IEEE innovative smart grid technologies (ISGT), pp 1–6, 2012
140. Initiative on Plug-in Electric Vehicles (2010) Commonwealth edison company, initial assessment of the impact of the introduction of plug-in electric vehicles on the distribution system. Technical report, Illinois Commerce Commission (15 Dec 2010)
141. Bae S, Kwasinski A (2012) Spatial and temporal model of electric vehicle charging demand. IEEE Trans Smart Grid 3(1):394–403
142. Etezadi-Amoli M, Choma K, Stefani J (2010) Rapid-charge electric-vehicle stations IEEE Trans Power Delivery 25(3):1883–1887
143. Green R, Wang L, Alam M (2010) The impact of plug-in hybrid electric vehicles on distribution networks: a review and outlook. In: IEEE power and energy society general meeting, pp 1–8, 2010
144. Raghavan S, Khaligh A (2012) Impact of plug-in hybrid electric vehicle charging on a distribution network in a smart grid environment. In: IEEE innovative smart grid technologies (ISGT), pp 1–7, 2012

145. Farmer C, Hines P, Dowds J, Blumsack S (2010) "Modeling the impact of increasing PHEV loads on the distribution infrastructure," 43rd Hawaii International Conference on System Sciences (HICSS), 5-8 Jan 2010 pp 1–10
146. Dyke K, Schofield N, Barnes M (2010) The impact of transport electrification on electrical networks. IEEE Trans Ind Electr 57(12):3917–3926
147. Pieltain Fernandez L, Gomez San Roman T, Cossent R, Domingo C, Frias P (2011) Assessment of the impact of plug-in electric vehicles on distribution networks. IEEE transactions on power systems 26(1):206–213

# Chapter 12
# Coordinating Distributed Energy Resources During Microgrid Emergency Operation

**C. Gouveia, D. Rua, C. L. Moreira and J. A. Peças Lopes**

**Abstract** The development of the Smart Grid (SG) concept is the pathway for assuring flexible, reliable and efficient distribution networks while integrating high shares of Distributed Energy Resources (DER): renewable energy based generation, distributed storage and controllable loads such as Electric Vehicles (EV). Within the SG paradigm, the Microgrid (MG) can be regarded as a highly flexible and controllable Low Voltage (LV) cell, which is able to decentralize the distribution management and control system while providing additional controllability and observability. A network of controllers interconnected by a communication system ensures the management and control of the LV microgrid, enabling both interconnected and autonomous operation modes. This new distribution operation philosophy is in line with the SG paradigm, since it improves the security and reliability of the system, being able to tackle the technical challenges resulting from the large scale integration of DER and provide the adequate framework to fully integrate SG new players such as the EV. By exploiting the MG operational flexibility and controllability, this chapter aims to provide an extended overview on MG self-healing capabilities, namely on its ability of operating autonomously from the main grid and perform local service restoration. The MG hierarchical management and control structure is revisited and adapted in order to exploit the flexibility of SG new players, like the EV and flexible loads and integrate smart

C. Gouveia · D. Rua · C. L. Moreira (✉) · J. A. Peças Lopes
INESC TEC-INESC Technology and Science,
Rua Dr. Roberto Frias, 378 4200–465 Porto, Portugal
e-mail: cmoreira@inescporto.pt

C. Gouveia
e-mail: cstg@inescporto.pt

D. Rua
e-mail: drua@inescporto.pt

J. A. Peças Lopes
e-mail: jpl@fe.up.pt

J. Hossain and A. Mahmud (eds.), *Renewable Energy Integration*,
Green Energy and Technology, DOI: 10.1007/978-981-4585-27-9_12,

metering infrastructures. The implementation of the MG architecture and communication infrastructure in a laboratorial facility is also presented and used to validate the MG self-healing capabilities.

**Keywords** Electric vehicle · Demand response · Frequency control · Islanding operation · Microgeneration · Microgrids · Smart grid

## 12.1 Introduction

The Microgrid (MG) is considered to be a Low Voltage (LV) distribution grid incorporating local generation, storage devices and responsive loads [1–3]. Through the future change of paradigm in the mobility sector, Electric Vehicles (EV) will become another resource to be integrated in the MG system, which can behave both as flexible loads or as mobile energy storage devices [4, 5]. In order to be flexible and controllable, the MG power infrastructure is supported by a communication and information system constituting the MG technical management and control system [1–3, 6, 7]. The local intelligence is responsible for coordinating and controlling local resources and enable innovative self-healing operating strategies [8].

MG applications can be divided in two major types:

- The MG operates as a Medium Voltage (MV) customer, being the main operational objective to meet the consumer's technical and economic requirements, while being capable of providing services to the upstream grid. This concept can be applied to a diversity of commercial, industrial or even military consumers.
- The MG is implemented as an extension of the distribution network management system. In this case, the MG main operational objective is to ensure the technical management and control of the LV network, when considering the integration of large amounts of microgeneration technologies, EV and responsive loads. However, the microgeneration, EV charging and loads control options will be limited by commercial agreements between the consumers and the electricity providers [5].

In the majority of the applications referred above, reliability is considered a major benefit of the MG system due to its ability to operate both interconnected to the main grid or autonomously as a physical island—emergency operation. In fact, case a disturbance occurs in the upstream MV network or in case of planned actions, the MG can be transferred to islanded operating conditions, supplying local loads with the MG generation capacity supported by fast acting storage units [3, 6]. However, if the islanding fails or a general blackout occurs, the MG flexibility also allows performing local service restoration exploiting locally available generation and energy storage units [8]. Ensuring a successful autonomous operation is quite challenging, since the sudden islanding of the MG may cause high unbalances between local load and generation, which have to be compensated by fast acting storage devices in coordination with the microgeneration units and load shedding mechanisms [4, 9–11].

The fact that EV are expected to be connected to the LV network during long periods of time for charging purposes along with the predictability of its power consumption, makes them good candidates for the development of dedicated load control strategies. When regarded as flexible loads or distributed storage, they can contribute to increase the global MG storage capacity [4, 5]. Nevertheless, the increase of regulation capacity will depend on the number of EV connected to the system and on the state of charge of their batteries.

Complementing EV control strategies with innovative emergency demand response schemes for MG islanded operation will improve the security of operation during emergency conditions and minimize the consumers discomfort during outages [9–14]. Under such cases, innovative control functionalities need to be developed in order to promote an adequate coordination of the available resources.

The roll out of smart meters with bidirectional communication capabilities will provide a set of valuable measurements essential to perform an adequate analysis of the MG operating conditions and identify the most appropriate actions during MG emergency operation, including islanding and restoration procedures. Updating the emergency operation strategies according to the actual operating state of the MG will allow a better coordination of the available resources, consequently improving the MG resilience and success of the autonomous operation [14]. However, the new functionalities to be developed have to take into consideration the limitation of different communication solutions and technologies currently available, namely the use of private or public networks.

The feasibility of the MG concept integrating EV has been the focus of several research projects around the world. Following the track record on MG research, significant demonstration activities have been exploited worldwide [15–20], dedicated to the three main areas of knowledge involved in the MG operation, namely power systems, power electronics and information and communication technologies. Within this line, INESC Porto has been implementing a MG laboratorial infrastructure in order to exploit distinctive control and management solutions for key DER such as EV, loads and Renewable Energy Sources (RES) based microgeneration [20]. The laboratorial infrastructure aims to demonstrate the feasibility of the MG operation both in interconnected and autonomous mode while exploiting the specificities of different communication technological solutions that can be envisioned for a SG environment.

## 12.2 MG Architecture and Communication Infrastructure

The basic architecture of a MG consists in a hierarchical structure composed by a network of local controllers which are connected to each MG element and are headed by a central manager agent installed at the MV/LV substation—the MG

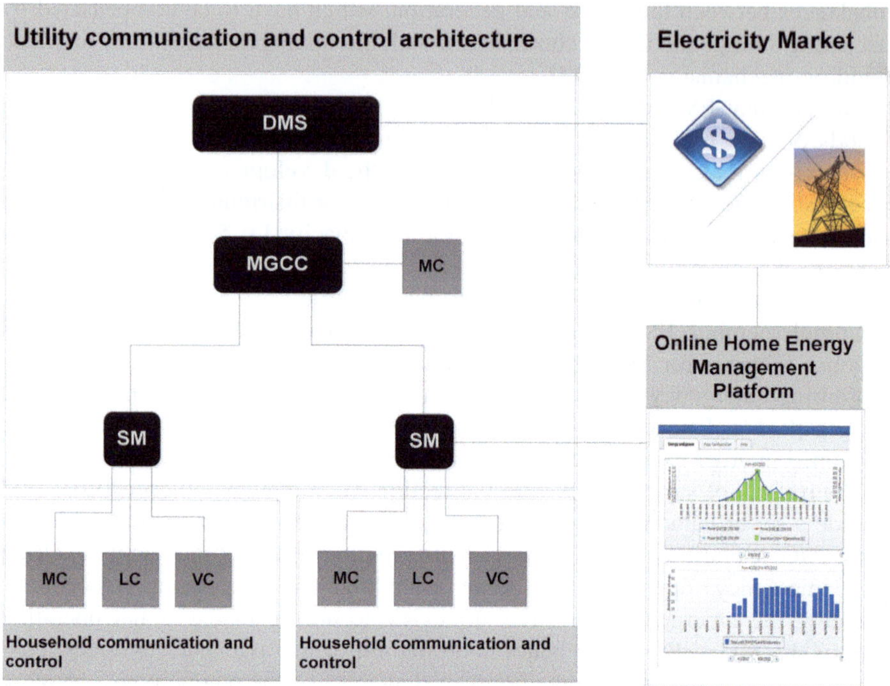


**Fig. 12.1** Smart distribution network architecture based on MG concept

Central Controller (MGCC). Given the different characteristics of the MG resources, three types of local controllers can be defined: the Microsource Controller (MC), EV Controller (VC) and the Load Controller (LC) [3–5]. As shown in Fig. 12.1, the MG architecture can be easily adapted to integrate smart metering infrastructures as well as the associated processing capabilities to manage the data collected from the meters. At the consumer premises, the Smart Meter (SM) is expected to support bidirectional communications between the local controllers and the MGCC. Acting also as a local gateway it is able to aggregate the information related with power consumption, generation, service interruptions, voltage and other relevant data.

Communications systems and technologies have long been developed to support a diverse number of applications, many of them related with electric power system. However, a particular interest has been recently raised with the introduction of the SG concept, which intends, among others, to support enhanced control strategies in the distribution grids. This new distribution system paradigm introduces challenges for communications systems, namely because in these networks the need for information exchange, was until recently, very modest.

Different reference models and architectures have been defined, namely by NIST, IEEE and by the European Standards Organizations (ESO) with the purpose of establishing the main actors, data networks and domains of applications.

Transversal to all of them is the highlighted importance of distribution and customer domains where the need of incorporating a considerable number of different potentially participating players is higher. There are also different functional and logical models that have been recently defined within the context of smart grids, establishing different information flows and requirement to be met to support specific types of information exchange.

## *12.2.1 Smart Grid Communication Technologies and their Applicability to MG Operation*

A considerable amount of communication technologies that suite the needs of different applications targeted for SG is available. However, economical aspects along with technical limitation factors are among the main reasons for the use of particular technologies in detriment of others. The candidate technologies for the electric distribution networks can be associated with the use of public or private network. They can be typically distinguished between wired and wireless technologies.

The use of public communications network is commonly associated with telecommunication operators that make use of their infrastructure, based for instance on copper, fiber optics, or cellular networks, to provide services to their customers. The use of such networks is regarded by electric system operators with some scepticism due to the limitations that telecommunication operators have in guaranteeing the security, availability and reliability indexes usually associated with the electric sector.

The deployment of private networks in the back-haul or backbone of utilities has been very common, using several technologies in order to meet the different requirements associated with the applications to support. The use of these types of networks is therefore also regarded as preferred solution for the more demanding scenarios and applications, making way for the use of public network in less restrictive cases.

The potential for wired technologies envisioned by utilities are somehow narrowed down to the use of power lines as a communications medium, mainly due to the technical and economic constraints of deploying a cabled alternative, like copper or fiber. This has led to the exploitation of wireless alternatives compatible with the SG context.

### 12.2.1.1 Power Line Technologies

Power Line Communication (PLC) technologies explore the use of electric power lines as a medium to convey a bidirectional data exchange. In recent years, SG activities, in-building communications and home automation have brought a lot of attention to PLC as an alternative to unwanted or impractical wiring to setup communication networks. Power lines are inherently exposed to time–frequency varying noise, unmatched loads and interference from electrical or communication devices. The fact that in electric distribution networks power lines vary from grounded to overhead lines introduces further challenges in using this type of medium to exchange data. National legislation in different countries has introduced different limits in the transmission power and accepted frequency bands. For these reasons a large number of PLC technologies have been developed, each one addressing different applications and thus targeting different throughput values, frequency bands and channel access mechanisms. Implementations of PLC can be globally divided into narrow and broadband versions.

### Narrowband PLC

The Narrowband PLC (NB-PLC) is usually associated with low and high speed versions. The latter is also sometimes referred to as medium speed to distinguish it from the broadband implementations. Different transmission frequency ranges are defined as well as a maximum transmission power in each range that must be respected either by a standard or by the legal disposition of each country. In Europe CENELEC is the entity responsible for the standardization and regulation of NB-PLC, whereas in the USA the Federal Communications Commission (FCC) and in Japan the Association of Radio Industries and Businesses (ARIB) are the legally assigned entities [21].

The first generation of NB-PLC made use of a single or double band carrier transmission schemes with simple modulation schemes like Phase-Shift Keying (PSK) and Frequency Shift Keying (FSK) to achieve a bandwidth of a few kbps, targeting remote metering applications. The second generation of PLC is associated with G3 and PRIME technologies, which were developed by different consortia with the purpose of supporting smart metering communication scenarios. Despite the use of Orthogonal Frequency-Division Multiplexing (OFDM) modulation schemes in G3 and PRIME, subtle physical layer details make them to differ slightly [22]. As a general rule, PRIME achieves higher data rates while G3 incorporates more powerful error correction coding for higher reliability of information. Another similar variation, G.hem, was introduced by the International Telecommunication Union (ITU) sharing some features with the previous two.

The IEEE 1901.2 [23] is an emerging narrowband standard under development since 2009. Defined as a low frequency OFDM narrowband for SG applications, it is set to use frequencies below 500 kHz and data rates up to 500 kbps supporting both indoors and outdoors communications. In the outdoors context this standard targets MV and LV distribution feeders. It is expected that coexistence mechanisms are defined to enable the interaction with G3 and PRIME systems.

Broadband PLC

The initial developments of Broadband PLC (BB-PLC), also known as Broadband Powerline (BPL), were started in the in-building environment, namely as a replacement for conventional Ethernet infrastructure as an alternative, or complementary solution for wireless solutions. In the last years BB-PLC has been explored as a communication infrastructure outside the building context, namely as a candidate technology for last-mile communications. Wide frequency bands, generally in the range of 2–30 MHz, are used worldwide, except in Japan where BB-PLC is not permitted. Higher bandwidths are potentially available in some countries but they are generally bounded due to TV broadcast systems. One of the main promoters of BB-PLC has been the HomePlug Alliance that targeted domestic environments to implement this type of networks. A particular implementation of HomePlug standard was introduced in 2010 and designated as Green PHY (GP) for in-home automation as a support for SG applications. It is prepared to coexist with other HomePlug implementations.

The IEEE 1901 [24] is another BPL standardized implementation which uses some of the strategies defined in the HomePlug variants. Unlike the latter, it is designed to be deployed both inside and outside the building environment. Different PHY layers are defined along with different modulation schemes and error correction coding mechanisms. Some of the approaches used in IEEE 1901 are expected to be included in the IEEE 1901.2 NB-PLC.

The ITU has also defined a BPL standard designated G.hn, with the purpose of supporting SG applications such as Advanced Metering Infrastructure (AMI) and energy management including EV. It allows the use of power line communications inside buildings, as well as in the last-mile.

#### 12.2.1.2 Wireless Technologies

Wireless communication systems are another type of technologies that can also be explored in the context of SG, either as an alternative or as a complementary solution. Some of the generally highlighted advantages of wireless technologies in this scope are the ease of installation, maintenance and modularity that allow a higher degree of flexibility. Another advantage of wireless implementations, when associated to battery systems, is the potential capability to continue operating despite the occurrence of disturbances that lead to the disconnection of parts of a feeder, thus compromising communications over power lines. The potential low cost is also pointed out as an advantage when well established and mature solutions are employed. Even so, there are some general challenges associated with the use of wireless technologies, namely the need for careful planning due to the adverse channel propagation conditions, the limitation in the use of free or licensed Radio Frequency (RF) bands, the presence of obstacles, and the shared nature of the communication medium.

### IEEE 802.11/Wi-Fi

One of the most known and disseminated IEEE standards is the 802.11 that defines the features of a wireless local area network as an alternative to cabled Ethernet. Its commercial designation, Wi-Fi, is promoted by the Wi-Fi Alliance, which has contributed for the widespread availability and implementation of this technology, making it a ubiquitous solution used in different communication infrastructures. In SG its use has been explored inside and outside the building environment, in applications as varied as distribution substation automation, DER monitoring, control LAN, among others. The IEC 61850 defines some of the applications that can be support by Wi-Fi networks [25].

### IEEE 802.16/WiMAX

The IEEE 802.16 was released by IEEE as a standard for wireless metropolitan area networks, targeting both Line of Sight (LOS) and non-LOS applications. The WiMAX is the designation of the wireless devices based on this standard. It is promoted by the WiMAX Forum that certified the compatibility and interoperability of the devices from different manufacturers. Although WiMAX was created with other type applications in mind, the fact is that SG applications like automatic meter reading, real-time pricing, outage detection and restoration are also targeted to be supported by WiMAX [25].

### IEEE 802.15.4/ZigBee

A family of standards for wireless personal area networks was introduced by IEEE with the 802.15.4. It defines low power, low rate and low complexity networks to convey small amount of information over relatively small distances. The ZigBee stack is responsible for the design and implementation of applications over IEEE 802.15.4. It originally targeted applications such as building automation, embedded sensing or wireless sensor networks. Recently, SG applications like the control of domestic appliances or even direct load control by a system operator or utility, allowing smart meters to interact with a home area network manager, have been considered to be supported by this type of networks [26].

### Emerging Wireless Solutions and Opportunities

The use of conventional wireless technologies, typically single-hop, for SG application has been subject to a general discussion and acceptance in specific scenarios. As an alternative, multi-hop technologies can also be considered in this context. One of such cases can be found in the recent IEEE 802.15.4 g, which defines a mesh network of flat smart meters to support smart metering. Nevertheless

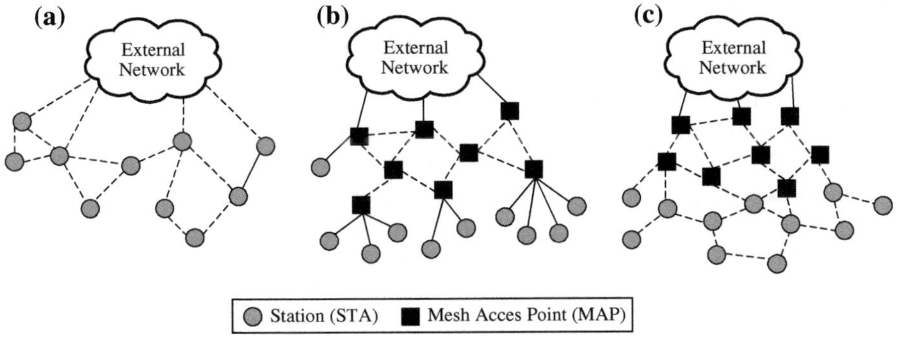


**Fig. 12.2** Wireless mesh network topologies

this type of solution is not tailored for last-mile implementations, found in distribution networks, where node distances can vary significantly [27].

Given the characteristics of the distribution networks, wireless mesh networks can be used as a means to extend the coverage of a communications network allowing access to an infrastructure network that can be wired or wireless. As shown in Fig. 12.2, several topologies can be explored for this purpose that range from a completely flat network where all nodes can relay information to more complex and tiered approached where mesh access point nodes are responsible for relaying the information. Nevertheless, hybrid strategies can also be implemented.

Wireless Mesh Networks (WMN) can potentially offer higher levels of redundancy and robustness in terms of data communications, in the event of temporarily or permanent node or link loss. Routing algorithms can be adapted to dynamic network topologies to deal with channel varying conditions. However, the use of WMN typically aggravates some of the limitations of conventional single-hop wireless communications. Associated challenges like fairness [28], scheduling in terms of fair-sharing [29] or resource optimization [30] and cross-layer mechanisms like scheduling with congestion control are already addressed by the research community. The use of WMN in distribution networks is facilitated given the scope in which they have to operate. Most devices are not expected to have mobility characteristics and planning is made usually in a medium/long term perspective. Power constraints are very low and the use of battery support may allow the survival of such systems in the event of power failures or even in more severe cases where blackout occurs. The geographic span of last mile SG communication networks may not require a high number of mesh nodes, namely relays, to ensure a proper coverage. Besides, data aggregation, typically upstream, may be used to allow mode flexible scheduling schemes.

Solutions like Wi-Fi network Infrastructure eXtension (WiFIX) allow the extension of wired communication networks using simple and efficient WMN [31]. This solution is regarded as an alternative to the IEEE 802.11 s which defines mesh networking within the 802.11 family. The original WiFIX system has evolved [31, 32] and it consists of a simple and efficient tree-based algorithm for stub WMN that runs over 802.11 legacy MAC. Being a system that runs above the MAC layer it can be implemented in wireless technologies other than IEEE 802.11, making it suitable of SG last-mile scenarios.

An opportunity for wireless communications in SG can be found in the disconnection of the analog TV broadcast services and consequent transition to digital terrestrial TV. This can potentially allow frequency bands to be available for SG applications. These are in particular desirable bands since they are associated with transmission ranges which are compatible with the last-mile segment. Even when using multi-hop solutions these frequencies could allow a similar level of coverage but with less number of hops. Within this RF reusability spirit, TV white spaces communications have also been considered as potential solution for the support of SG services, although current regulation restrictions limit the reuse of these bands [26].

## 12.3 Microgrid Control and Emergency Functionalities

The stability of the conventional power system relies in the regulation of the system's voltage and frequency, which is achieved by the control of large synchronous units. However, the majority of MG systems are inverted dominated being inertialess, due to the inexistence of rotating masses directly coupled to the grid [3].

MG components such as Microsource (MS), storage units and EV require a power electronic interface (DC/AC or AC/DC/AC) in order to be connected and exchange power with the AC distribution grid. The inverters control system is responsible for controlling the active and reactive power exchanged with the grid, ensure high power quality and grid synchronization functionalities. Besides, some additional services may be required by the grid operator, namely voltage and frequency regulation, harmonic compensation or even active filtering [33, 34]. The two most common control strategies used for the inverters connected to the power systems are [3, 6, 33]:

- PQ inverter control: the inverter operates by injecting a given active and reactive power set point into the grid. The control is implemented as a current-controlled voltage source, with the objective of exporting a controlled amount of power (active and reactive) to the grid (these inverters are usually referred as grid tied inverters).
- Voltage Source Inverter (VSI) control: the inverter is able to control the frequency and magnitude of its output voltage. A VSI, also referred as a grid forming unit, defines the grid voltage and frequency reference by assuring a fast response in order to continuously balance power generation and loads.

In interconnected mode, the MG voltage and frequency reference is given by the upstream network. In these operating conditions the inverters can be operated

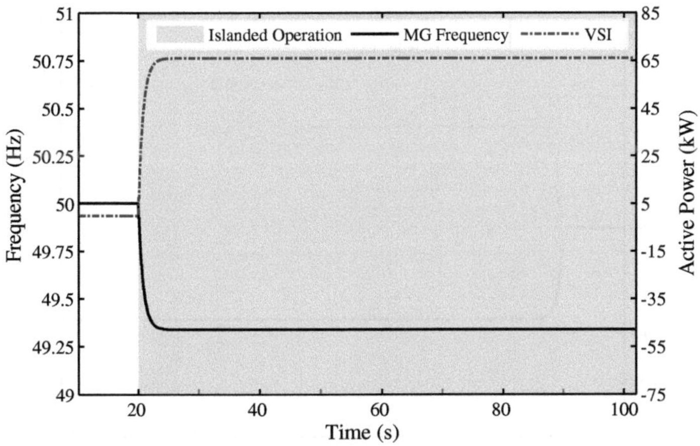


**Fig. 12.3** SMO strategy: MG frequency and active power injected by storage unit

with a PQ control strategy, being the set-points for P and Q provided either by the MGCC or by other forms of local control. However, when the MG disconnects from the main grid, due to a fault upstream or a planned outage, it is necessary to establish the voltage in magnitude and frequency. Similarly to what has been defined for large power systems, the MG requires specific primary and secondary frequency control strategies, in order to maintain power balance between generation and loads and restore frequency to adequate values [3, 6]:

- Primary Control, which is performed locally to continuously assure the balance between load and generation.
- Secondary Control, which ensures voltage and frequency regulation in order to maintain these electrical variables within acceptable limits. This control level also includes the coordination of synchronization actions to reconnect the MG to the main grid after islanding operation. It is usually performed at the central controller such as the MGCC.

### *12.3.1 Primary Voltage and Frequency Regulation*

In conventional power systems, synchronous generators share load variations in accordance with their droop characteristics. This general principle of conventional power systems can be applied to inverters in order to change its output frequency as a function of the output power. This principle is also important for parallel operation of variable frequency AC voltage sources, similarly to what happens in conventional power systems. In order to provide voltage and frequency regulation, the storage unit grid coupling inverter is controlled as a VSI device with external droop control loops, as in Eq. (12.1) [3, 6],

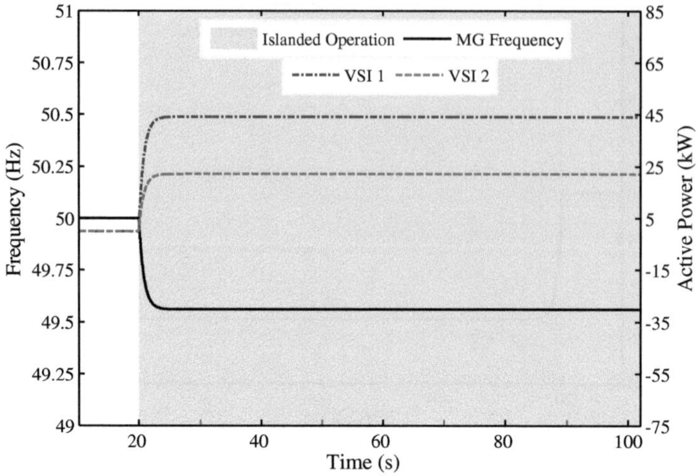


**Fig. 12.4** MMO strategy: MG frequency and active power injected by the two storage units (VSI 1 and VSI 2)

$$\begin{aligned} \omega &= \omega_{grid} - k_P \times P \\ V &= V_{grid} - k_Q \times Q \end{aligned} \tag{12.1}$$

where $P$ and $Q$ are the inverter active and reactive power outputs, $k_P$ and $k_Q$ are the droop slopes (positive values) and $\omega_0$ and $V_0$ are the idle values of the angular frequency and voltage (at no load conditions).

Figure 12.3 shows a MG frequency response after an unplanned islanding at t = 20 s. The power unbalance between the MG generation and load will cause a frequency variation in accordance to the active power-frequency (P-*f)* droop, similarly to the response of the conventional synchronous machines. In this case the MG was importing about 66 kW prior to islanding, which was compensated by the storage unit. This control principle allows the VSI to react to system disturbances based only on information available at its terminals, not requiring fast communications between the MC controllers and the MGCC.

When the MG operates with a single VSI, the MG is said to operate in Single Master Operation (SMO) mode. However, the MG can operate with more than one VSI controlled through droop functions, in a Multi-Master Operation (MMO) strategy. In this case, when a disturbance occurs, the storage units will respond to the power imbalance and share the power variations according to their droop characteristics, as in Eq. (12.2).

$$\Delta P = \sum_{i=1}^{n} \Delta P_i \tag{12.2}$$

where $\Delta P$ is the unbalance between the MG generation and load and $\Delta P_i$ is the active power injected by the i-th VSI connected to the MG.

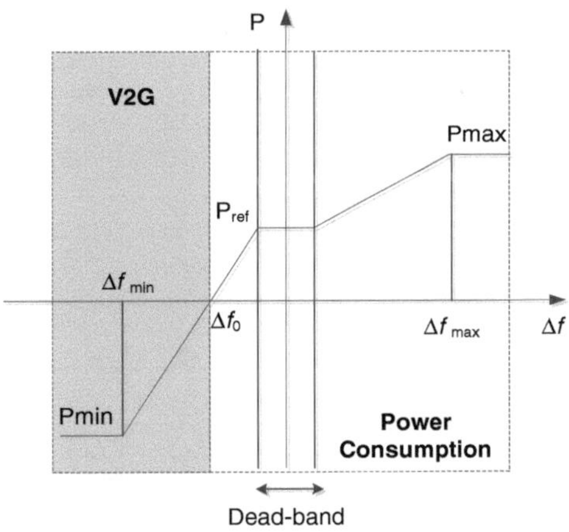


**Fig. 12.5** EV frequency-droop characteristic

The MG frequency deviation and the power sharing among the VSI can be determined by the system of equations represented in Eq. (12.3).

$$\begin{bmatrix} 1 & k_{P1} & 0 & \cdots & 0 \\ 1 & 0 & k_{P2} & \cdots & 0 \\ \cdots & \cdots & \cdots & \ddots & \vdots \\ 1 & 0 & 0 & \cdots & k_{Pn} \\ 0 & 1 & 1 & \cdots & 1 \end{bmatrix} \times \begin{bmatrix} \omega' \\ \Delta P_1 \\ \Delta P_2 \\ \vdots \\ \Delta P_n \end{bmatrix} = \begin{bmatrix} \omega_{grid} \\ \omega_{grid} \\ \vdots \\ \omega_{grid} \\ \Delta P \end{bmatrix} \tag{12.3}$$

where $\omega'$ is the post-disturbance MG angular frequency and $\omega_{grid}$ is the pre-disturbance MG angular frequency.

Figure 12.4 shows the MG frequency response after an islanding when adopting an MMO strategy. The power disturbance is shared by the two units as a function of the storage unit capacity. In this case VSI 2 has half of VSI 1 capacity. Comparing to the SMO strategy in Fig. 12.3 the resulting frequency deviation is smaller, since the power injected by each unit is also lower. MMO strategy is more robust than SMO, because the voltage and frequency control relies in several (distributed) VSI units [3, 6].

## *12.3.2 EV Contribution to Primary Frequency Support*

The deployment of EV within the new mobility paradigm introduces significant power requirements for battery charging purposes which impacts the operation of the LV network by affecting voltage profiles and line flows. In order to mitigate the resulting impacts, the implementation of smart charging strategies has been shown to be an effective way of dealing with the mentioned problems. Additionally, during MG autonomous operation, EV can also be exploited in order to reduce the

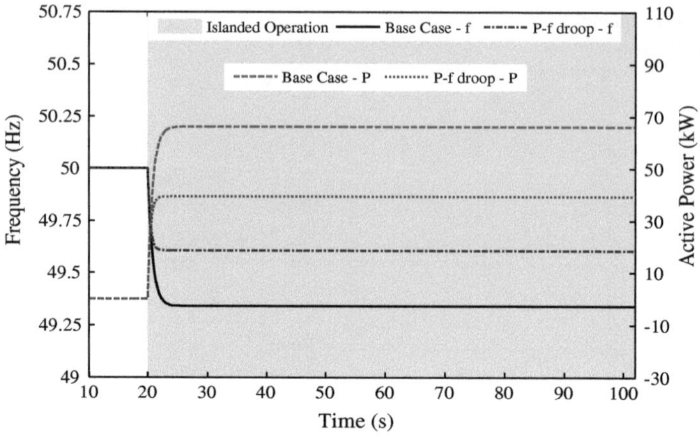


**Fig. 12.6** MG islanding with EV control: MG frequency and VSI active power response

**Fig. 12.7** MG frequency and EV active power response

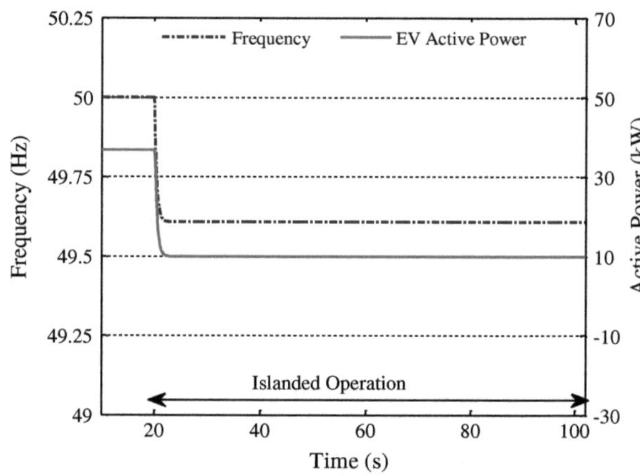


central storage capacity required to assure MG system robustness during islanding conditions [4, 5, 14]. A P-*f* droop control strategy implemented at the EV charger has been proposed in [4, 5], where the EV will modify the power exchanged with the LV grid based on the MG frequency.

As shown in Fig. 12.5, for frequencies around the nominal value the EV will charge the battery at a pre-defined charging rate ($P_{ref}$). If a disturbance occurs and the frequency drops below the dead-band minimum, the EV reduces its power consumption, reducing the load of the system. When the MG frequency overpasses the frequency dead-band maximum, the EV can also increase its power consumption. For large disturbances, causing the frequency to go below the zero-crossing frequency ($f_0$), the EV starts to inject power into the grid (Vehicle-to-Grid—V2G functionality). When the MG frequency becomes out of the pre-defined frequency range the vehicle will inject/absorb a fixed power ($P_{max}$ or $P_{min}$).

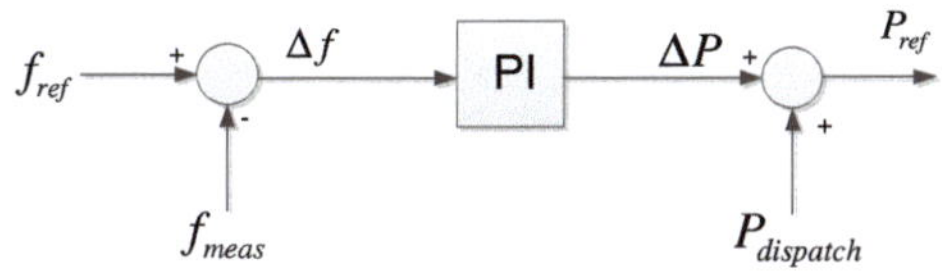


**Fig. 12.8** Local secondary frequency control of PQ controlled MS

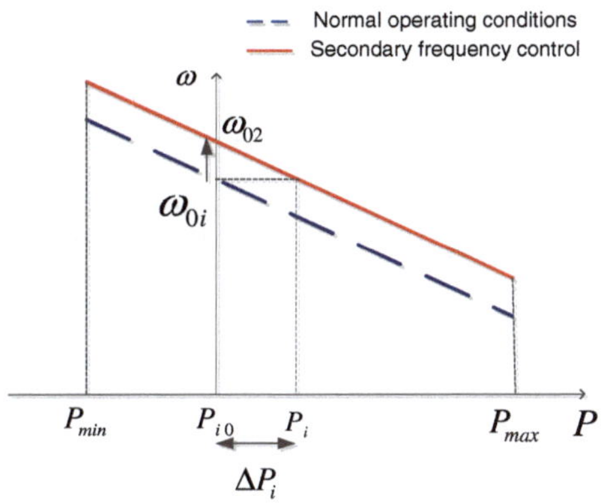


**Fig. 12.9** Local secondary frequency control for VSI controlled MS

The definition of the EV control parameters will depend on the EV charger characteristics and on the willingness of EV owners to participate in such services. These parameters may differ from grid to grid and can be changed remotely by the MGCC in order to promote adequate coordination with the MG frequency regulation mechanisms (load shedding schemes, availability of energy storage devices and their state of charge).

Figure 12.6 compares the MG frequency and the VSI active power response after an islanding at t = 20 s when a SMO control strategy is adopted. As shown in Fig. 12.7 the EV controlled through P-f droop have reduced their power consumption from 36.75 to 10 kW, reducing the MG load and consequently the power injected by the VSI.

### *12.3.3 Secondary Load–Frequency Control*

The secondary frequency control ensures that after a disturbance the frequency returns to its nominal value, by increasing the power output of controllable MS, such as microturbines and/or fuel cells. As shown in Figs. 12.3 to 12.6, during a transient the power balance is assured by the MG energy storage devices, which would keep on injecting or absorbing active power until the frequency is restored to the idle value.

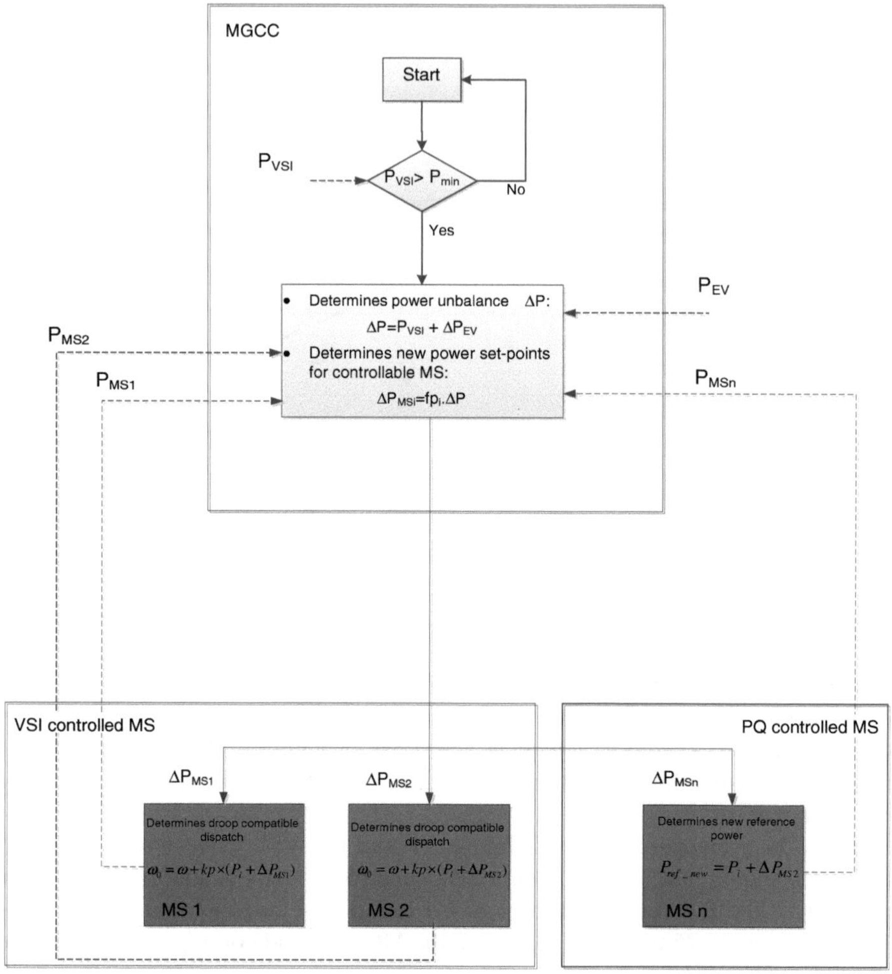


**Fig. 12.10** Centralized secondary control algorithm

In [35] two secondary load–frequency control strategies are identified: one implemented locally at each controllable MS and another centralized and mastered by the MGCC. Local secondary control is implemented as an external control loop of the active power control of controllable MS, as shown in Fig. 12.8. If the MS inverter is controlled with a PQ strategy, the reference active power ($P_{ref}$) to be injected by the MS is determined by a PI controller based on MG frequency error. If the MS inverter is controlled as VSI, local secondary control can be performed by updating the idle frequency $\omega_{0i}$ of the MS P-*f* droop function, as represented in Fig. 12.9. As proposed in [3, 35] a PI controller can be used to determine the new idle frequency ($\omega_{02}$) based on the MG frequency deviation in order to update the

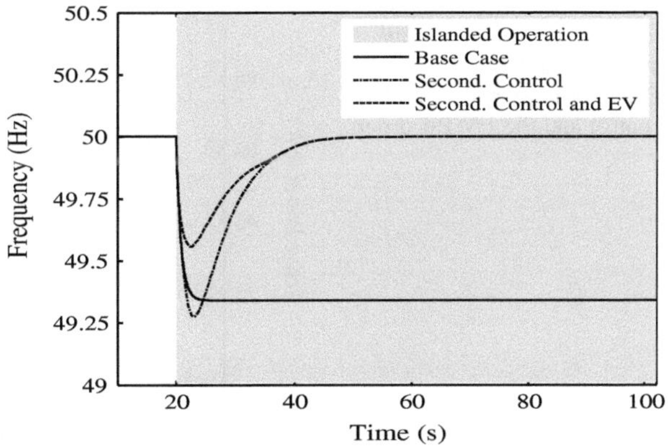


**Fig. 12.11** MG frequency response considering: only primary regulation (*Base Case*), primary and secondary control and the EV P-f droop

active power droop characteristic. The main advantage of local secondary frequency control is that it only relies on local measurement to define the new reference power. However, the MS active power response will also depend on the PI controller parameters.

Centralized secondary control determines the new MS active power set-points at the MGCC and then sends them to the respective local controllers. Based on the MG frequency deviation and on an economic dispatch algorithm, the MS new set-points are determined according to participation factors, which are based on pre-defined cost functions or on the MS capacity [35, 36].

Taking advantage of the data provided by the smart meters a centralized secondary control strategy based on active power unbalance can be designed. The algorithm shown in Fig. 12.10 is implemented at the MGCC and performs the MS power dispatch when the power injected by the storage unit installed at the MV/LV substation surpasses a defined limit.

As shown in Fig. 12.10, the MG power unbalance (ΔP) is determined based on the power injected by the MG main storage unit ($P_{VSI}$) and on the EV injected power ($P_{EV}$). The contribution of each controllable MS will be defined through a participation factor ($f_{pi}$), which is determined (for unit *i*) by the ratio between its reserve and the total controllable MS reserve, as in Eq. (12.4),

$$\begin{cases} fp_i = \dfrac{R_i}{R} \\ \Delta P_{MS_i} = \Delta P \times fp_i \\ \sum_{i=1}^{n} \Delta PMS_i = \Delta P \end{cases} \tag{12.4}$$

where $\Delta P_{MSi}$ is the emergency active power step of unit *i*, $\Delta P$ is the active power unbalance following a disturbance and $R_i$ is the reserve capacity of unit *i*. The resulting set-points are then sent from the MGCC to local MC. In order to be

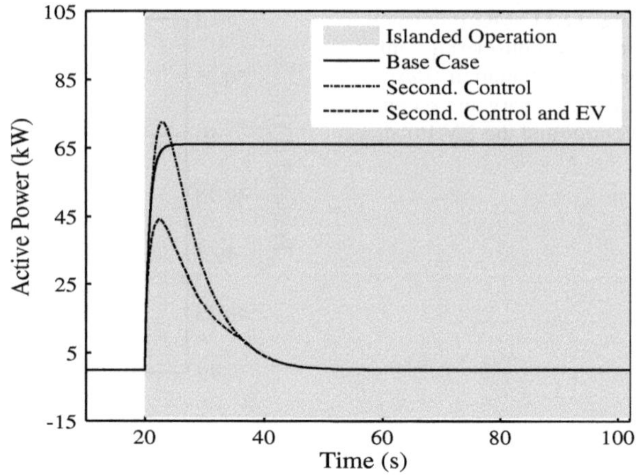


**Fig. 12.12** VSI active power response considering: only primary regulation (*Base Case*), primary and secondary control and the EV P-f droop

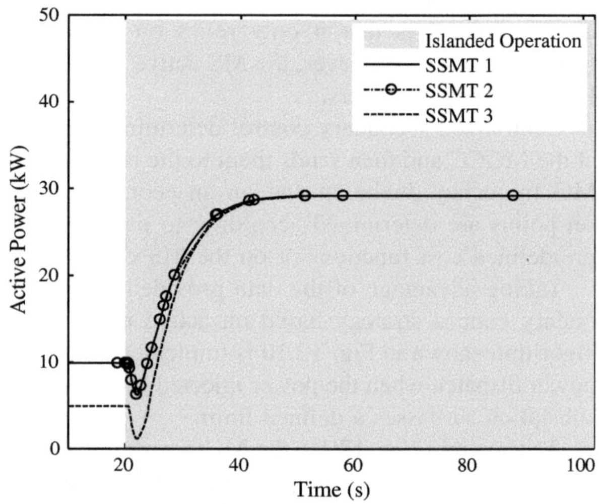


**Fig. 12.13** Controllable MS power dispatch

compatible with both SMO and MMO modes of operation, in the case of MS controlled as VSI, the algorithm determines new idle values for the angular frequency of the MS droop characteristics ($\omega_0$), in order to obtain the desired power adjustment ($\Delta P_{MSi}$). For PQ controlled MS, the new power set-point is directly sent to the MC.

Figure 12.11 compares the MG frequency when considering primary and secondary frequency response with and without the participation of EV through P-*f* droop characteristic (MG islanding takes place at t = 20 s).

As shown in Figs. 12.12 and 12.13, secondary control reduces the power injected by the storage unit, recovering the frequency to the nominal value (in this case 50 Hz). In this case, in order to compensate the 66 kW of power imported from the MV network, it was assumed that the MG comprises also three Single Shaft Microturbines (SSMT) running on gas that were dispatched to their maximum capacity (30 kW in this scenario) after islanding. The typical response of this type of units is presented in Fig. 12.13.

## 12.3.4 Demand Response

Controlling the MG loads may be required when operating autonomously for the two distinct situations:

- MG with insufficient generation reserve. In this case, the load will have to be curtailed until the system reconnects to the upstream network or the generation increases.
- MG with sufficient reserve. Although the MG has sufficient reserve, it may happen that, for a certain disturbance, the storage units State of Charge (SOC) is not able to ensure a secure operation or the frequency deviation expected for a given disturbance surpasses admissible limits. The implementation of a temporary load curtailment will help improve the MG frequency regulation capacity and reduce the solicitation from the storage unit(s).

The MG load shedding schemes are usually based on under-frequency or on rate-of-change-frequency relays [3]. However, if the MG has sufficient generation capacity, some demand response strategies may help dealing with transient disturbances resulting from load or variable generation, compensating the slow response of MS to power control signals.

In [13] an Adaptive Hill Climbing (AHC) load control method was proposed in order to improve the MG voltage and frequency regulation. The AHC determines the minimum percentage of responsive loads (electric water heaters) which should be disconnected/connected based on the MG frequency deviation. The commands are then sent to the electric water heaters controllers. The method was tested considering a small system with a diesel generator, which ensured secondary frequency control. Still, in the absence of synchronous machines, the MG frequency response will depend on the storage unit and controllable MS response to control signals.

Several frequency responsive load control methods based on local frequency measurements have been proposed in literature. The Frequency Adaptive Power Energy Reschedule (FAPER) method [10] proposes the control of consumers loads based on a local frequency measurement. In case of a sudden load increase or generation loss causing a severe frequency disturbances, the loads will be disconnected during the time of response of the secondary control. In [11], a Dynamic Demand Control (DDC) method is proposed in order to control refrigerators. An alternative controller to the conventional thermostat will connect/disconnect the refrigerator based on the frequency in addition to the freezer air temperature. The results presented have shown that the DDC method potentially improves power

**Fig. 12.14** Priority load shedding algorithm

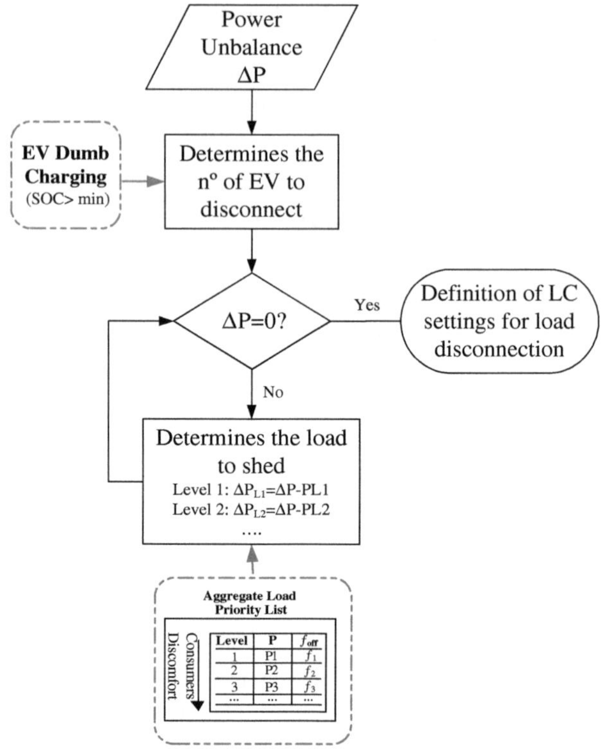


system stability, when facing the sudden loss of generation, increase of load or dealing with a large penetration of variable generation. In [12] a load reduction algorithm is proposed considering the local frequency measurement from smart meters. The load control method aggregates typical domestic appliances according to the level of discomfort caused by their disconnection.

Based on the frequency load control methods proposed in [10–12, 14] a selective load control scheme is proposed in order to control the MG loads, where the consumers will be able to prioritize the availability of its home appliances to participate in the MG load control scheme. At the MGCC an online tool periodically characterizes the MG operating state and evaluates the MG resilience in the event of a disturbance affecting MG load and generation balance. Based on the load priority list aggregated at the MGCC and on the amount of load which is required to be disconnected ($\Delta P$), the load shedding algorithm shown in Fig. 12.14 defines a set of actions to be parameterized at the local control level, namely: load disconnection in LC, EV disconnection or droop parameterization at VC.

The activation/parameterization of the controls is performed previously to the disturbances. The MGCC will send to the SM a frequency threshold for the activation of the different load priority levels, which is then communicated to LC and VC. When there is available reserve capacity, load reconnection takes place

Fig. 12.15 MG frequency response with EV control and demand response

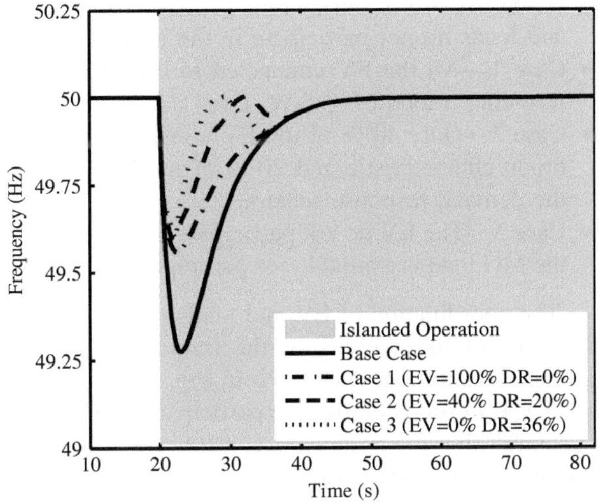


Fig. 12.16 VSI active power output with EV control and demand response

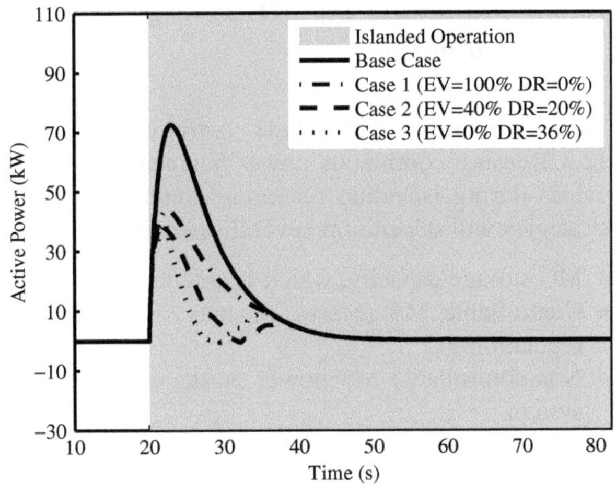


when the frequency recovers to a value close to the nominal one (for example, at least 49.9 Hz). In order to avoid simultaneous load reconnections, a random time delay is considered for load reconnection after the moment frequency effectively reaches the pre-defined value.

Figure 12.15 shows the MG frequency behavior when the load participates in the MG frequency regulation following islanding at t = 20 s. Four scenarios were tested:

- Base Case—considering only primary and secondary frequency control. The EV and loads do not participate in the MG frequency regulation.
- Case 1—All the EV connected to the MG are controlled with P-*f* droop, representing a total of 49 kW. Load control is not considered.
- Case 2—Only 40 % of the EV connected to the system are controlled with P-f droop characteristic and 20 % of the MG load is available for participating in the demand response scheme.
- Case 3—The EV do not participate in the MG frequency regulation and 36 % of the MG load is available for participating in an active demand response scheme.

The coordination of EV and load control strategies demonstrates great benefits in terms of the severity of the frequency disturbance and consequently of the storage solicitation. As shown in Fig. 12.16, similarly to the results obtained for the EV P-f droop control, the participation of load provides spinning reserve to the system during the secondary control response time. Further improvements on the reconnection of loads can be performed in order to minimize the transients.

### *12.3.5 Managing the MG Energy Balance Following Islanding*

The MG frequency and voltage control mechanisms described in Sects. 12.3.1 and 12.3.3 ensure continuous power balancing and frequency restoration to nominal values during islanding operating conditions. However, the effectiveness of the strategies will depend on several operating conditions, such as:

- MG storage capacity, which is essential to ensure primary frequency regulation.
- Controllable MS reserve capacity, in order to perform secondary frequency regulation.
- Non-controllable MS power production, which acts as a negative load in the system.
- MG flexible load, including the EV connected to the LV network.
- MG load in comparison to total generation.

The information sent by the smart meters to the MGCC can be used to characterize the MG operating conditions and coordinate the available resources in order to improve the MG resilience to severe disturbances. Updating the MG emergency operation strategies according to the MG actual operating state will allow:

- Minimizing the amount of load to be disconnected.
- Minimizing the time during which loads are disconnected.
- Ensure that the MG has sufficient storage capacity to ensure primary frequency control following a given disturbance.
- Maintain frequency excursions within admissible limits.

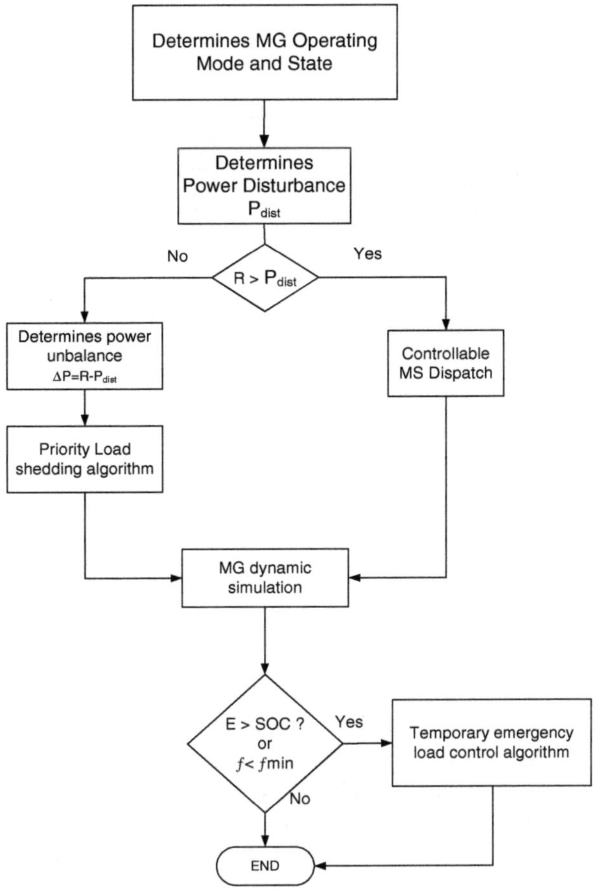


**Fig. 12.17** Algorithm for defining the MG emergency operation strategy

The algorithm presented in Fig. 12.17 will run online at the MGCC in order to analyze the MG operating conditions and then identify the most appropriate actions to ensure a secure islanding. It consists in four main steps, namely:

(1) *Characterize the MG operating state* based on the information sent by smart meters. The algorithm determines the MG storage capacity, microgeneration reserve capacity, power flow between the MG and the MV network and load (divided between the controllable and non-controllable load).
(2) *Determine the severity of the disturbance*. If the MG is operating interconnected to the main grid, the occurrence of an unplanned islanding will cause an active power unbalance ($P_{dist}$) equal to the power exchanged with the MV network. If the MG is operating islanded, the power unbalance will result from the changes in loads or generation. It is important to notice that planned islanding events are not a key concern, since adequate control action can be taken in order to balance the MG load and generation prior to islanding, thus minimizing the associated transient phenomena.

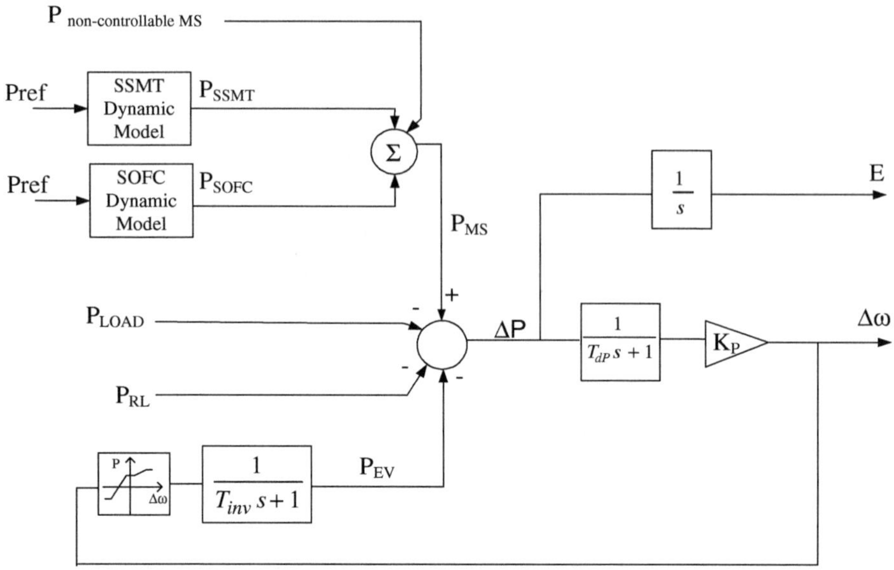


**Fig. 12.18** MG simplified dynamic model to run at the MGCC

(3) *Determine the amount of load to shed*. In case the MG does not have enough reserve capacity, it is necessary to exploit emergency responsive loads and shed some part of the MG load in order to ensure power balance.

(4) *Evaluate the security of the MG during an unplanned islanding*. The algorithm determines if the MG has enough storage capacity to ensure power balance during the time required for restoring the frequency to nominal values or if the frequency goes out of admissible limits. Otherwise, it might be necessary to shed some load temporarily.

In Step (3) the amount of load to shed ($\Delta P_{shed}$) is determined as in Eq. (12.5), based on the MG reserve capacity (R) and on the active power unbalance ($P_{dist}$) estimated in Step (2),

$$\Delta P_{shed} = R - P_{dist} \tag{12.5}$$

During islanding conditions, if the MG lacks energy to ensure power balance, the system will collapse. The energy balance in the storage unit ($E$) can be determined by Eq. (12.6).

$$E = E_{\Delta P} - E_{MS} - E_{VE} - E_{RL} \tag{12.6}$$

where $E_{\Delta P}$ is the energy resultant from the power unbalance after the MG disturbance, $E_{MS}$ is the energy provided by the MS (both controllable and non-controllable) and $E_{VE}$ is the energy resultant from the response of the EV to the frequency deviation. $E_{RL}$ results from: (1) the energy not supplied to responsive loads, which were disconnected due to the emergency state of the MG or (2) the energy to be supplied to dump loads, in case the disturbance leads to an excess of generation regarding the MG load.

Based on Eq. (12.6) it is possible to evaluate the energy required to ensure secure islanded operation. However, the energy injected/absorbed by the MG storage unit will depend on the response of the controllable MS such as SSMT or fuel cells (for example Solid Oxide Fuel Cells—SOFC). Since these units present a non-linear power response, the energy required to ensure MG robust operation following a disturbance cannot be accurately estimated using simplified linear models.

In order to overcome this difficulty, a simplified MG dynamic model that is represented in Fig. 12.18 was adopted. The model consists on a single equivalent bus, considering only the load, generation and storage resources. The model adopted neglects the presence of the LV network and the dynamics of the power converters, which are faster than the MG dynamic behavior. The storage unit and EV are represented by their external P-$f$ control loop. $T_{dP}$ is the delay related to the response of voltage source inverters and $T_{inv}$ is the delay of the EV grid coupling inverter [13]. Loads and non-controllable MS are represented as constant active power.

The load shedding and the generation emergency dispatch are used as inputs of the MG dynamic model in order to evaluate the energy balance within the MG and the expected frequency deviation for a given period. As outputs, the model provides the total energy injected by the storage units and the MG frequency response. Based on these values, the algorithm then verifies if the MG storage units have sufficient capacity to ensure power balance and if the minimum frequency does not violate the admissible frequency limits ($f_{\min}$). If these conditions are violated, it is possible to follow an iterative procedure to determine an additional amount of load to be disconnected and compensate the slow response of some MS to power control signals. The frequency threshold for its activation can also be identified, based on the frequency response obtained from the MG simplified model.

The proposed approach is intended to support MG islanding operation during short periods of time (i.e. less than 1 h). For larger time frames of operation in islanding conditions, complementary approaches need to be considered, involving forecasting of loads with different degrees of flexibility (including EV), as well as forecasts for renewable based microgeneration.

## 12.4 Microgrid Service Restoration Procedure

The occurrence of general blackouts is a rare event but has severe economic and social impacts, as a consequence of the long restoration times resulting from complex procedures followed by system operators [37–39]. The unique characteristics of each electric power systems preclude the definition of general restoration plans. The restoration procedures usually consist on pre-established guidelines and operating routines, involving transmission and distribution operators. For each step, system operators have to coordinate generation and load to maintain system stability, ensure reactive power balance to keep system voltages within acceptable limits and limit switching transient voltages, amongst others [37–39].

When compared to conventional power system restoration, the MG Black Start (BS) procedure will benefit from a considerable problem size reduction, and hence the reduction of the number of controllable variables. However, as discussed previously, the MG is an inverter dominated grid, thus requiring very specific voltage and frequency regulation strategies exploring the MG energy storage and the MS generation capacity [8].

The MG restoration procedure is triggered by the MGCC when a general or local blackout occurs, or when the MV network is not able to restore MG operation after a pre-defined time interval. Similarly to conventional power systems, the MG restoration procedure consists on a sequence of actions to be checked and performed by MG local controllers in coordination with the MGCC. However, the procedure should be fully automatic without requiring the intervention of distribution network operators.

In order to perform service restoration at the MG level, it is assumed that the MG is equipped with [8]:

- MS with black start capability, capable of communicating to the MGCC their generation availability and operational status.
- LV switches to disconnect the MG feeders, loads and MS in case a generalized fault occurs.
- Communication infrastructure powered by dedicated auxiliary power units, in order to ensure the communication between the MGCC and the local controllers.
- Adequate protection equipment in order to protect MS and the LV grid from the fault currents and to isolate the faulted area. Since the BS procedure involves a step-by-step connection of MS to the LV grid, the short-circuit power at the point where protection devices are installed will change. Thus, under such protection strategy, it is assumed that during the restoration procedure the MGCC has the ability to change the protection devices settings in order to efficiently detect and isolate MG faults.

In order to provide BS capability, controllable MS such as SSMT and or fuel cells require additional DC storage capacity to power the ancillary equipment required when starting the MS. Also, in order to supply local loads, the MS coupling inverter has to be controlled as a VSI. Therefore, during the MG restoration procedure a Multi-Master Operation (MMO) strategy could be adopted increasing the MG robustness during the procedure. When the MG stabilizes before synchronizing to the upstream network, the MS grid-coupling inverters can switch its control strategy to PQ.

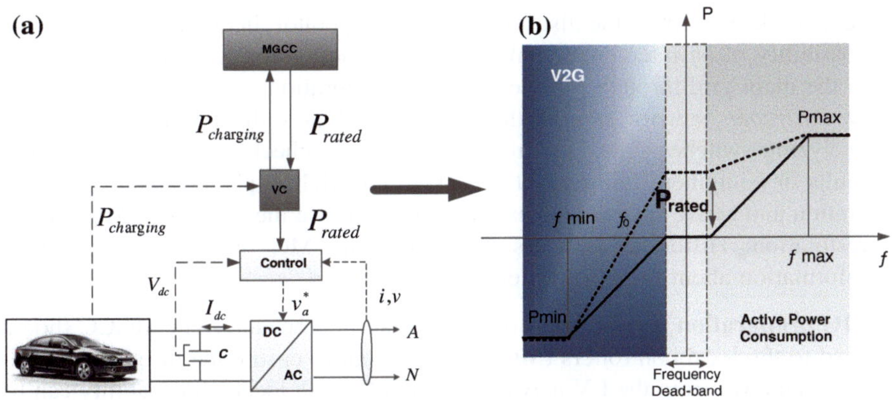


**Fig. 12.19** EV control and interaction with the MGCC for the MG service restoration

## 12.4.1 Participation of EV in the MG Restoration Procedure

The EV can be integrated in the MG restoration procedure as grid supporting units. The EV available to provide this service will be connected to the MG during the first steps of the restoration procedure. They won't be allowed to charge at the nominal frequency. Instead, the EV should be synchronized to the grid with a zero power set-point, as shown in Fig. 12.19. As a result, during the first steps of the procedure and whenever the frequency remains within the EV P-f droop frequency dead-band, the EV will not exchange power with the MG. However, when the loads and MS begin to be reconnected, the EV will inject or absorb power in order to reduce the resulting frequency disturbance, consequently reducing the solicitation from the main storage unit and providing frequency support to the MG.

## 12.4.2 BlackStart Methodology

After a general blackout, the MGCC will trigger the BS procedure, based on the information about the MG pre-disturbance load scenario. Following the MG restoration strategy proposed in [8], the overall procedure can be organized in the following sequence of events:

1. **MG status determination**. The MGCC evaluates the network status both upstream and downstream:

a. *Upstream network status*—the MGCC only activates the restoration procedure if there is no alternative to reconnect to the main grid. Before activating the procedure, the MGCC waits for a confirmation of the occurrence of local or general blackout from the distribution system operator, in order to eliminate the possibility of an interconnection switch malfunction, or possible reconnection to the main grid through MV network reconfiguration.

b. *Downstream network status*—the MGCC evaluates the LV network status, analyzing switches status and alarms, in order to check the existence of local faults or equipment failures. At this stage the MGCC also evaluates the generation and active load resources, in order to ensure the successful MG service restoration. Historical data resulting from the MG operation can provide information about the priority loads to be restored.

2. **MG preparation** in order to start the restoration procedure. The MGCC should send to the local controllers (MC, LC), a signal to ensure the disconnection of loads and MS from the LV network. Then, MS with local BS capability can be restarted and be used in order to power some of the local loads. This procedure ensures that the MS with storage capacity providing back power to their local loads are not energizing larger parts of the LV network.
3. **MG energization**, connecting the MG central storage unit at the MV/LV substation and closing the substation LV feeder switches. The connection of the storage unit in no load/generation conditions ensures that the MG is operating at nominal frequency and voltage.
4. **Synchronization of the running MS to the MG**. The synchronization is enabled by the MGCC, being the necessary conditions—such as phase sequence, frequency and voltage differences (both in phase and amplitude) checked by the MC, through synchro-check relays.
5. **Connection of EV**. The MGCC will send a control signal to all smart meters in order to connect the EV chargers with a zero power set-point and enable the P-f droop functionality.
6. **Coordinated reconnection of loads and non-controllable MS**, considering the available storage capacity and local generation, in order to avoid large frequency and voltage deviations during load and non-controllable generation connection.
7. **Enable EV charging**. After restoring priority loads and generation, if the MG remains with sufficient reserve capacity the MGCC will enable the EV charging. In order to avoid further disturbances, the MGCC will send new reference power to the smart meters and dispatch the controllable MS to meet the new load.
8. **MG synchronization with the main grid**, when the service is restored at the MV network. The MGCC should receive a confirmation from the distribution network operator to start the synchronization with the upstream network. The synchronous conditions are checked locally through a synchro-check relay.

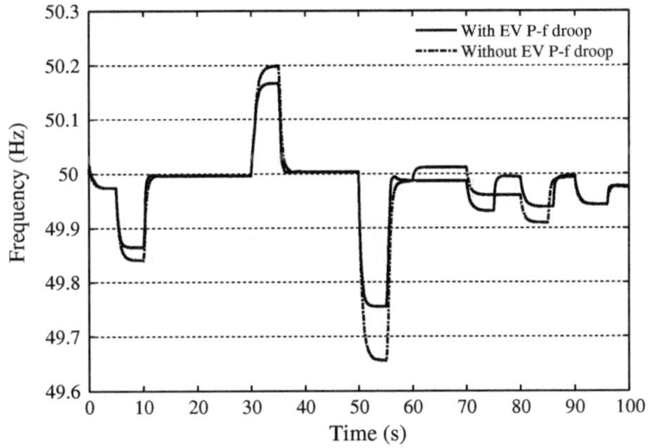


**Fig. 12.20** MG frequency response during the MG rebuilding

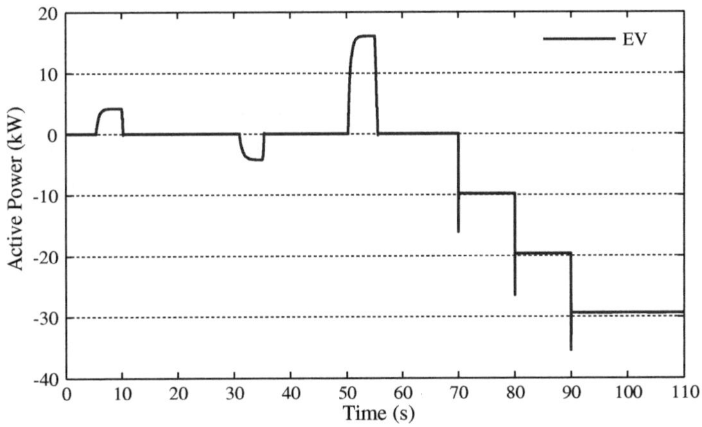


**Fig. 12.21** EV total power output during MG rebuilding

### *12.4.3 Analyzing the Benefits of EV Participation on the MG Rebuilding*

Figure 12.20 compares the MG frequency behavior during the reconnection of loads and non-controllable MS when considering both non-controlled charging and V2G service. At t = 5 s the MGCC enables the reconnection of the first load block causing a frequency deviation of 1.5 Hz, which is then corrected by the secondary control by increasing the controllable generation output. The load reconnection is interleaved with the reconnection of the non-controllable MS at t = 30 s, followed by the reconnection of the second block of loads at t = 50 s. After the restoration

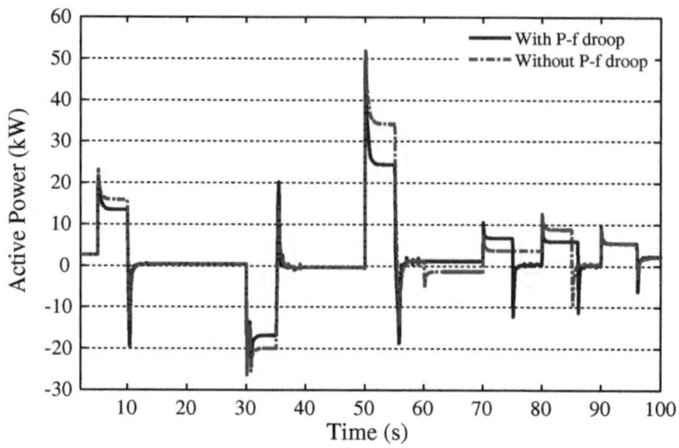


**Fig. 12.22** Active power injected by the MG main storage unit

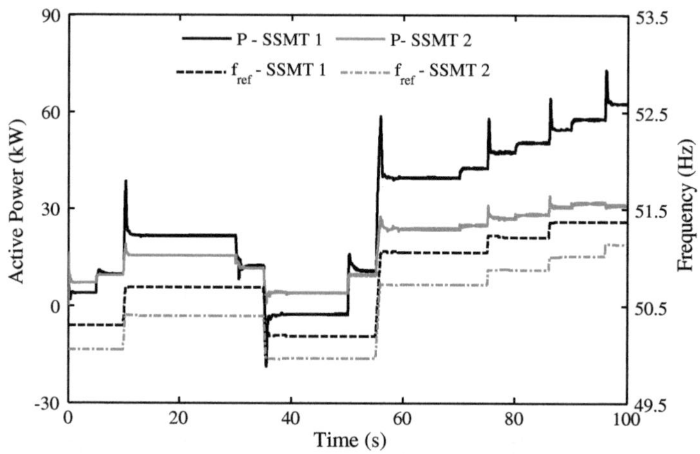


**Fig. 12.23** SSMT active power response to the MGCC centralized secondary frequency control

of load there is still sufficient reserve capacity and the MGCC enables the increase of the EV charging power, which is divided in three steps: 20 % of their nominal power at t = 70 s, 40 % at t = 80 s and 60 % at t = 90 s.

As it can be observed in Figs. 12.20 and 12.21, the active participation of the EV in the frequency regulation reduces the frequency deviation caused by the reconnection of loads and the non-controllable MS. When the frequency deviation falls outside of the P-f droop frequency dead-band (from 49.9 to 50.1 Hz), the EV inject power during the reconnection of loads and increase their consumption during the reconnection of the non-controllable MS. The participation of EV in frequency regulation becomes more significant for large frequency disturbances, according to their droop characteristics. Following the MG frequency stabilization (t = 65 s) and since the MG had enough reserve capacity, the P-f droop parameter $P_{rated}$ is increased gradually, in order to allow the EV to charge their batteries.

Figure 12.22 shows the MG main storage power output for both case studies. As demonstrated, when considering the V2G operation mode there is a reduction of the active power required from the MG main storage unit.

The active power response of the SSMT to the centralized secondary load frequency control is shown in Fig. 12.23. The SSMT change their power output considering the frequency set-points sent by the MGCC. During the restoration procedure the SSMT coupling inverters are controlled as VSI with a MMO strategy. The DC coupled storage provides immediate power response to the MGCC frequency control signals, compensating the SSMT slower power response.

## 12.5 Experimental Validation of MG Emergency Operation

Following the track record on MG research, significant demonstration activities have been exploited worldwide [15–20]. In Europe, the Association of European Distributed Energy Resources Laboratories (DERlab) clusters European DER laboratories in order to widen the access to the available testing facilities. In the United States, the CERTS Microgrid Laboratory Test Bed aims to demonstrate peer-to-peer and plug-and-play functionalities for different DER, while exploiting the MG islanding operation mode. Large scale demonstration projects can also be found in Japan, with a strong emphasis on the integration of RES in the grid, namely on the development and test of adequate control solutions involving different energy storage technologies to balance the RES variability. Recently, the Zhejiang Electric Power Test and Research Institute implemented a cluster of multiple distributed generation and energy storage technologies capable of operating in different conditions and involving the testing of control strategies for grid connected and islanding mode, while dealing with protection and power quality issues.

Envisioning the development of advanced experimental infrastructures for feasibility demonstration of MG operation in both grid connected and autonomous mode, INESC Porto has been implementing a laboratorial infrastructure in order to exploit distinctive control and management solutions for key DER, such as EV and loads. At the same time, a distinct feature of the MG experimental set-up relies on the integration of both commercially available solutions and in-house developed prototypes. The laboratorial infrastructure aims also to evaluate the specificities of communication requirements for a MG environment and to assess the impact of different technological solutions on the electric power system operational performance [20].

**Fig. 12.24** Laboratory renewable energy sources and commercial inverters

**Fig. 12.25** Commercial EV and commercial charger

## *12.5.1 MG and EV Laboratory: Electric Infrastructure*

The main building block of this laboratory includes micro-generation technologies, energy storage, controllable loads, EV and the associated chargers as well as the LV grid cable simulators. As shown in Fig. 12.24, the RES based microgeneration installed at the laboratory includes 15.5 kWp of photovoltaic installed capacity and a 3 kW micro-wind turbine emulator. The RES based microgeneration units can be connected to the electric network either through single-phase state of the art commercial inverters or through single-phase inverter prototypes developed in-house. A 20 kW 4-quadrant back-to-back inverter is also available in the laboratory. The inverter is remotely controlled in terms of injected or absorbed active power, thus allowing the emulation of different operation conditions either as a generation unit or as a load.

Regarding storage units, two Flooded Lead-Acid (FLA) battery banks (50 V, 20 kWh @ 10 h) were integrated in the laboratorial infrastructure, being connected to two three-phase groups of SMA Sunny Islands inverters (15 kW, 400 V each). These inverters are mainly used for the electrification of remote areas, being able to operate autonomously in isolated systems, by managing storage and local generation (renewable based and/or small backup generators). However, the inverters can also operate in parallel with an existing grid, while providing a smooth transition to autonomous (islanded) operation [40].

Regarding electric mobility, the laboratory has two commercial plug-in EV, as shown in Fig. 12.25, which can be charged by a commercial single-phase home charger. Since the commercial available chargers do not allow controlling the EV charging power, a single-phase DC/AC bi-directional inverter prototype was also developed and connected to a bank of lithium-ion batteries (128 cells, 3 V per cell, 40 Ah).

Two three-phase-four wire LV cables emulators (100 and 50 A) enable the development of different network testing scenarios, considering the resistive nature of LV feeders. Different unbalanced scenarios can also be implemented, distributing the laboratory electric devices by the three-phases of the system.

All the devices available in the laboratory are connected to an electric panel with three 400 V busbars with the possibility of being sectionalized (thus allowing a configuration with six semi-busbars). From the installed busbars, a set of output feeders was installed, allowing the connection of the different laboratory equipment. The configuration of the electric panel allows the composition of different MG setups through the use of a contactor based system allowing the selection of the busbar to which a certain feeder should be connected. The laboratory supervision and automation is carried out by a SCADA system, which supports all the laboratorial operations and ensures the electrical network configuration and monitoring. Each feeder and busbar is equipped with a universal metering device, which provides a large set of electrical values and power quality indicators. The interface and field bus capabilities (Modbus) of the universal metering devices enables the communication of measurement data and its incorporation into the laboratory SCADA system, as well as in the software routines used to operate and control the MG system.

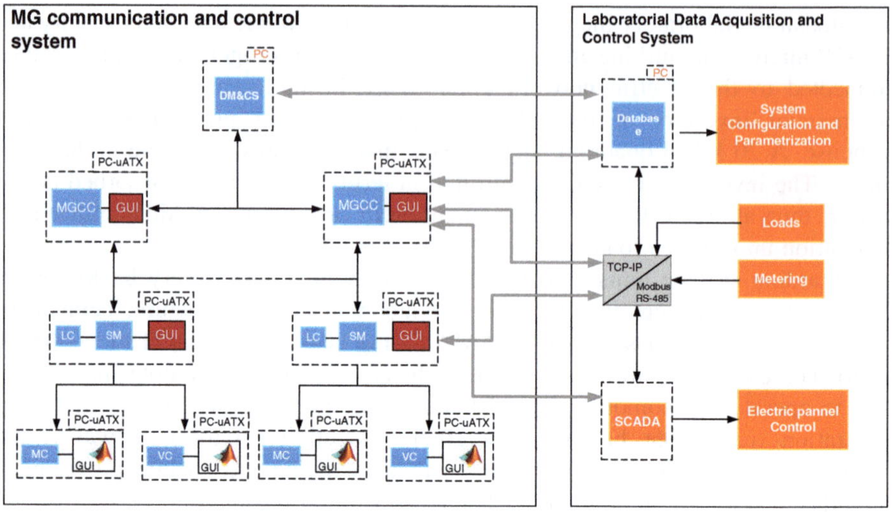


**Fig. 12.26** Smart grid laboratorial architecture

### *12.5.2 Microgeneration Power Converters and Bi-directional EV Charger Prototypes*

As discussed in Sect. 12.3, the MG management and control strategies have to take into account the specificities of power electronic interfaces coupling microgeneration and EV to the network. In order to develop and test new control strategies and since the commercial inverters are relatively closed technologies, two single-phase microgeneration and an EV charger prototypes were developed [41–43].

The power electronic converters architecture is based on a modular structure, enabling the configuration of different types of converters. The converters are composed by half-bridge assemblies, including IGBT switches and hybrid gate drivers, as well as passive components, such as protection devices, voltage and current sensors and the control hardware.

The inverters prototypes have similar architectures. However, considering their distinct applications they present different control and hardware features. As an illustrative example, the EV bidirectional charger prototype is mentioned. The charger can be divided in two stages with independent control schemes, namely: a grid tied Full-Bridge (FB) inverter that controls the power flow between the DC bus and the LV grid and a Dual Active Bridge (DAB) that regulates the current flowing from/to the batteries and assures the galvanic isolation between the grid and the battery pack. The FB inverter regulates the DC bus voltage to 400 V, in order to ensure adequate supply to the DAB input stage. The inverter is controlled with a PQ strategy within a limit of ± 3,680 W, using a proportional-integrative controller implemented in a synchronous reference frame and with additional harmonic compensation. The reference power will be determined according to the droop characteristic described in Sect. 12.3.2.

### 12.5.3 Communication and Control Architecture

As represented in Fig. 12.26, the laboratory electric infrastructure is supported by an information and communication system, enabling the management and control of the MG. In terms of communications and control, the laboratory comprises two major building blocks: the MG communication and control system and the laboratory data acquisition and control system. A communication network consisting of an Ethernet infrastructure was deployed and it is used to interlink the previously mentioned building blocks.

At the MG communication and control system, the Ethernet communication system ensures hierarchical and bidirectional communications in line with the MG concept previously discussed and it allows a controllable communication medium interconnecting the different devices and entities (e.g., MGCC, SM, local controllers). Additionally, it was also envisioned the possibility of considering a layer dedicated to the distribution network operator, which is represented by the Distribution Management and Control System (DM&CS), which will allow the implementation of coordinated operations between the MG cell and the upstream distribution grid.

The MG controllers are emulated in conventional computers integrating the different software and hardware modules responsible for the acquisition and processing of data, which is then used as input to the MG high level management and control functionalities. The information received from the SM includes: power generation, load, EV, responsive loads and power quality indicators, which are processed and aggregated according to the system operator needs. The information received is then used by local software modules, responsible for managing the MG technical operation, during normal and emergency conditions. At the lower control layer, the SM is able to receive set-points from the MGCC and forward them downstream to the respective local controllers. The SM has also a local processing capability, in order to integrate preferences of customers regarding the participation in grid support services and remote management of load and generation.

Regarding the MG communication infrastructure, a Medium Behavior Controller (MBC) was developed. It allows emulating different communication technologies. The Ethernet network is used as a controllable communication medium where distributed MBC are able to impose controlled bandwidth values to the different communication interfaces and define different communication profiles considering data packet delays and losses.

With respect to the laboratory data acquisition and control system, it is intrinsically related to the SCADA system and to the universal metering devices which are used as a platform for data collection for detailed experiment analysis

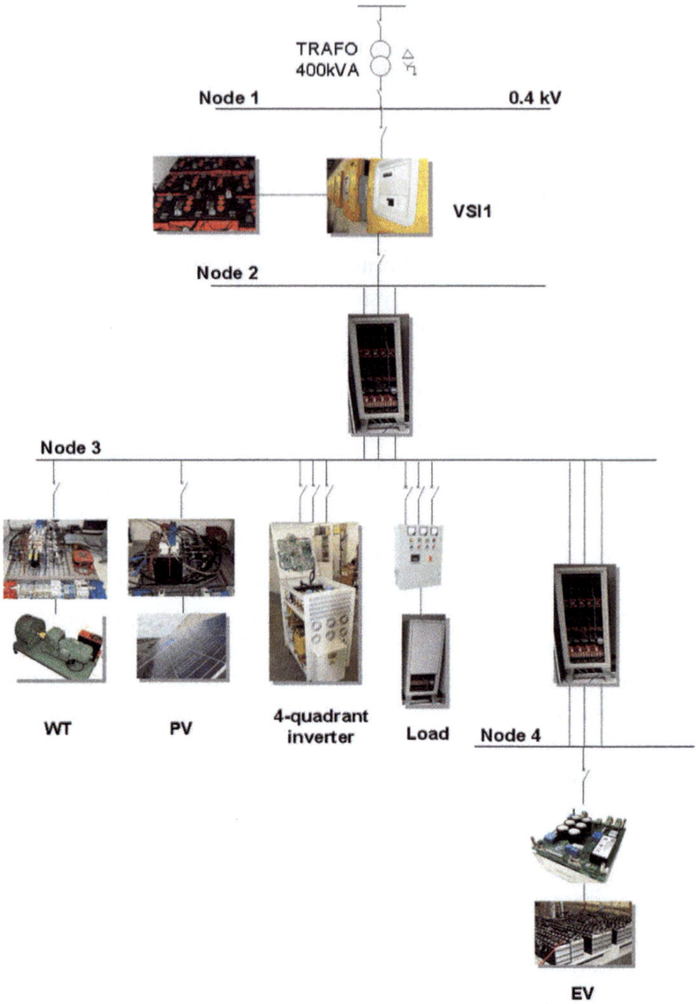


**Fig. 12.27** MG test system for MG islanded operation experiment

and to support MG software modules. The Modbus interface of the universal metering devices enables the communication of measurement data and its incorporation into the laboratory SCADA system. Simultaneously, a Modbus to TCP-IP protocol conversion platform is also used in order to share the measurement data with the MG control system and with a general database for storing experimental results. Additionally, laboratory load banks are remotely connected through Ethernet remote I/O devices that allow its remote control and operation.

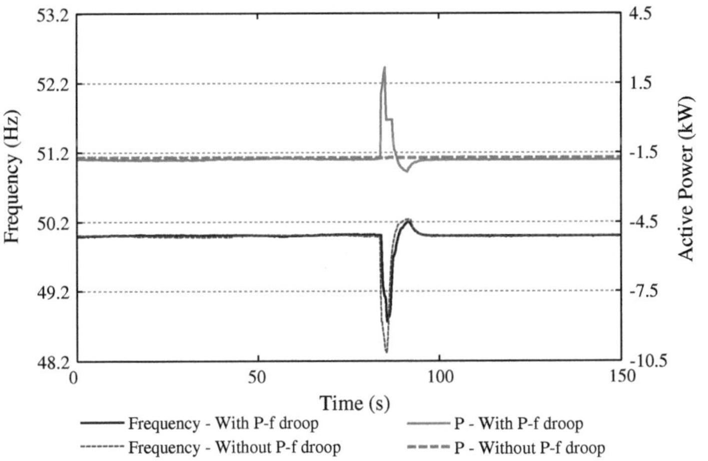


**Fig. 12.28** MG frequency and EV active power response to the MG islanding

## *12.5.4 Experimental Validation of MG Autonomous Operation*

The experimental MG topology represented in Fig. 12.27 was adopted in order to validate the MG islanded operation. The system includes two PV strings connected to the DC/AC solar power converter prototype, the wind-turbine emulator, the EV bidirectional charger prototype, two 27 kW resistive banks and the 4-quadrant inverter emulating a gas microturbine with DC coupled storage. The communication architecture, illustrated in Fig. 12.26, was used to support the data control exchanged during the different phases of the scenarios considered.

The three-phase group of VSI inverters—SMA Sunny Island—is connected to node 1. When operating in normal conditions, loads are supplied directly from the main grid. When the grid switch is disconnected the Sunny island inverter coupled to the FLA storage units ensures power balance and the MG voltage and frequency reference. Primary control is assured by the Sunny Island inverter; however, these power converters intrinsically include a fast additional secondary frequency control strategy, which allows restoring the grid frequency and voltage to the nominal values [40, 44]. The solar and micro-wind turbine inverter prototypes are operated with fixed power injection, while the 4-quadrant inverter is fully regulated in terms of active power injection. In this case, the 4-quadrant inverter is dispatched to provide secondary regulation, compensating the power injected by the battery bank.

***Experiment 1** -MG islanded operation with EV*

As discussed in previous sections, one of the most important features of the MG is the possibility of operating autonomously, through the exploitation of adequate primary and secondary frequency and voltage regulation strategies. Figure 12.28

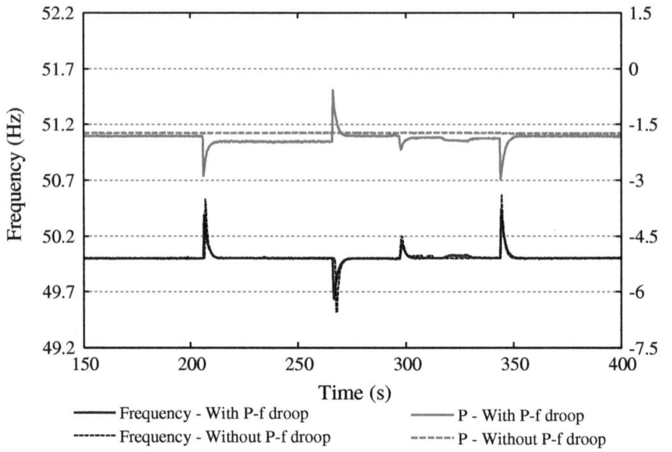


**Fig. 12.29** MG frequency and EV active power response during load following

**Fig. 12.30** Active power injected by the SMA Sunny Island inverter

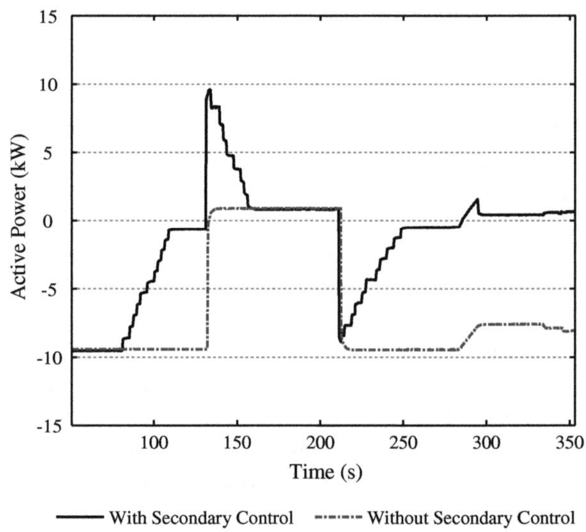


compares the MG frequency in the moments subsequent to islanding when considering both the non-controlled and P-$f$ droop EV charging strategies. The MG frequency and EV active power were measured with a power quality analyzer Fluke 1760 (negative values of EV active power corresponds to power absorption for battery charging). Similarly to the results obtained in [4] and [5], the participation of EV in the MG frequency regulation reduces the MG frequency drop during the islanding transient, since it actively responds to the MG frequency by starting to reduce the absorbed power and even reversing it (operation in the V2G

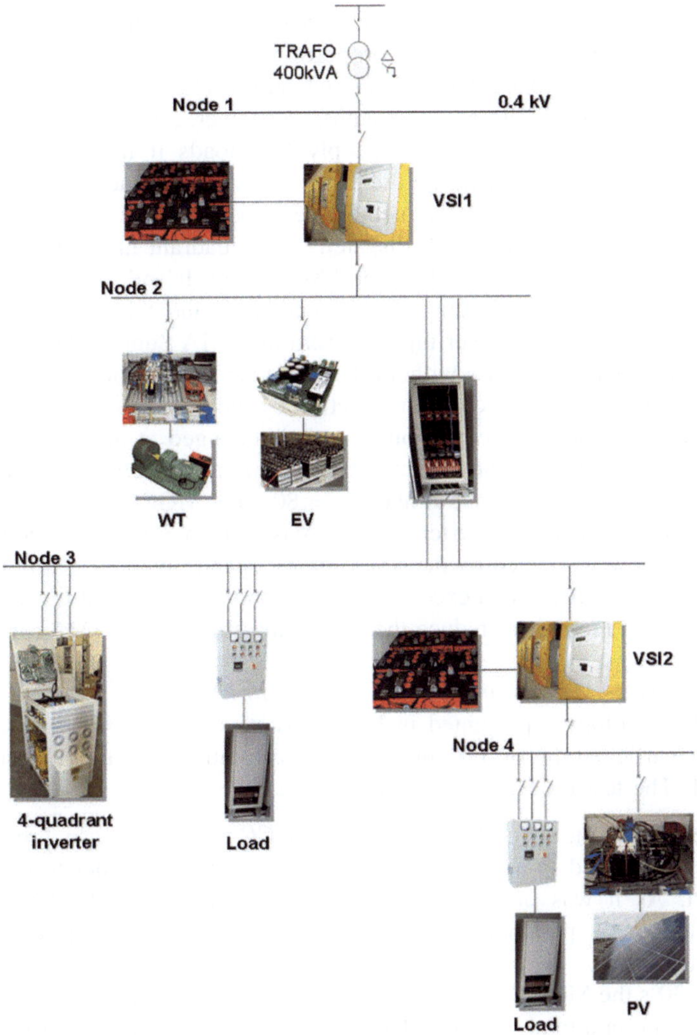


**Fig. 12.31** MG test system for the MG restoration procedure

mode). This autonomous behavior implemented in the EV charger prototypes contributes to reduce the power unbalance in the islanded MG and therefore provides an active contribution to MG frequency regulation.

Following MG islanding, the MG laboratory setup allows also to evaluate the ability of the EV charge prototype to participate in load following conditions. As shown in Fig. 12.29, the EV controlled through P-f droop actively participates in load following during MG islanding operation, by increasing or decreasing its charging power as a response to the different load steps connection (frequency decrease) and disconnection (frequency increase). During this process it can also be observed that EV participation is proportional to the MG frequency behavior.

As previously mentioned, SMA Sunny Island inverters are able to provide both primary and secondary frequency regulation. However, after correcting frequency deviation, the inverter continues to supply MG loads if no other MS can be dispatched. In order to compensate the energy provided by the FLA battery bank, the secondary frequency control method described in Sect. 12.3.3 was implemented at the MGCC, which will dispatch the 4-quadrant inverter. Figure 12.30 depicts the active power injected by the SMA Sunny Island inverters both when the secondary regulation functionality is activated and when it is deactivated (negative power refers to power injected from the SMA Sunny Island inverters to the MG). Considering the case without the secondary regulation, in Fig. 12.30 it is possible to observe the active power injection by the SMA Sunny Island inverters starting from an operating conditions which is changed at about t = 140 s and t = 210 s by a load disconnection and reconnection, respectively. When considering the secondary regulation, at about t = 80 s the MGCC activates secondary control and the 4-quadrant inverter starts increasing its power output, consequently reducing the power injected by the SMA Sunny Island inverters. On the contrary, at t = 140 s when there is an excess of generation the 4-quadrant inverter reduces its power output in order to reduce the power absorbed by the SMA Sunny Island inverters.

***Experiment 2*** *-MG restoration procedure with EV*

The MG topology represented in Fig. 12.31 was adopted in order to demonstrate the participation of EV in the MG restoration procedure discussed in Sect. 12.4. The test procedure adopted was the following:

1. Before starting the rebuilding phase, the MGCC enables the synchronization of the EV connected to node 2 to the MG (at t = 25 s), in order to provide grid support. As it was previously explained in Sect. 12.4.1, the EV P-f droop characteristic is parameterized by the MGCC with a zero power charging reference.
2. At t = 30 s the MGCC enables the reconnections of the loads installed in node 3, representing a total power of 10 kW and waits until the system frequency is recovered to nominal values.
3. At t = 44 s the MGCC enables the reconnection of PV panels connected to node 4.
4. At t = 56 s the MGCC reconnects the controllable generation represented by the 4-quadrant converter connected to node 3 and activates the secondary control.
5. At t = 80 s, a single-phase (phase A) load step in node 4 is allowed to be connected to the islanded MG, representing a total of 3 kW.
6. Since the system has enough reserve capacity, at t = 136 s the MGCC enables the increase of the EV charging power in three steps of 20, 40 and 60 % of the EV charging power (3 kW).

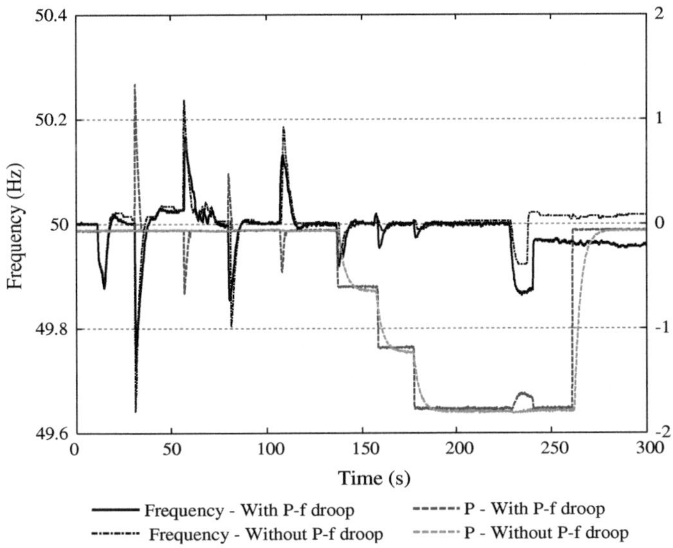


**Fig. 12.32** MG frequency response and EV active power during BS with and without P-f droop control

**Fig. 12.33** Secondary control response of the 4-Q inverter—delays

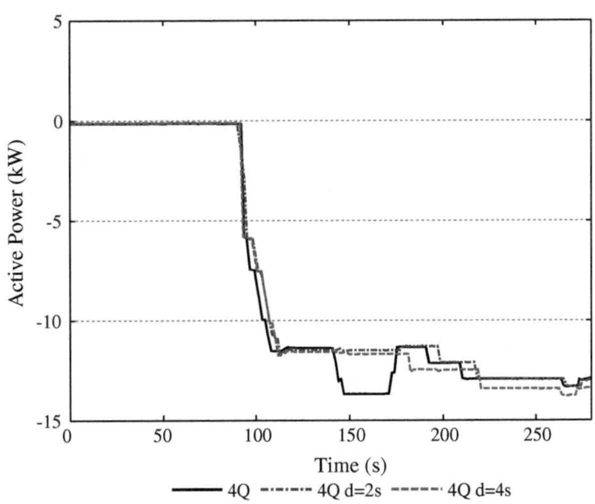


7. At t = 228 s the MV network becomes available and the synchronization process is enabled. Synchronization is performed at the LV side of the MV/LV transformer.
8. At t = 239 s the MG is successfully reconnected to the MV network.

Figure 12.32 compares the MG frequency response and the EV active power output during the restoration procedure (negative power represents EV charger

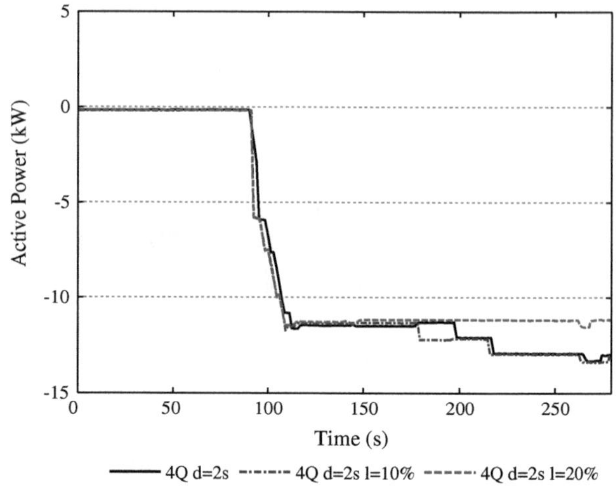


**Fig. 12.34** Response of the 4-quadrant converter—impact of communication delays and losses

prototype power abortion from the MG). Similarly to the results obtained for the islanded operation, the participation of EV reduces the frequency disturbances resulting from the reconnection of loads and generation. For example, at t = 30 s when the load at node 3 is reconnected, the EV active power injection of 1.3 kW and it contributes to reduce the frequency excursion from 49.64 to 49.73 Hz. On the contrary, when the 4-quadrant converter is reconnected at t = 56 s, the increase of the EV power consumption to 600 W reduced the frequency excursion from 50.24 to 50.1 Hz.

In order to evaluate the influence of the communication system uncertainty on the restoration procedure previously developed, the experiment was repeated considering the action of the MBC (described in Sect. 12.5.3) through the emulation of different scenarios of information delays and losses. The introduction of delays and losses is based on probabilistic process over which average values and deviations can be established. Figure 12.33 depict the active power output of the 4-quadrant inverter in the presence of delays associated with the information exchange. When the secondary control is enabled by the MGCC, in this case at t = 90 s, the secondary control response depends on the MG communication system uncertainties, which are considered to be the delays for information exchange that are changed from a base case (zero delays) to 2 and 4 s (mean delay values) with a jitter of 500 ms. As shown in Fig. 12.33, an increase in the average delay time, associated with each set-point exchange, delays the overall response of the 4-quadrant converter.

In Fig. 12.34 it is possible to notice the effect of set-point losses over the same experiment, in the particular case of the 4-quadrant converter. A mean delay of approximately 2 s was considered and different values of information loss were defined. The impact resulting from the communication system uncertainties can be evaluated with the introduction of losses. The impact is visible when comparing the base case with no losses with the remainder cases where losses occur. With a loss rate of 10 % a set-point loss occurs around t = 180 s; with a loss rate of 20 %, failures are visible t = 175 s and t = 220 s.

## 12.6 Final Remarks

Future power system planning and operation will focus on the LV distribution networks since the majority of small scale DER—EV, MS, storage units and flexible loads—are connected to it. Within the Smart Grid paradigm, the MG concept has been pointed out as an efficient solution to deal with DER integration acting either as a local DER control and management system or as an aggregation cell that is able to provide a coordinate response to the upstream distribution grid.

This new distribution operation philosophy embodies the active distribution network concept, being able to tackle the technical challenges resulting from the large scale integration of DER and EV and at the same time take advantage of the local resources flexibility to improve the security and reliability of the system. The coordination of MG local resources endows the LV system with sufficient autonomy to operate interconnected to the upstream network or islanded, or even perform local restoration procedures in case of generalized disturbances.

Following a worldwide track record on MG research activities, significant demonstrations have been implemented. INESC Porto experience on implementing a MG laboratorial infrastructure was presented aiming the feasibility demonstration of MG operation in grid connected and autonomous mode, while exploiting distinctive control and management solutions for DER. The infrastructure is composed of commercially available components, which are complemented by hardware and software prototypes developed in accordance with strategies developed for DER active integration in MG operation. Additionally, the specificities of communication requirements for a Smart Grid environment and the evaluation of its impact on the MG operational performance, namely due to the associated uncertainty, were addressed given its key relevance for MG operation and control. Furthermore, the presence of distributed control schemes allows the MG to react even when the uncertainty introduced by communication systems is considered.

The industrial development of Smart Grid related products and technologies are growing, enhancing the importance of experimental demonstrations which take into consideration real world conditions. The development of laboratorial facilities, such as the one described in this chapter, plays an important role in the consolidation of innovative solutions that are the key for a successful development of the smart grid paradigm. The results described in this chapter reinforce the fact that the coordination between the MG centralized and local control strategies provides additional resources to deal with the increasing integration of DER.

## References


1. Lasseter RH (2011) Smart distribution: coupled microgrids. Proc IEEE 99(6):1074–1082
2. Lasseter B (2001) Microgrids [distributed power generation]. IEEE Power Eng Soc Winter Meet 2001 1:146–149
3. Peças Lopes JA, Moreira CL, Madureira AG (2006) Defining control strategies for MicroGrids islanded operation. IEEE Trans Power Syst 21(2):916–924
4. Peças Lopes JA, Silvan Polenz A, Moreira CL, Cherkaoui R (2010) Identification of control and management strategies for LV unbalanced microgrids with plugged-in electric vehicles. Electr Power Syst Res 80(8):898–906
5. Peças Lopes JA, Soares FJ, Almeida PMR (2011) Integration of electric vehicles in the electric power system. Proc IEEE 99(1):168–183
6. Guerrero JM, Vasquez JC, Matas J, de Vicuna LG, Castilla M (2011) Hierarchical control of droop-controlled AC and DC microgrids—a general approach toward standardization. IEEE Trans Industr Electron 58(1):158–172
7. Katiraei F, Iravani MR (2006) Power management strategies for a microgrid with multiple distributed generation units. IEEE Trans Power Syst 21(4):1821–1831
8. Moreira CL, Resende FO, Lopes JAP (2007) Using low voltage microgrids for service restoration. IEEE Trans Power Syst 22(1):395–403
9. Strbac Goran (2008) Demand side management: benefits and challenges. Energy Policy 36(12):4419–4426
10. Schweppe FC, Tabors RD, Kirtley JL, Outhred HR, Pickel FH, Cox AJ (1980) Homeostatic utility control. IEEE Trans Power Apparatus Syst PAS-99(3):1151–1163
11. Short JA, Infield DG, Freris LL (2007) Stabilization of grid frequency through dynamic demand control. IEEE Trans Power Syst 22(3):1284–1293
12. Samarakoon K, Ekanayake J, Jenkins N (2012) Investigation of domestic load control to provide primary frequency response using smart meters. IEEE Trans Smart Grid 3(1):282–292
13. Pourmousavi SA, Nehrir MH (2012) Real-time central demand response for primary frequency regulation in microgrids. IEEE Trans Smart Grid 3(4):1988–1996
14. Gouveia C, Moreira J, Moreira CL, Pecas Lopes JA (2013) Coordinating storage and demand response for microgrid emergency operation. IEEE Trans Smart Grid PP(99):11
15. Hatziargyriou N, Asano H, Iravani R, Marnay C (2007) Microgrids. IEEE Power Energ Mag 5(4):78–94
16. European Distributed Energy Resources Laboratories. http://www.der-lab.net/
17. Lasseter RH, Eto JH, Schenkman B, Stevens J, Vollkommer H, Klapp D, Linton E, Hurtado H, Roy J (2011) CERTS microgrid laboratory test bed. IEEE Trans Power Delivery 26(1):325–332
18. Zhao B, Zhang X, Chen J (2012) Integrated microgrid laboratory system. IEEE Trans Power Syst 27(4):2175–2185
19. Schmitt L, Kumar J, Sun D, Kayal S, Venkata SSM (2013) Ecocity upon a hill: microgrids and the future of the European City. IEEE Power Energ Mag 11(4):59–70
20. Project REIVE – Smart Grids with Electric Vehicles (2013) http://reive.inescporto.pt/en Accessed Jan 2013
21. Shaver D (2009) Low frequency, narrowband PLC standards for smart grid. Presentation
22. Hoch M (2011) Comparison of PLC G3 and PRIME. In: 2011 IEEE international symposium on Power line communications and its applications (ISPLC), pp 165–169, Apr 2011
23. IEEE P1901.2. http://standards.ieee.org/develop/project/1901.2.html
24. IEEE standard for broadband over power line networks: medium access control and physical layer specifications. IEEE Std 1901-2010, pp 1–1586, 2010
25. Parikh PP, Kanabar MG, Sidhu TS (2010) Opportunities and challenges of wireless communication technologies for smart grid applications. In: 2010 IEEE power and energy society general meeting, pp 1–7, Jul 2010

26. Chin-Sean S, Harada H, Kojima F, Lan Z, Funada R (2011) Smart utility networks in TV white space. IEEE Commun Mag 49(7):132–139
27. IEEE Standard for Local and Metropolitan Area Networks - Part 15.4: Low-Rate Wireless Personal Area Networks (LR-WPANs) Amendment 3: Physical Layer (PHY) Specifications for Low-Data- Rate, Wireless, Smart Metering Utility Networks. IEEE Std 802.15.4 g-2012 (Amendment to IEEE Std 802.15.4-2011), pp 1–252, 27 2012
28. Gambiroza V, Sadeghi B, Knightly EW (2004) End-to-end performance and fairness in multihop wireless backhaul networks. In: Proceedings of the 10th Annual international conference on mobile computing and networking, mobiCom'04, New York, NY, USA, 2004. ACM, pp 287–301
29. Ernst JB, Denko MK (2011) The design and evaluation of fair scheduling in wireless mesh networks. J Comput Syst Sci 77(4):652–664. JCSS IEEE AINA 2009
30. Gupta GR, Shroff NB (2010) Practical scheduling schemes with throughput guarantees for multi-hop wireless networks. Comput Netw 54(5):766–780
31. Campos R, Duarte R, Sousa F, Ricardo M, Ruela J (2011) Network infrastructure extension using 802.1D-based wireless mesh networks. Wirel Commun Mob Comput 11(1):67–89
32. Campos R, Oliveira C, Ruela J (2011) WiFIX + : a multicast solution for 802.11-based wireless mesh networks. In: 2011 Eighth international conference on wireless on-demand network systems and services (WONS), pp 179–186, Jan 2011
33. Green TC, Prodanovic M (2007) Control of inverter-based micro-grids. Electr Power Syst Res 77(9):1204–1213
34. Blaabjerg F, Teodorescu R, Member SS, Liserre M, Timbus AV (2006) Overview of control and grid synchronization for distributed power generation systems. IEEE Trans Ind Electron 53(5):1398–1409
35. Madureira A, Moreira CL, Peças Lopes JA (2005) Secondary load-frequency control for microgrids in islanded operation. In: Proceedings of ICREPQ'05 - international conference on renewable energies and power quality, Zaragoza, Spain, Março, 2005
36. Jong-Yul K, Jin-Hong J, Seul-Ki K, Changhee C, June Ho P, Hak-Man K, Kee-Young N (2010). Cooperative Control strategy of energy storage system and microsources for stabilizing the microgrid during Islanded operation. IEEE Trans Power Electron 25(12):3037–3048
37. Adibi MM, Fink LH (1994) Power system restoration planning. IEEE Trans Power Syst 9(1):22–28
38. Pham Thi Thu Ha, Besanger Y, Hadjsaid N (2009) New challenges in power system restoration with large scale of dispersed generation insertion. IEEE Trans Power Syst 24(1):398–406
39. Adibi MM, Martins N (2008) Power system restoration dynamics issues. In: 2008 IEEE power and energy society general meeting - conversion and delivery of electrical energy in the 21st Century, pp 1–8
40. SMA Report, "Technology Compedium 2-Solar Stand-Alone Power and Backup Power Supply", SMA 2009. Available at via http://files.sma.de/dl/10040/INSELVERSOR-AEN101410.pdf
41. Miguel Rodrigues J, Resende F (2012) Using photovoltaic systems to improve voltage control in low voltage networks. In: ISGT2012 - third IEEE PES innovative smart grid technologies Europe conference, Berlin, Germany, Oct 2012
42. Rodrigues J, Resende FO, Moreira CL (2011) Contribution of PMSG based small wind generation systems to provide voltage control in low voltage networks. In: ISGT 2011 - IEEE PES ISGT 2011 Europe, Manchester, UK
43. Varajao D, Araujo RE, Moreira C, Lopes JP (2012) Impact of phase-shift modulation on the performance of a single-stage bidirectional electric vehicle charger. In: IECON 2012 - 38th annual conference on IEEE industrial electronics society, pp 5215–5220, 25–28 Oct 2012
44. Engler A (2005) Applicability of droops in low voltage grids. DER J 1:1–5

# Chapter 13
# A Novel Aggregation Technique Using Mechanical Torque Compensating Factor for DFIG Wind Farms

**M. A. Chowdhury**

**Abstract** A novel aggregated model for wind farms consisting of wind turbines equipped with doubly fed induction generators (DFIG) is proposed in this paper. In the proposed model, a mechanical torque compensating factor (MTCF) is integrated into a full aggregated wind farm model to deal with the nonlinearity of wind turbines in the partial load region and to make it behave as closely as possible to a complete model of the wind farm. The MTCF is initially constructed to approximate a Gaussian function by a fuzzy logic method and optimized on a trial and error basis to achieve less than 10% discrepancy between the proposed aggregated model and the complete model. Then, a large scale offshore wind farm comprising 72 DFIG wind turbines is used to verify the effectiveness of the proposed aggregated model. The simulation results show that the proposed model is able to approximate collective dynamic responses at the point of common coupling with significant reduction in the simulation computation time.

**Keywords** Doubly fed induction generator · Wind farm · Aggregated model · Mechanical torque compensating factor · Fuzzy logic

## 13.1 Introduction

Wind power has been the fastest growing energy source since the last decade due to its inherent attribute of the reproducible, resourceful and pollution-free characteristics. Wind power capacity reached 215 GW (3 % of global electricity consumption) worldwide with a growth rate of 22.9 % in 2010. With this growth rate, wind power capacity will be doubled every 3 years. Based on this accelerated

M. A. Chowdhury (✉)
Faculty of Engineering and Industrial Sciences, Swinbune University of Technology, Internal Mail H38, PO Box 218, Hawthorn, VIC 3122, Australia
e-mail: achowdhury@swin.edu.au

J. Hossain and A. Mahmud (eds.), *Renewable Energy Integration*, Green Energy and Technology, DOI: 10.1007/978-981-4585-27-9_13,

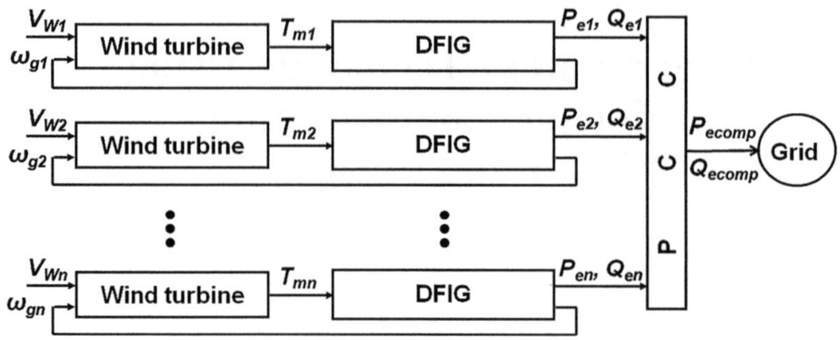


**Fig. 13.1** Block diagram of a complete DFIG wind farm model

development and further improved policies, 12 % of global electricity demand (1900 GW) is predicted to be provided by wind energy systems by the year 2020 [1].

Wind farms of 50 MW ratings or more are integrated into high voltage transmission networks [2]. With the increasing amount of wind power penetration in power systems, wind farms begin to influence power systems. This justifies the need for adequate models for wind farms in order to represent overall power system dynamic behavior of grid-connected wind farms during both normal operations and grid disturbances. A wind farm may consist of tens to hundreds of wind turbines. This leads to model complexity and computation burden [3, 4]. Figure 13.1 shows a complete wind farm model with *n* number of wind turbines equipped with doubly-fed induction generator (DFIG).

To simplify the complete wind farm model, an aggregated wind farm model is required to reduce the size of the power system model, the data requirement and the simulation computation time [5–7], where this aggregated model can (1) represent the behavior (active and reactive power exchanged with the power system at the point of common coupling (PCC)) of the wind farm during normal operation, characterized by small deviations of the grid quantities from the nominal values and the occurrence of wind speed changes and (2) represent the behavior of the wind farm during grid disturbances, such as voltage drops and frequency deviations.

Two types of wind farm aggregation techniques have been proposed: the full aggregated and the semi aggregated techniques. Figure 13.2 shows the full aggregated and semi aggregated wind farm models. The full aggregated model consists of one equivalent wind turbine and one equivalent generator for a wind farm with one operating point at an average wind speed for all the wind turbines in the wind farm [7–12]. The semi aggregated model consists of all the wind turbines in the wind farm and one equivalent generator [13, 14].

For a wind farm consisting of DFIG wind turbines, the ability of the full or semi aggregated model to approximate the complete model depends on the operating region of the DFIG wind turbines. The operating regions of the DFIG wind turbine adopted in this work is shown in Fig. 13.3, which can be segmented into two parts: a partial load region, where the wind speed ranges between 4.5 and 14.5 m/s and a

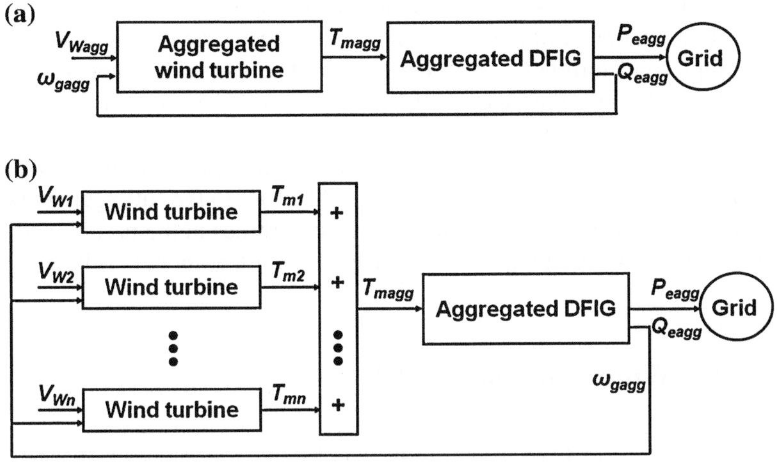


**Fig. 13.2** Block diagram of **a** full aggregated and **b** semi aggregated DFIG wind farm models

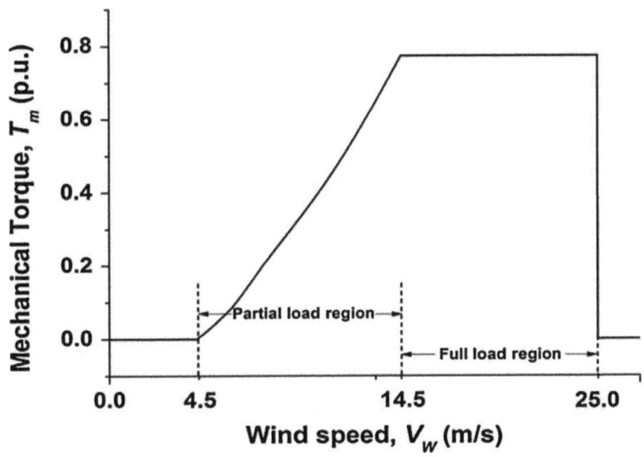


**Fig. 13.3** Operating regions of the DFIG wind turbine at the turbine rotor speed of 1 p.u

full load region, where the wind speed ranges between 14.5 and 25 m/s).The DFIG wind turbine is stopped when wind speed is less than 4.5 m/s or greater than 25 m/s.

The full or semi aggregated model can represent the complete model when DFIG wind turbines in the wind farm operate in the full load region regardless of the differences in the operating points of the wind turbines in the wind farm. This is due to the fact that all generators produce the same current at its maximum rating in this region.

But, the full aggregated model cannot provide an accurate approximation of a complete model when DFIG wind turbines in the wind farm operate in the partial load region. This is due to the fact that the full aggregated technique does not consider the operating points of all corresponding wind turbines in the wind farm and a nonlinear relationship between wind speed ($V_W$) and mechanical torque ($T_m$) as shown in Fig. 13.3.

The semi aggregated model, on the other hand, improves the approximation of a complete model in the partial load region by considering the operating points of all corresponding wind turbines in the wind farm. The use of an average generator rotor speed ($\omega_g$) for all of the wind turbines still contributes to discrepancies in the magnitude of mechanical torque and consequently electromagnetic torque.

This chapter thus proposes a new aggregation technique with the incorporation of a mechanical torque compensation factor (MTCF) into the full aggregated wind farm model to deal with the nonlinearity of wind turbines in the partial load region and to make it behave as closely as possible to a complete model of the wind farm.

## 13.2 DFIG Wind Turbine Model

The DFIG wind turbine is modelled in terms of behavior equations of each of the subsystems, mainly the turbine, the drive train, the induction generator and the control system (Fig. 13.4).

The aerodynamics of the wind turbine is characterized by $Cp$-$\lambda$-$\beta$ curve. $Cp$ is the power coefficient, which corresponds to maximum mechanical power extraction from wind for its maximum value, and is a function of the tip-speed ratio ($\lambda$) and the pitch angle ($\beta$), which is given by [15]

$$C_p(\lambda, \beta) = 0.22\left(\frac{116}{\lambda_i} - 0.4\beta - 5\right)e^{\frac{-12.5}{\lambda_i}} \tag{13.1}$$

where

$$\frac{1}{\lambda_i} = \frac{1}{\lambda + 0.08\beta} - \frac{0.035}{\beta^3 + 1} \tag{13.2}$$

For a given $Cp$, the mechanical torque ($T_m$) produced by the wind turbine is given by [16]

$$T_m = \frac{\rho A C_p V_W^3}{2\omega_t} \tag{13.3}$$

where $\rho$ is the air density, $A$ is the sweep area of the blades, $V_W$ is the wind speed and $\omega_t$ is the turbine rotor speed.

For modeling drive train, the rotor is treated as two lumped masses, i.e., turbine mass and generator mass are connected together by shaft with a certain damping and stiffness coefficient values [17].

After simplifications by neglecting the turbine and generator self-damping, shaft stiffness and torsional oscillations, the mathematical equation can be expressed

$$2H\frac{d\omega_g}{dt} = T_m - T_e \tag{13.4}$$

Fig. 13.4 Configuration of a DFIG wind turbine

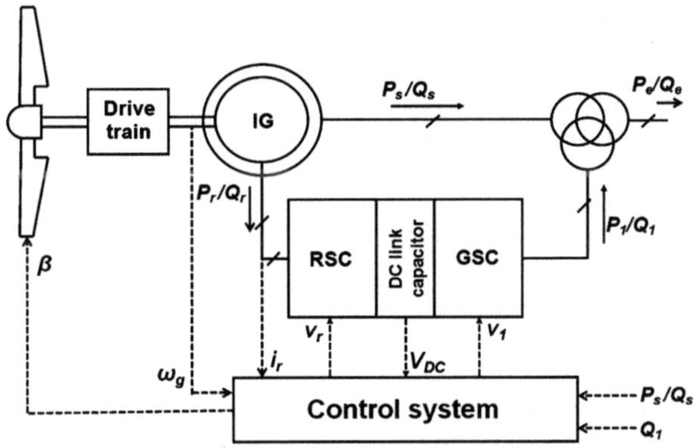


where $H$ is the inertia constant, $\omega_g$ is the generator rotor speed and $T_e$ is the electromagnetic torque.

For modeling induction generator, a synchronously rotating *d-q* reference frame is chosen, which is rotating with the same speed as the stator voltage. Stator and rotor voltages in this reference frame are given by

$$v_{ds} = -R_s i_{ds} - \omega_s \phi_{qs} + \frac{d\phi_{ds}}{dt} \quad (13.5)$$

$$v_{qs} = -R_s i_{qs} + \omega_s \phi_{ds} + \frac{d\phi_{qs}}{dt} \quad (13.6)$$

$$v_{dr} = -R_r i_{dr} - s\omega_s \phi_{qr} + \frac{d\phi_{dr}}{dt} \quad (13.7)$$

$$v_{qr} = -R_r i_{qr} + s\omega_s \phi_{dr} + \frac{d\phi_{qr}}{dt} \quad (13.8)$$

where $v$ is the voltage, $i$ is the current, $R$ is the resistance, $\omega_s$ is the synchronous speed, $\phi$ is the flux and $s$ is the slip. Suffix $s$, $r$, $d$ and $q$ denote stator, rotor, d-axis component and q-axis component, respectively.

The electromagnetic torque ($T_e$) is expressed in [18] as

$$T_e = \phi_{ds} i_{qs} - \phi_{qs} i_{ds} \quad (13.9)$$

where $p$ is the number of poles.

The output active and reactive powers ($P_e$ and $Q_e$) are then calculated as

$$P_e = v_{ds} i_{ds} + v_{qs} i_{qs} + v_{dr} i_{dr} + v_{qr} i_{qr} \quad (13.10)$$

$$Q_e = v_{qs} i_{ds} - v_{ds} i_{qs} + v_{qr} i_{dr} - v_{dr} i_{qr} \quad (13.11)$$

**Table 13.1** DFIG wind turbine parameters

| Parameter | Symbol | Value | Unit |
|---|---|---|---|
| Nominal mechanical output power | $P_{mec}$ | 1.5 | MW |
| Nominal electrical power | $P_{elec}$ | 1.5/0.9 | MW |
| Nominal voltage (L–L) | $V_{nom}$ | 575 | Volt |
| Stator resistance | $R_s$ | 0.00706 | p.u. |
| Stator leakage inductance | $L_r$ | 0.171 | p.u. |
| Rotor resistance | $R_r$ | 0.0058 | p.u. |
| Rotor leakage inductance | $L_r$ | 0.156 | p.u. |
| Magnetizing inductance | $L_m$ | 2.9 | p.u. |
| Base frequency | $f$ | 60 | Hz |
| Inertia constant | $H$ | 1 | s |
| Friction factor | $F$ | 0.01 | p.u. |
| Pair of poles | $p$ | 3 | – |

The control system comprises of two external power electronic converter controllers and a pitch angle controller. These controllers are modelled by the conventional PI controllers [19]. The converter controllers generate the voltage command signal $v_r$ and $v_l$ for the rotor side converter (RSC) and grid side converter (GSC), respectively in order to control the DC voltage and the reactive power or the voltage at the grid terminals. The pitch angle controller generates pitch angle command signal ($\beta$) for temporary reduction of mechanical power when the DFIG wind turbine operates in the full load region. Their detail explanation can be found in [20]. The parameters of the DFIG wind turbine used in this work are shown in Table 13.1 [21].

## 13.3 Formation of a Complete DFIG Wind Farm Model

The model of a wind farm with all of its electrical networks is presented in this section, which is a modified version of a 120 MVA offshore wind farm model implemented by 'NESA Transmission Planning' of Denmark for power stability investigations [8] as shown in Fig. 13.5. The wind farm model comprises of 72 DFIG wind turbines with the parameters specified in Table 13.1. Each WTG is connected to the cable sections through 0.67/30 kV transformer (LV/MV) and a line impedance of $0.08 + j0.02$ p.u. The wind farm is connected to the power grid through a 30/132 kV tertiary transformer (MV/HV) and then through a high voltage (132 kV) transmission network (HVTN) with the impedance value of $1.6 + j3.5$ p.u.

The internal and external electrical networks including electric lines, transformers and cables are represented by constant impedances [13]. Short circuit capacity viewed from the PCC into the HVTN is around 1500 MVA. The power grid is modeled by an infinite bus with the MVA-rating of 1000 MVA.

A pair of indices identifies the WTG within the wind farm, where the first index (from 1 to 6) denotes the number of the group (the 30 kV sea cable) and the second

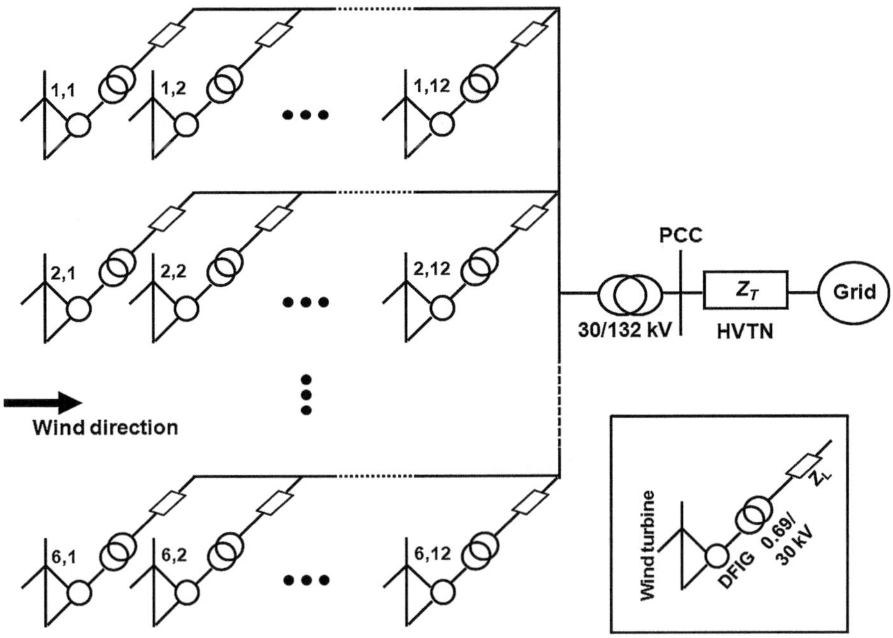


**Fig. 13.5** A 120 MVA offshore DFIG wind farm model

**Table 13.2** DFIG wind farm parameters

| Parameter | Symbol | Value | Unit |
|---|---|---|---|
| *Internal electrical network* | | | |
| Base power | $S_{WTG}$ | 1.5/0.9 | MVA |
| Base voltage | $V_{WTG}$ | 575 | V |
| LV/MV transformer | – | 0.69/30 | kV |
| | $S_T$ | 2 | MVA |
| | $\varepsilon_{cc}$ | 6 | % |
| Line impedance | $Z_L$ | 0.08 + j0.02 | p.u. |
| *External electrical network* | | | |
| MV/HV transformer | – | 30/132 | kV |
| | $S_T$ | 150 | MVA |
| | $\varepsilon_{cc}$ | 8 | % |
| HVTN impedance | $Z_T$ | 1.6 + j3.5 | p.u. |
| Short circuit capacity of the PCC | $S_{PCC}$ | 1500 | MVA |
| X/R ratio of PCC | $(X/R)_{PCC}$ | 20 | p.u. |
| Short circuit capacity of the grid | $S_G$ | 1000 | MVA |

index (from 1 to 12) denotes the number of the WTG within the group. The parameters of the DFIG wind farm used in the simulation are shown in Table 13.2 [8, 13].

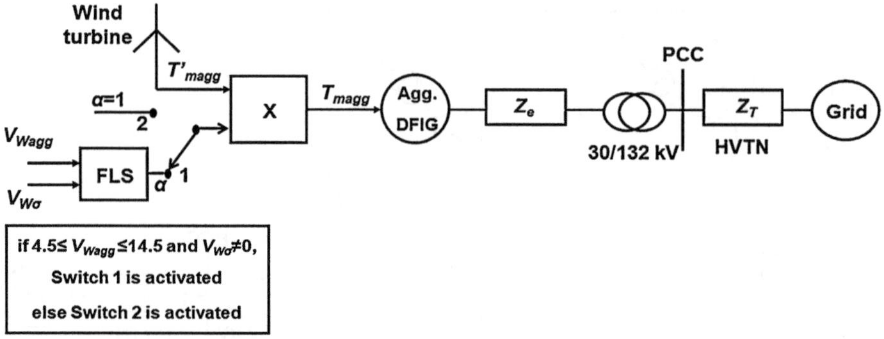


**Fig. 13.6** Block diagram of the proposed aggregated DFIG wind farm model

## 13.4 Proposed Aggregated DFIG Wind Farm Model

Figure 13.6 shows the proposed aggregated DFIG wind farm model that consists of a mechanical torque compensating factor (MTCF) incorporated into a traditional full aggregated model. The MTCF ($\alpha$) is a multiplication factor to the mechanical torque ($T'_{magg}$) of the full aggregated model that minimizes this inaccuracy in approximation. The mechanical torque ($T_{magg}$) of the proposed aggregated DFIG wind farm model is thus calculated by

$$T_{magg} = T^{'}_{magg} * \alpha \tag{13.12}$$

The proposed model also involves the calculation of an equivalent internal network and the simplification of the power coefficient ($C_p$) function.

### *13.4.1 Full Aggregated DFIG Wind Farm Model*

The full aggregated DFIG wind farm model converts all DFIG wind turbines in the wind farm into one equivalent unit with the same per unit value of mechanical and electrical parameters in the voltage, flux linkage and motion equations [13], which is driven by an average wind speed ($V_{Wagg}$) [14]

$$V_{Wagg} = \frac{1}{n}\sum_{i=1}^{n} V_{Wi} \tag{13.13}$$

where $n$ is the number of WTGs in the wind farm and suffix *agg* denotes the aggregated wind farm model.

This gives the mechanical torque as

$$T'_{magg} = \frac{\rho A C_p V_{Wagg}^3}{2\omega_{tagg}} \tag{13.14}$$

where $\omega_{tagg}$ is the average turbine rotor speed calculated from $V_{Wagg}$.

## 13.4.2 Basis of MTCF Calculation

As stated earlier, the full aggregated model can provide an approximation of the complete model when DFIG wind turbines operate in the full load region. Thus, in this region the MTCF takes a value equal to 1.

When the wind turbines operate at different wind speeds and thus different operating points, the adoption of an average operating point for the DFIG wind turbines causes the discrepancies between the complete and full aggregated models. Figure 13.7 shows that the torque of the full aggregated model is generally lower than that of the complete model in the partial load region. Thus, in this region the MTCF takes a value more than 1.

It means that the MTCF increases from the value 1 as $V_{Wagg}$ increases from 4.5 m/s or $V_{Wagg}$ decreases from 14.5 m/s, which implies that the MTCF may take its maximum value between 4.5 and 14.5 m/s. On the other hand, the MTCF maintains a proportional relation with the wind speed deviation ($V_{W\sigma}$) and it takes a value equal to 1 when the operating points of the DFIG wind turbines in the wind farm are identical (i.e., $V_{W\sigma} = 0$). Thus, the MTCF is a function of $V_{Wagg}$ and $V_{W\sigma}$ and may be 'approximated' by an ideal Gaussian function (see Fig. 13.8) in the partial load region:

$$\alpha = 1 + le^{\frac{-(V_{Wagg} - V_{W\mu})^2}{2\sigma^2}} V_{W\sigma} \tag{13.15}$$

According to Eq. 13.15, the maximum value of $\alpha$ is $(1 + l)$ when the wind speed is equal to $V_{W\mu}$ (in this work, $V_{W\mu} = 9.5$ m/s) and $\sigma$ is the standard deviation from $V_{W\mu}$.

From empirical rule of the central limit theorem, it is known that 99.993 % of data lie within four standard deviation from their mean value [22]. It gives the value of $\sigma$ and $l$.

$$4\sigma = 5 \tag{13.16}$$

$$l = \frac{1}{\sigma\sqrt{2\pi}} = 0.32 \tag{13.17}$$

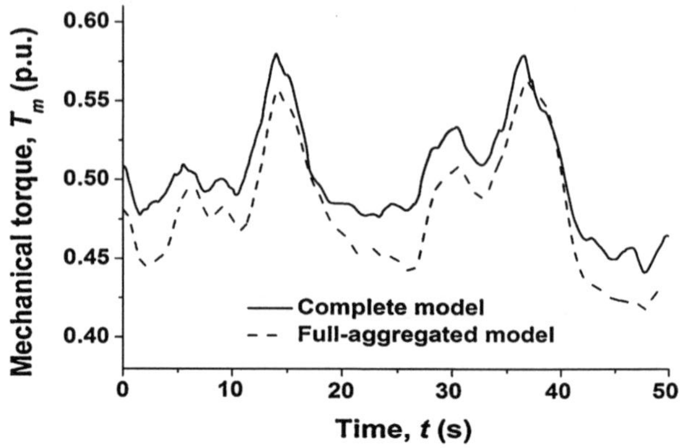


**Fig. 13.7** Torque curves of the complete and full aggregated model in the partial load region

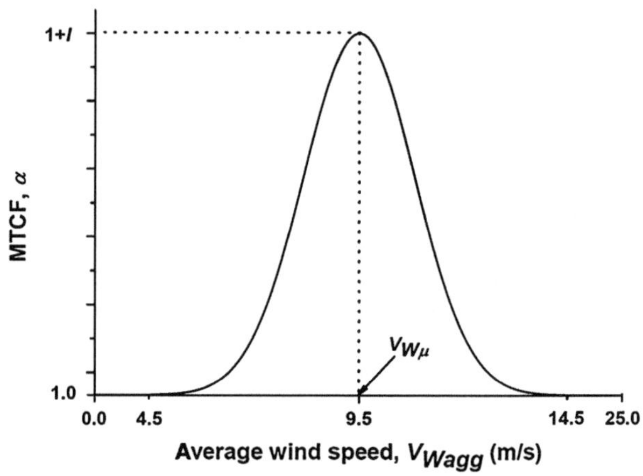


**Fig. 13.8** Gaussian distribution of MTCF ($\alpha$) with respect to average wind speed ($V_{Wagg}$)

### *13.4.3 MTCF Calculation by Fuzzy Logic System*

Due to the complex nonlinear relationship and ambiguous dynamics of the wind energy generation system, it is difficult to find the mathematical model for the input–output relationship for the calculation of the MTCF. However, based on the expert (operator's) knowledge, a human operator can express the input–output relationship of a MTCF by using linguistic rules without knowing the exact mathematical relationship and this expert knowledge described by linguistic rules can be used to design the fuzzy logic system (FLS), which makes the FLS a very good candidate to compute the MTCF. Thus, the FLS is adopted to calculate the MTCF in this work.

The FLS is initially constructed by assigning overlapped triangular membership functions for the fuzzy sets and setting fuzzy rules based on the ideal Gaussian function. Triangular membership functions are easy to implement, quicker to process and give more sensitivity, especially as variables approach to zero.

The design is optimized by making possible changes in membership functions for the fuzzy sets and fuzzy rules on trial and error basis to achieve less than 10 % discrepancy between the proposed aggregated model and the complete model.

The FLS takes two inputs: average wind speed ($V_{wagg}$) and wind speed deviation ($V_{W\sigma}$). In the design of the FLS, $V_{wagg}$ ranges between 4.5 and 14.5 m/s. $V_{W\sigma}$ ranges between 0 and its maximum possible value. The value of $V_{W\sigma}$ is the maximum when wind speeds received by the wind turbines are equally spaced within the specified range of $V_{Wagg}$. For 72 DFIG wind turbines, the maximum value of $V_{W\sigma}$ is found by the following calculation:

$$V_{W\sigma\max} = \sqrt{\frac{1}{72}\sum_{i=1}^{72}\left(\frac{14.5-4.5}{72-1}i - 9.5\right)^2} = 5.25 \tag{13.18}$$

Then, according to Eq. 13.15, the MTCF ($\alpha$) is 2.7 when $V_{Wagg} = V_{W\mu}$ and $V_{W\sigma} = 5.25$ (takes its maximum value), thus the range of the MTCF should be between 1 and 2.7.

Figure 13.9 shows that triangular membership functions are assigned to each input or output variable. It has been selected 7 membership functions for $V_{Wagg}$, 7 for $V_{W\sigma}$ and 8 for output $\alpha$. Overall 49 (i.e., 7 $\times$ 7) rules are built by crossing the fuzzy sets, as shown in Table 13.3.

The $i$th fuzzy rule is expressed as [23]

Rule $i$: if $V_{Wagg}$ is $A_a$ and $V_{W\sigma}$ is $B_b$,

$$\text{then}\,\alpha(n)\ \text{is}\, C_c. \tag{13.19}$$

$a = 1, 2, \ldots., 7;\ b = 1, 2, \ldots., 7;\ c = 1, 2, \ldots., 49$

where $A_a$ and $B_b$ denote the antecedents and $C_c$ is the consequent part.

The FLS gives the values of the MTCF ($\alpha$) by applying the center of gravity method [23]

$$\alpha(n) = \sum_{i=1}^{49}\omega_i C_c / \sum_{i=1}^{49}\omega_i \tag{13.20}$$

where $\omega_i$ denotes the grade for the antecedent, which is the product of grade for the antecedents of each rule.

### 13.4.4 Equivalent Internal Electrical Network

The aggregated wind farm must operate at an equivalent internal electrical network. Thus, the internal electrical network of each individual DFIG wind turbine in the complete model is required to be replaced by equivalent impedance in the proposed aggregated wind farm model. The short circuit impedance of the aggregated wind

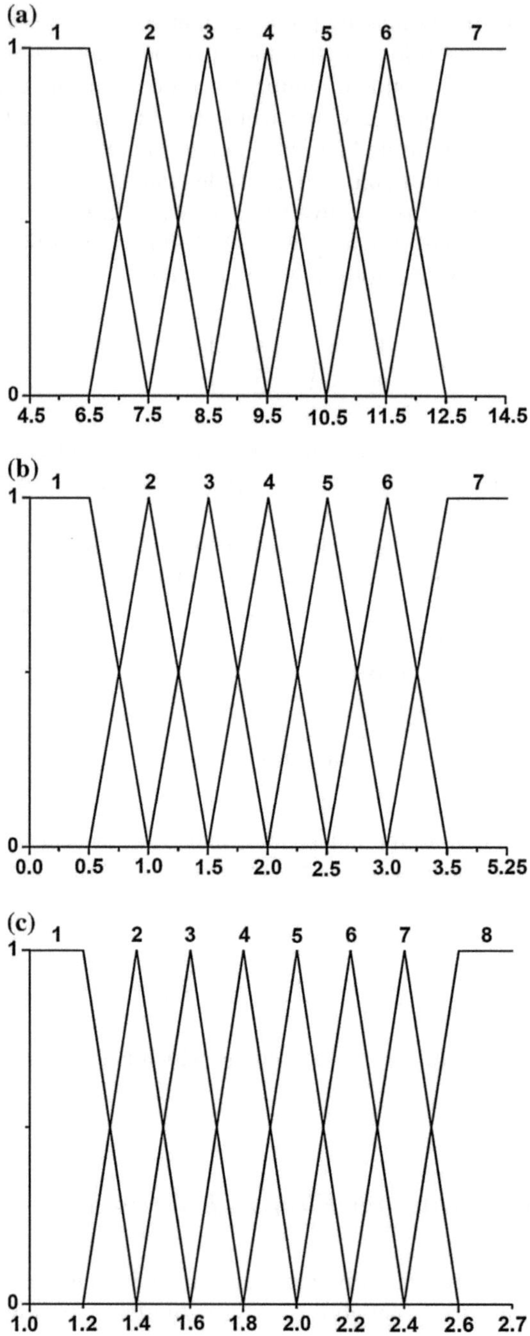


**Fig. 13.9** Membership functions: **a** $V_{Wagg}$, **b** $V_{W\sigma}$ and **c** $\alpha$

Table 13.3 Rules of the FLS

| $\alpha$ | | $V_{W\sigma}$ | | | | | | |
|---|---|---|---|---|---|---|---|---|
| | | 1 | 2 | 3 | 4 | 5 | 6 | 7 |
| $V_{Wagg}$ | 1 | 1 | 1 | 1 | 2 | 3 | 3 | 4 |
| | 2 | 1 | 1 | 2 | 3 | 3 | 4 | 5 |
| | 3 | 1 | 2 | 3 | 3 | 5 | 6 | 7 |
| | 4 | 2 | 3 | 4 | 5 | 6 | 7 | 8 |
| | 5 | 1 | 2 | 3 | 4 | 5 | 6 | 7 |
| | 6 | 1 | 1 | 2 | 3 | 4 | 5 | 6 |
| | 7 | 1 | 1 | 2 | 3 | 3 | 4 | 5 |

farm must be equal to that of the complete wind farm, which gives the calculation of the equivalent impedance ($Z_e$) of the aggregated wind farm [13]

$$Z_e = Z_{awt} - \frac{Z_{wt}}{n} \tag{13.21}$$

where $Z_{awt}$ is the equivalent impedance of the internal electrical network of each individual DFIG wind turbine in the complete model, $Z_{wt}$ is the impedance of DFIG wind turbine.

### 13.4.5 Model Simplification

The detailed representation of wind farms with DFIG wind turbines is quite complex. However, it can be simplified assuming that the power coefficient ($C_p$) is always equal to the maximum value because the control mechanism of the DFIG wind turbine maintains its power-speed characteristics such that $C_p$ is always tracking its maximum value (in this work, $C_{pmax} = 0.48$) [24]. Due to the adoption of the maximum power coefficient, the complicated $C_p(\lambda, \beta)$ characteristics from the model (Eq. 13.1) is replaced by the transfer characteristics by a first order approximation (see Fig. 13.10).

### 13.4.6 Simulation Results

Both the proposed aggregated model and the full aggregated model are simulated to obtain the dynamic responses at the PCC under the following two conditions: (1) normal operation and (2) grid disturbance. The variables considered for the comparison are the active ($P_e$) and reactive power ($Q_e$) exchange between the wind farm and power system. The reactive power is taken into the calculation of the proposed aggregated model because the reactive power does not solely depend on active power generation in the DFIG wind turbine, where the reactive power and

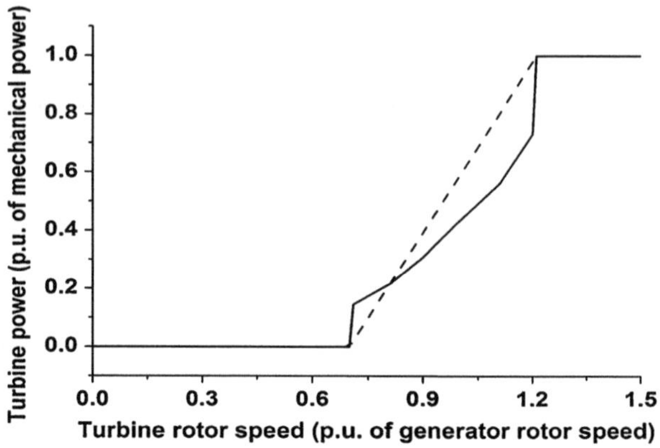


**Fig. 13.10** First order approximation (*dashed line*) of transfer characteristic (*solid line*) of the DFIG wind turbine

active power are independently regulated by the converter controllers in the exchange of reactive power with the grid. In addition, the operation in the dynamic speed range could demand lower reactive power output due to increased active power output at higher and gusty wind conditions [25].

Figure 13.11 shows the speed of the wind received by the first DFIG wind turbine in each group. The time delay and wake effect are accounted for approximating wind speed for the following DFIG wind turbines in each corresponding group.

Any changes in wind speed upstream have effects on the wind speed downstream after a certain time delay due to the wind speed transport. The delay is a function of distance and wind speed. The transport time delay of wind speed ($t_{delay}$) passing between two successive columns can be roughly estimated using [26]

$$t_{delay} = \frac{d}{\overline{V_W}} \tag{13.22}$$

where $d$ is the distance between the two successive turbine columns and $\overline{V_W}$is the average wind speed passing the first DFIG wind turbine.

Power extraction on wind flow passing the turbine creates a wind speed deficit in the area behind the turbine. This phenomenon is known as 'wake effect'. As a consequence, the turbines that are located downstream obtain lower wind speed than those that are located upstream. The deficit in wind speed due to the wake effect depends on several factors, such as the distance behind turbine, turbine efficiency and turbine rotor size. Wind speed in the wake at a distance $x$ behind the turbine rotor can be calculated as [27]

$$V_W(x) = V_o\left[1 - \left(\frac{R}{k_w x + R}\right)^2 \left(1 - \sqrt{1 - C_T}\right)\right] \tag{13.23}$$

where $V_o$ is the incoming free-stream wind speed, $C_T$ is the turbine thrust coefficient whose value is adopted from [28] and $k_w$ is the wake decay constant.

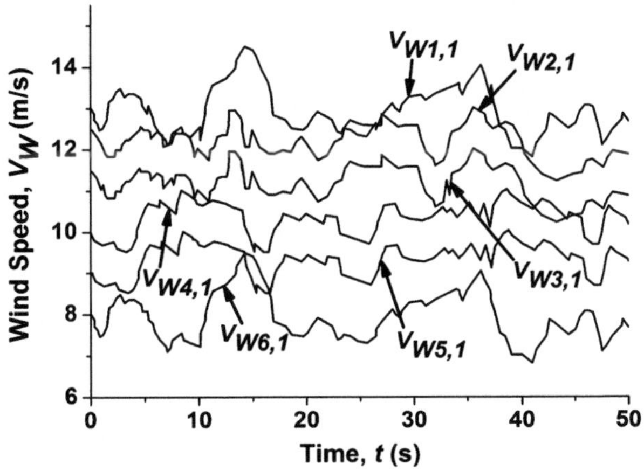


**Fig. 13.11** Wind speed received by the first DFIG wind turbine in each group

## 13.4.7 Normal Operation

The collective responses of the complete, the full aggregated and the proposed aggregated wind farm models at the PCC during normal operation are shown in Fig. 13.12.

The proposed aggregated model has a higher correspondence in approximating active power (see Fig. 13.12a). Comparing with the complete model, it has the maximum and average discrepancy of 2.94 and 2.35 %, respectively, while the full aggregated model has the maximum and average discrepancy of 8.23 and 6.58 %, respectively. The multiplication factor MTCF, dynamically produced by a well-tuned FLS, manipulates the mechanical torque to compensate the existing non-linearities in the wind farm in order to have a better approximation in the proposed aggregated model. The proposed aggregated model cannot respond to the high frequency fluctuations of wind speed as compared to the complete model, as can be seen during periods of time between 23 and 26.5 s and between 37.5 and 50 s.

The proposed aggregated model has a higher correspondence in approximating reactive power as well (see Fig. 13.12b). Comparing with the complete model, it has the maximum and average discrepancy of 5.45 and 4.36 %, respectively, while the full aggregated model has the maximum and average discrepancy of 9 and 8.14 %, respectively. The reactive power for DFIG wind turbine depends on the active power and the generation voltage. These variables differ in each DFIG wind turbine when the incoming winds are different. Therefore, it leads to different converter controller action for each DFIG wind turbine. This is not accounted for in the aggregated model resulting in lesser accuracy in the approximation of reactive power at the PCC. The manipulation of mechanical torque in the proposed aggregated model enables it to provide a better performance. However, poor approximation of reactive power is observed during the periods between 13 and 18 s and between 34.5 and 38 s.

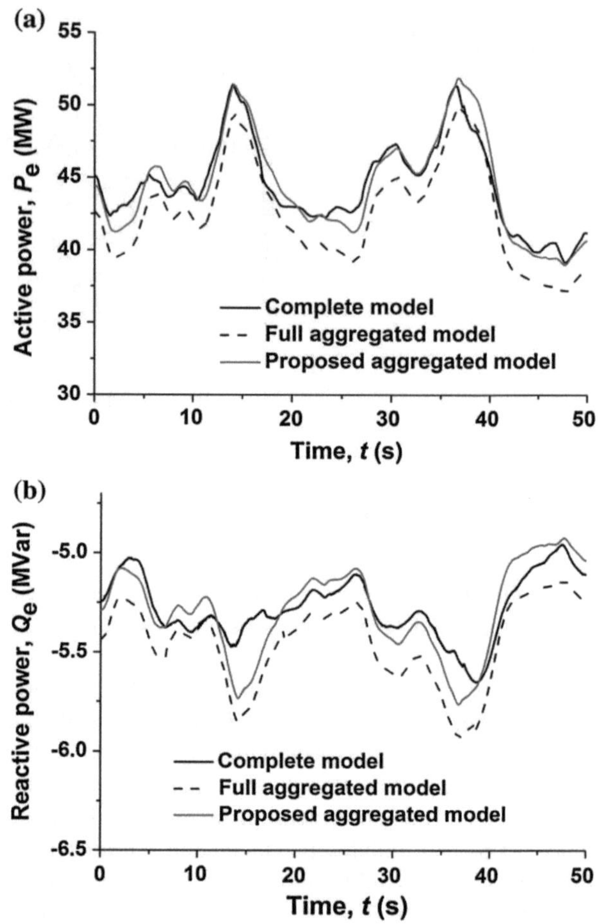


**Fig. 13.12** Evaluation of the proposed aggregated wind farm model during normal operation at the PCC: **a** Active power and **b** Reactive power

## *13.4.8 Grid Disturbance*

A voltage sag of 50 % lasting for 0.1 s is originated at the PCC at $t = 1$ s to evaluate the proposed aggregated wind farm model during grid disturbances, the collective responses of the complete, the full aggregated and the proposed aggregated wind farm models at the PCC are shown in Fig. 13.13.

Figure 13.13 shows that the active power produced by the wind farm reduces and it goes to negative values for a short time (i.e. the grid supplies active power to the DFIG to keep it spinning) during grid disturbances. On the other hand, the reactive power which is normally negative (which means the wind farm takes reactive power from the grid) changes sign and increases during the disturbance. This means the wind farm supplies reactive power to the grid during the disturbance caused by the voltage sag.

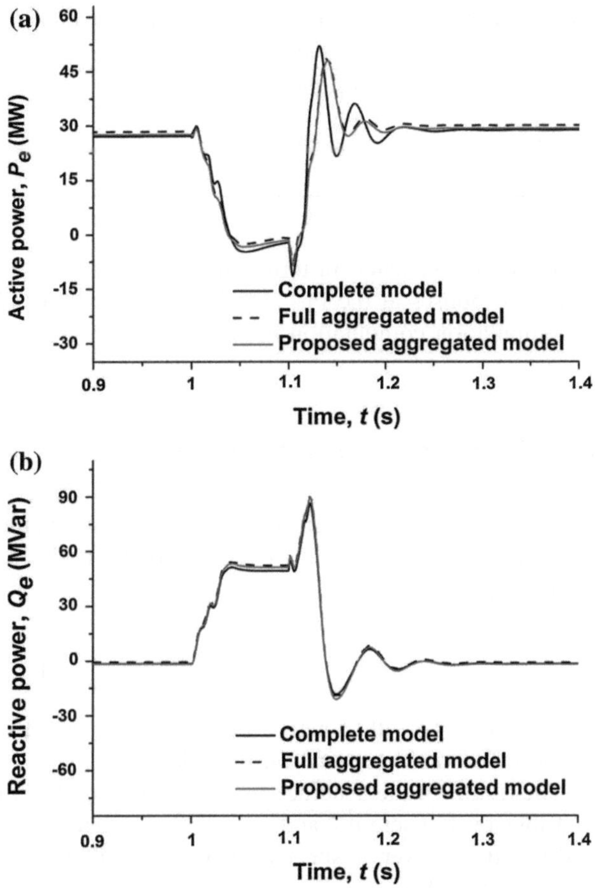


**Fig. 13.13** Evaluation of the proposed aggregated wind farm model during grid disturbance at the PCC: **a** Active power and **b** Reactive power

It also shows a high correspondence among the collective responses at the PCC of the complete, the full aggregated and the proposed aggregated wind farm models with negligible discrepancies on the active and reactive power. However, the active power ($P_e$) slightly mismatches in both aggregated models right after clearing the fault when different parameters start retaining their normal values. This high level of correspondence is partly due to the fact that the grid disturbances are much faster than the wind speed variations [13] and, therefore, the discrepancies during normal operations are unimportant during grid disturbances.

## 13.5 Evaluation of the Proposed Aggregated Technique

In previous section, the good agreement of the collective responses at the PCC between the proposed aggregated model and the complete model verifies the stability of the aggregated model. In the following, the proposed aggregated

**Table 13.4** Accuracy in approximating the collective responses at the PCC

| Operation type | Full aggregated model | | Proposed aggregated model | |
|---|---|---|---|---|
| Normal operation | $n_{Pe}$ (%) | 91.3 | $n_{Pe}$ (%) | 100 |
| | $n_{Qe}$ (%) | 87.5 | $n_{Qe}$ (%) | 100 |
| Grid disturbance | $n_{Pe}$ (%) | 95 | $n_{Pe}$ (%) | 95 |
| | $n_{Qe}$ (%) | 100 | $n_{Qe}$ (%) | 100 |

technique is evaluated in terms of the accuracy in the approximation of the collective responses at the PCC, such as active power ($P_e$) and reactive power ($Q_e$) and simulation computation time.

### 13.5.1 Accuracy in Approximation

The discrepancy between any instantaneous output power of the proposed aggregated model and that of the complete model can be calculated by the following equation [29]

$$\Delta x = \left| \frac{x_{comp} - x_{agg}}{x_{comp}} \right| \tag{13.24}$$

where $x$ can be either active power ($P_e$) or reactive power ($Q_e$). Suffix *comp* denotes the complete wind farm model.

The results of the accuracy in approximating the collective responses are shown in Table 13.4, where $n_{Pe}$ and $n_{Qe}$ are the number of instantaneous values of active and reactive power, respectively. It shows that less than 10 % discrepancy has been achieved between the proposed aggregated model and the complete model. It can be seen as well that in normal operation the proposed aggregated model approximates active power ($P_e$) and reactive power ($Q_e$) more accurately than the full aggregated model by 8.7 and 12.5 %, respectively. However, in grid disturbance both models show the same level of accuracy.

### 13.5.2 Simulation Computation Time

The comparison of computation time for the complete and both aggregated wind farm models are made and the results are shown in Table 13.5. The simulations are carried out on a personal computer with the following specifications: Intel (R) Pentium (R) Dual CPU E2200, 2.20 GHz, 1.96 GB of RAM.

It can be seen that the proposed aggregated wind farm model has higher simulation computation time than the full aggregated wind farm model by 2.38 and 3 % during normal operation and grid disturbance, respectively. A slight increase

**Table 13.5** Comparison of simulation computation time

| Operation type | Simulation computation time (s) | | | Reduction in simulation time (%) | |
|---|---|---|---|---|---|
| | Complete model | Full aggregated model | Proposed aggregated model | Full aggregated model | Proposed aggregated model |
| Normal operation | 1476 | 110 | 142 | 92.5 | 90.3 |
| Grid disturbance | 2283 | 235 | 298 | 89.7 | 87 |

in the computation time is caused by the additional computing block with the FLS to generate the MTCF. However, it has significantly reduced the simulation computation time by 90.3 and 87 %, respectively, comparing with the complete model during normal operation and grid disturbance.

## 13.6 Conclusions

This chapter describes the development of a novel aggregated technique with the incorporation of a mechanical torque compensation factor (MTCF) into the full aggregated wind farm model to obtain dynamic responses of a wind farm at the point of common coupling. The aim is to simulate the dynamic responses of the wind farm with an acceptable level of accuracy while reducing the simulation time considerably by using the aggregation technique. The MTCF is a multiplication factor to the mechanical torque of the full aggregated wind farm model that is initially constructed to approximate a Gaussian function by a fuzzy logic method and optimized on a trial and error basis to achieve less than 10 % discrepancy between the proposed aggregated model and the complete model. The proposed aggregated model is then applied to a bigger 120 MVA offshore wind farm comprising of 72 DFIG wind turbines. Simulation results show that the proposed aggregated wind farm model has the average discrepancy in approximating active power ($P_e$) and reactive power ($Q_e$) of 2.35 and 4.36 %, respectively, during normal operation as compared to the complete model. But it has 8.7 and 12.5 % more approximation capability of $P_e$ and $Q_e$, respectively, than the full aggregated model. However, the proposed aggregated model can mimic $P_e$ and $Q_e$ with negligible discrepancy during grid disturbance. Computational time of the proposed aggregated model is slightly higher than that of the full aggregated model but much faster than the complete model by 90.3 % during normal operation and 87 % during grid disturbance.

## References


1. Executive Summary (2011) World wind energy report 2011. http://www.wwindea.org/home/index.php. Accessed 15 Nov 2011
2. [2] Li L, Jing Z, Yihan Y (2009) Comparison of pitch angle control models of wind farm for power system analysis. In: IEEE power and energy society general meeting, Calgary, pp 1-7 July 2009
3. Rodriguez-Amenedo JL, Arnalte S, Burgos JC (2002) Automatic generation control of a wind farm with variable speed wind turbines. IEEE Trans Energy Convers 17:279–284
4. Tapia A, Tapia G, Ostolaza JX, Saenz JR, Criado R, Berasategui JL (2001) Reactive power control of a wind farm made up with doubly fed induction generators. In: IEEE Power Tech Proceedings, Porto, p 6, September 2001
5. Feijóo A, Cidrás J, Carrillo C (2000) A third order model for the doubly-fed induction machine. Elect Power Syst Res 56:121–127
6. Ekanayake JB, Holdsworth L, XueGuang W, Jenkins N (2003) Dynamic modeling of doubly fed induction generator wind turbines. IEEE Trans Power Syst 18:803–809
7. Nunes MVA, Lopes JAP, Zurn HH, Bezerra UH, Almeida RG (2004) Influence of the variable-speed wind generators in transient stability margin of the conventional generators integrated in electrical grids. IEEE Trans Energy Convers 19:692–701
8. Akhmatov V, Knudsen H (2002) An aggregate model of a grid-connected, large-scale, offshore wind farm for power stability investigations—importance of windmill mechanical system. Int J Electr Power Energy Syst 24:709–717
9. Usaola J, Ledesma P, Rodriguez JM, Fernandez JL, Beato D, Iturbe R et al. (2003) Transient stability studies in grids with great wind power penetration. Modelling issues and operation requirements. In: IEEE power engineering society general meeting, Toronto, p 1541, July 2003
10. Ledesma P, Usaola J, Rodríguez JL (2003) Transient stability of a fixed speed wind farm. Renewable Energy 28:1341–1355
11. Slootweg JG, Kling WL (2003) Aggregated modelling of wind parks in power system dynamics simulations. In: IEEE power tech conference proceedings, Bologna, p 6, June 2003
12. Fernández LM, García CA, Saenz JR, Jurado F (2009) Equivalent models of wind farms by using aggregated wind turbines and equivalent winds. Energy Convers Manage 50:691–704
13. Fernández LM, Jurado F, Saenz JR (2008) Aggregated dynamic model for wind farms with doubly fed induction generator wind turbines. Renewable Energy 33:129–140
14. Poller M, Achilles S (2003) Aggregated wind park models for analyzing power system dynamics. In: Proceedings of the international workshop on large scale integration of wind power and transmission networks for offshore wind farms, Billund, Oct 2003
15. Perdana A, Carlson O, Persson J (2004) Dynamic response of grid-connected wind turbine with doubly fed induction generator during disturbances. In: Proceedings of the IEEE Nordic workshop on power and industrial electronics, Trondheim, June 2004
16. Sedaghat A, Mirhosseini M (2012) Aerodynamic design of a 300 kW horizontal axis wind turbine for province of Semnan. Enerergy Convers Manage 63:87–94
17. García-Gracia M, Comech MP, Sallán J, Llombart A (2008) Modelling wind farms for grid disturbance studies. Renewable Energy 33:2109–2121
18. Ghennam T, Berkouk EM, Francois B (2009) Modeling and control of a Doubly Fed Induction Generator (DFIG) based wind conversion system. In: International conference on power engineering, energy and electrical drives, Lisbon, March 2009
19. Wei Q, Venayagamoorthy GK, Harley RG (2009) Real-Time implementation of a STATCOM on a wind farm equipped with doubly fed induction generators. IEEE Trans Ind Appl 45:98–107
20. Salman SK, Badrzadeh B (2004) New approach for modelling doubly-fed induction generator (DFIG) for grid-connection studies. European wind energy conference and exhibition, London, November 2004

21. SimPowerSystems (2010) – model and simulate electrical power systems. User's guide. Natick (MA): The Mathworks Inc
22. Voelker DH (2001) Statistics. John Wiley and Sons Inc, New York
23. Senjyu T, Sakamoto R, Urasaki N, Funabashi T, Fujita H, Sekine H (2006) Output power leveling of wind turbine generator for all operating regions by pitch angle control. IEEE Trans Energy Convers 21:467–475
24. Slootweg JG (2003) Wind power: modelling and impact on power system dynamics. Dissertation, Delft University of Technology
25. Engelhardt S, Erlich I, Feltes C, Kretschmann J, Shewarega F (2011) Reactive power capability of wind turbines based on doubly fed induction generators. IEEE Trans Energy Convers 26:364–372
26. Magnusson M, Smedman AS (1999) Air flow behind wind turbines. J Wind Eng Ind Aerodyn 80:169–189
27. Rudion K (2008) Aggregated modelling of wind farms. Dissertation, Otto-von-Guericke University
28. Perdana A (2008) Dynamic models of wind turbines. Dissertation, Chalmers University of Technology
29. Trilla L, Gomis-Bellmunt O, Junyent-Ferre A, Mata M, Sanchez Navarro J, Sudria-Andreu A (2011) Modeling and validation of DFIG 3-MW wind turbine using field test data of balanced and unbalanced voltage sags. IEEE Trans Sustain Energy 2:509–519

[illegible] ... and sensitive electrical power ... [illegible]
[illegible] The Mathworks Inc
[illegible]
[illegible]
[illegible]
[illegible]
[illegible]
[illegible]
[illegible]

# Chapter 14
# DC Grid Interconnection for Conversion Losses and Cost Optimization

**R. K. Chauhan, B. S. Rajpurohit, S. N. Singh and F. M. Gonzalez-Longatt**

**Abstract** The rapid increment of DC compatible appliances in the buildings, emerge the photovoltaic energy as the fastest growing source of renewable energy and expected to see continued strong growth in the immediate future. The photovoltaic power is, therefore, required to be provided with a certain reliability of supply and a certain level of stability. Motivated by the above issues, many grid operators have to develop DC micro-grids, which treat photovoltaic power generation in a special manner. The interconnection of different voltage rating distributed generation, storage and the load to the DC micro-grid requires large number of converters, which will increase the conversion power losses and the installation cost. Different types of Low Voltage Direct Current (LVDC) grid and their topologies are helpful in understanding the interconnection of distributed generation with consumers end. The connections of LVDC distribution system is discussed in this chapter. Optimization of LVDC grid voltage may reduce the conversion stage and the power loss in DC feeders. A multi-objective technique is discussed, in this chapter, to design a minimum power loss and low cost DC micro-grid.

---

R. K. Chauhan (✉) · B. S. Rajpurohit
School of Computing and Electrical Engineering, Indian Institute of Technology Mandi, Mandi, Himachal Pradesh 175001, India
e-mail: rkchr_iitm@yahoo.com

B. S. Rajpurohit
e-mail: bsr@iitmandi.ac.in

S. N. Singh
Department of Electrical Engineering, Indian Institute of Technology Kanpur, Kanpur, Uttar Pradesh 208016, India
e-mail: snsingh@iitk.ac.in

F. M. Gonzalez-Longatt
Coventry University, Faculty of Engineering and Computing EC3-32, Priory Street, Coventry CV1 5FB, UK
e-mail: francisco.gonzalez-longatt@coventry.ac.uk

J. Hossain and A. Mahmud (eds.), *Renewable Energy Integration*, Green Energy and Technology, DOI: 10.1007/978-981-4585-27-9_14,

**Keywords** DC-DC converter · DC load · Hybrid electric vehicle · LVDC grid · Photovoltaic

## 14.1 Introduction

In the last decade, micro-grids have become a more attractive option for rural electrification compared with the extension of electrical transmission infrastructure to connect rural areas to the centralized grid. DC power transmission and its significance have been explained in detail [1]. According to world energy outlook 2011, around 70 % rural area will be connected to a micro-grid or standalone off grid solution while only 30 % rural areas may be connected to the main grid [2]. Rapid increment in DC appliances and the integration of Distributed Generation (DG), Hybrid Electrical Vehicles (HEV) leads to change in the structure of AC power systems as well as the direction of power flow from a single direction to bidirectional in the distribution systems [3, 4]. In [5], a review has been given about the demand and economics of Renewable Energy Resources (RESs). The DC power of RESs such as Photovoltaic (PV), and the fuel cell has to converted to AC for feeding to AC grid, while AC power have to be convert into DC power at storage and load ends, resulting in more conversion losses, and poor efficiency. For high power quality distributed resources have been used in combination with DC distribution system [6].

The consumption of DC power by the DC appliances eliminates the conversion stages required in the present AC distribution system. Moreover, the DC systems are free from inductance, capacitance effects and skin effect, resulting in a lesser voltage drop, power loss and line resistance. The DC system also requires less amount of insulation than an AC system because of lesser potential stress for same working voltage. Moreover, the multilevel voltage transformation ability of DC system promotes to interconnect the different voltage level and power rating DG sources and battery bank. It will go to increase the stability and reliability of the distribution system.

The DC load is supplied by DC power generated from load site DC sources and stored energy dramatically improves the system efficiency, reduces energy costs and environmental impact. On the other hand, the optimization of the voltage level of the DC distribution system is the most important factor for reduction in conversion losses and stages. A power-sharing method presents the control modes of micro-grid when multiple distributed generators have been connected [7]. Some literature explains the power sharing for hybrid power system [8]. Moreover, the conversion of existing AC distribution systems into the DC distribution systems is the greatest challenge for electrical engineers.

In the present scenario, Low Voltage Direct Current (LVDC) distribution systems do not have standard voltage ratings as in the case of AC distribution systems. The non-standardized voltage rating of the DC distribution systems again requires more number of DC-DC conversion stages at the load, storage and the generation ends, resulting in an increase in the conversion stages, losses and

Table 14.1 Losses based on partial AC and DC loading of power systems [16]

| Loads | AC (Watt) | DC (Watt) | DC (Watt) | DC (Watt) |
|---|---|---|---|---|
| AC % / DC % | | 95 %[a] | 97 %[a] | 95 %[a] |
| 100/0[b] | 412 | 2317 | 1811 | 1119 |
| 50/50[b] | 1679 | 1982 | 1313 | 583 |
| 0/100[b] | 2872 | 1653 | 1008 | 329 |

[a] Represents DC–DC converter efficiency
[b] Represents ratio of AC/DC loads

system cost. However, some literatures are available on the DC-DC converter and AC-DC converter for renewable energy systems [9–15]. Table 14.1 shows the conversion losses based on partial AC and DC loading of power system [16]. Three efficiencies viz. 95, 97, and 99.5 % were assigned to the DC-DC converters for their comparative study.

The $I^2R$ loss can be represented by varying the voltages of the DC system. From this Table, the difference in loss compared to the AC become small for 95 % DC-DC converter, if 50 % of the loads are AC and 50 % of the loads are DC.

This chapter includes the topics which deals with not only the AC-DC-AC-DC, DC-AC-DC conversion stages, but also the topics related to the reduction of the DC-DC conversion stages, losses including system cost strategy in Indian environment. Moreover, the conversion of wiring such as conversion of the existing AC distribution system i.e. three-phase three-wire system, three-phase four-wire system, etc. into multi-pole DC distribution system without any change in the hardware is also discussed. The present chapter discusses the conversion losses in the DC and AC distribution systems. Different topologies have been discussed which may be helpful in reducing converter stages i.e. decreasing the conversion losses. Total cost of the DC system has been calculated and some methods for cost optimization have been explained.

## 14.2 Topologies of Different LVDC Grids

### *14.2.1 Topologies of Mini-Grid*

> "Mini-grid having the more expansion of serving. It is an electrical power system that having the span between two points to fulfil the requirements of more than one building" [17].

The converters and DC links are the main parts of DC distribution system. The AC-DC conversion occurs near to the Medium Voltage (MV) line in each topology. However, the points of DC-AC conversion may be at any locations. The location, where DC-AC conversion is located at consumer range, is called the wide LVDC distribution and the High Voltage Direct Current (HVDC) link may be DC-AC conversion locations [18, 19]. Figure 14.1 represents the wide LVDC

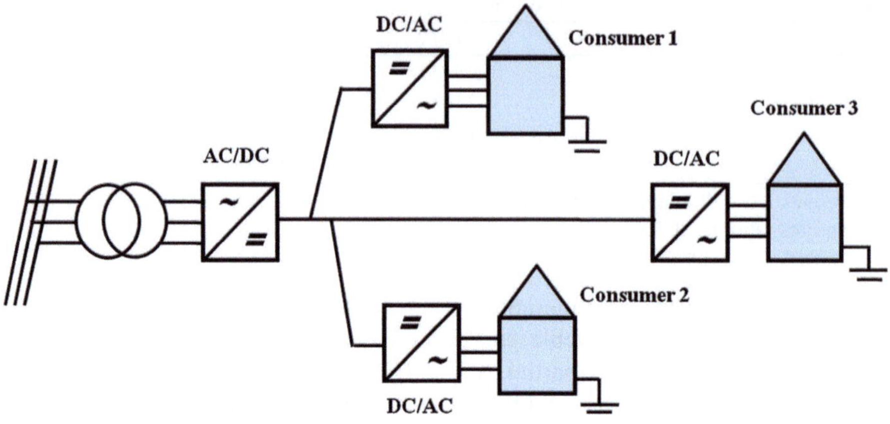


**Fig. 14.1** Wide LVDC distribution system

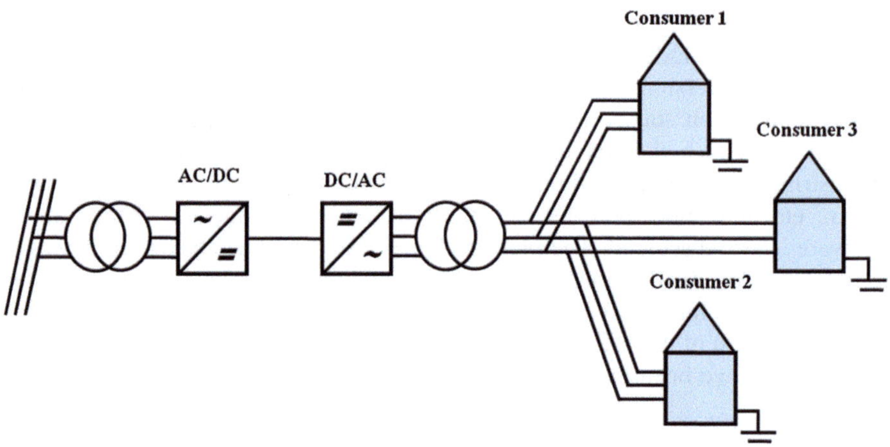


**Fig. 14.2** HVDC link distribution systems

distribution system. The three-phase three-wire AC system is replaced here by single DC line [20].

In Fig. 14.2, a DC line showing a HVDC link based DC distribution system interconnects two separate AC networks. To make a suitable connection with existing AC system, DC links are connected using transformers.

### *14.2.2 Topologies of Micro-Grid*

> "A DC micro-grid is a group of many distributed energy resources which have the capability to fulfill a specified load demand by a sufficient energy continuously" [17].

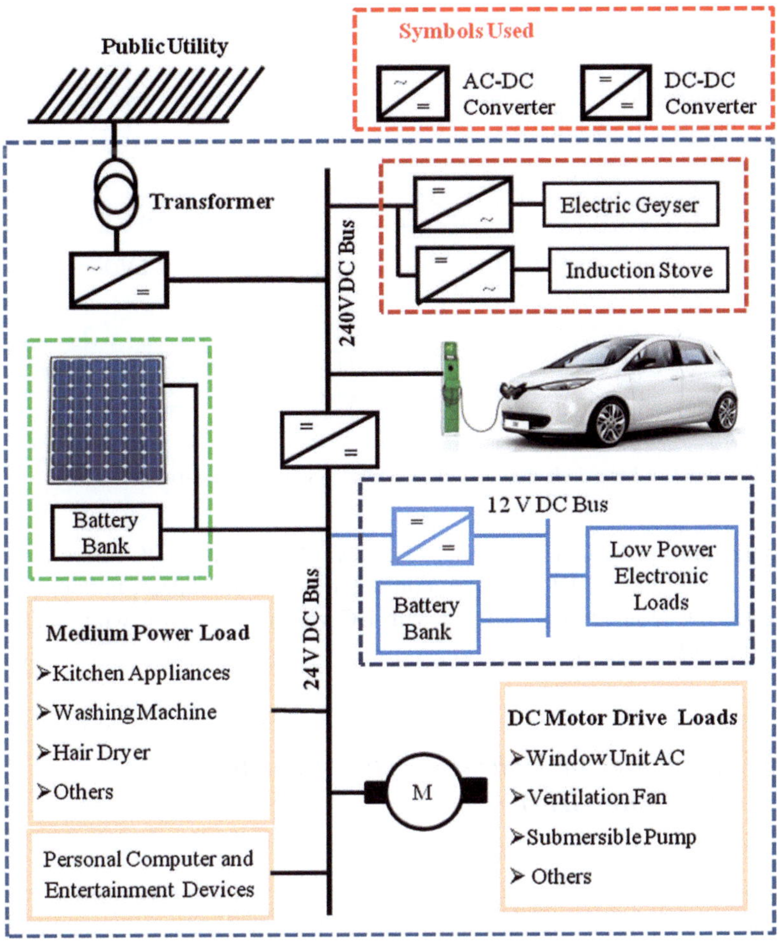


**Fig. 14.3** Layout of hybrid home power system (DC Micro-Grid)

The layout of a DC micro-grid is shown in Fig. 14.3. This micro-grid structure is a combination of three 240 volt (HVDC), 24 and 12 volt (LVDC) buses to draw the three different types of loads viz. 240 volt, 24 and 12 volt loads in a same building. The power fed to the HVDC bus (240 volt DC) is supplied by external AC system via AC-DC converter while the photovoltaic plant and battery bank are connected to the LVDC bus (24 volt DC). The Buck-boost converter is used to interconnect the HVDC and LVDC buses. The HVDC bus is responsible to supply the AC compatible appliances via DC-AC converter and the hybrid car. The photovoltaic panel, battery bank and other appliances of 24 volt voltage rating has been directly connected to LVDC bus and is used to supply the power to the low power rating equipments such as DC motor drive loads, personal computer, medium power load etc. While the 12 volt LVDC bus is designed for low power

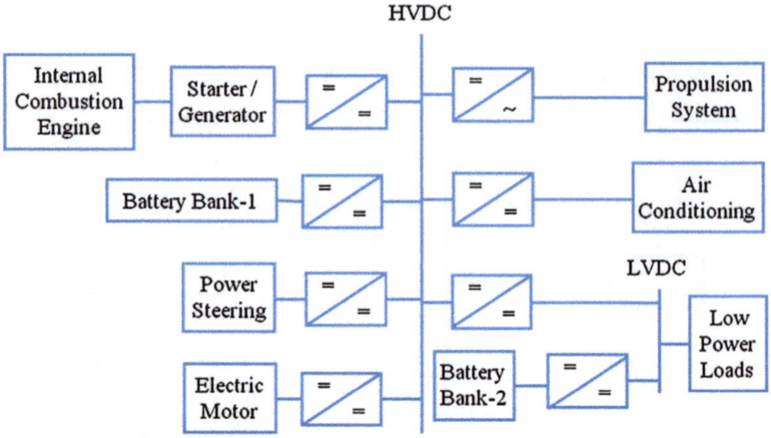


**Fig. 14.4** Layout of hybrid electrical vehicle power system (Nano-Grid)

electronic appliances such as mobile. The power balancing in DC micro-grid could be achieved by the AC-DC network converter and photovoltaic, battery bank and storage DC-DC converters equipped with DC voltage regulator, which adjusts the voltage in all the buses.

### *14.2.3 Topologies of Hybrid Electric Vehicle (DC Nano-Grid)*

> "Nano-grid is an electrical power system, defined between two universal points for fulfilling the demand of a particular space in any building" [17].

Hybrid Electric Vehicle (HEV) designs is made up of a gasoline engine, a fuel tank, an electric motor, a generator, batteries and a transmission system. The electric motor, generator and batteries all work to gather on behalf of the hybrid concept. Motors not only help power the car, but can also act as generators. It means they can draw energy from the batteries, as well as return energy to the batteries. The generator itself works solely to produce electrical power, while the batteries store the energy to electric motors. Fig. 14.4 shows the layout of DC hybrid electric vehicle (nano-grid) including distributed battery bank. According to the load, the HEV power system is divided into HVDC and LVDC bus. The power fed to the HVDC bus is supplied by the generator and battery bank-1 via DC-DC converter. The HVDC bus is also connected to the propulsion system via DC-AC converter. The air conditioning, power steering, and electric motor have different operating voltage level and they are connected to HVDC bus via its own DC-DC converter. The DC-DC converters interconnect the HVDC and battery bank-2 to LVDC bus is responsible for power balancing in LVDC bus.

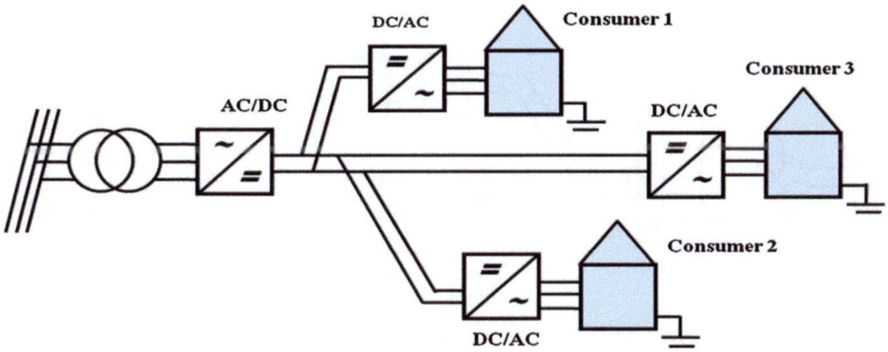


**Fig. 14.5** Unipolar LVDC distribution system

## 14.3 LVDC System Connections

The LVDC distribution system may be categorized in two parts according to the polarities of conductor.

### *14.3.1 Unipolar LVDC System*

The construction of a unipolar system has two conductors with neutral and positive polarity. The energy in the unipolar system is transmitted via single voltage level, by which all consumers are connected. Fig. 14.5 shows a unipolar DC system.

### *14.3.2 Bipolar LVDC System*

Bipolar system is a combination of two series connected unipolar system. The consumers may be connected between different voltage levels as shown in Fig. 14.6. The consumer connections 1 and 2 can lead unsymmetrical loading situations between DC poles in system. The overvoltage is consequence of current superposition at neutral wire. The possible overvoltage can be restricted with cable cross-section selection and the equal load balancing. The load balancing may be achieved by placing the load between positive pole and neutral, negative pole and neutral, positive and negative poles, and positive and negative poles with neutral connection. The connections 1 and 2 are chosen to be used in studied of $\pm 120$ volt DC bipolar system. The main lines of system contain all three conductors but at consumer end 2-wire cables connected between positive or negative pole. Therefore, the consumer supply voltage is either +120 VDC or –120 VDC [20, 21].

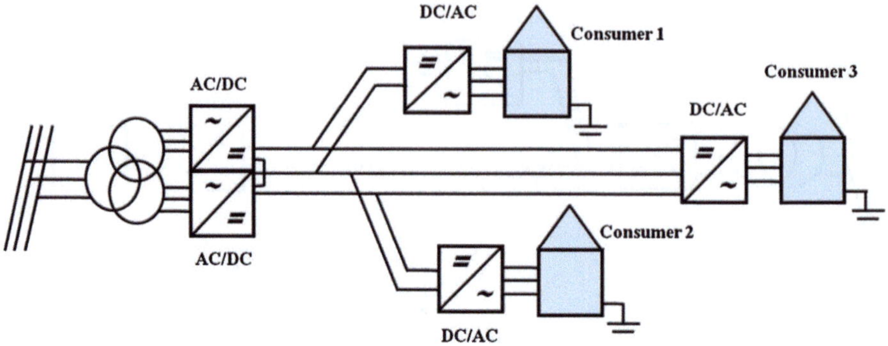


**Fig. 14.6** Bipolar LVDC distribution system with different consumer connection

## 14.4 DC-DC Converter Efficiency

In this section, the formulation of minimum required efficiency of DC-DC converter of DC distribution system is discussed. This formulation has the ability to make the efficiency of the DC distribution system near to any AC distribution system. A step-by-step approach is used to derive the technique.

- Step1

In order to match the overall efficiency of DC system with the AC distribution system, under the constraints of same load, the total power taken in $P_{totalin}$ should be same [22]. The system total power input can be expressed as:

$$P_{totalin} = \delta P_{dli} + P_{closses} \tag{14.1}$$

where, $\delta$ is the number of DC/DC converters distribution levels in the distribution system, $P_{closses}$ the losses in the distribution feeders transporting power from the bulk power source to the distribution level converters.

In the DC distribution system, individual distribution level DC-DC converter power $P_{dli}$ can be obtained by an iterative process. A seed value is required in the beginning of the iterative process. $P_{dli}$ is considered as the seed value and can be found as:

$$P_{dli} = \frac{(\gamma \cdot P_{bulin} + P_{Closses})}{\eta_{dl}} \tag{14.2}$$

where $\gamma$, the number of residential buildings served by one distribution level DC/DC converter, $\eta_{dl}$, the efficiency of the converter, $P_{bulin}$ the power input in one building. It can be concluded that:

$$P_{dli} > \gamma \cdot P_{bulin > \gamma \cdot (P_I + P_{AC} + P_{DC})} \tag{14.3}$$

**Table 14.2** Description of categories with percentage loading

| Category | Description | Relative Percentage |
|---|---|---|
| AC | Loads utilizing AC power | 33.26 |
| DC | Loads utilizing DC power | 66.00 |
| I | Loads that can use both (Independent loads) | 0.74 |

where $P_I$, $P_{AC}$ and $P_{DC}$ represent power requirements of the three categories as listed in Table 14.2.

In the present scenario three types of load such as AC compatible, DC compatible and AC-DC compatible load can be found in the building. The relative percentage of these loads has been described in Table 14.2.

Due to power losses in the house hold converters, $P_{dli}$ may be assumed slightly higher than the total power consumed in the loads. The iterative process solves for voltage and current values at each distribution level converter, starting from the source side. Once the far end is reached, the following equality is tested,

$$V_{cal} \times I_{cal} = P_{cal} = P_{dli\_assumed} \tag{14.4}$$

where $V_{cal}$ the calculated voltage, $I_{cal}$ the calculated current, $P_{cal}$ the calculated total power input to the building, $P_{dli\_assumed}$ is the assumed total power input to the building.

If the equality holds, then $P_{dli\_assumed}$ is the actual value of power input for the system. Otherwise, iteration is performed again with a modified value of $P_{dli}$. The modification depends upon the results of the previous iteration. Specifically, if

$$V_{cal} \times I_{cal} < P_{dli\_assumed} \tag{14.5}$$

- Step 2

The individual building input power $P_{bulin}$ can be expressed as:

$$P_{bulin} = P_I + \frac{(P_{AC} + P_{DC})}{\eta_{PEC}} \tag{14.6}$$

where, $\eta_{PEC}$ is the Power Electronics Converter (PEC) used in appliances.

Substituting expressions of $P_{bulin}$ and cable losses in Eq. (14.2), and choosing a suitable value for $\eta_{dl}$, the resultant can be simplified to produce the following equation.

$$P_{dli} = a.\eta_{PEC}^{-2} + b.\eta_{PEC}^{-1} + \rho \tag{14.7}$$

where,

$$a = \frac{R \cdot (P_{AC} + P_{DC})^2}{V_b^2} \times \frac{\gamma}{\eta_{dl}} \tag{14.8}$$

$$b = \left[ (P_{AC} + P_{DC}) + \frac{2P_I \cdot (P_{AC} + P_{DC})}{V_b^2} \right] \times \frac{\gamma}{\eta_{dl}} \tag{14.9}$$

$$\rho = \left[ P_I + \frac{P_I^2}{V_b^2} \times R \right] \times \frac{\gamma}{\eta_{dl}} \tag{14.10}$$

where $V_b$ the voltage at each housing unit, $R$ the resistance of power conductor from distribution converter to individual buildings. Equation (14.7) can be further simplified as

$$a'.\eta_{PEC}^{-2} + b'.\eta_{PEC}^{-1} + \lambda' = 0 \tag{14.11}$$

$a'$, $b'$ are remain the same as $a$, $b$ while

$$\lambda' = \left[ P_I + \frac{P_I^2}{V_b^2} \times R \right] \times \frac{\gamma}{\eta_{dl}} - P_{dli} \tag{14.12}$$

## 14.5 Loss Calculation in DC System

### *14.5.1 Cable Loss*

A DC system has only active power. If the load power consumption and cable type are same as AC then current and corresponding power losses in a DC system can be expressed as [23]:

$$I_{dc} = \frac{P}{V_{dc}} \tag{14.13}$$

where, $I_{dc}$ is current in DC system, $P$ total power consumption in load and $V_{dc}$ the dc voltage:

$$\Delta P_{dc} = 2.r.l.I_{dc}^2 = 2.r.l.\frac{P^2}{V_{dc}^2} \tag{14.14}$$

$\Delta P_{dc}$ power losses in DC system, $r$ the cable resistance per unit length, $V_{dc}$ the DC voltage and $l$ the cable length.

### *14.5.2 Transformation and Conversion Losses*

There are AC-DC conversion losses and DC-DC conversion losses present at different voltage levels in any multilevel DC system [24]. Main losses are conduction and switching losses. Due to partial distribution of current between IGBT

**Table 14.3** Conduction losses parameters of IGBT 2—generation [27]

| Parameters | Magnitude |
|---|---|
| $u_{T0}$ | 1.1 V |
| $u_{D0}$ | 0.832 V |
| $r_T$ | 11.0 mΩ |
| $r_D$ | 3.55 mΩ |

**Table 14.4** Switching losses parameters of IGBT2—generation [27]

| Parameter | Magnitude |
|---|---|
| $a_{0Eon}$ | 0.0794 |
| $a_{1Eon}$ | 0.0060 |
| $a_{2Eon}$ | $2 \times 10^{-5}$ |
| $a_{0Eoff}$ | 0.493 |
| $a_{1Eoff}$ | 0.0239 |
| $a_{2Eoff}$ | $-1 \times 10^{-5}$ |
| $a_{0Erec}$ | 0.6369 |
| $a_{1Erec}$ | 0.029 |
| $a_{2Erec}$ | –6E – 5 |

and anti-parallel diode, the accurate finding for the conduction losses is becoming difficult. Consideration is made by connecting a series resistance in series to create a voltage drop. According to [25], IGBT module with series resistance exhibits properties very close to an on-state resistance diode, and for this reason, conduction losses can be calculated in a simplified manner.

### 14.5.3 AC-DC Converter Losses

The power electronic converter is a combination of transistor and diode. The power losses in a power electronic converter can be divided into conduction and switching losses [25]. The conduction losses parameters of IGBT 2—generation has been shown in Table 14.3. The average conduction losses in the transistor or diode can be expressed as:

$$\bar{P}_{c.loss,T/D} = u_{T_0/D_0}\hat{} \left(\frac{1}{2\pi} \pm \frac{\alpha \cos\phi}{8}\right) + r_{T/D}\hat{I}^2\left(\frac{1}{8} \pm \frac{\alpha \cos\phi}{3\pi}\right) \tag{14.15}$$

where $\overline{P}_{c.loss,T/D}$ the average conduction losses for transistor or diode, $u_{T0}$, the threshold voltage of transistor, $u_{D0}$ threshold voltage of diode, $r_T$, differential resistance of transistor, $r_D$ differential resistance of diode.

The switching losses in the power electronic converters occur during the turn on and turn off the power semiconductor and depend on the current blocking voltage, the chip temperature and the current which are represented in characteristic maps

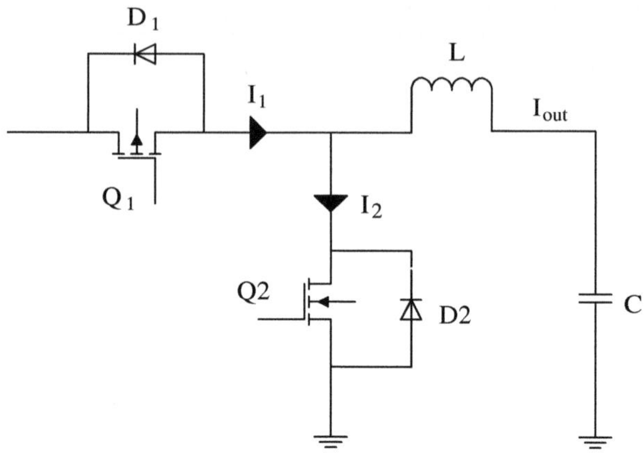


**Fig. 14.7** Basic topology of buck converter

of the energy losses per switch delivered by the manufacturer [26]. The switching losses parameters of IGBT 2—Generation is shown in Table 14.4. The power losses in switching are calculated by the following equation [27].

$$\bar{P}_s = f_s \left[ \begin{array}{l} \frac{1}{2\pi}\int_0^{2\pi} E_{onT}(i_T)d\phi + \frac{1}{2\pi}\int_0^{2\pi} E_{offT}(i_T)d\phi \\ + \frac{1}{2\pi}\int_0^{2\pi} E_{rec}(i_D)d\phi \end{array} \right] \frac{U_{dc}}{U_{dcref}} \tag{14.16}$$

where $E_{onT}$, the switching on losses of the transistor, $E_{offT}$ the switching off losses of the transistor, and $E_{rec}$, the switching off losses of the diode.

### 14.5.4 DC-DC Converter

The power losses in a power electronic DC-DC converter can be divided into conduction and switching losses, where conduction losses consist of inductor conduction losses and MOSFET conduction losses [28] (Fig.14.7).

- Inductor Conduction Losses

Inductor conduction losses is as follows

$$P_L = I_L^2 \times R_L \tag{14.17}$$

where $R_L$ is the DC-Resistance of the inductor,

The inductor rms current ($I_L$):

$$I_L = I_o^2 + \frac{\Delta I^2}{12} \tag{14.18}$$

where $I_o$ the output current and $\Delta I$ the ripple current.

Typically, $\Delta I$ is about 30 % of the output current. Therefore, the inductor current can be calculated as:

$$I_L = I_o \times 1.00375 \tag{14.19}$$

Because the ripple current contributes only 0.375 % of $I_L$, it can be neglected. The power dissipated in the inductor now can be calculated as:

$$P_L = I_o^2 \times R_L \tag{14.20}$$

- Power Dissipated in the MOSFETs

The power dissipated in the high-side MOSFET is given by:

$$P_{Q1} = I_{rms\_Q1}^2 \times R_{DSON1} \tag{14.21}$$

where $R_{DSON1}$ is the on-time drain-to-source resistance of the high-side MOSFET.

$$P_{Q1} = \frac{V_o}{V_{IN}}\left(I_o^2 + \frac{\Delta I^2}{12}\right)R_{DSON1} \tag{14.22}$$

The power dissipated in the low-side MOSFET is given by:

$$P_{Q2} = I_{rms\_Q2}^2 \times R_{DSON2} \tag{14.23}$$

where $R_{DSON2}$ is the on-time drain-to-source resistance of the low-side MOSFET.

$$P_{Q2} = \left(1 - \frac{V_o}{V_{IN}}\right)\left(I_o^2 + \frac{\Delta I^2}{12}\right)R_{DSON2} \tag{14.24}$$

The total power dissipated in both MOSFET's is given by:

$$P_{FET} = P_{Q1} + P_{Q2} \tag{14.25}$$

$$P_{FET} = \left(I_o^2 + \frac{\Delta I^2}{12}\right)\left[\left(1 - \frac{V_o}{V_{IN}}\right)R_{SDON2} + \frac{V_o}{V_{IN}}R_{SDON1}\right] \tag{14.26}$$

where

$$\Delta I = \frac{\left(V_{IN} \times V_o - V_o^2\right)}{L \times f \times V_{IN}} \tag{14.27}$$

and:
$L$ = Inductance (H)
$f$ = Frequency (Hz)

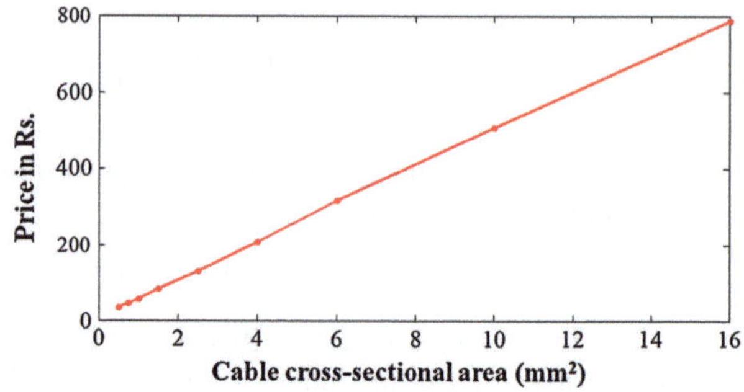


**Fig. 14.8** Cable size versus prices in Indian scenario

**Table 14.5** Investigated appliances, appliance ratings and estimated on time per day

| Product name | Quantity | Ratings | | | Appliance on time in a day |
|---|---|---|---|---|---|
| | | Power rating (W) | Current rating (A) | Voltage Rating (V) | |
| Light bulb LED | 4 | 7 | 0.60 | 12 DC | 10 |
| Light bulb CFL | 2 | 12 | 1.00 | 12 DC | 10 |
| Microwave oven | 1 | 235 | 10.00 | 24 DC | 1 |
| Induction stove | 1 | 2000 | 10.00 | 230 AC | 2 |
| Electric geyser | 1 | 1500 | 8.00 | 230 AC | 1 |
| Sandwich maker | 1 | 550 | 23.00 | 24 DC | 0.5 |
| Coffee maker | 1 | 135 | 11.00 | 12 DC | 0.5 |
| Refrigerator (DC) | 1 | 72 | 3.00 | 24 DC | 12 |
| Water purifier | 1 | 11 | 0.50 | 24 DC | 1 |
| Ventilation fan | 4 | 20 | 0.90 | 24 DC | 5 |
| Submersible pump | 1 | 240 | 10.00 | 24 DC | 0.5 |
| Washing machine | 1 | 70 | 3.00 | 24 DC | 0.5 |
| Vacuum cleaner | 1 | 95 | 8.00 | 12 DC | 0.25 |
| Window unit AC | 2 | 800 | 33.30 | 24 DC | 12 |
| Laptop | 1 | 65 | 3.34 | 19.5 DC | 7 |
| Perdonal computer | 1 | 170 | 14.00 | 12 DC | 5 |
| External modem | 1 | 5 | 0.43 | 12 DC | 24 |
| 15.6"LCD television | 1 | 30 | 2.50 | 12 DC | 5 |
| Ceiling fan | 4 | 20 | 1.70 | 12 DC | 4 |
| Hair dryer | 1 | 425 | 15.00 | 24 DC | 0.5 |
| Cell phone | 4 | 4 | 0.30 | 12 DC | 5 |
| Hybrid car | 1 | 3000 | 12.50 | 240 DC | 10 |

$V_{IN}$ = Input voltage (V)
$V_o$ = Output voltage (V)

- MOSFET Conduction Losses

**Table 14.6** Cable size, power loss and energy consumption for DC system

| Product Name | Cable Size ($mm^2$) | Cable length (meter) | Energy Loss in feeder (kWh/yr) | Total Energy Consumption (kWh/yr) |
|---|---|---|---|---|
| Light Bulb LED | 4 | 40 | 0.23 | 102.43 |
| Light Bulb CFL | 4 | 30 | 0.47 | 88.07 |
| Microwave oven | 6 | 10 | 1.05 | 86.82 |
| Induction Stove | 1.5 | 20 | 16.74 | 1476.74 |
| Electric Geyser | 1.5 | 30 | 8.04 | 555.54 |
| Sandwich Maker | 35 | 20 | 0.95 | 101.32 |
| Coffee Maker | 25 | 20 | 0.30 | 24.94 |
| Refrigerator DC | 4 | 20 | 3.39 | 318.75 |
| Water Purifier | 1.5 | 20 | 0.02 | 4.04 |
| Ventilation fan | 1.5 | 30 | 0.51 | 146.51 |
| Submersible Pump | 6 | 10 | 0.52 | 44.32 |
| Washing Machine | 6 | 25 | 0.12 | 12.89 |
| Vacuum Cleaner | 16 | 15 | 0.09 | 8.76 |
| Window Unit AC | 25 | 10 | 33.42 | 7041.42 |
| Laptop | 10 | 40 | 1.96 | 168.04 |
| Personal Computer | 25 | 15 | 3.69 | 313.94 |
| External Modem | 1.5 | 30 | 0.56 | 44.36 |
| 15.6" LCD Television | 6 | 20 | 0.65 | 55.40 |
| Ceiling Fan | 4 | 20 | 0.36 | 117.16 |
| Hair Dryan | 16 | 15 | 0.66 | 78.22 |
| Cell Phone | 1.5 | 20 | 0.04 | 29.24 |
| Hybrid Car | 1.5 | 10 | 65.40 | 11015.40 |
| Total Energy in kWh/year | | | 139.17 | 21834.31 |

For typical buck power supply designs, the inductor's ripple current, $\Delta I$, is less than 30 % of the total output current, so the contribution of $\Delta I_2/12$ to the is negligible and can be dropped to get:

$$P_{FET} = I_o^2 \left[ \frac{V_o}{V_{IN}} (R_{SDON1} - R_{SDON2}) + R_{SDON2} \right] \tag{14.28}$$

Note that when $R_{SDON1} = R_{SDON2}$, then:

$$P_{FET} = I_o^2 \times R_{SDON2} \tag{14.29}$$

The power dissipated in the MOSFET is independent of the output voltage. By using Eq. (14.29) the conduction losses of MOSFET can be calculated at any output voltage. On the other side, inductor conduction losses and switching losses etc. are independent of output voltage and remains constant with change in output voltage [22].

Hence, $P_D$ now can be computed as:

$$P_D = P_L + P_{FET} + otherlosses \tag{14.30}$$

There are some other types of losses such as the MOSFET switching losses, quiescent current etc. At any output voltage, the overall efficiency can be calculated by the known total power supply losses and power supply output power.

$$\eta = \frac{P_o}{P_o + P_D} \tag{14.31}$$

## 14.6 Total Cable Cost

The voltage drop across the feeder cable and current is high for the 12 V DC systems. The dependency of these losses is on the household appliances as well as on the cable size (length and cross sectional area). For the 12 V DC systems, the current is double for the appliances of same power rating compared to the 24 V DC systems. Hence, power losses and voltage drops will decrease by increasing the voltage rating. If the wire resistance is reduced then the power losses across the cable can be reduced, as the resistance is inversely proportional to the cross section of the wire. The losses in the cable can be reduced by increasing its cross-sectional area. For example in a load of 500 W the power losses reduces 40 % if a 2.5 $mm^2$ cable is used instead of a 1.5 $mm^2$ cable. However, when the cross-sectional area increases it will definitely increase the copper cost used in the cable. The total cost of the cable can be minimized by minimizing the cross sectional area of the cable [29].

The total cost of the cable is calculated as the sum of the investment cost of the cable and the cost of the losses in the cable.

The total annual cost is calculated as:

$$C_t = C_c + \frac{W_{fl}}{year} \times N \times C_e \tag{14.32}$$

where $C_t$ the total cost, $C_c$ the cable cost, $W_{fl}$ feeder energy loss, $N$ life time and $C_e$ energy cost. The life time is assumed to be 25 years and the energy cost Rs.1 / kWh. From Fig. 14.8, it is seen that the copper wire cost increases almost linearly with the cross section of wire.

The annual cable cost can be expressed as:

$$C_c = (C_1 + C_2A)\text{Rs.} \tag{14.33}$$

The annual cost of energy waste can be calculated as:

$$C_{ew} = \frac{C_3}{A}\text{Rs.} \tag{14.34}$$

Total annual cost

$$C_t = \left(C_1 + C_2 A + \frac{C_3}{A}\right) l \tag{14.35}$$

where C1, C2, C3 are constant and A is the cross sectional area of conductor.

$$\begin{aligned} &C_1 = 2.662 \text{ Rs./m}, \quad C_2 = 1.959 \text{ Rs./mm}^2.\text{ m, and} \\ &C_3 = P_r \cdot N.\rho \cdot E_{ON} \cdot \frac{I_{on}}{V_{dc}} \text{ Rs. mm}^2/\text{m} \end{aligned} \tag{14.36}$$

The optimum area that minimizes the total cost can be calculated as

$$\frac{dC_t}{dA} = C_2 - \frac{C_3}{A^2} = 0 \tag{14.37}$$

$$A = \sqrt{\frac{3.75 N \rho E_{on} I_{on}}{1.959 V_{dc}}} \tag{14.38}$$

The on time in a typical day of different ratings appliances in a building is shown in Table 14.5.

The description of cable size (area of cross section and length) to connect particular appliances to the supply is based on Table 14.6. The energy losses in the feeder and the total energy consumption in each appliances of building for a year is also shown.

## 14.7 Conclusions

In this chapter, DC grids and Hybrid Electric Vehical (HEV) architechture has been discussed. Topologies and wiring system discussed here are very helpful to understand the most efficient way of interconnection. Tabular data of energy dissipiated in DC appliences and the cable cost data representing a easy way to understand the correlation of different parameters associated with losses. Graphical relation of cable cost vs. cable cross-sectional area representing a recent study of indian power economics. The relation is almost linear in nature with some exceptions. So the cost calculation can be done by standard formulation. Overall chapter giving a brief knowledge about DC grid topologies, losses and cost optimization, which are the main requirements to design a DC system.

**Acknowledgments** This work was supported by the Department of Science and Technology, Government of India, under the Science and Engineering Research Board Fast Track Scheme for Young Scientists (SERC/ET-0123/2012).

## References

1. Uhlmann E (1975) Power transmission by direct current. Springer, Berlin, 289
2. International Energy Agency, World Energy Outlook (2011) Paris, France IEA publications
3. Garbesi K, Vossos V, Shen H (2012) Catlog of dc appliances and power system. http://efficiency.lbl.gov/sites/all/files/catalog_of_dc_appliances_and_power_systems_lbnl-5364e.pdf
4. Chiu HJ, Huang HM, Lin LW, and Tseng MH (2005) A multiple input DC-DC converter for renewable energy systems. In: Proceedings of 2005 IEEE international conference on industrial technology, pp 1304–1308
5. Khanna M, Rao ND (2009) Supply and demand of electricity in the developing world. Annu Rev Resource Econ 1:567–596
6. Noroozian R, Abedi M, Gharehpetian GB, Hosseini SH (2010) Distributed resources and DC distribution system combination for high power quality. Int J Electr Power Energy Syst 32(7):769–781
7. Solero L, Lidozzi A, Pomilio JA (2005) Design of multiple-input power converter to hybrid vehicles. IEEE Trans Power Electron 20(5):1007–1016
8. Jiang W, Zhang Y (2011) Load sharing techniques in hybrid power systems for dc micro-grids. In: Proceedings of 2011 IEEE power and energy engineering conference, pp 1–4
9. Ahmad Khan N (2012). Power loss modeling of isolated AC–DC converter. Dissertation, KTH
10. Hayashi Y, Takao K et al (2009) Fundamental study of high density DC/DC converter design based on sensitivity analysis. In: IEEE telecommunications energy conference (INTELEC), pp 1–5
11. Kang T, Kim C et al (2012) A design and control of bi-directional non-isolated DC-DC converter for rapid electric vehicle charging system.In: Twenty-seventh IEEE annual applied power electronics conference and exposition (APEC), pp 14–21
12. Sizikov G, Kolodny A et al (2010) Efficiency optimization of integrated DC-DC buck converters. In: 17th IEEE international conference on electronics circuits and systems (ICECS), pp 1208–121
13. Vorperian V (2010) Simple efficiency formula for regulated DC-to-DC converters. IEEE trans Aerosp Electron 46(4):2123–2131
14. Wens M, Steyaert M (2011) Basic DC-DC converter theory. Design and implementation of fully-integrated inductive DC-DC converters in standard CMOS, Springer
15. Zhang F, Du L et al. (2006) A new design method for high efficiency DC-DC converters with flying capacitor technology.In: IEEE twenty-first annual applied power electronics conference and exposition (APEC), pp 92–96
16. Starke M, Tolbert L M, Ozpineci B (2008) AC vs. DC distribution: a loss comparison. In proceedings. of transmission and distribution conference and exposition, pp 1–7
17. Savage P, Nordhaus R et al (2010) DC microgrids: benefits and barriers. published for Renewable Energy and International Law (REIL) project. Yale School of Forestry and Environmental Studies, pp 0–9
18. Pellis J, P J I et al (1997). The DC low-voltage house. Dissertation, Netherlands Energy Research Foundation ECN
19. Chauhan R K, Rajpurohit B S, Pindoriya N M (2012) DC power distribution system for rural applications. In: Proceedings of 8th national conference on indian energy sector, pp 108–112
20. Salonen P, Kaipia T, Nuutinen P, Peltoniemi P, Partanen J (2008) An LVDC distribution system concept. Nordic Workshop on Power and Industrial Electronics. A3-1–A3-16
21. Hossain MJ, Pota HR, Ugrinovskii V, Ramos RA (2009) Robust STATCOM control for the enhancement of fault ride-through capability of fixed speed wind generators. IEEE Control Applications (CCA) and Intelligent Control (ISIC),pp.1505–1510
22. Dastgeer F, Kalam A (2009) Efficiency comparison of DC and AC distribution systems for distributed generation. In: IEEE power engineering conference, pp 1–5

23. Nilsson D, Sannino A (2004) Efficiency analysis of low-and medium-voltage dc distribution systems. In: IEEE power engineering society general meeting, pp 2315–2321
24. Peterson A Lundberg S (2002) Energy efficiency comparison of electrical systems for wind turbines, Nordic workshop on power and industrial electronics (NORPIE)
25. Schroeder D(2008) Leistungselektronische Schaltungen (Power Electronic Circuits), Springer, Berlin
26. Muehlbauer K, Gerling D (2012) Experimental verification of energy efficiency enhancement in power electronics at partial load. In: 38th annual conference on IEEE industrial electronics society (IECON), pp.394–397. doi: 10.1109/IECON.2012.6388788
27. Muehlbauer K, Bachl F, Gerling D (2011) Comparison of measurement and calculation of power losses in AC/DC-converter for electric vehicle drive. In: International conference on electrical machines and systems (ICEMS), pp 1–4
28. Raj A (2010) PMP-DCDC controllers calculating efficiency: application report. http://www.ti.com/lit/an/slva390/slva390.pdf. Accessed February 2010
29. Amin M, Arafat Y et al. (2011) Low voltage DC distribution system compared with 230 V AC. In: IEEE electrical power and energy conference (EPEC), pp 340–345

# Chapter 15
# Interconnected Autonomous Microgrids in Smart Grids with Self-Healing Capability

**Farhad Shahnia, Ruwan P. S. Chandrasena, Sumedha Rajakaruna and Arindam Ghosh**

**Abstract** In order to minimize the number of load shedding in a Microgrid during autonomous operation, islanded neighbour microgrids can be interconnected if they are on a self-healing network and an extra generation capacity is available in Distributed Energy Resources (DER) in one of the microgrids. In this way, the total load in the system of interconnected microgrids can be shared by all the DERs within these microgrids. However, for this purpose, carefully designed self-healing and supply restoration control algorithm, protection systems and communication infrastructure are required at the network and microgrid levels. In this chapter, first a hierarchical control structure is discussed for interconnecting the neighbour autonomous microgrids where the introduced primary control level is the main focus. Through the developed primary control level, it demonstrates how the parallel DERs in the system of multiple interconnected autonomous microgrids can properly share the load in the system. This controller is designed such that the converter-interfaced DERs operate in a voltage-controlled mode following a decentralized power sharing algorithm based on droop control. The switching in the converters is controlled using a linear quadratic regulator based state feedback which is more stable than conventional proportional integrator controllers and this prevents instability among parallel DERs when two microgrids are interconnected.

---

F. Shahnia (✉)
Center of Smart Grid and Sustainable Power Systems, Department of Electrical and Computer Engineering, Curtin University, Perth, Australia
e-mail: farhad.shahnia@curtin.edu.au

R. P. S. Chandrasena · S. Rajakaruna
Department of Electrical and Computer Engineering, Curtin University, Perth, Australia
e-mail: ruwan.chandrasena@curtin.edu.au

S. Rajakaruna
e-mail: s.rajakaruna@curtin.edu.au

A. Ghosh
School of Electrical Engineering and Computer Science Queensland University of Technology, Brisbane, Australia
e-mail: a.ghosh@qut.edu.au

J. Hossain and A. Mahmud (eds.), *Renewable Energy Integration*,
Green Energy and Technology, DOI: 10.1007/978-981-4585-27-9_15,

The efficacy of the primary control level of DERs in the system of multiple interconnected autonomous microgrids is validated through simulations considering detailed dynamic models of DERs and converters.

**Keywords** Interconnected microgrids · Self-healing network · Power sharing · DER · DSTATCOM

## 15.1 Introduction

The ever increasing energy demand, the necessity of cost reduction and higher reliability requirement are driving the modern power systems towards using Distributed Energy Resources (DER) as an alternative to the expansion of large centralized generating stations [1, 2].

Microgrid is a cluster of DERs and loads which can operate in grid-connected mode as well as in autonomous (islanded) mode during planned (network maintenance) or unplanned (network fault) conditions [3]. A survey on microgrid control strategies is presented in [4]. General introduction on microgrid basics, including the architecture, protection and power management is given in [5, 6]. To deliver high quality and reliable power, the microgrid should appear as a single controllable unit that responds to changes in the system [7]. A review of ongoing research projects on microgrid in US, Canada, Europe and Japan is presented in [6, 8].

In grid-connected mode, the grid dictates voltage and frequency of the network and DERs operate at their nominal (rated) capacities. For the renewable energy based DERs, the rated capacity is dynamically changing depending on the prevailing weather conditions. In this mode of operation, the desired reference for DERs output can be derived using a constant PQ control strategy [9]. Alternatively, in autonomous mode, if the DERs generation capacity is higher than local load demand, they will be sharing the loads. In this operation, voltage and frequency droop control strategy is proposed in [10] for deriving the reference for the DER converters, for power sharing among parallel converter-interfaced DERs similar to the droop control technique used for power sharing among generators in a conventional power system. Later, voltage and angle droop is proposed in [11] to be used instead of the voltage and frequency droop for converter-interfaced DER applications. For improving the system response, the modified droop control is proposed in [12]. For improving the small signal stability, an arctan power frequency droop is proposed in [13] instead of the conventional droop method. Recently, intelligent power sharing algorithms such as adaptive droop control [14] and intelligent droop control [15] are proposed to remove the dependency of droop control on line parameters. In addition, in [16], a potential function based method is utilized instead of the droop control to adjust common set points required for power sharing among the DERs in a microgrid. Stability of the microgrid system with different power sharing algorithms is also investigated in [17–19]. Once the references for the DERs are assigned, the DER converters can be controlled either in voltage-controlled [20, 21] or current-controlled [22, 23] strategies.

In grid-connected mode, the grid dictates the network voltage; however, in autonomous mode, the network voltage can be indirectly regulated by the DERs based on the droop control. Although in both modes, the voltage along the low voltage network is maintained within the acceptable limits, it is desired to hold the voltage to the nominal value of 1 pu. This can be achieved if one of the DERs in microgrid regulates the network voltage to 1 pu, referred to as Master DER in [4]. However, the DERs in residential low voltage networks are owned by customers and are not responsible for the network voltage support. This is because utilizing the converter of a DER to generate reactive power, for supporting the network voltage profile, will reduce the active power generation capacity of the converter which is not desired by their owners. Alternatively, a Distribution Static Compensator (DSTATCOM) can be used in each microgrid to regulate its voltage at Point of Common Coupling (PCC) [24].

Another important issue in microgrids is islanding detection and resynchronization. Islanding is referred to the isolating of microgrid from the grid. Different islanding detection methods are presented in [25]. In [25], it is indicated that a communication based method can be utilized to send the circuit breaker status from the circuit breaker to the DER converters. This method is preferred among other islanding detection methods since it has no None Detection Zones (NDZ). Resynchronization is referred to the reconnection process of a DER to microgrid or a microgrid to the grid [26]. If a DER operates in current-controlled strategy, no resynchronization is required when connecting to microgrid. However, if a DER operates in voltage-controlled strategy, proper resynchronization is required. Reconnection should only take place once the voltage magnitude and voltage phase differences across the respective circuit breaker are zero or lower than a very small specified value [26]. Inappropriate reconnection may cause high current fluctuations which can damage the network assets or result in system instability. Different resynchronization methods are proposed in [27–29].

If DERs generation capacity in an autonomous microgrid is less than the local load demand, load shedding has to be carried out in order to maintain the voltage and frequency in the microgrid [30, 31]. In [32, 33] based on the microgrid and power market concepts and availability of control and communication infrastructures, a model of distributed autonomous microgrids is proposed. In such a model, if the DERs generation capacity in an autonomous microgrid is less than the local load demand but there is surplus generation capacity in DERs of neighbour autonomous microgrids, interconnecting these two microgrids can reduce the load shedding in the microgrid with generation deficiency. In this model, the DERs in interconnected microgrids should be properly controlled to share the total load demand in the interconnected system. With the increased interest in Smart Grid and self-healing networks, there is a possibility of interconnecting autonomous microgrids in the near future [32].

Smart Grid is a term referred to the improved condition of existing electric networks into more reliable, efficient, sustainable and customer-interactive status by properly adding advanced metering, protection and communication infrastructures [34, 35]. A general summary of the available smart grid definitions, standards, protection and management plans and the required information technology, energy and communication infrastructures are given in [34]. Among various anticipated smart features, self-healing is a key attribute in smart grids. It is mainly driven by the requirement to improve system reliability [35–37]. In a self-healing network, it is expected that the network can continuously detect, analyze and respond to faults and restore feeders with minimum human intervention. Therefore, in case of a fault in the network, the normal operation can be restored in different feeder sections by properly isolating only the faulted subsections such that the amount of the affected loads is minimized. The feasibility study carried out in [38] concludes that integration of self-healing capabilities to the future smart grids will bring high financial benefits to both utilities and customers by reducing the number of affected customers as well as the amount of unsupplied energy.

For a self-healing network, the intelligent agents are required to adapt the system operation conditions. These agents are then utilized for analyzing and maintaining the system reliability in real-time. A framework, required to implement autonomous agents throughout an interconnected system, is proposed in [39]. Such a framework can be utilized to support a self-healing smart grid through the system monitoring and controlling. Some self-healing reconfiguration techniques are proposed in [40, 41] to divide the network into isolated grids while minimizing the number of effected loads. Some utilities in US have already started implementing self-healing projects [42].

In a smart grid with self-healing capability, the following assumptions are required when interconnecting two autonomous microgrids:

- Availability of communication infrastructure among DERs, protection devices and circuit breakers,
- Self-healing capability in the network,
- Surplus generation capacity in DERs of at least one of the microgrids,
- Possibility of bypassing some of the current technical requirements for DER interconnection.

This book chapter is on the dynamic performance of DER converters and possibility of power sharing among the DERs in such systems. This is referred to as the primary control level of the system of interconnected microgrids. An automatic supply restoration algorithm is required for interconnecting the neighbor microgrids. It is to be noted that developing and validating the control philosophy for interconnecting neighbour microgrids and the required hardware are not discussed in this chapter.

## 15.2 A Network with Self-Healing Capability

Let us assume a medium voltage feeder with self-healing capability with the single line diagram as shown in Fig. 15.1. The assumed network contains 10 microgrids which can be interconnected at certain times to fulfil the power demand requirement of each other. Each microgrid is a combination of several DERs and loads. The network is divided into three zones by proper installation and coordination of circuit breakers $CB_G$, $CB_{M1}$, $CB_{M2}$ and $CB_{M3}$, as shown in this figure. This zone forming can be realized by deploying the required protection and communication infrastructure, which is not the scope of this chapter.

In case of a fault within any microgrid, the faulty microgrid can be isolated from rest of the network by the proper operation of the low voltage circuit breaker, located in the secondary side of the distribution transformer in the relevant microgrid. Let us assume a fault is occurred on the medium voltage feeder. Based on the fault location, three separate cases can be identified as below:

Case A: The fault is on the medium voltage feeder within Zone 2.
Case B: The fault is on the medium voltage feeder within Zone 1.
Case C: The fault is in the grid (i.e. upstream of Zone 1).

For Case A, when the fault is within Zone 2, it is expected that, based on the protection and circuit breaker coordination and self-healing process, $CB_{M2}$ to be opened while all other circuit breakers are closed. In such a case, the DERs in MG-5, MG-6 and MG-7 have to independently supply their local load demand and a load shedding is required if their generation capacity is less than their load demand.

For Case B, when the fault is within Zone 1, it is expected $CB_G$ and $CB_{M1}$ to be opened while all other circuit breakers are closed. In such a case, the DERs in MG-1–MG-4 have to independently supply their local load demand. However, as Zone 2 and Zone 3 microgrids are interconnected, they can share the load demand together and this will prevent or reduce the load shedding for the loads of these two zones.

For Case C, when the fault is in the upstream of Zone 1, it is expected only $CB_G$ to be opened while all other circuit breakers are closed. In such a case, the MGs in all three zones are interconnected and their DERs will share the load demand altogether. Hence, the load shedding for the loads of these three zones will be prevented or reduced. If this interconnection is not possible, then load shedding is inevitable for these MGs.

## 15.3 Distribution Network: Configuration and Model

Let us consider a subsection of the medium voltage feeder in Fig. 15.1 where it is connected to two low voltage microgrids, namely MG-1 and MG-2, through distribution transformers $T_1$ and $T_2$, as shown in Fig. 15.2. Assume that MG-1 has

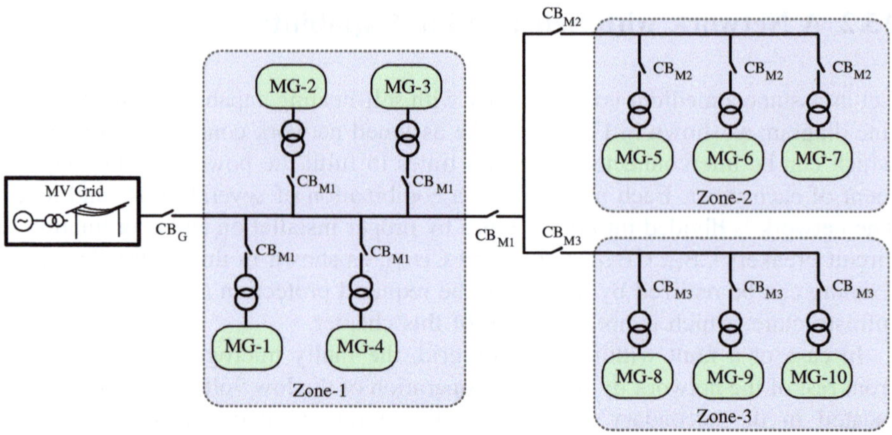


**Fig. 15.1** Schematic diagram of the large medium voltage feeder with self-healing capability containing several microgrids

three DERs (i.e. $DER_1$ to $DER_3$) and five loads while MG-2 has two DERs (i.e. $DER_4$ to $DER_5$) and four loads. A DSTATCOM is installed at the secondary side of the distribution transformer in each microgrid to regulate the voltage at its PCC. The loads are assumed to be residential loads and all DERs are assumed to be converter-interfaced DERs.

During grid connected mode, all DERs are operating at their rated capacities or the capacities desired by the economic analyses, subjected to limitations posed by weather conditions in the case of renewable energy sources. During planned maintenance or unplanned fault situations on the medium voltage feeder, the microgrids will work in autonomous mode. In this mode, when $CB_{M1}$ and $CB_{M2}$ are open, the DERs of each microgrid will be sharing the loads of that microgrid separately. Now, if the load demand in each microgrid is higher than the power generation capacity of the DERs in that microgrid, load shedding must be applied to some of the (non-critical) loads in that microgrid.

Now, let us assume a scenario in which the power generation in MG-2 is less than its demand; while the power generation capacity in DERs of MG-1 is higher than its demand. Assuming the network has self-healing and automatic supply restoration capability, $CB_{M1}$ and $CB_{M2}$ can be closed while $CB_G$ remains open. In this way, the two microgrids will be interconnected together. Hence, the DERs in MG-1 can share some of the loads in MG-2 and prevent/reduce load shedding in MG-2.

Sections below present a brief explanation on modeling the network.

## *15.3.1 DER, Converter Structure and Modeling*

The considered DERs in this chapter are photovoltaic cells (PV), fuel cells and batteries, connected to the microgrid through Voltage Source Converters (VSC).

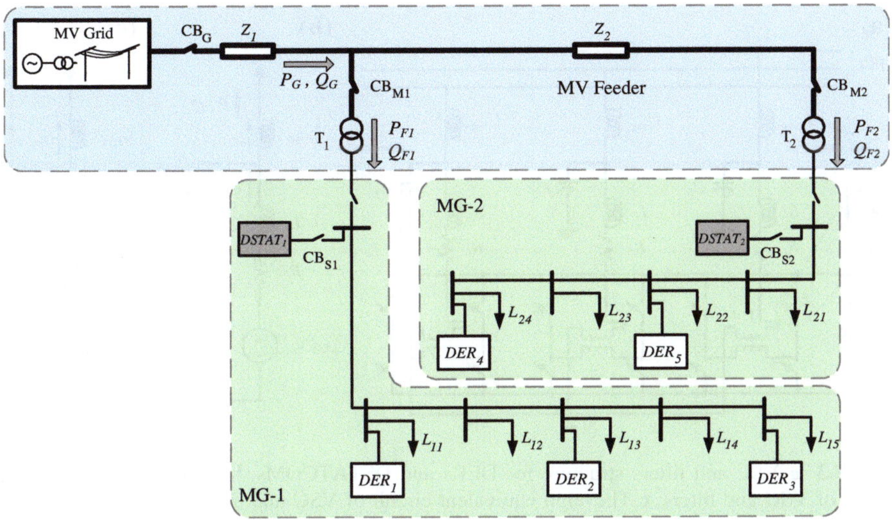


**Fig. 15.2** Schematic diagram of the considered network and microgrid structure

Detailed dynamic models of these micro sources are utilized in this study. These models are shown in Appendix A.

The considered DERs have a VSC consisting of three single-phase H-bridges, as shown in Fig. 15.3a. This VSC structure has better controllability and dynamic performance under unbalanced conditions in the network compared to VSCs with three-phase three-leg or four-leg configurations; since in this configuration, each phase can be controlled individually.

Each H-bridge of VSCs is composed of IGBTs with proper parallel reverse diode and snubber circuits. The outputs of each H-bridge are connected to a single-phase transformer, with 1 : $a$ ratio, and the three transformers are star-connected. The transformers provide galvanic isolation as well as voltage boosting. In this figure, the resistance $R_f$ represents the switching and transformer losses, while the inductance $L_f$ represents the leakage reactance of the transformers. The filter capacitor $C_f$ is connected to the output of the transformers to bypass the switching harmonics.

## 15.3.2 Network Voltage Regulation

As mentioned in the Introduction, in grid connected mode, the grid dictates the network voltage and in autonomous mode, the DERs indirectly regulate the network voltage based on the droop control. Although in both modes, the voltage along the low voltage network will be within acceptable limits, it is desired to hold the network voltage to its nominal value of 1 pu. For this, in this research, voltage

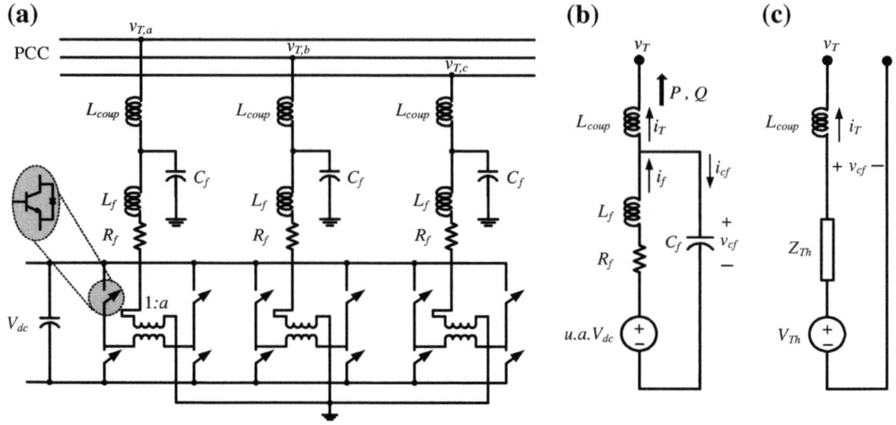


**Fig. 15.3** **a** VSC and filters structure for DERs and DSTATCOMs, **b** single-phase equivalent circuit of VSC and filters, **c** Thevenin equivalent circuit of VSC and filters

regulation in the microgrids is achieved by a DSTATCOM installed at the secondary side of the distribution transformer in each microgrid, which regulates its PCC voltage to a desired value, by exchanging reactive power with the network. The implemented DSTATCOM has the same converter structure as those of the DERs and its control is discussed in Sect. 15.4.4.

### 15.3.3 MICROGRID Resynchronization

As mentioned in the Introduction, resynchronization of a DER to microgrid and a microgrid to grid is required as DERs are operating in voltage-controlled strategy. For resynchronization, the voltage magnitude and angle are measured on both sides of the circuit breaker. A Phase-Locked Loop (PLL) is utilized for measuring the voltage angle [26]. The circuit breaker closes once the voltage angles and magnitudes are the same on both sides. However, based on network load and parameters, resynchronization can be slow and may take from several milliseconds to minutes [28].

It is to be noted that resynchronization can be a complicated process depending on the network configuration, existence of systems of interconnected microgrids and their location along the medium voltage feeder [43]. Therefore, a sophisticated resynchronization, self-healing and automatic supply restoration algorithm is required for this purpose. This controller will utilize communications between the network circuit breakers.

## 15.4 Control Strategy of Multi Microgrid System

A three-level hierarchical control system shown in Fig. 15.4, is required for proper operation of a microgrid within a network, where the interconnection of neighbour microgrids are considered [16, 44–46].

The primary (lowest) control level consists of inner-loop and outer-loop controls. Inner-loop control is responsible for appropriate switchings in the converter such that a proper tracking of the desired reference in DER converter output is achieved. This control is based on the references determined by outer-loop control and the local current and voltage measurements in DER converter output. Outer-loop control is responsible for proper output power control of DERs in the microgrid. This control generates the proper references for the inner-loop control and is different for grid-connected and autonomous modes. The dynamic performance of DER converters in the system of interconnected microgrids depends on the primary control level which is the main focus of this chapter and is discussed further in Sects. 15.3.1–15.3.3.

The secondary control is the central controller of the microgrid. This controller sends the desired (reference) output power of each DER converter to them. In autonomous mode, this controller sends the desired output power of each DER converter based on monitoring the network voltage and frequency. However, in grid-connected mode, the desired output power of each DER converter is received from the tertiary controller. This controller runs in a slower time frame compared to that of the primary control [16].

The tertiary (highest) control communicates with the central controllers in each microgrid and the protection devices and circuit breakers of the network. In general, this controller can utilize load forecasting, electricity market and demand response information for optimal power flow of the network and microgrids [47]. Additionally, in case of a network with self-healing capability, this controller utilizes the information received from the circuit breakers to define the network configuration and status. In this case, microgrids coordination agent decides whether interconnection of neighbour microgrids is required and if any, identifies those microgrids. Then, using the self-healing and supply restoration agent, it decides which circuit breakers should open/close and sends the proper commands to them. This controller also monitors the network data and decides whether two interconnected microgrids should be isolated and sends the proper commands to the relevant circuit breakers to operate.

As mentioned before, designing the secondary and tertiary control with its sub agents requires an extensive study and is not discussed in this chapter.

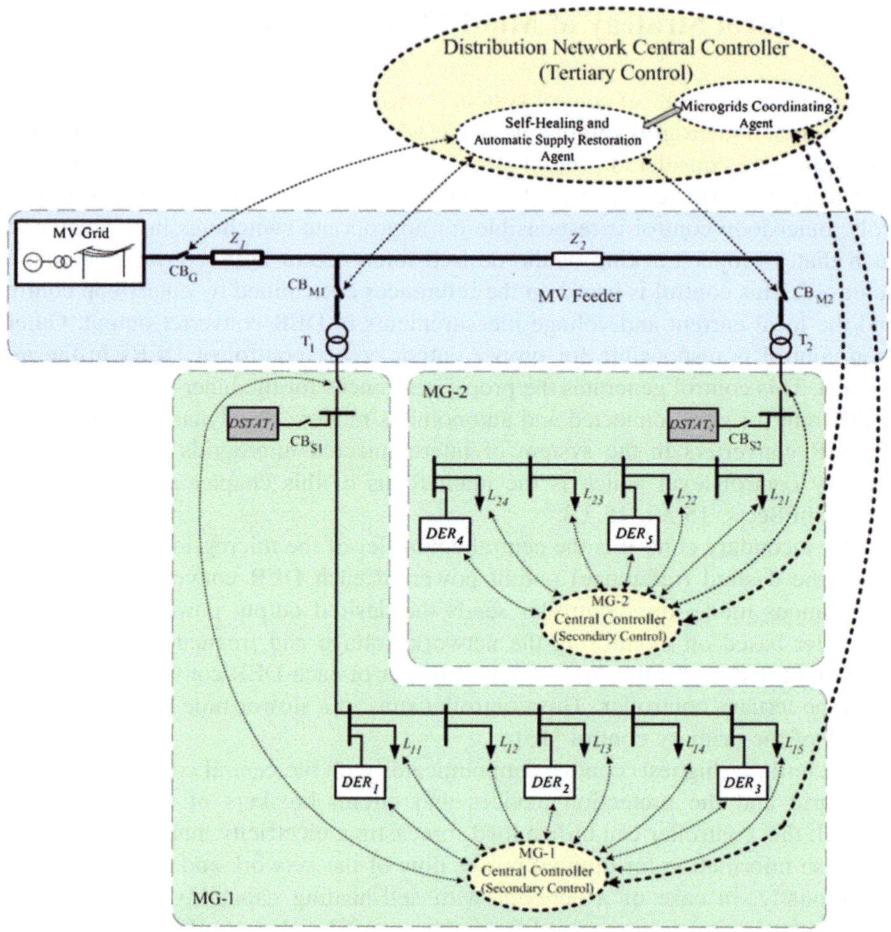


**Fig. 15.4** Hierarchical control structure of the network with microgrids

## *15.4.1 Outer-Loop Control in Grid-Connected Mode*

Let us consider the DERs have a VSC and filter structure as shown in Fig. 15.3a with its equivalent single-phase circuit as shown in Fig. 15.3b. This equivalent circuit can also be represented by its Thevenin equivalent parameters (i.e. $V_{Th}$ and $Z_{Th}$) as shown in Fig. 15.3c. From Fig. 15.3c, it can be seen that the instantaneous active power ($p$) and reactive power ($q$) supplied from DER-$i$ to its PCC can be expressed as [48]

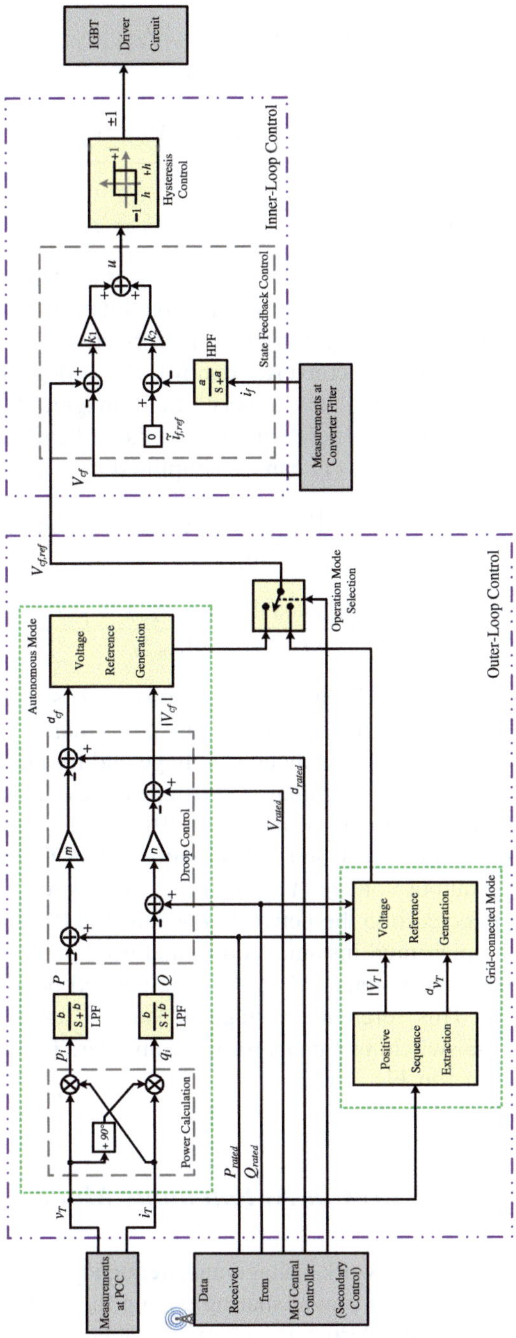


**Fig. 15.5** The block diagram of primary control level (outer- and inner-loop controls) of the DER converters in the microgrid

$$p_i = \frac{\left(\left|V_{T,i}\right|\left|V_{Th,i}\right|\cos\phi_i - \left|V_{Th,i}\right|^2\right)\cos\theta_i + \left|V_{T,i}\right|\left|V_{Th,i}\right|\sin\phi_i\sin\theta_i}{\left|Z_{Th,i} + Z_{coup,i}\right|}$$
$$q_i = \frac{\left(\left|V_{T,i}\right|\left|V_{Th,i}\right|\cos\phi_i - \left|V_{Th,i}\right|^2\right)\sin\theta_i - \left|V_{T,i}\right|\left|V_{Th,i}\right|\sin\phi_i\cos\theta_i}{\left|Z_{Th,i} + Z_{coup,i}\right|} \quad (15.1)$$
$$\phi_i = \delta_{Th,i} - \delta_{T,i} \qquad \& \qquad \theta_i = \angle\left(Z_{Th,i} + Z_{coup,i}\right)$$

where $V_T$ is the PCC voltage, $Z_{coup}$ is the coupling impedance and $V = |V| \angle\delta$ represents the phasor notation of $v(t)$. It is to be noted that the coupling impedance is dominantly inductive (i.e. $Z_{coup} \approx j\omega L_{coup}$). In Sect. 15.4.3, it will be shown that after applying the proposed converter control, $Z_{Th}$ is dominantly inductive in 50 Hz (i.e. $Z_{Th} \approx j\omega L_{conv}$) and has a very small magnitude. Therefore, it is expected that $V_{Th} \approx V_{cf}$ in Fig. 15.3c where $V_{cf}$ is the voltage across the capacitor $C_f$. Based on these assumptions, (1) can be simplified as

$$p_i = \frac{\left|V_{T,i}\right|\left|V_{cf,i}\right|\sin\left(\delta_{cf,i} - \delta_{T,i}\right)}{\omega L_{conv,i} + \omega L_{coup,i}}$$
$$q_i = \frac{\left|V_{T,i}\right|\left|V_{cf,i}\right|\cos\left(\delta_{cf,i} - \delta_{T,i}\right) - \left|V_{T,i}\right|^2}{\omega L_{conv,i} + \omega L_{coup,i}} \quad (15.2)$$

The average active power ($P$) and reactive power ($Q$) supplied by each DER can then be calculated from $p$ and $q$ using a low pass filter.

In grid-connected mode, the grid dictates voltage and frequency of the network and the DERs operate at their nominal (rated) capacities. This can be achieved using a constant PQ control mode of operation [9]. In this discussion, a voltage-controlled technique is utilized which will monitor the PCC voltage ($V_T$) to regulate DER converter output voltage ($V_{cf}$) such that the desired nominal active and reactive power are injected into the network. For this, the PCC voltage magnitude and angles should be measured instantaneously and used in (15.2) to calculate the converter output reference voltage ($V_{cf,ref}$) for the desired powers and the known coupling inductance. This reference voltage is later be used in the inner-loop control to generate the switching signals for the converter. The schematic diagram of this control is shown in Fig. 15.5.

## 15.4.2 Outer-Loop Control in Autonomous Mode

During autonomous mode, network voltage and frequency should be regulated by the DERs. In addition, a proper power sharing among the DERs is desired. Below is a detailed description of the designed and implemented power sharing algorithm based on droop control.

Let us now consider a microgrid with two parallel converter-interfaced DERs supplying a common load. DER-1 is connected to the load through a feeder impedance of $Z_{line,1}$ where DER-2 is connected to the load through a feeder impedance of $Z_{line,2}$. Let us also assume the feeder is highly inductive (i.e. $Z_{line} \approx j\omega L_{line}$). Assuming the voltage at load PCC is $V_{load}$, the active and reactive power supplied from each DER to the load can be expressed as

$$\begin{aligned} p_i &= \frac{|V_{load,i}|\,|V_{cf,i}|\sin(\delta_{cf,i} - \delta_{load,i})}{\omega L_{conv,i} + \omega L_{coup,i} + \omega L_{line,i}} \\ q_i &= \frac{|V_{load,i}|\,|V_{cf,i}|\cos\,(\delta_{cf,i} - \delta_{load,i}) - |V_{load,i}|^2}{\omega L_{conv,i} + \omega L_{coup,i} + \omega L_{line,i}} \end{aligned} \tag{15.3}$$

As the feeder is assumed to be highly inductive, the active and reactive powers are decoupled and a DC load flow analysis can be applied. In addition, the angle difference between $V_{cf}$ and $V_{load}$ is small. Based on DC load flow, the average active power supplied from each DER to the load is

$$P_i = \frac{|V_{load,i}|\,|V_{cf,i}|(\delta_{cf,i} - \delta_{load})}{\omega L_{conv,i} + \omega L_{coup,i} + \omega L_{line,i}} \tag{15.4}$$

To supply the load with an average active power of $P_i$, from (15.4), the angle of the voltage across the filter capacitor of DER-$i$ will be

$$\delta_{cf,i} = P_i(\Gamma_{conv,i} + \Gamma_{coup,i} + \Gamma_{line,i}) + \delta_{load} \tag{15.5}$$

where $\Gamma_i = \frac{\omega L_i}{|V_{cf,i}|\,|V_{load}|}$

Decentralized power sharing among several DERs in a microgrid can be achieved by changing the voltage magnitude and angle of DERs using the droop control as [15]

$$\begin{aligned} \delta_{cf,i} &= \delta_{rated,i} - m_i\left[\frac{X_{line}}{Z_{line}}(P_{rated,i} - P_i) - \frac{R_{line}}{Z_{line}}(Q_{rated,i} - Q_i)\right] \\ |V_{cf,i}| &= V_{rated,i} - n_i\left[\frac{R_{line}}{Z_{line}}(P_{rated,i} - P_i) + \frac{X_{line}}{Z_{line}}(Q_{rated,i} - Q_i)\right] \end{aligned} \tag{15.6}$$

where $V_{rated}$ and $\delta_{rated}$ are respectively the rated voltage magnitude and angle of the DER when supplying rated active power ($P_{rated}$) and reactive power ($Q_{rated}$). Reactive power–voltage and active power-angle droop coefficients are represented by $n$ and $m$, respectively. As the feeder lines are assumed to be inductive, (15.6) can be further simplified as

$$\begin{aligned} \delta_{cf,i} &= \delta_{rated,i} - m_i(P_{rated,i} - P_i) \\ |V_{cf,i}| &= V_{rated,i} - n_i(Q_{rated,i} - Q_i) \end{aligned} \tag{15.7}$$

This is monitored and controlled in the microgrid through the outer-loop control. Therefore, it is expected that the outer-loop control determines the desired $V_{cf}$

for each DER in microgrid. The block diagram of outer-loop control is shown in Fig. 15.5.

Let us assume when each DER changes its output active power from zero to its nominal (rated) capacity, the frequency of the DER voltage reduces by $\Delta\omega$. Based on this assumption, the active power-angle droop coefficient for each DER [49] is derived from

$$m_i = \frac{\Delta\omega}{P_{rated,i}} \tag{15.8}$$

Assuming $\Delta\omega$ to be constant for DERs $i$ and $j$ with different nominal capacities, we have

$$\frac{m_i}{m_j} = \frac{P_{rated,j}}{P_{rated,i}} \tag{15.9}$$

Now, let us assume that all DERs in the microgrid have the same power factor and when each DER changes its output reactive power from zero to its nominal (rated) capacity, the DER voltage reduces by $\Delta V$. Based on this assumption, the reactive power–voltage droop coefficient for each DER [49] is derived from

$$n_i = \frac{\Delta V}{Q_{rated,i}} \tag{15.10}$$

Assuming $\Delta V$ to be constant for DERs $i$ and $j$ with different nominal capacities, we have

$$\frac{n_i}{n_j} = \frac{Q_{rated,j}}{Q_{rated,i}} \tag{15.11}$$

Now, for the microgrid in steady state condition, let us assume DER-$i$ and DER-$j$ have the same $\delta_{rated}$. Therefore, from (15.7) and (15.9) we have

$$\begin{aligned} \delta_{cf,i} - \delta_{cf,j} &= \left(\delta_{rated,i} - \delta_{rated,j}\right) - m_i\left(P_{rated,i} - P_i\right) + m_j\left(P_{rated,j} - P_j\right) \\ &= m_iP_i - m_jP_j \end{aligned} \tag{15.12}$$

Replacing $\delta_{cf,i}$ and $\delta_{cf,j}$ from (15.5) in (15.12), we have

$$P_i\left(\Gamma_{conv,i} + \Gamma_{coup,i} + \Gamma_{line,i}\right) - P_j\left(\Gamma_{conv,j} + \Gamma_{coup,j} + \Gamma_{line,j}\right) = m_iP_i - m_jP_j \tag{15.13}$$

Therefore, the ratio of the active power supplied by the DERs is equal to

$$\frac{P_j}{P_i} = \frac{-m_i + \Gamma_{conv,i} + \Gamma_{coup,i} + \Gamma_{line,i}}{-m_j + \Gamma_{conv,j} + \Gamma_{coup,j} + \Gamma_{line,j}} \tag{15.14}$$

Equation (15.14) shows that output active power of each DER is inversely proportional to the sum of $\Gamma$ in its output. The three components of $\Gamma_{\text{conv}}$, $\Gamma_{\text{coup}}$

and $\Gamma_{\text{line}}$ are dependent on three inductances between the DER and the load. Since $\Gamma$ has a parameter of voltage square in its denominator, it is expected that

$$\Gamma_{conv,i} << \Gamma_{line,i} << \Gamma_{coup,i} << m_i \tag{15.15}$$

Therefore, (15.14) can be simplified further as

$$\frac{P_j}{P_i} \approx \frac{m_i}{m_j} \tag{15.16}$$

In a similar way, it can be shown that the ratio of the reactive power supplied by the DERs in the microgrid will be

$$\frac{Q_j}{Q_i} \approx \frac{n_i}{n_j} \tag{15.17}$$

Hence, based on the above assumption, from (15.9), (15.16) and (15.11), (15.17), it is expected that the output active and reactive power ratio among two DERs in the microgrid will be same as the ratio of their nominal active and reactive power capacities.

On the other hand, in parallel operation of converter-interfaced DERs in a microgrid, it is desired that, for all the DERs, the voltage angle difference across their coupling inductances (i.e. $\delta_{cf} - \delta_T$) in (15.2) to be constant. This voltage angle difference is preferred to be small [49] so that it lies on the linear section of sinusoidal $P$–$\delta$ characteristic of (15.2). Similarly, it is desired that the voltage drop across the coupling inductances (i.e. $|V_{cf}| - |V_T|$) in (15.2) to be constant, for all the DERs. This voltage drop is preferred to be small and in the range of 1–2 % [50]. For achieving these assumptions, the coupling inductances are designed inversely proportional to the nominal power ratio of DERs as

$$\frac{L_{coup,i}}{L_{coup,j}} = \frac{P_{rated,j}}{P_{rated,i}} = \frac{Q_{rated,j}}{Q_{rated,i}} \tag{15.18}$$

From (15.1)–(15.18), it is concluded that for an accurate power sharing among DERs in microgrid, we require

$$\begin{aligned} \frac{P_j}{P_i} &\approx \frac{m_i}{m_j} = \frac{L_{coup,i}}{L_{coup,j}} = \frac{P_{rated,j}}{P_{rated,i}} \\ \frac{Q_j}{Q_i} &\approx \frac{n_i}{n_j} = \frac{L_{coup,i}}{L_{coup,j}} = \frac{Q_{rated,j}}{Q_{rated,i}} \end{aligned} \tag{15.19}$$

### 15.4.3 Inner-Loop Control

As mentioned before, the outer-loop control regulates the output power of DERs by adjusting the proper references for the each DER from (15.2) or (15.7). The inner-loop control will calculate and apply proper switching signals for the IGBTs in the DER converter such that the desired voltage ($V_{cf,ref}$) is perfectly generated across the AC filter capacitor ($C_f$).

For this, let us consider the equivalent circuit of VSC as shown in Fig. 15.3b. In this figure, $u{\cdot}a{\cdot}V_{dc}$ represents the converter output voltage, where $u$ is the switching function. For a 2-level (bipolar) switching, $u$ can take on $\pm$ 1 value which will be used subsequently to turn ON/OFF the IGBTs.

Let us consider the DER converter and its output filter as system for which a controller is to be developed. The differential equations that describe the dynamic behavior of this system are given by

$$\begin{aligned} u_i.a.Vdc &= R_f i_{f,i} + L_f \frac{di_{f,i}}{dt} + v_{cf,i} \\ i_{f,i} &= C_f \frac{dv_{cf,i}}{dt} + i_{T,i} \end{aligned} \tag{15.20}$$

A closed-loop optimal linear robust controller based on state feedback is utilized to generate $u$. Let us assume the state vector $x(t)$ for each phase of the system is defined as

$$x_i(t) = [\, v_{cf,i}(t) \quad i_{f,i}(t)\,]^T \tag{15.21}$$

where $v_{cf}(t)$ represent the instantaneous voltage across AC filter capacitor, $i_f(t)$ is the current passing through filter inductor $L_f$ and $T$ is the transpose operator. Then, the equivalent circuit of this system can be represented with state space equation of

$$\dot{x}_i(t) = A\,x_i(t) + B_1\,u_{c,i}(t) + B_2\,i_{T,i}(t) \tag{15.22}$$

where

$$A = \begin{bmatrix} 0 & \frac{1}{C_f} \\ -\frac{1}{L_f} & -\frac{R_f}{L_f} \end{bmatrix} \quad B_1 = \begin{bmatrix} 0 \\ \frac{a.Vdc}{L_f} \end{bmatrix} \quad B_2 = \begin{bmatrix} -\frac{1}{C_f} \\ 0 \end{bmatrix} \tag{15.23}$$

In (15.22), $u_c(t)$ is the continuous time version of switching function $u$ and $i_T$ represents the network load change effects on this system; hence it is assumed as a disturbance for the controller.

In the control systems, the desired values for each control parameter in steady state condition must be known. However, it is rather difficult to determine the reference for $i_f$ in (15.21). Nevertheless, it is desired that $i_f$ has only low frequency components. Therefore, instead of using $i_f$ as a control parameter, its high frequency components ($\tilde{i}_f$) can be used in the control system. Based on this assumption, $\tilde{i}_f$ can be described and expanded in Laplace domain as [51]

$$\tilde{i}_f(s) = \frac{s}{s+\alpha} i_f(s) = \left(1 - \frac{\alpha}{s+\alpha}\right) i_f(s) = i_f(s) - \hat{i}_f(s) \tag{15.24}$$

where $\alpha$ is the cut-off frequency of this high-pass filter and $\hat{i}_f$ is the low frequency components of $i_f$ which is given by

$$\hat{i}_f(s) = \frac{\alpha}{s+\alpha} i_f(s) \tag{15.25}$$

Equation (15.25) can be expressed in differential equation form as

$$\frac{d\hat{i}_f(t)}{dt} = \alpha\left(i_f(t) - \hat{i}_f(t)\right) \tag{15.26}$$

Now, let us define a new state vector for the system, which includes $\hat{i}_f$, as [51]

$$x'_i(t) = \begin{bmatrix} v_{cf,i}(t) & i_{f,i}(t) & \hat{i}_{f,i}(t) \end{bmatrix}^T \tag{15.27}$$

In this case, the system can be represented with new state space equation of

$$\dot{x}'_i(t) = A'\, x_i(t) + B'_1\, u_{c,i}(t) + B'_2\, i_{T,i}(t) \tag{15.28}$$

where

$$A' = \begin{bmatrix} 0 & \frac{1}{C_f} & 0 \\ -\frac{1}{L_f} & -\frac{R_f}{L_f} & 0 \\ 0 & \alpha & -\alpha \end{bmatrix} \quad B'_1 = \begin{bmatrix} B_1 \\ 0 \end{bmatrix} \quad B'_2 = \begin{bmatrix} B_2 \\ 0 \end{bmatrix} \tag{15.29}$$

Equation (15.28) can be represented in discrete-time domain as [52]

$$x'_i(k+1) = Fx'_i(k) + G_1 u_{c,i}(k) + G_2 i_{T,i}(k) \tag{15.30}$$

where

$$F = e^{A'.T_s}, \quad G_1 = \int_0^{T_s} e^{A't} B'_1\, dt, \quad G_2 = \int_0^{T_s} e^{A't} B'_2\, dt \tag{15.31}$$

where $T_s$ is the sampling time. From (15.30), $u_c(k)$ can be computed, using a suitable state feedback control law, as

$$u_{c,i}(k) = -K[x'_i(k) - x'_{ref,i}(k)] \tag{15.32}$$

where $K$ is a gain matrix and $x'_{ref}(k)$ is the desired state vector for (15.27), in discrete-time mode.

It is to be noted that the desired reference value for $v_{cf}(t)$ for each DER will be determined by the outer-loop control. This value is calculated from (15.2) when microgrid operates in grid-connected mode and from (15.7) when microgrid operates in autonomous mode. As mentioned before, the desired reference value for $\tilde{i}_f(t)$ for all DERs is always set to zero to minimize the high frequency components of the current flowing through $L_f$. Therefore, $i_f(t)$ must only contain low frequency components (i.e. $i_f(t) = \hat{\imath}_f(t)$). Now, the reference vector, $X'_{ref}$ can be defined for each DER as

$$X'_{ref,i} = [V_{cf,ref,i} \quad \hat{I}_f \quad \hat{I}_f]^T = [V_{cf,i}\angle\delta_{cf,i} \quad \hat{I}_f \quad \hat{I}_f]^T \tag{15.33}$$

As the system behavior in steady-state is interested and assuming a full control over $u_c(k)$, an infinite time Linear Quadratic Regulator (LQR) [52] can be designed for this problem to define $K$. This controller is more stable than Proportional Integrator (PI) based controls and prevents the instability among parallel DERs when two autonomous microgrids interconnect together.

In a discrete LQR problem, an objective function $J$ is chosen as

$$J_i(k) = \sum_{k=0}^{\infty}\left[\left(x_i'(k) - x'_{ref,i}(k)\right)^T Q_i(k)\left(x_i'(k) - x'_{ref,i}(k)\right) + u_i(k)^T R_i(k) u_i(k)\right] \tag{15.34}$$

where $R$ is the control cost matrix, $Q$ is the state weighting matrix which reflects the importance of each controlling parameter in $x$ and $J(\infty)$ represents the objective function at infinite time (steady-state condition) for the system. Equation (15.34) is then minimized to obtain the optimal control law $u(k)$ through solution of steady state Riccati equations while satisfying system constraints in (15.30) [52]. The LQR method ensures the desired results for the system while the variations of system load and source parameters are within acceptable limits of reality.

Equation (15.32) shows the total tracking error of each DER converter. The tracking error can be minimized by limiting this error within a very small bandwidth (e.g. $h = +10^{-4}$). Now, from (15.32), for each DER, the switching function $u$ (i.e. which pairs of IGBTs to turn ON/OFF) is generated using a hysteresis control based on the error level as

$$\begin{aligned} &\mathit{If}\ u_{c,i}(k) > +h \quad \mathit{then} \quad u_i = +1 \\ &\mathit{If} -h \le u_{c,i}(k) \le +h \quad \mathit{then} \quad u_i = \mathit{previous}\ u_i \\ &\mathit{If}\ u_{c,i}(k) < -h \quad \mathit{then} \quad u_i = -1 \end{aligned} \tag{15.35}$$

More detail on the discussed converter control is given in [24].

To achieve a good tracking of the output voltage, a state feedback control was utilized for each DER. Therefore, the controller expression can be presented as

$$u_i.a.Vdc = v_{cf,ref,i} - K(X_i - X_{ref,i}) = v_{cf,ref,i} - k_1\left(v_{cf,i} - v_{cf,ref,i}\right) - k_2\left(\tilde{\imath}_{f,i} - 0\right) \tag{15.36}$$

where $K = [k_1\ k_2\ k_2]$ is the gain matrix in (15.32).

Replacing (15.36) in (15.20) and representing that in Laplace domain, we have

$$V_{cf,i}(s)\left[L_fC_fs^2+R_fC_fs+1\right]+I_{T,i}(s)\left[L_fs+R_f\right]$$
$$=V_{cf,ref,i}(s)[k_1+1]-k_1V_{cf,i}(s)-\frac{k_2s}{s+\alpha}I_{f,i}(s) \quad (15.37)$$

From (15.37), the parameters of the Thevenin equivalent circuit of the DER converter and filter including the developed state feedback control are given by

$$V_{cf,i}(s)=G_1(s)V_{cf,ref,i}(s)+G_2(s)I_{T,i}(s) \quad (15.38)$$

where $G_1(s)$ and $G_2(s)$ are defined as

$$G_1(s)=\frac{(k_1+1)\,s+\alpha\,(k_1+1)}{L_fC_fs^3+C_f\left(R_f+k_2+\alpha L_f\right)s^2+\left(k_1+1+\alpha R_fC_f\right)s+\alpha\,(k_1+1)}$$

$$G_2(s)=\frac{L_fS^2+\left(R_f+\alpha L_f-k_2\right)S+\alpha R_f}{L_fC_fs^3+C_f\left(R_f+k_2+\alpha L_f\right)s^2+\left(k_1+1+\alpha R_fC_f\right)s+\alpha\,(k_1+1)} \quad (15.39)$$

and $G_1(s)\ V_{cf,ref}\ (s)\ =\ V_{Th}$ and $G_2(s)\ =\ -Z_{Th}$.

The comparison of the performance of the state feedback control versus the open loop control of the DER converter and filter system can be studied from frequency domain analysis. For this, the Bode diagram of $G_1(s)$ and $G_2(s)$ are obtained and shown in Fig. 15.6a and b, respectively. From Fig. 15.6a, it can be seen that at the frequency range of interest (i.e. 50 Hz), $G_1(s)$ has the property of a unity gain with zero phase shift (i.e. $V_{Th}\approx V_{cf}$). This helps us to simplify (15.1) into (15.2) in Sect. 15.4.1. On the other hand, in Fig. 15.6b, it can be seen that at frequency range of interest, $G_2(s)$ magnitude is relatively high in open loop condition. The high value takes the network load and $i_T$ change effects into the DER converter control. This magnitude is reduced effectively by properly designing the state feedback gains. It is to be noted that $G_2(s)$ is relatively inductive around 50 Hz. This validates the assumption that $Z_{Th}$ to be pure inductive in (15.1) in Sect. 15.4.1.

### *15.4.4 DSTATCOM Control*

As discussed in Sect. 15.3.2, voltage regulation in the microgrid is achieved by a DSTATCOM installed at the secondary side of the distribution transformer, which regulates its PCC voltage to the desired value. The utilized DSTATCOM in this research has an LCL filter in its converter output and voltage regulation is achieved by exchanging the reactive power with the network. Voltage control strategy in DSTATCOM is based on directly controlling the PCC voltage. In this method, the difference between the PCC RMS voltage ($V_{T-DSTAT}$) and its desired

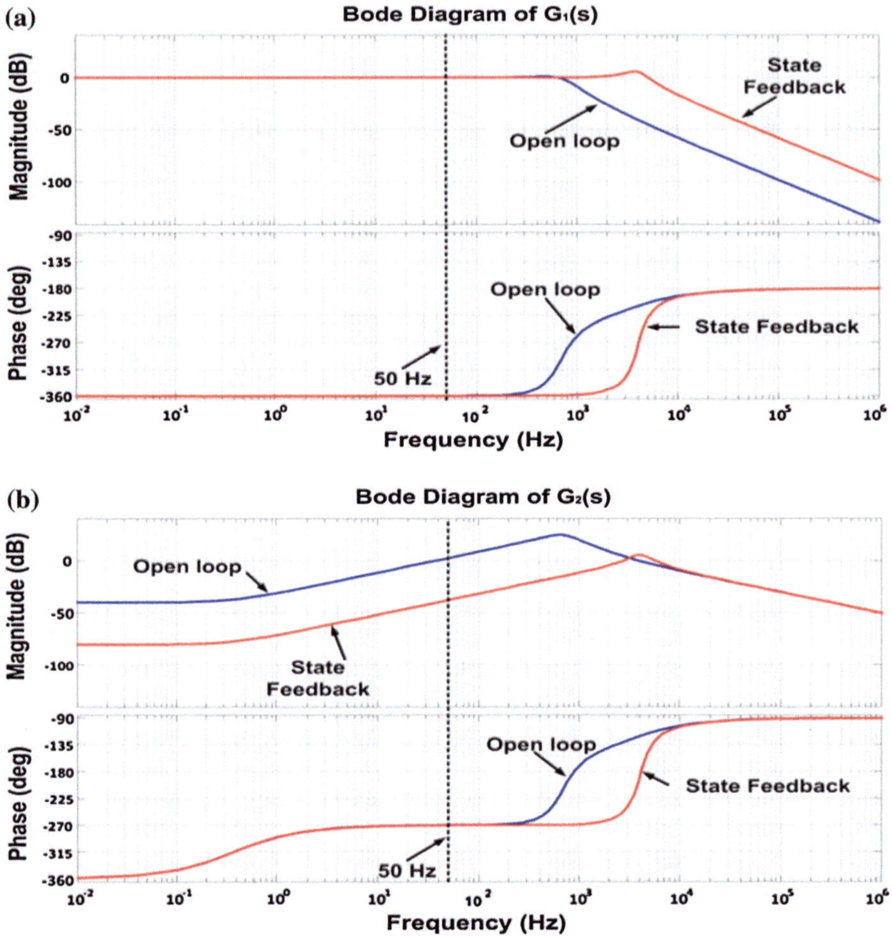


**Fig. 15.6** **a** Open loop and closed loop Bode diagram of $G1(s)$, **b** open loop and closed loop Bode diagram of $G2(s)$

value ($V_{T-DSTAT,desired}$) is utilized to generate the required voltage magnitude across the AC filter capacitor in DSTATCOM as

$$\left|V_{cf-DSTAT}\right| = V_{cf-DSTAT,ref} + \left(K_P + \frac{K_I}{s}\right)\left(V_{T-DSTAT,desired} - V_{T-DSTAT}\right) \quad (15.40)$$

where $V_{cf-DSTAT,ref}$ is the reference value for this voltage, $K_P$ and $K_I$ are PI controller parameters and the suffix *DSTAT* represents DSTATCOM.

The DC capacitor voltage ($V_{dc}$) in DERs is stabilized by their DC sources; however, there is no such a DC source in DSTATCOMs. $V_{dc}$ in DSTATCOM can be kept equal to its reference value ($V_{dc,ref}$) when the AC system does not exchange any power with the DC capacitor [53]. This can be reassured if the AC system replenishes the DSTATCOM converter losses. For this, the angle of the voltage across the AC filter capacitor ($\delta_{cf}$) must be varied with respect to the DC capacitor voltage variations as

$$\delta_{cf-DSTAT,ref} = \left(K_P'' + \frac{K_I''}{s}\right)\left(V_{dc-DSTAT,ref} - V_{dc-DSTAT}\right) \tag{15.41}$$

## 15.5 Verification and Simulation Studies

For verification of the proposed primary control level for the DERs in the system of interconnected autonomous microgrids, several case studies are presented below using PSCAD/EMTDC software [44, 54]. The technical data for the network, DER converters and the droop control parameters are provided in Appendix B.

### *15.5.1 Case 1: Microgrid Operation with Three-Phase Balanced Loads*

Let us consider the system shown in Fig. 15.2 to investigate the DERs operation during grid-connected and autonomous modes of microgrid operation. In grid-connected mode, each DER will generate its rated power and the extra load demand will be supplied by the grid or the DER extra generation will flow back to the grid. In autonomous mode, the load demand of each microgrid is shared among the DERs of that microgrid proportional to their rating.

The individual operation of MG-1 is investigated in this section. For this, let us assume in the system shown in Fig. 15.2, $CB_G$ and $CB_{M1}$ are closed while $CB_{M2}$, $CB_{S1}$ and $CB_{S2}$ are open. Let us also assume all loads are three-phase balanced. The system is assumed to be in steady state condition at $t = 0$ s and all the DERs are running at their rated conditions. The DSTATCOM is not connected to the system.

At $t = 1$ s, the grid is disconnected (i.e. $CB_G$ is opened) and MG-1 starts to operate in autonomous mode. Therefore, the DERs increase their output power to satisfy the load demand within the microgrid. At $t = 2$ s a 25 % load increase (i.e. 3 kW) and at $t = 3$ s a load decrease of 25 % are applied in the microgrid. It can be seen that all DERs are sharing the load change proportional to their ratings. At $t = 4$ s, it is desired that microgrid reconnect to the grid. In order to prevent fluctuations in the current and power, the resynchronization method described in Sect. 15.3.3 is used. The resynchronization is achieved at $t = 10.1$ s and $CB_G$ closes. It can be seen the system reaches to the steady state condition after each change within five cycles. Figure 15.7a shows the active power dispatch of grid and three existing DERs in MG-1 between 0 and 11 s in the above mentioned network while Fig. 15.7b shows the reactive power dispatch within the same time period.

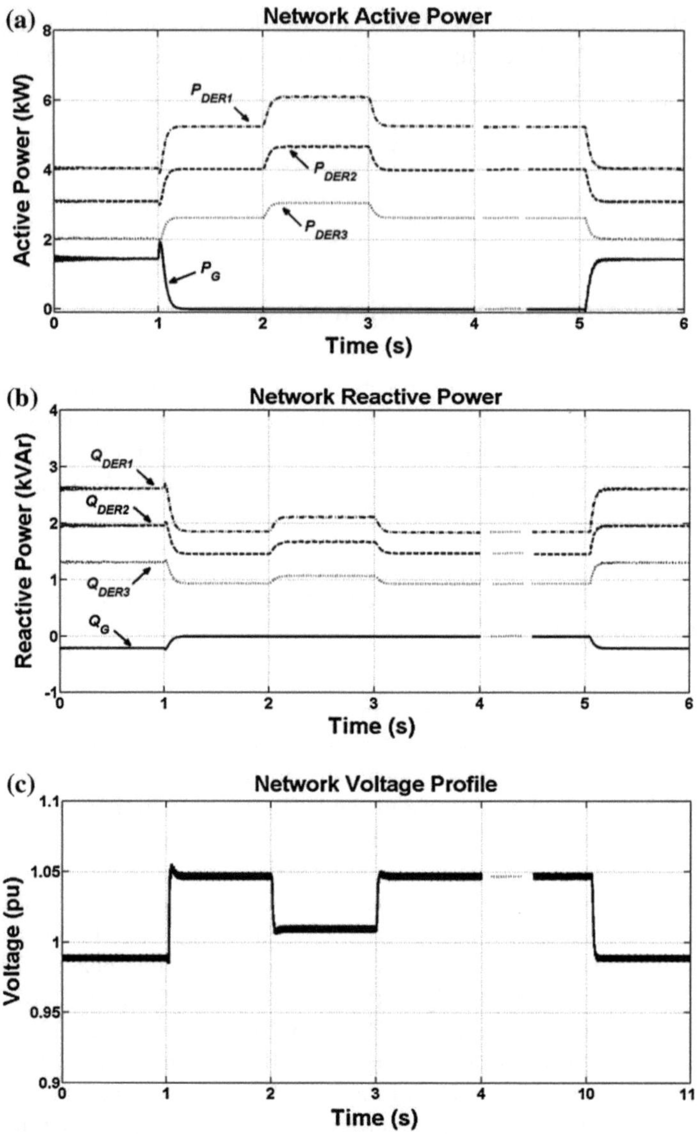


**Fig. 15.7** Simulation results for MG-1 without DSTATCOM: **a** active power dispatch of grid and DERs, **b** reactive power dispatch of grid and DERs, **c** network voltage profile monitored in the secondary side of distribution transformer

The voltage profile of the network is also shown in Fig. 15.7c. As it can be seen in this figure, during the autonomous mode, there might be an uncontrolled voltage drop/rise in the network, as none of the DERs are regulating the network voltage. In this study, the DSTATCOM is not utilized to regulate the network voltage. To investigate the efficacy of DSTATCOM, let us now assume that in Fig. 15.2, $CB_{S1}$ is closed and DSTATCOM-1 is connected. For this condition, the active and reactive power dispatch of grid and DERs are shown in Fig. 15.8a and b, respectively. The voltage profile of the network is now highly improved and it is closer to the desired value of 1 pu, as shown in Fig. 15.8c. The difference in the active powers in Figs. 15.7a and 15.8a is due to the fact that voltage is varying before DSTATCOM installation. The voltage profile of DSTATCOM DC capacitor is shown in Fig. 15.8d which has a negligible drop of 1 % in autonomous mode. The output active and reactive power of DSTATCOM is also shown in Fig. 15.8e. The oscillations in the results of Fig. 15.8 are due to the dynamic characteristics of DSTATCOM. It is also to be noted that the resynchronization mechanism is similar to the previous case and not shown here.

### 15.5.2 Case 2: Microgrid Operation with Unbalanced and Harmonic Loads

The simulation results presented in Fig. 15.8 represent the microgrid operation in the presence of three-phase balanced loads. However, the networks are always supplying single-phase loads and nonlinear loads. To study the effects of these loads on microgrid operation, let us assume in the system of Fig. 15.2, at $t = 0$ s $CB_G$, $CB_{M1}$ and $CB_{S1}$ are closed and the system is in steady state condition. At $t = 0.5$ s, $CB_G$ is opened to disconnect the grid hence microgrid will work in autonomous mode. It is to be noted that distribution transformer is still connected to the low voltage side.

At $t = 1.5$ s a new single-phase 2 kW load is connected to phase-A. Later, at $t = 2.5$ s another single-phase 2 kW load is connected to phase-C. Figure 15.9a shows the active power dispatch of grid and the 3 DERs in the microgrid between 0 and 3.5 s in the above mentioned network. From this figure, it can be seen that, the total amount of active power generation among DERs are kept based on the desired power sharing ratio among them.

The active power output of each phase of one of the DERs (e.g. DER-1) is shown in Fig. 15.9b. From this figure, it can be seen that for $t < 1.5$ s, all three phases of DER-1 have an equal amount of generated active power. However, at $t = 1.5$ s when a single-phase load is connected to phase-A, the active power output of all phases of the DER increase. This increase is slightly higher for phase-A. This slight difference is further discussed in [54]. In a similar way, at $t = 2.5$ s when another single-phase load is connected to phase-C, the output active power of all phases of the DER increases but this increase is more for phase-C. Similar results are observed in all the DERs of the microgrid. From this figure it can be seen that single phase load power demand is shared among the three phases of the DERs. Figure 15.9c shows the active power output of all DERs in their phase-A for the studied case. From this figure, it can be seen that for a load change in the

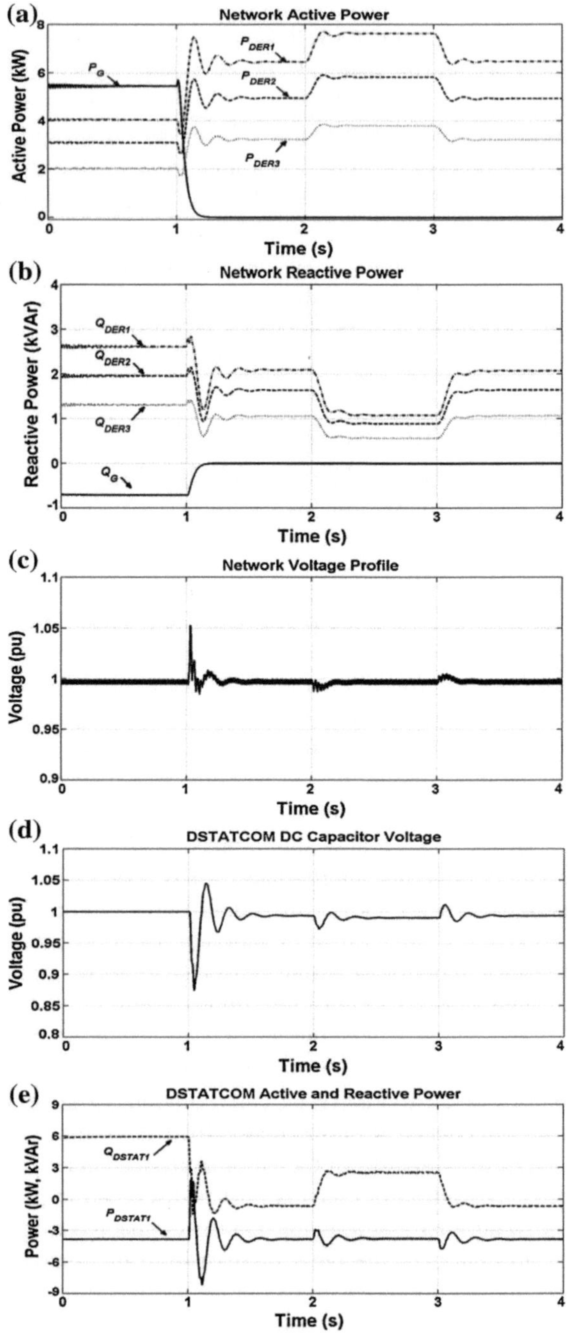


**Fig. 15.8** Simulation results of MG-1 in with DSTATCOM connected: **a** active power dispatch of grid and DERs, **b** reactive power dispatch of grid and DERs, **c** network voltage profile monitored in the secondary side of distribution transformer, **d** DSTATCOM output active and reactive power injection, **e** voltage profile of DSTATCOM DC capacitor

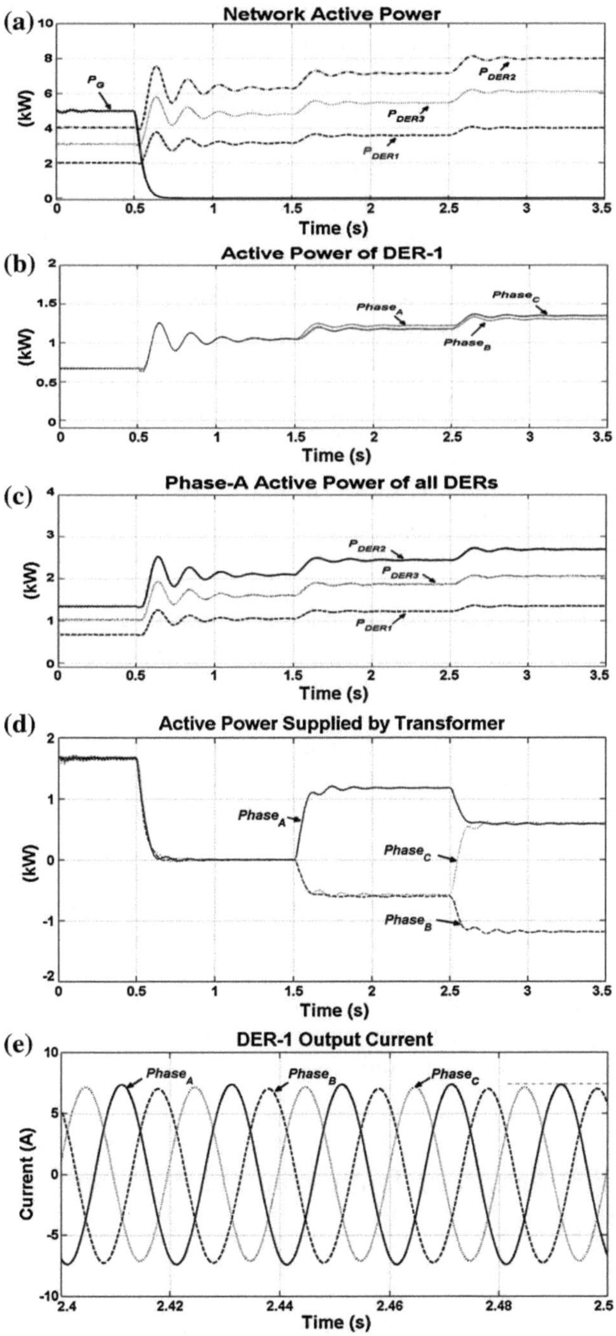


**Fig. 15.9** Simulation results of microgrid operation in presence of unbalanced loads: **a** active power dispatch of grid and three DERs, **b** active power output of DER-1, **c** active power output of all DERs in their phase-A, **d** active power supplied by the distribution transformer, **e** three-phase instantaneous current output of a DER-1

network, the contribution level of each phase of the DERs in power generation is also based on the desired sharing ratio among them.

It is to be noted that, all phases of the DERs contribute to a single-phase (or unbalanced) load change in the network since there is a possibility of power circulation from one phase of microgrid to the other two phases through the windings of the distribution transformer. Figure 15.9d shows the active power supplied by the distribution transformer in the studied case. From this figure, it can be seen that the distribution transformer output power is zero when the microgrid is working in autonomous mode with balanced loads (i.e. $0.5 < t < 1.5$ s). However, it can be seen that for $1.5 < t < 2.5$ s there is a negative active power flow into the distribution transformer in phase-B and C which is circulated and retuned to phase-A where the single phase load is connected. In the same way, for $2.5 < t < 3.5$ s there is a negative active power flow into the distribution transformer in phase-B which is circulated and retuned to phase-A and C where the single phase loads are connected. The three-phase instantaneous current output of a sample DER (e.g. DER-1) is shown in Fig. 15.9e which shows the DERs are generating unbalanced current in their output.

Now let us consider the microgrid of Fig. 15.2 with a harmonic load. Assume that the system is in steady state condition at $t = 0$ s and at $t = 0.5$ s, the grid is disconnected (i.e. $CB_G$ is opened) and microgrid operates in autonomous mode. At $t = 1.5$ s a new three-phase harmonic load of 3 kW is connected to the network. At $t = 2.5$ s, its demand is increased to 6 kW. The harmonic load has a Total Harmonic Distortion (THD) of 25.5 % with the Fast Fourier Transform (FFT) Spectrum as shown in Fig. 15.10a. Figure 15.10b shows the active power dispatch of grid and three existing DERs in the microgrid between 0 and 3.5 s in the above mentioned network. The single-phase active power output of one of the DERs (e.g. DER-1) is shown in Fig. 15.10c. This is the same for all phases of A, B and C. The active power output in phase-A of all three DERs are also shown in Fig. 15.10d. From these figures, it can be seen that as the load is supplied from the three-phase network, its demand is shared equally among the three phases of each DER. However, the DERs share the extra demand based on their desired power sharing ratio. The three-phase instantaneous current output of a sample DER (e.g. DER-1) is shown in Fig. 15.10e which shows the output current of the DERs are distorted as required by the network harmonic load.

### 15.5.3 *Case 3: System of Interconnected Autonomous Microgrids*

Let us assume that in system of Fig. 15.2, at $t = 0$ s, circuit breakers $CB_G$, $CB_{M1}$, $CB_{M2}$, $CB_{S1}$ and $CB_{S2}$ are closed and the network is at steady state condition. Now, let us also assume, due to a fault in the medium voltage grid, after self-healing process, $CB_G$ is open while $CB_{M1}$ and $CB_{M2}$ are closed at $t = 1$. As

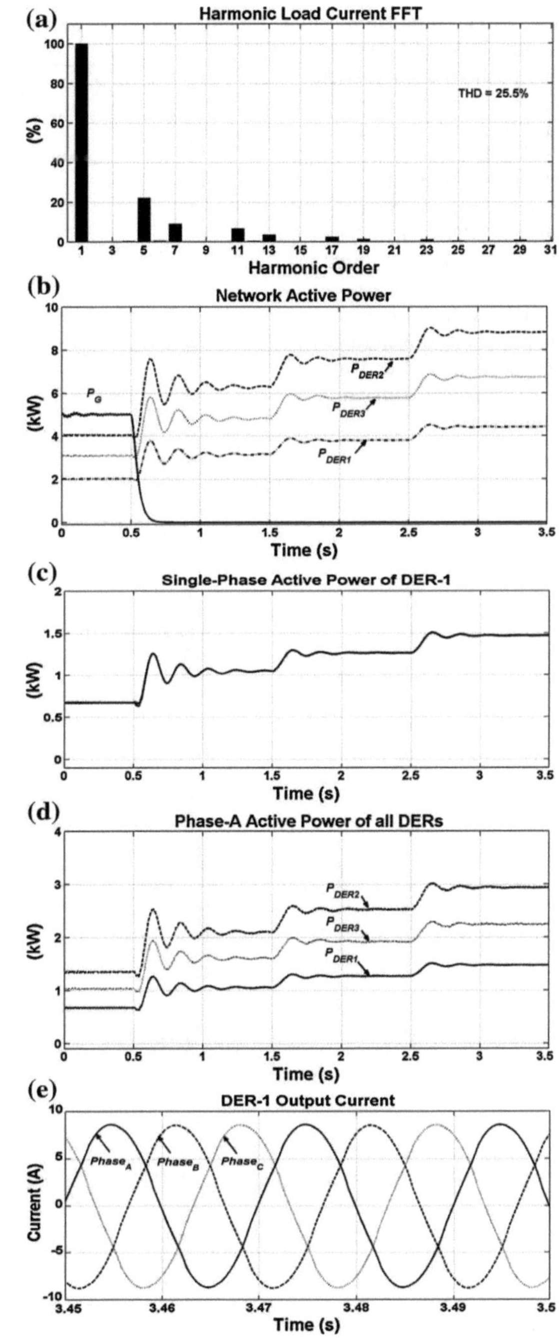


**Fig. 15.10** Simulation results of microgrid operation in presence of harmonic load: **a** FFT Spectrum of harmonic load current, **b** active power dispatch of grid and three DERs, **c** single-phase active power output of DER-1, **d** active power output of all DERs in their phase-A, **e** three-phase instantaneous current output of a DER-1

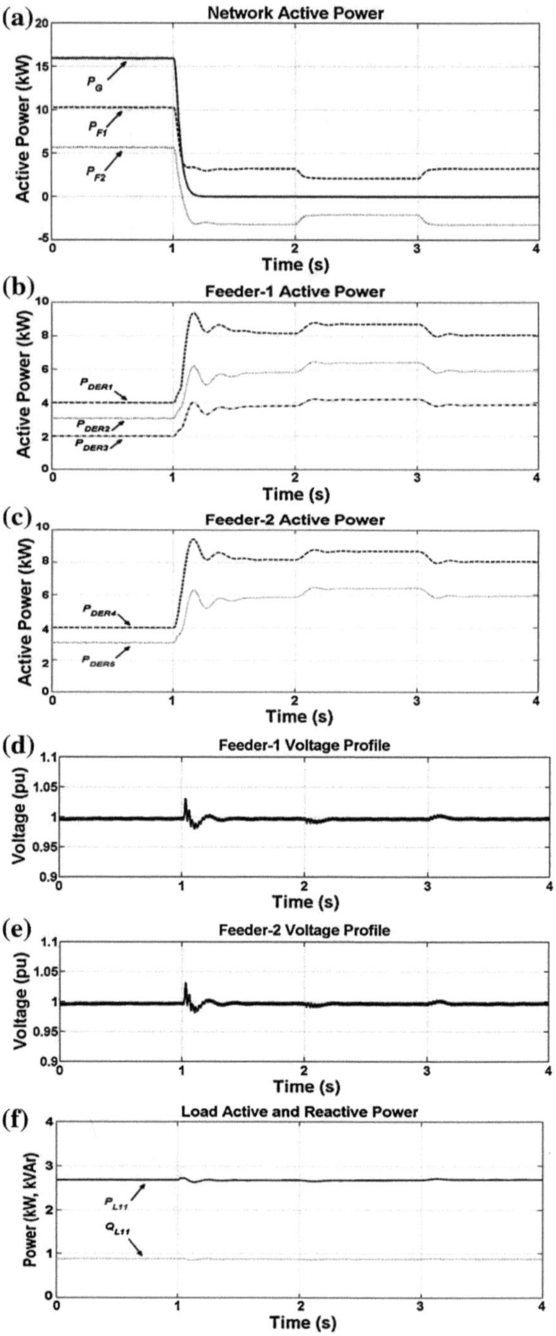


**Fig. 15.11** Simulation results for system of interconnected microgrids: **a** active power supplied from grid into each microgrid feeder, **b** active power output of each DER in MG-1, **c** active power output of each DER in MG-2, **d** MG-1 voltage profile monitored in transformer secondary side, **e** MG-2 voltage profile monitored in transformer secondary side, **f** active and reactive power demand of a sample load

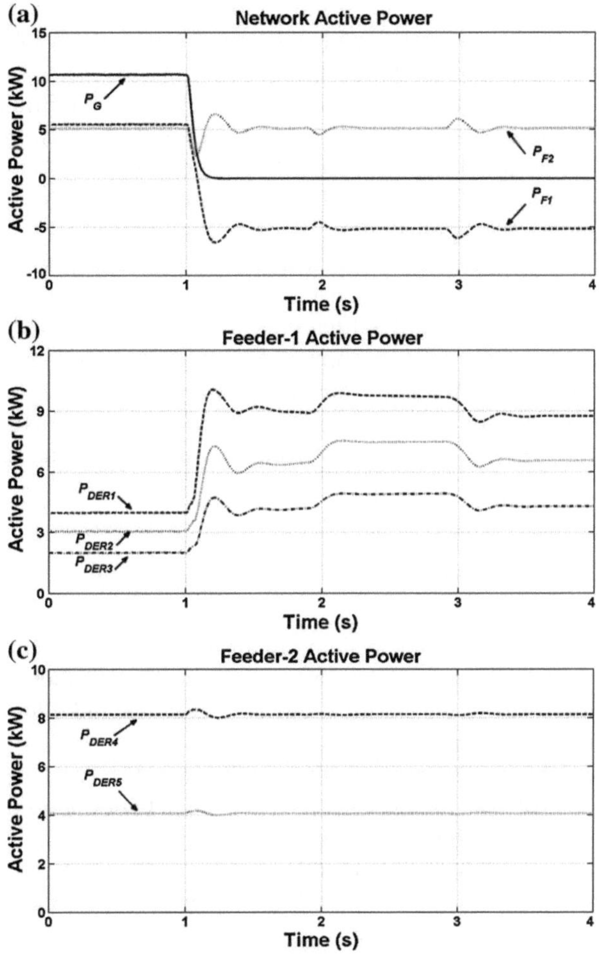


**Fig. 15.12** Simulation results for the system of interconnected MG-1 and MG-2 when MG-2 DERs operate in their maximum capacity: **a** active power supply from grid and active power flow into each microgrid feeder, **b** active power output of each DER in MG-1, **c** active power output of each DER in MG-2

mentioned before, the operation of the protection devices and circuit breakers and resynchronization of interconnecting microgrids during self-healing process is beyond the scope of this discussion. In this study, this transition period is not considered. At $t = 2$ s a load increase of 25 % and at $t = 3$ s a load decrease of 25 % are also applied in MG-1.

The total active power supply from the grid in addition to the active power flow into each microgrid is shown in Fig. 15.11a. From this figure, it can be seen that MG-2 has a negative power flow and is delivering, approximately, 20 % of the load demand in MG-1. The active power output of all the DERs in MG-1 and MG-2 are shown separately in Fig. 15.11b and c, respectively. It can be seen that all DERs in MG-1 and MG-2 are sharing the load demand of the system of interconnected microgrids proportional to their ratings. The voltage profile in the secondary side of the distribution transformers in each microgrid is regulated to the desired value of 1 pu as shown in Fig. 15.11d and e. The active and reactive power drawn by a sample load in the network, with a 2.7 kW demand and power factor of 0.95, is also shown in Fig. 15.11f.

Now, let us assume a case in which the two DERs in MG-2 are running in their maximum capacity (i.e. 4 and 8 kW, respectively). Let us also assume at $t = 1$ s, due to a fault in the medium voltage grid and after self-healing process, $CB_G$ is open while $CB_{M1}$ and $CB_{M2}$ are closed. A 25 % load increase and decrease in the network is applied at $t = 2$ s and 3 s, respectively.

For this case, the total active power supply from the grid in addition to the active power flow into each microgrid is shown in Fig. 15.12a. From this figure, it can be seen that MG-1 has a negative power flow and is delivering, approximately, 30 % of the load demand in MG-2. The active power output of all the DERs in MG-1 and MG-2 are shown separately in Fig. 15.12b and c, respectively. It can be seen that all DERs in MG-2 are running in their maximum rating at all times while the DERs in MG-1 are sharing the rest of the load proportional to their ratings.

## 15.6 Conclusions

To minimize load shedding in a Microgrid during autonomous operation, islanded neighbour microgrids can be interconnected if they are on a self-healing network and an extra generation capacity is available in DERs in one of the microgrids. The primary control level for the DER converters in a system of interconnected neighbour autonomous microgrids was discussed in this chapter. A hierarchical control system was described for the network and microgrid levels. The concepts of self-healing and supply restoration algorithm, protection devices and their coordination and communication belong to the tertiary control level and were not the focus of this chapter. The chapter focused on the primary control level which guarantees a proper power sharing among all the DERs in the system of interconnected microgrids. Control algorithms were developed to manage the parallel operation of converter-interfaced DERs at grid-connected and autonomous operation modes in the microgrids while the network voltage was regulated by the DSTATCOMs installed in each microgrid. Using the developed inner loop and outer loop control techniques, an acceptable dynamic performance of DER converters in transients and steady-state conditions can be achieved. The proposed method prevents any undesired voltage, power and frequency variations in the network. Based on the described primary control level, a DER with extra generation capacity can successfully share some of the loads in other microgrids, once the other microgrids have a shortage of generation.

# Appendix A

As described in Sect. 15.3.1, the DERs considered in this chapter were modeled in detail. The technical data for these models are summarized below.

## *A.1 Fuel Cell*

Based on experimental validations, a typical Proton exchange membrane fuel cell (PEMFC) has an output V–I characteristic of

$$V(i) = 371.3 - 12.38\log(i) - 0.2195i - 0.2242e^{0.025i} \tag{A.1}$$

reported and utilized in [55].

## *A.2 Photovoltaic Cell (PV)*

In [55], the simplified equivalent circuit of a PV was utilized where its output voltage was a function of its output current and its output current was a function of load current, ambient temperature and radiation level. In this model, the voltage output of PV is calculated by

$$V_{PV} = \frac{AkT_C}{e} Ln\left(\frac{I_{ph} + I_o - I_c}{I_o}\right) - R_s I_c \tag{A.2}$$

where

$A$ Constant value for curve fitting
$e$ Electron charge (1.602 × 10¯19 C)
$k$ Boltzmann constant (1.38 × 10¯23 J/ºk)
$I_c$ Output current of PV cell
$I_{ph}$ Photocurrent (1 A)
$I_o$ Diode reverse saturation current (0.2 mA)
$R_s$ Series resistance of PV cell (1 mΩ)
$V_{PV}$ Output voltage of PV cell
$T_c$ PV cell reference temperature (25 °C).

A Maximum Power Point Tracking (MPPT) method was also used to achieve maximum power from the PV based on the load or ambient condition changes. The MPPT algorithm was presented in [55].

**Table B.1** Technical data of the considered network

| Medium voltage feeder | 11 kV L–L RMS, 50 Hz |
|---|---|
| | $R = 0.2\ \Omega$, $L = 10$ mH |
| Low voltage feeder | 410 V L–L RMS, 50 Hz |
| | $R = 0.02\ \Omega$, $L = 1$ mH |
| Transformers | 30 kVA, 11 kV/410 V, three-phase, Δ/Y-grounded, $Z_l = 5\ \%$, |
| Impedance loads | 7 three-phase RL load each $P = 3$ kW, $PF = 0.95$ |
| Induction motors | 2 three-phase each $P = 1.5\ kW$, $PF = 0.8$ |
| DER VSCs and filters | $R_f = 0.1\ \Omega$, $L_f = 0.4$ mH, $C_f = 50\ \mu$F, $V_{dc} = 150$ V, $a = 3.33$, $h = 10^{-4}$ |
| DSTATCOM VSCs and filters | $R_f = 0.1\ \Omega$, $L_f = 4$ mH, $C_f = 25\ \mu$F, $V_{dc} = 1$ kV, $a = 1$, $h = 10^{-4}$ |

**Table B.2** Technical data of the DERs and droop control coefficients

| DER type | DER rating [kW] | $L_{coup}$ [mH] | $m$ [rad/kW] | $n$ [V/kVAr] |
|---|---|---|---|---|
| Fuel cell (DER-1, 4) | 4 | 5.61 | 1.5708 | 4.5 |
| PV (DER-2, 5) | 3 | 7.48 | 2.0944 | 6.0 |
| Battery (DER-3) | 2 | 11.22 | 3.1416 | 9.0 |

### *A.3 Battery*

The battery is assumed to be a constant voltage source with fixed amount of energy and modeled as a constant DC voltage source with series internal resistance [55].

## Appendix B

The technical data (Tables B.1, B.2) of the microgrid network under consideration in Fig. 15.2 is provided.

## References

1. Kroposki B, Pink C, DeBlasio R, Thomas H, Simões M, Sen PK (2010) Benefits of power electronic interfaces for distributed energy systems. IEEE Trans Energy Convers 25(3):901–908
2. Senjyu T, Nakaji T, Uezato K, Funabashi T (2005) A hybrid power system using alternative energy facilities in isolated island. IEEE Trans Energy Convers 20(2):406–414
3. Hatziargyriou N, Asano H, Iravani R, Marnay C (2007) Microgrids. IEEE Power and Energy Magazine 5(4):78–94
4. Huang W, Lu M, Zhang L (2011) Survey on microgrid control strategies. Energy Procedia 12:206–212

5. Kroposki B, Lasseter R, Ise T, Morozumi S, Papatlianassiou S, Hatziargyriou N (2008) Making microgrids work. IEEE Power Energy Magazine 6(3):40–53
6. Katiraei F, Iravani R, Hatziargyriou N, Dimeas A (2008) Microgrids management. IEEE Power Energy Magazine 6(3):54–65
7. Lasseter RH (2002) Microgrids. IEEE Power Eng Soc Winter Meet 1:305–308
8. Barnes M, Kondoh J, Asano H, Oyarzabal J, Ventakaramanan G, Lasseter R, Hatziargyriou N, Green T (2007) Real–world micro grids–an overview. In: IEEE international conference on systems engineering, pp 1–8
9. Lopes JAP, Moreira CL, Madureira AG (2006) Defining control strategies for microgrids islanded operation. IEEE Trans Power Syst 21(2):916–924
10. Chandorkar MC, Divan DM, Adapa R (1993) Control of parallel connected inverters in standalone AC supply systems. IEEE Trans Ind Appl 29(1):136–143
11. Majumder R, Ghosh A, Ledwich G, Zare F (2009) Angle droop versus frequency droop in a voltage source converter based autonomous microgrid. In: IEEE power engineering society general meeting, pp 1–8
12. Majumder R, Shahnia F, Ghosh A, Ledwich G, Wishart M, Zare F (2009) Operation and control of a microgrid containing inertial and non–inertial micro sources. In: IEEE region 10 conference (TENCON), pp 1–6
13. Rowe CN, Summers TJ, Betz RE, Cornforth DJ, Moore TG (2013) Arctan power–frequency droop for improved microgrid stability. IEEE Trans Power Electron 28(8):3747–3759
14. Rokrok E, Golshan MEH (2010) Adaptive voltage droop scheme for voltage source converters in an islanded multibus microgrid. IET Gener Transm Distrib 4(5):562–578
15. Bevrani H, Shokoohi S (2013) An intelligent droop control for simultaneous voltage and frequency regulation in islanded microgrids. In: Accepted in IEEE transactions on smart grid, vol 99, pp 1–9
16. Sanjari MJ, Gharehpetian GB (2013) Small signal stability based fuzzy potential function proposal for secondary frequency and voltage control of islanded microgrid. Electric Power Compon Syst 41(5):485–499
17. Johnson B, Davoudi A, Chapman P, Sauer P (2011) A unified dynamic characterization framework for microgrid systems. Electric Power Compon Syst 40(1):93–111
18. Dou CX, Liu DL, Jia XB, Zhao F (2011) Management and control for smart microgrid based on hybrid control theory. Electric Power Compon Syst 39(8):813–832
19. Majumder R (2013) Some aspects of stability in microgrids. In: Accepted in IEEE Transactions on power systems, vol 99
20. Majumder R, Chaudhuri B, Ghosh A, Majumder R, Ledwich G, Zare F (2010) Improvement of stability and load sharing in an autonomous microgrid using supplementary droop control loop. IEEE Trans Power Syst 25(2):796–808
21. Nian H, Zeng R (2011) Improved control strategy for stand-alone distributed generation system under unbalanced and non-linear loads. IET Renew Power Gener 5(5):323–331
22. Katiraei F, Iravani MR (2006) Power management strategies for a microgrid with multiple distributed generation units. IEEE Trans Power Syst 21(4):1821–1831
23. Yazdani A, Iravani R (2006) A unified dynamic model and control for the voltage-sourced converter under unbalanced grid conditions. IEEE Trans Power Delivery 21(3):1620–1629
24. Ghosh A, Ledwich G (2002) Power quality enhancement using custom power devices. Kluwer Academic Publishers, Dordrecht
25. Teodorescu R, Liserre M, Rodriguez P (2011) Grid converters for photovoltaic and wind power systems. Wiley, New Jersey
26. Blaabjerg F, Teodorescu R, Liserre M, Timbus AV (2006) Overview of control and grid synchronization for distributed power generation systems. IEEE Trans Ind Electron 53(5):1398–1409
27. Rocabert J, Azevedo G, Candela I, Teoderescu R, Rodriguez P, Etxebarria-Otadui I (2010) Microgrid connection management based on an intelligent connection agent. In: IEEE 36th annual conference on industrial electronics (IECON), pp 3028–3033

28. Majumder R, Ghosh A, Ledwich G, Zare F (2008) Control of parallel converters for load sharing with seamless transfer between grid connected and islanded modes. In: IEEE power engineering society general meeting, pp 1–7
29. Vandoorn TL, Meersman B, De Kooning JDM, Vandevelde L Transition from islanded to grid–connected mode of microgrids with voltage-based droop control. In: Accepted in IEEE transactions on power systems, vol 99, p 1
30. Hong YY, Hsiao MC, Chang YR, Lee YD, Huang HC, Multiscenario underfrequency load shedding in a microgrid consisting of intermittent renewables. In: Accepted in IEEE transactions on power delivery, vol 99, p 1
31. Seethalekshmi K, Singh SN, Srivastava SC (2011) A synchrophasor assisted frequency and voltage stability based load shedding scheme for self–healing of power system. IEEE Trans Smart Grid 2(2):221–230
32. Lasseter RH (2011) Smart distribution: coupled microgrids. In: Proceedings of the IEEE vol 99, no 6, pp 1074–1082
33. Shahnia F, Chandrasena RPS, Rajakaruna S, Ghosh A (2013) Autonomous operation of multiple interconnected microgrids with self-healing capability. In: IEEE power engineering society general meeting, pp 1–5
34. Moslehi K, Kumar R (2010) A reliability perspective of the smart grid. IEEE Trans Smart Grid 1(1):57–64
35. Fang X, Misra S, Xue G, Yang D (2012) Smart grid—the new and improved power grid: a survey. IEEE Commun Surv Tutorials 14(4):944–980
36. Liu H, Chen X, Yu K, Hou Y (2012) The control and analysis of self-healing urban power grid. IEEE Trans Smart Grid 3(3):1119–1129
37. Kezunovic M (2011) Smart fault location for smart grids. IEEE Trans Smart Grid 2(1):11–22
38. Moslehi K, Kumar ABR, Hirsch P (2006) Feasibility of a self-healing grid—part II: benefit models and analysis. In: IEEE power engineering society general meeting, pp 1–8
39. Zidan A, El–Saadany EF (2012) A cooperative multiagent framework for self-healing mechanisms in distribution systems. In: IEEE transactions on smart grid, vol 3, no. 3, pp 1525–1539, Sept 2012
40. Arefifar SA, Mohamed YAI, EL–Fouly THM (2012) Supply-adequacy-based optimal construction of microgrids in smart distribution systems. In: IEEE transactions on smart grid, vol 3, no 3, pp 1491–1502, Sept 2012
41. Košt'álová A, Carvalho PMS (2011) Towards self-healing in distribution networks operation: Bipartite graph modeling for automated switching. In: Electric power systems research, vol 81, Issue 1, pp 51–56, Jan 2011
42. Yinger RJ (2012) Self-healing circuits at Southern California Edison. In: IEEE transmission and distribution conference and exposition, pp 1–3, May 2012
43. Spitsa V, Ran X, Salcedo R, Martinez JF, Uosef RE, de Leon F, Czarkowski D, Zabar Z (2012) On the transient behavior of large-scale distribution networks during automatic feeder reconfiguration. IEEE Trans Smart Grid 3(2):887–896
44. Shahnia F, Chandrasena RPS, Rajakaruna S, Ghosh A (2013) Primary control level of parallel DER converters in system of multiple interconnected autonomous microgrid with in self-healing networks. In: IET generation transmission and distribution, under review
45. Justo JJ, Mwasilu F, Lee J, Jung JW (2013) AC-microgrids versus DC-microgrids with distributed energy resources: a review. Renew Sustain Energy Rev 24:387–405
46. Guerrero JM, Vasquez JC, Matas J, de Vicuna LG, Castilla M (2011) Hierarchical control of droop-controlled ac and dc microgrids—a general approach toward standardization. IEEE Trans Ind Electron 58(1):158–172
47. Katiraei F, Iravani R, Hatziargyriou N, Dimeas A (2008) Microgrids management. IEEE Power Energy Magazine 6(3):54–65
48. Vasquez JC, Mastromauro RA, Guerrero JM, Liserre M (2009) Voltage support provided by a droop-controlled multifunctional inverter. IEEE Trans Ind Electron 56(11):4510–4519
49. Lasseter RH, Piagi P (2006) Control and design of microgrid components. Final project report, Power Systems Engineering Research Center, University of Wisconsin–Madison

50. Salamah AM, Finney SJ, Williams BW (2008) Autonomous controller for improved dynamic performance of AC grid, parallel-connected, single-phase inverters. IET Gener Transm Distrib 2(2):209–218
51. Ghosh A, Ledwich G (2010) High bandwidth voltage and current control design for voltage source converters. In: 20th Australasian University power engineering conference (AUPEC), pp 1–6
52. Tewari A (2002) Modern control design with Matlab and Simulink. Wiley, New York
53. Ghosh A, Ledwich G (2003) Load compensating DSTATCOM in weak AC systems. IEEE Trans Power Delivery 18(4):1302–1309
54. Chandrasena RPS, Shahnia F, Rajakaruna S, Ghosh A (2013) Control, operation and power sharing among parallel converter-interfaced DERs in a microgrid in the presence of unbalanced and harmonic loads. In: 23rd Australasian University power engineering conference (AUPEC), pp 1–6, Sep/Oct 2013
55. Shahnia F, Majumder R, Ghosh A, Ledwich G, Zare F (2010) Operation and control of a hybrid microgrid containing unbalanced and nonlinear loads. Electric Power Syst Res 80(8):954–965

# Chapter 16
# Agent-Based Smart Grid Protection and Security

**Md Shihanur Rahman and H. R. Pota**

**Abstract** The vision of a smart grid is to provide a modern, resilient, and secure electric power grid as it boasts up with a highly reliable and efficient environment through effective use of its information and communication technology (ICT). Generally, the control and operation of a smart grid which integrate the distributed energy resources (DERs) such as, wind power, solar power, energy storage, etc., largely depends on a complex network of computers, softwares, and communication infrastructure superimposed on its physical grid architecture facilitated with the deployment of intelligent decision support system applications. In recent years, multi-agent system (MAS) has been well investigated for wide area power system applications and specially gained a significant attention in smart grid protection and security due to its distributed characteristics. In this chapter, a MAS framework for smart grid protection relay coordination is proposed, which consists of a number of intelligent autonomous agents each of which are embedded with the protection relays. Each agent has its own thread of control that provides it with a capability to operate the circuit breakers (CBs) using the critical clearing time (CCT) information as well as communicate with each other through high speed communication network. Besides physical failure, since smart grid highly depends on communication infrastructure, it is vulnerable to several cyber threats on its information and communication channel. An attacker who has knowledge about a certain smart grid communication framework can easily compromise its appliances and components by corrupting the information which may destabilize a system results a widespread blackout. To mitigate such risk of cyber attacks, a few innovative counter measuring techniques are discussed in this chapter.

---

M. S. Rahman (✉) · H. R. Pota
School of Engineering and Information Technology, The University of New South Wales,
PO Box 7916, Canberra, ACT 2610, Australia
e-mail: md.rahman3@student.adfa.edu.au

H. R. Pota
e-mail: h.pota@adfa.edu.au

J. Hossain and A. Mahmud (eds.), *Renewable Energy Integration*,
Green Energy and Technology, DOI: 10.1007/978-981-4585-27-9_16,

**Keywords** Power system · Smart grid · Multi-agent system · Protection relay · Cyber security and stability

## 16.1 Introduction

A smart grid is a complex electrical power network that provides efficient integration of real-time monitoring, advanced sensing and communication infrastructure superimposed on the physical grid. It utilizes estimations and controls to enable dynamic flows of information and energy from a generation source to end-users that ensures an economically efficient, reliable and sustainable power system environment for the generation and distribution of energy with low losses, high quality and security of energy supply. A smart grid has a role to play in the process of integrating various distributed energy resources (DERs), such as wind power or solar power or both, to supply all the critical loads in a network in either a gird-connected or standalone mode. Safety and reliability of supply are increased by using an intelligent multi-agent system (MAS) approach to avoid frequent occurrences of cascading failures due to the inability of the system to operate in an autonomous mode. For reliable and secure operation of a smart grid both physical and cyber influences are significant for protection and security. Recent research studies have indicated that an MAS technology provides a more flexible way of increasing both the resilience and efficiency of a smart grid by combining top–down and bottom–up intelligent autonomous decision-making in its communication infrastructure and control architectures [1].

According to Russell and Norvig [2] "the notion of an agent is meant to be a tool for analysing systems, not an absolute characterization that divides the world into agents and non-agents," being able to distinguish agent systems from existing systems is important. The autonomous agents and multi-agent technology, differ from the existing systems and system engineering approaches as they exhibits a distributed characteristics along with the dynamic adaptability and flexibility. As a matter of fact, the potential advantages gained through this difference motivated the power engineers to explore the applications of MAS to power engineering problems. From power engineering point of view, some existing systems could be classified as agents i.e., agents can be assumed to be embedded with the power system components. For example, a protection relay may be considered as an agent present in an *environment*, such as a power system. It reacts to any changes in it environment, i.e., changes to voltage or current due to fault or sudden generator load change. It can also exhibit a degree of *autonomy*. Similar arguments can be made for software agent systems such as detection and mitigation of cyber threats for smart grid security.

In this chapter, we discuss about the physical protection and cyber security issues in smart grids by means of an agent-based approach. As smart grids are quite large, complex and extensively interconnected, they are vulnerable to several types of disturbances, such as three-phase faults and sudden generator load changes, which may cause loss of synchronism and a risk of widespread blackouts. The problem of assessing the transient stability has been dealt by several methods, including a few applications of agent-based techniques such as, an agent-based turbine valve control [3–5], a multi-agent-based real-time wide-area power system stabilizer using reinforcement learning (RL) [6], an agent-based technique for instability prediction and control [7], a multi-agent based decentralized coordinated control [8], etc. To date, the transient stability enhancement considering a dynamic evaluation of the critical clearing time (CCT) with high wind power penetration using a multi-agent concept has not been studied. Therefore, a MAS architecture for smart grid protection is proposed in this chapter for well-coordinated protection framework, in which autonomous agents embedded with the protection relays which co-operate and communicate with each other to provide real-time coordination of system protection devices with CCT information. By using a continual streaming of this CCT information, it can precisely detect and isolate a fault before its CCT and reclose its circuit breakers (CBs) for the resumption of normal operation to promote the scalability and capability of the system, hence ensuring the transient stability improvement.

Besides physical failures or disturbances, a smart grid, being a critical cyber-physical system (CPS), is also vulnerable to various cyber threats to its data management system and communication layers. A cyber attack can potentially impact on the smart grid's dynamic states, appliances and its components by manipulating or corrupting its dynamic data and tampering with its information, thereby causing its control centre to make wrong decisions regarding its control operations which can lead to a widespread blackout in the grid. A few innovative counter measuring techniques within an agent-based smart grid infrastructure are discussed in this chapter which can be incorporated to detect and mitigate such a cyber attack which could cause abnormal physical behavior of the system.

## 16.2 Protection and Security Requirement for Smart Grid

In a smart grid, a protection system is, in general, set to be adequate a certain operating condition. Therefore, if the operating condition has changed, incorrect operation of the protection relays can result in a sequence of CB tripping, which may cause cascading failure results into a widespread blackout in the smart grid [9]. To ensure the safe and reliable operation of a smart grid, a few basic protection requirements are necessary for protection relay coordination. Since a smart grid is being a major infrastructure of a CPS, so its security requirement is also necessary for its safe and resilient operation. In general, cyber security requirements for a smart grid include three main security properties, such as confidentiality which prevents an unauthorized user from obtaining private information, integrity which prevents an unauthorized user from modifying the information and availability which ensures that the resources can be used when requested [10].

### 16.2.1 Smart Grid Protection Requirements

The most important requirements for protecting an agent-based smart grid are discussed below.

- **Reliability:** It is the ability of a protection system protection system in a smart grid which operates only when it is needed to be operated rather than remain in normal operation.
- **Sensitivity:** It is the ability of a protection system to detect even a small disturbance or a small change in a power network. It is an important issue that ensures the detection of high-impedance faults or the reduced contribution to faults from small, dispersed generators [11].
- **Speed:** It is the ability of a protection system to operate with a very short time after a fault begins, but must trip the CBs before the CCT, which is significant for preserving the system stability, reducing the equipment damage and hence improving the power quality.
- **Selectivity:** It is the ability of a protection system to isolate a fault occurring on a grid with a shortest possible time along with least disconnection of system equipments.

### 16.2.2 Smart Grid Security Requirements

In this section we discuss about the smart grid cyber security requirements which are important to protect it with respect to the main security properties [10].

- **Confidentiality:** It is important as the power system information provides it with a usage patterns for individual smart appliances. Confidentiality of price information, control commands and software should not critical since the security of the system does not rely on the public knowledge and the secrecy of the software rather than only the secrecy of the keys [12].
- **Integrity:** Integrity of information, control commands and software is critical. It may alter the price information which results electricity utilization wrong billing information to the customers. It can cause a control center to make wrong decision by corrupting the information obtained from PMU results a significant loss to the utility grid. Integrity of software is critical as compromised software can control any device and grid component [10].
- **Availability:** Availability of power system information is a key factor of which Denial-of-service (DoS) attacks can send fake or delayed information to a server or a network, and distributed DoS (DDoS) attack can be accomplished by utilizing distributed attacking source such as, compromised smart meter and appliances [10].

## 16.3 Smart Grid System Model

Smart grids are large-scale interconnected, distributed and complex systems consisting of four major units which are, (i) power generation: the electric power is retrieved from the coal, water dams, wind, nuclear reaction, solar radiation, etc., which can be synchronous generator, induction generator or photovoltaic generator; (ii) transmission: transporting electrical power through a high voltage (HV) transmission lines over a long distance; (iii) distribution: the transmitted energy is distributed to residents over a medium or low voltage distribution system; and (iv) consumption: energy consumption at the customer device level. A node of typical physical smart grid architecture is shown in Fig. 16.1 where each node may consist of a generator with necessary supporting control equipments and a load or a generator along with a protection relay. Each node of this system is capable of independent control action.

In this chapter, we have considered a smart micro-grid system [13] as shown in Fig. 16.2 with some modifications. This system consists of two DGs, a conventional synchronous generator with nominal capacity of 7.3 MW–6.9 kV connected to bus 1 using a 6.9/66 kV, 8 MVA transformer and an induction generator-based wind farm model consists of 10 wind generators for 500 kW each i.e., 5 MW wind turbines are connected at bus 7 through a transmission line of impedance, $z = 0.0374 + j0.3741$ pu using a 12.5/66 kV, 10 MVA transformer. Two electrical loads, load-1 and load-2, are connected at bus 4 and bus 6, respectively. Power is delivered to load-1 of 3.94 MW, 0.95 Mvar by a 66/12.5 kV, 5 MVA transformer and load-2 of 2.82 MW, 0.6 Mvar by a 66/12.5 kV, 4 MVA transformer. Bus 2 and bus 5 are connected through a transmission, TL1, of impedance, $z = 0.0416 + j0.0663$ pu. A 66 kV, 1,000 MVA utility grid, considered as infinite bus, is connected to bus 3 through a transmission line of impedance, $z = 0.02 + j0.032$ pu. The modelling of synchronous generator and induction generator-based wind farm are discussed in the following sections.

### *16.3.1 Modelling of Synchronous Generators*

In power system, a synchronous generator is one of the most important components that generate power delivered to the consumer end through transmission and distribution system. With some typical assumptions, a classical third-order nonlinear dynamic model of the synchronous generator can be represented by the following set of differential equations [14, 15]:

- **Generator electrical dynamics:**

$$\dot{E}'_q = -\frac{1}{T'_{d0}}E'_q + \frac{1}{T_{d0}}\left(\frac{x_d - x'_d}{X'_d}\right)V_s\cos\delta + \frac{1}{T_{d0}}E_f \tag{16.1}$$

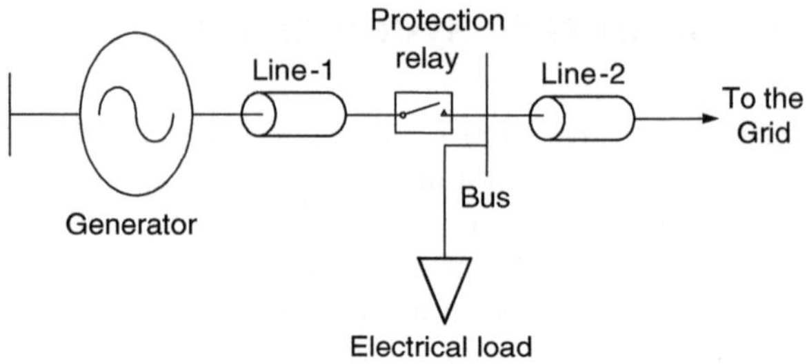


**Fig. 16.1** A typical node of a smart grid

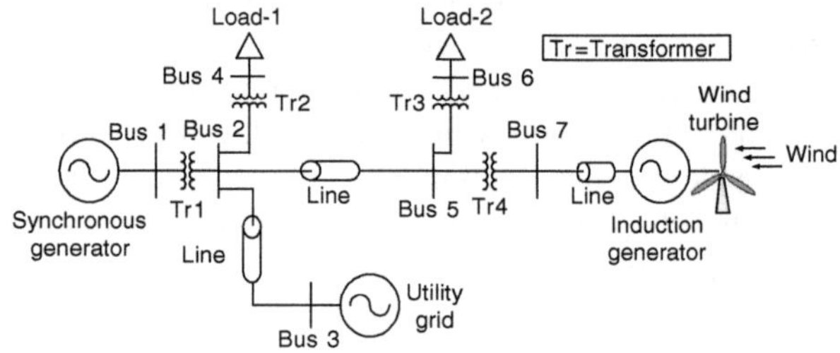


**Fig. 16.2** *Single-line* diagram of the smart micro-grid system

- **Generator mechanical dynamics:**

$$\dot{\delta} = -\omega_s(\omega - 1) \tag{16.2}$$

$$\dot{\omega} = \frac{\mathrm{D}}{2\mathrm{H}}\omega + \frac{\mathrm{P_m}}{2\mathrm{H}} - \frac{1}{2\mathrm{H}}\frac{\mathrm{V_s E'_q}}{\mathrm{x'_d}}\sin\delta \tag{16.3}$$

where $E'_q$ is the quadrature-axis transient voltage of the generator, $T'_{d0}$ the direct-axis open-circuit transient time constant of the generator, $X_d$ the direct-axis synchronous reactance, $X'_d$ the direct axis transient reactance, $V_s$ the infinite bus voltage, $\delta$ the power angle of the generator, $E_f$ the equivalent voltage in the excitation coil, $\omega$ the rotor speed with respect to synchronous reference, $D$ the damping constant of the generator, $H$ the inertia constant of the generator, $P_m$ the mechanical input power to the generator which is assumed to be constant.

### *16.3.2 Modelling of Wind Power Generation*

In this chapter, the nonlinear model the wind turbines is based on a static model of the aerodynamics, and a second order model of the induction generator. With some typical assumptions, the mathematical model of the induction generator-based wind farm can be represented by the following set of equations [16, 17].

#### 16.3.2.1 Wind Farm Model

A fixed-speed wind turbine (FSWT) model, in which the wind energy is transformed into mechanical energy through a mechanical drive train, is considered in this chapter. The wind speed and mechanical power extracted from the wind are related as [17],

$$P_{\mathrm{wt}_i} = \frac{\rho_i}{2} A_{wt_i} c_{p_i}(\lambda_i, \theta_i) V_{w_i}^3 \tag{16.4}$$

where $c_{p_i}$ is approximated by the following relation [17]:

$$c_{p_i} = (0.44 - 0.0167\theta_i)\sin\left[\frac{\pi(\lambda_i - 3)}{15 - 0.3\theta_i}\right] - 0.00184(\lambda_i - 3)\theta_i \tag{16.5}$$

where $i = 1, \cdots, n$, and $n$ is the number of wind turbines, $P_{wt_i}$ the power extracted from the wind turbine depends on the wind speed, $V_{w_i}$(m/s), the air density, $\rho_i$ (kg/m$^3$), and the swept area, $A_{wt_i}$(m$^2$), $c_{pi}$ the performance/power coefficient which depends on the pitch angle of the blade, $\theta_i$, and the ratio between the speed of the blade tip and the wind speed, denoted tip-speed ratio, $\lambda_i = \frac{\omega_{m_i} R_i}{V_{w_i}}$, $R_i$ the wind turbine radius (m), $\omega_{m_i}$ the wind turbine rotational speed (rad/s).

#### 16.3.2.2 Induction Generator Model

The dynamic equations for the induction generator used in this chapter in a structure similar to the synchronous generator is written as,

$$\dot{\delta} = -(\omega_{\mathrm{s}} - \omega_{\mathrm{r}}) - \frac{\mathrm{r_r X_m}}{\omega_{\mathrm{s}} \mathrm{X_r X_s}} \frac{\mathrm{V}_\alpha \sin\delta}{\lambda_{\mathrm{dr}}} \tag{16.6}$$

$$J\left(\frac{2}{P}\right)\dot{\omega} = \left(\frac{3}{2}\right)\left(\frac{\mathrm{P}}{2}\right)\frac{\mathrm{X_m}}{\mathrm{X_r}}\left(-\lambda_{\mathrm{dr}}\frac{\mathrm{V}_\alpha \sin\delta}{\omega_{\mathrm{s}} \mathrm{X}'_{\mathrm{s}}}\right) - \mathrm{T_L} \tag{16.7}$$

where $\delta$ is the power angle between the d-axis and phase-a peak of the induction generator, $\omega_s$ the synchronous speed, $\omega_r$ the electrical angular speed of the rotor, $r_r$ the rotor winding resistance, $X_m$ the magnetizing reactance, $X_r$ the leakage reactance of the rotor winding, $X'_s$ the leakage reactance of the stator winding, $V_\alpha$ the infinite bus voltage, $\lambda_{dr}$ the rotating flux of the d-axis, $J$ the inertia co-efficient, $P$ the pole of the induction machine and $T_L$ the load torque.

## 16.4 Multi-agent System

A MAS framework is a distributed and coupled network composed of intelligent software agents within an environment which work together to achieve a particular goal by solving a certain problem in a particular domain. Agent's intelligence may

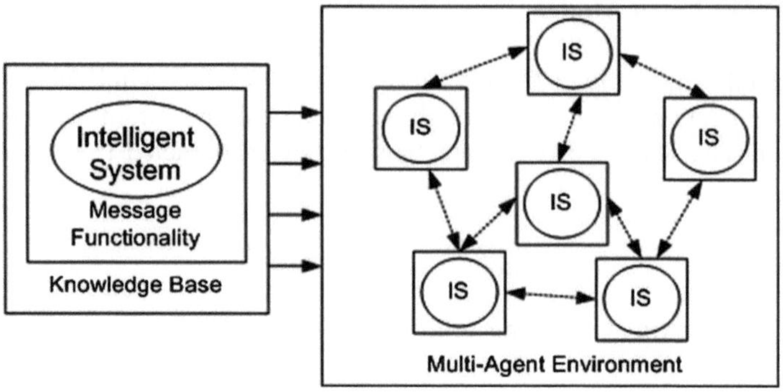


**Fig. 16.3** Intelligent system wrapped in a multi-agent system

include a few methodic, functional, procedural or algorithmic search, find and processing approach [18]. According to Wooldridge [19], an agent is merely "a software entity that is presented in some *environment* and is able to *autonomously* react to changes in that *environment*", which must be observable to, or alterable by, each intelligent agent. The environment may be a physical system e.g., a power system, and therefore it is observable through sensors using a phasor measurement unit (PMU), or it may be the computing environment e.g., data or computing resources by messaging. An agent may alter the environment by taking control actions through actuators either physically such as, opening and re-closing a circuit breaker to reconfigure a power network, or storing diagnostic information in a database for others to access [20].

The key component of the MAS is the communication principle. According to Wooldridge's definitions, intelligent agents must have social ability and therefore must be capable of communicating with each other, where an agent encapsulates a particular task or a set of functionality, in a similar way to object-oriented programming [20]. If any agents need to cooperate and negotiate with neighbouring agents, they use messaging with a standard agent communication language (ACL) [21] along with an additional capability of autonomous control action. A multi-agent technology provides an ideal means of achieving systems integration by 'wrapping' disparate systems as 'intelligent systems' which commonly forms a MAS as shown in Fig. 16.3 with rules and knowledge enabling it to react within its environment and automate its internal function and reasoning.

By giving an agent this autonomy, it is possible to overcome the problems associated with intermittent communications between tools [22]. The intelligent autonomous agents have two basic criteria, one is *knowledge base*, i.e., each and every agent must have knowledge about other agents and the whole environment; another one is the *message functionality*, i.e., they use their communication capability to exchange the message or information among each other to achieve their common goal. A MAS has significant wide area applications in power engineering research such as, strategic power infrastructure defence system, fault analysis and diagnosis, power system protection, secondary voltage control, energy-management system (EMS), wide-area control, power system restoration and reconfiguration, transformer condition monitoring system, energy resource scheduling, power markets analysis, etc.

### *16.4.1 Characteristics of Intelligent Agents*

The agents in MAS have several important characteristics along with them which are given as follows.

- **Autonomy:** The agents are at least partially independent and autonomous, hence they have an ability to schedule action based on environment observations [23].
- **Decentralization:** There is no designated controlling agent or the system is effectively reduced to a monolithic system [24].
- **Reactivity:** It is an ability to react any changes in its environment and takes action based on its perception.
- **Social ability:** It is an ability to interact, negotiate and cooperate with each other by simply passing information ACL for conversation.
- **Pro-activeness:** It is an ability to take initiatives to exhibit its goal-directed behavior [20], through which it can dynamically change its behavior according to the changes in the system.

### *16.4.2 Engineering Benefits of MAS*

From engineering perspective, a MAS has numerous benefits over various engineering problems, by providing a solution in decentralized manner. MASs are used in a wide range of power system applications requiring flexibility, extensibility and adaptability with a rapidly changing environment due to their distributed nature, modularity, and ease of implementation which are beneficial [25]. Some benefits of MAS are discussed in this chapter as below.

- **Protocol:** MAS provides an infrastructure that specifies the communication and interaction protocols with all other intelligent agents.
- **Decentralize:** MAS are decentralized in their structures to solve a complex problem in an environment autonomously.
- **Cooperative:** MAS contains agents that are autonomous and distributed, and may be self-interested and cooperative.
- **Flexibility:** MAS has the ability to respond properly to the dynamic situations, and support for replication in variable situations [20].
- **Extensibility:** MAS has the ability to easily add a new feature to a system, augmenting or upgrading any existing functionality by sensing any change in the environment.

### 16.4.3 Interaction of MAS with Smart Grid

A smart grid is operated locally and distributed by heterogeneous agents that can range from simple devices to intelligent entities [26]. A few meta-heuristic techniques such as, genetic algorithm (GA), particle swarm optimization (PSO), simulated annealing (SA), and tabu search (TS) have been extensively used in power engineering research. As a matter of fact, a key disadvantage of these techniques is that most of these approaches are centralized and they mainly depend on a central computing facility to handle huge amount of data using a high speed communication capability. If the central computing facility fails then these methods could not be able to resolve the problems. Since MAS technology is an application of distributed intelligence dealing with problems in decentralized manner, so as to provide it with high flexibility and extensibility. Zhang et al. [27] have suggested that the MAS technology would be the best solution for the problems associated with computing, communications, and data integration in power systems because of its distributed nature and dynamic adaptability. A hierarchical agent-based framework is used in [28] for smart grid infrastructure. Inspired by this work, we have proposed a decentralized hierarchical structured multi-agent-based smart grid architecture in this chapter.

A multi-agent-based smart grid architecture is shown in Fig. 16.4, where the various node agents are: Through Node (TN) agent for end power consumption devices, ZIP Node (ZIPN) agent for ZIP (Z-impedance, I-constant current, P-constant power) loads which aggregates house-level TN, Dynamic Node (DN) agent for devices such as generators and motors with associated equipments, Root Node (RN) agent which provide a reference value for Dynamic Nodes, and Super Node (SN) agent that communicates with all RNs. A TN agent sends information of its on–off status and ratings of real and reactive powers to its cluster ZIP node agent. The ZIP node agents send the aggregated data to its DN agent which will collect this data and use this for simulating the system model given by system's dynamic equations. The RN agents provide the reference value down to its cluster of DN agents. The RN agent can also send its advice of pricing and load shedding down to the TN agents. Each subsystem in this architecture communicates through their RN agents, and can export power outside and receive injections of external power into them.

When intelligent autonomous software agents in a MAS environment are embedded or integrated with major power system components, for example protection devices, then they have their own thread of control over the protection relays to coordinate them after experiencing a disturbance on power system through a high-speed communication network perhaps via optical fibre link or wide-area network (WAN) used among protection relay agents. Researchers are still investigating the MAS technology as a way of developing a novel protection scheme which has an ability to be fault tolerant and self-healing if any agents fails to operate.

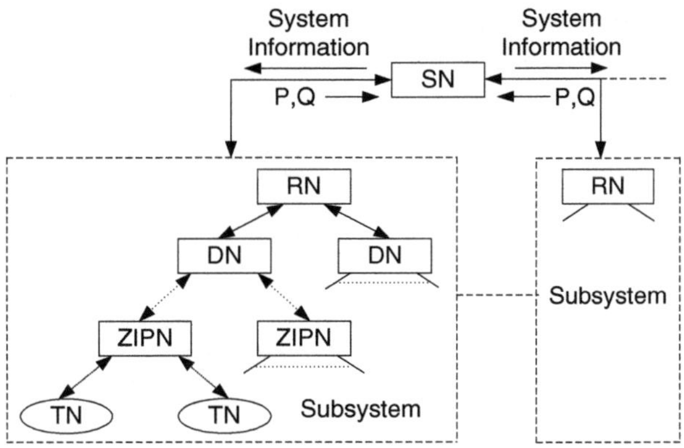


**Fig. 16.4** MAS system framework

## 16.5 Multi-agent Framework for Smart Grid Protection

Generally, a MAS is characterized as a group of individual intelligent agents running and co-operating with one another to solve problems in dynamic and unpredictable domains such as power systems which are inherently nonlinear and undergo a wide range of transient conditions. Since a smart grid is interconnected, distributed and a complex network, hence it is well suited for decentralized multi-agent technology. The basic operating requirement of a smart grid is that its synchronous generators remain in synchronism, and hence the ability of it to retain this synchronism after a large disturbance such as, three-phase fault or sudden generator load change is referred as transient stability. A smart grid will become unstable, if the operating time of its circuit breaker (CB) is longer than the CCT. Therefore, a better coordination among its protection devices is necessary to operate with the corresponding CCT information to regain the stability after experiencing a fault on it. Recent blackouts in different countries have illustrated the significance and vital need for investigating the transient stability of power systems, therefore the process of fault detection, isolation and reclosing is indispensible for the assessment of it.

A multi-agent based protection scheme is discussed in this chapter, where each intelligent autonomous agent is assumed to be embedded with each of the protection relays of the smart grid. A MAS-based protection scheme is shown in Fig. 16.5 in which the agents are able to coordinate the protection relays through an algorithm developed by a combination of MATLAB and Java. In this framework, agents can continuously monitor the present system status from the power system network topology. Once a fault occurs in a smart grid, using the algorithm, MASs precisely detect the fault with its location by measuring the fault current and hence obtain the CCT for the actual system condition. Based on their perceptions, they communicate and cooperate with each other through inter-agent communication to autonomously decide for coordinating the protection relays to trip the CB to isolate the fault before the CCT and reclose it automatically as soon as the fault

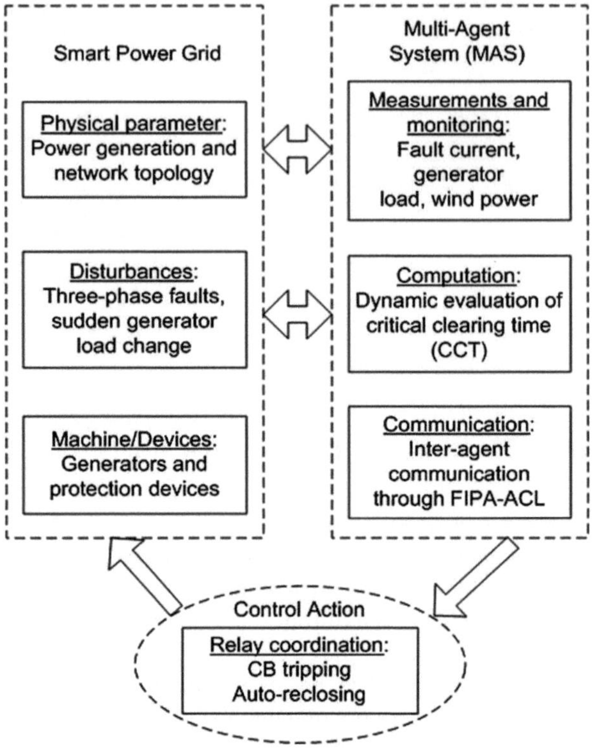


**Fig. 16.5** MAS-based protection system framework

is cleared in order to improve the transient stability of the system by maintaining a synchronism among the generators.

The conventional protection systems only respond to faults or abnormal events in a fixed and predetermined manner based on some certain assumptions made about the power system information e.g., considering a fixed generation with a certain critical fault clearing time. In fact, the variation in generation or high wind power integration into the grid may adversely affect the dynamic evaluation of the CCT which is an important key factor for assessing the transient stability of the system. However, these MASs can dynamically adapt the on-line measurement capabilities when the system conditions vary, and have the full flexibility to reconfigure themselves in an autonomous way i.e., they are able to dynamically determine the new stability limit each time the operating condition has changed due to the increasing penetration level of wind power of the smart micro-grid system as shown in Fig. 16.2 and flexibly obtain the corresponding new CCT to coordinate the protection relay operation. This means that this MAS has the full flexibility to adapt to the stability margin of the system for the assessment of the transient stability issue.

### 16.5.1 MAS-Based Protection Scheme

In the developed agent-based protection scheme, all the agents are able to simultaneously work together to update the system information at each integration step and have the capability to store pre-fault, fault and post-fault values. For fault detection, a threshold value of the current ($I_{th}$) is set for each protection device agent. When a fault occurs on the power system, agents detect that fault when fault current ($I_f$) flows above of its rated values and at the onset of the fault, agents dynamically check the wind power penetration level into the grid and for this penetration, the agents obtain the corresponding CCT which is a maximum value of the fault clearing time for a given fault for which the post-fault system is stable. If the fault occurring on smart grid is cleared within this CCT, the system will remain stable otherwise it will lose its stability.

In fact, the determination of the CCT is very important for protection scheme as it is a complex function of pre-fault system conditions, such as operating condition, network topology, system parameters, fault structure, such as its type and location and post fault conditions that themselves depend on the protective relaying plan employed [16].

- **CCT calculation:** The CCT is first estimated by using the following equations and then determined from the time-domain simulation approach. The swing equation of the induction generator can be written as,

$$\dot{\omega} = \frac{T_{\mathrm{L}} - T_e}{\mathrm{J}}. \tag{16.8}$$

In terms of speed it can be represented by,

$$\dot{s} = \frac{1}{2H}[T_{\mathrm{L}} - T_e]. \tag{16.9}$$

Integrating both sides of the above equation, it can be written as,

$$s = \int_0^t \frac{1}{2H}(T_{\mathrm{L}} - T_e) + s_0. \tag{16.10}$$

After solving this equation, the CCT of the machine can be written as,

$$t_{cr} = \frac{1}{T_m} 2H(s_{cr} - s_0) \tag{16.11}$$

where $s_{cr}$ is the critical slip, which is given by the intersection between the torque-speed curve for the specified system and $T_m$ the mechanical torque [29].

The torque-slip curve from Eq. 16.11 can be used to determine the slip of induction generator beyond which the generator become unstable which is called unstable slip or critical slip, $s_{cr}$. Its corresponding speed is called unstable speed or

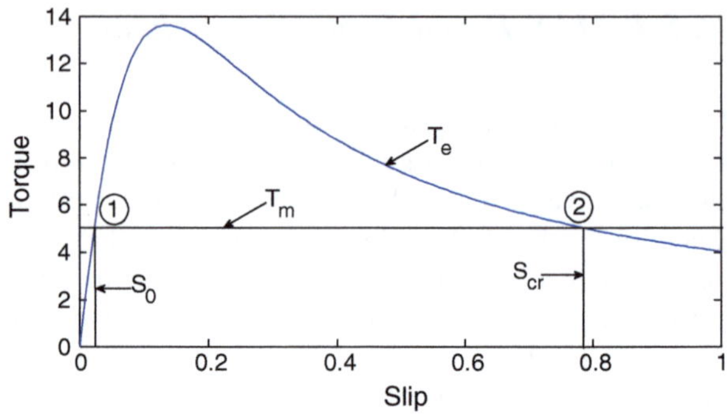


**Fig. 16.6** Torque-slip curve of an induction generator

critical speed and its corresponding time is called CCT, $t_{cr}$. To better understand, let us consider a Fig. 16.6, which shows the torque versus slip curve superimposed on the generator input mechanical torque, $T_m$. At steady-state condition, it is assumed that the operating point is located at point 1 in which the generator operating at a steady-state constant slip, $s_0$. From Fig. 16.6, it can be seen that, for a certain circumstances e.g., three-phase short circuit fault, the generator operates at any point beyond point 2 where the input mechanical torque is higher than the electromagnetic torque developed inside the generator. Hence, the generator would accelerate to possibly a dangerous speed. Therefore, this point is considered as the unstable operating region, which also represents the stability limit of induction generator. So, the slip at this point is called unstable slip or critical slip, $s_{cr}$.

On the other hand, the agents can continuously monitor the induction generator based wind power penetration ($P_{WG}$) and dynamically determine the new stability limit for increasing wind power penetration level by simultaneously checking it, hence obtain the new corresponding CCT for the new wind generation ($P'_{WG}$). Agents use the following simple logic to dynamically sense the wind power penetration.

$$
\begin{aligned}
&P_{WG} = P_{WG}^{0} \text{ (for normal operation)}\\
&\mathit{if}\ P_{WG} > \mathit{or} < P_{WG}^{0}(P_{WG}^{0} \pm \%\text{penetration level change})\\
&P'_{WG} = P_{WG} \text{ (for new penetration level)}\\
&\mathit{else}\ P_{WG} = P_{WG}^{0} \text{ (for normal penetration level)}
\end{aligned}
$$

where $P_{WG}^{0}$ is the initial wind power generation extracted from the wind turbine. Depending on the actual CCT information obtained, agents in the multi-agent platform communicate, cooperate and negotiate with each other to coordinate protection devices for opening of the CBs before the corresponding CCT is reached and reclose the breakers as soon as the fault is cleared. For CB operation, each agent uses a simple control logic signal $c_i(t)$ for breaker i, where i = 1,2,3,……n, where n is the number of CBs installed in a smart grid. The control logic signal for opening and reclosing of CB can be initiated as,

$$C_i(\mathrm{t}) = \begin{cases} 0, & \text{for opening of CB at time t} \\ 1, & \text{for closing of CB at time t} \end{cases}$$

Initially this control logic signal is set to 1 for normal operation when there is no fault occurring on the system. Once a fault occurs and agents obtained the CCT information, they initiate the control signal $c_i(t) = 0$ to trip the corresponding CBs at any time *t* to remove the faulted line and when the fault is cleared they initiate $c_i(t) = 1$ for reclosing the CBs to reconnect the line again. In this way agents cooperate with each other to enhance the transient stability of a smart grid through proper protection device coordination.

### 16.5.2 Smart Grid Protection Strategy

In the current implementation, the agents are primarily responsible for the protection of the transmission line of a smart grid. In the developed multi-agent based smart grid protection scheme, the agents use an algorithm which is developed through a combination of MATLAB and Java Agent Development Framework (JADE) that provides them with a capability to simultaneously monitor the system present status e.g., its present wind power penetration and measure the system parameters to store the values of voltage, current and power. In order to link these two platforms, they are integrated using the MATLAB control server through the MATLAB S-function. A flow chart of the proposed method is shown in Fig. 16.7.

When a disturbance like a three-phase short circuit fault occurs in a smart grid system, agents detect it by measuring the fault current (when $I_f > I_{th}$) whereas, if there is no disturbance then they take the decision for normal operation. After fault detection, the wind energy penetration level is checked according to the logic demonstrated in Sect. 16.5.1 and for the present wind power penetration the corresponding CCT is calculated. At the same time inter-agent communication takes place among the protection device agents to decide for tripping corresponding CBs to isolate the faulted portion from the rest of the system with the corresponding CCT information. On the other hand, when the fault is cleared, the agents take the decision to reclose the breakers as soon as possible for the resumption of the normal operation. At the same time, the agents can still perform their individual tasks within their own environments by continuously monitoring the system along with the appropriate control action.

The MAS framework developed in this chapter provides a scalability and capability for real-time coordination of protection devices with versatile system conditions as they autonomously decide to control the CBs using the knowledge of present system status. It can update the measured system information at each integration step with the corresponding CCT computation for the present wind power generation, hence the system relies on a continual streaming of on-line CCT information at any instant which enhances the on-line capability of the real-time

Fig. 16.7 Flow-chart of the agent-based smart grid protection algorithm

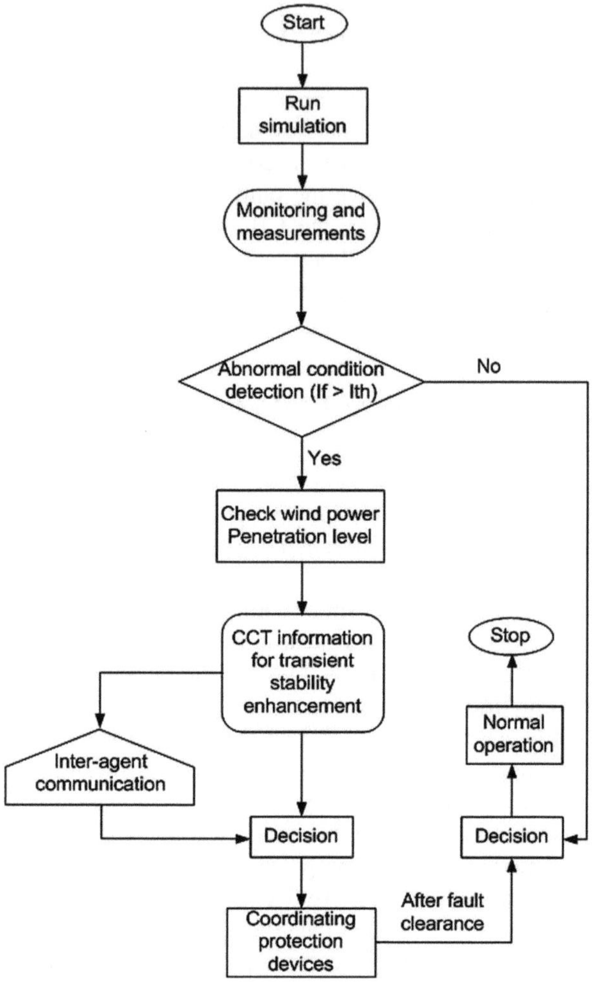


coordination of the protection devices for the secure and reliable operation of a smart grid through agents.

## 16.6 Illustrative Example and Results

To evaluate the performance of the MAS-based smart grid protection system for the impact of growing penetration of wind energy in the grid, a test system as shown in Fig. 16.2 is used for simulation under different wind power penetration level. In the following sections, we discuss about the function of individual autonomous agents which are responsible for coordinating the protection relays by detecting and isolating a fault with the corresponding CCT information.

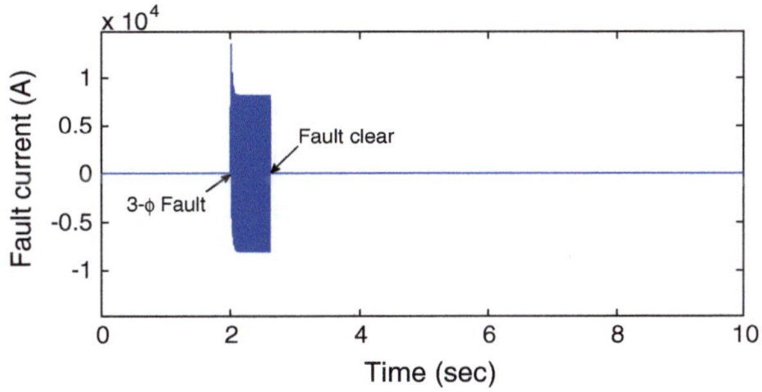


**Fig. 16.8** Fault current

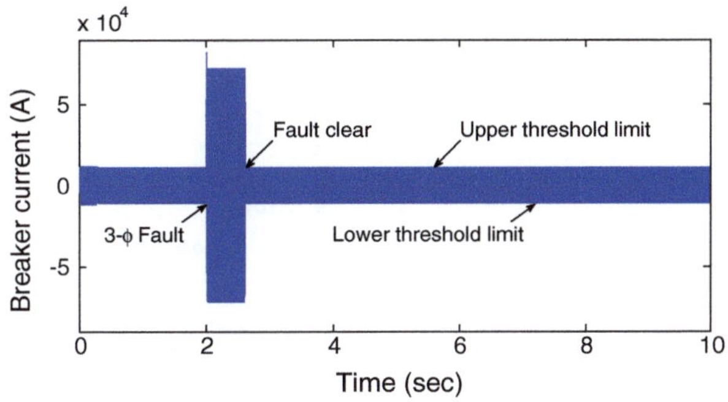


**Fig. 16.9** Breaker fault current

## *16.6.1 Fault Detection and Isolation*

In this section, we have described the capability of agents to coordinate the protection relays by operating CBs for precise fault detection and isolation with the corresponding CCT information. Under the given wind power extracted from the wind turbines i.e., 5 MW, a three-phase short circuit fault is applied at $t = 2.0$ s at line TL1 of the smart micro-grid system as shown in Fig. 16.2. The agents cooperate and communicate with each other to coordinate the protection relays for corresponding circuit breaker operation to detect and isolate the fault while the fault current exceeds its predefined threshold value. This fault is cleared by removing the line TL1 by tripping the corresponding CBs, where the fault current is shown in Fig. 16.8 and its flows in the breaker with their threshold limit is shown in Fig. 16.9 over a period of 10 s.

From Fig. 16.9 it is seen that the breaker agents can precisely detect that fault by sensing the fault current while it flows above its rated value i.e., upper and lower threshold limit. At the same time the agents dynamically evaluate the CCT using a transient simulation for tripping the breaker to isolate the fault from the rest of the system and hence reclose the breakers quickly to reconnect the line TL1 after the fault is cleared. The rotor angle of the synchronous generator and the rotor speed of the induction generator are shown in Figs. 16.10 and 16.11, respectively.

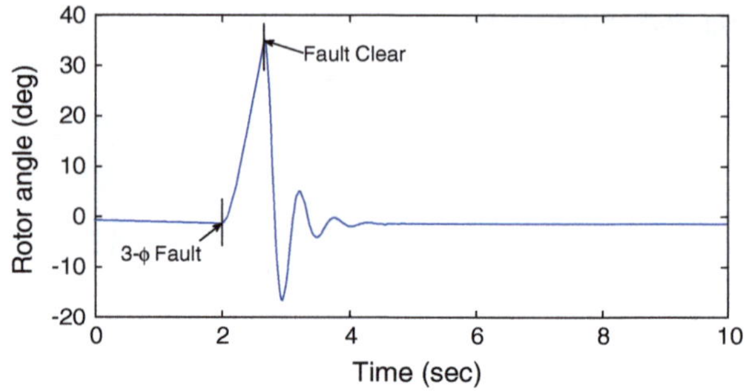


**Fig. 16.10** Synchronous generator rotor angle for fault at *line* TL1

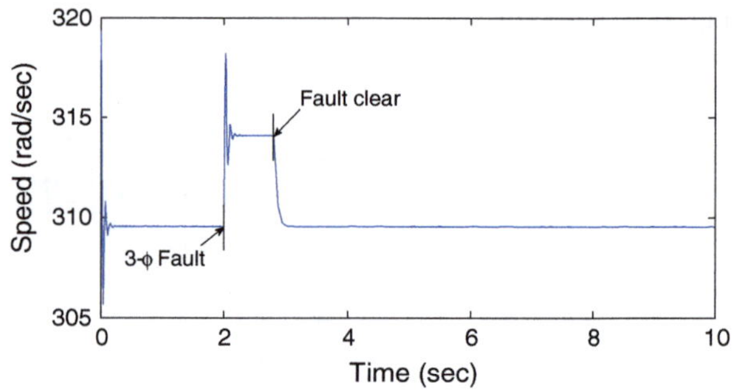


**Fig. 16.11** Induction generator rotor speed for fault at *line* TL1

From these figures it can be seen that, the synchronous generator rotor angle and induction generator rotor speed are transiently stable following the removal of the fault from the rest of the system by tripping the line TL1 before the corresponding CCT through agent-based protection relay coordination. This means that the agents successfully cooperate and communicate with each other to coordinate the corresponding CB operation with the corresponding CCT information in order to enhance the transient stability of the system.

## *16.6.2 Impact of Increasing Wind Power on Smart Grid*

The stability and security of a smart grid with a large amount of wind power penetration have recently become the major concerns, especially in terms of transient stability issue. Addressing this issue the impact of growing wind power generation on transient stability is evaluated in this section by applying a 3-phase short circuit fault at $t = 2.0$ s at line TL1. To dynamically evaluate the impact of this wind energy penetration on the transient stability of the smart micro-grid as shown in Fig. 16.2, the wind energy penetration level is gradually increased from 5.0 MW penetration to 6.25 MW penetration with a regular interval of 5 % penetration i.e., 0.25 MW. Table 16.1 summarize of the effect of increasing wind power penetration on the CCT of the system.

Table 16.1 Effect of increasing wind power penetration on CCT

| Fault | Location | CCT (s) | | | | | |
|---|---|---|---|---|---|---|---|
| | | 5 MW | 5.25 MW | 5.50 MW | 5.75 MW | 6 MW | 6.25 MW |
| 1 | TL1 | 0.611 | 0.593 | 0.554 | 0.512 | 0.475 | 0.457 |

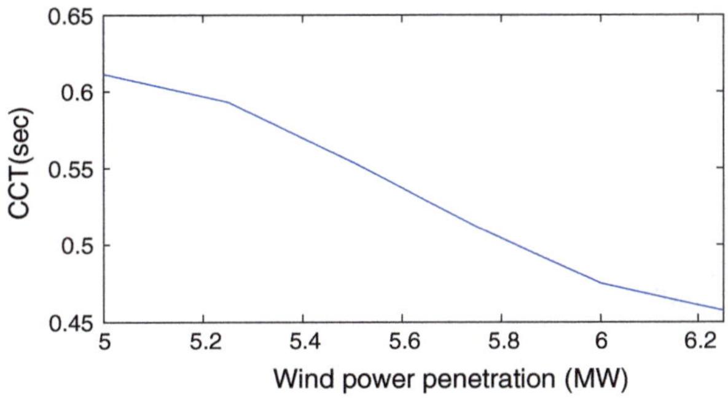


**Fig. 16.12** Relationship between wind power versus CCT

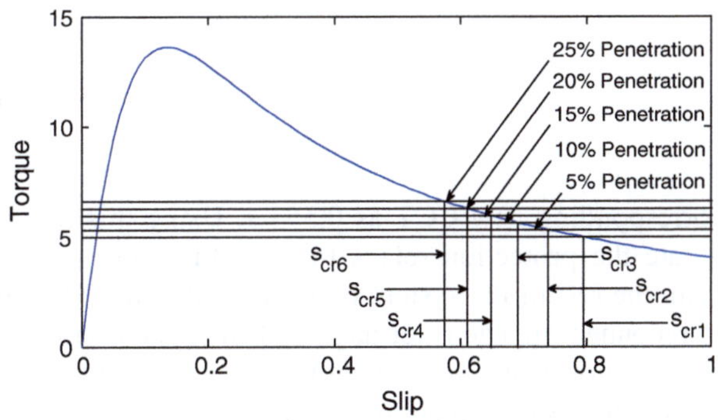


**Fig. 16.13** Torque-slip curve for different wind power penetration level

The relationship between the calculated CCTs and the wind power penetration levels is shown in Fig. 16.12 in which it is seen that increasing the wind power penetration level deteriorates the transient stability margin. On the other hand, a torque-slip curve of the induction generator is shown in Fig. 16.13 for operating at different level of wind power generation, in which it is seen that the critical slip decreases as the wind power penetration level increases i.e., the CCT decreases with growing wind power. Hence, a new stability limit or margin needs to be determined at each time the system operating condition has changed, which can be dynamically well achieved by the agents using logic developed in Sect. 16.5.1.

In this section, the angular stability of the synchronous generator and the rotor speed of the induction generator for wind power penetration levels of 5.5 MW is shown only to understand the effectiveness of the proposed approach for increasing wind power penetration. The wind power penetration level is increased from 5.0 to 5.5 MW, where for this penetration level, the agents dynamically adapt to this change using the logic demonstrated in Sect. 16.5.1 and hence obtains the

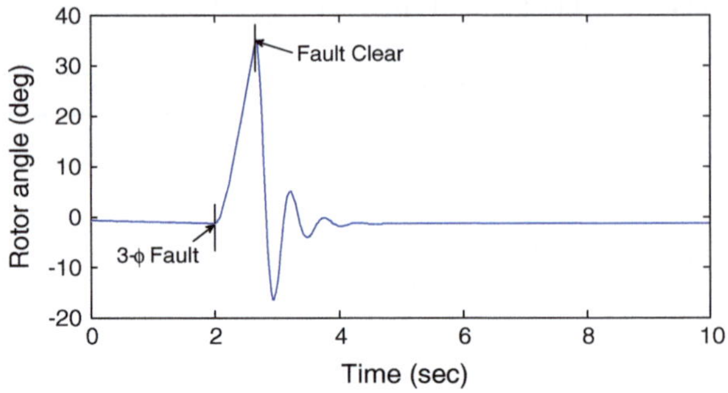


**Fig. 16.14** Synchronous generator rotor angle for 5.5 MW wind power penetration

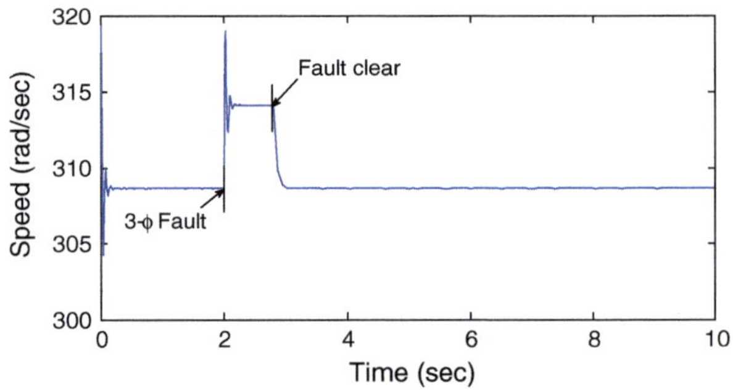


**Fig. 16.15** Induction generator rotor speed for 5.5 MW wind power penetration

corresponding new CCT as 0.554 s. Using this new CCT information they coordinate the protection relays to open the corresponding CBs to remove the fault from the rest of the system by tripping the line TL1 and reclose the breakers again to reconnect it. The rotor angles of the synchronous generator and the rotor speed of induction generator are shown in Figs. 16.14 and 16.15 for a fault at line TL1 for a wind power penetration level of 5.5 MW in which it can be seen that they are transiently stable following the removal of the fault by the tripping the line TL1 by agents before the corresponding CCT.

From the above analyses, it can be interpreted that, with a gradual increasing of the wind power penetration level into the grid, the CCT of the system decreases, i.e., the system's stability margin decreases. However, the agents can successfully adapt to this situation and properly coordinate the corresponding CB operations with the corresponding CCT information in order to improve the transient stability of the system.

## 16.7 Multi-agent Framework for Smart Grid Security

A CPS is a next-generation network system [28] which is often featured with tight interaction of the computing intelligence, networking function and physical processes of the system. A multi-agent based smart grid exhibits the important aspects of the behaviour of the next generation CPS infrastructure. CPS requires a capability of sensing and operation of the physical process locally while relevant observations and control messages are securely transferred in real-time over the network [30]. Recently CPS has been identified as the top research priority by the U.S. President's Council of Advisers on Science and Technology (PCAST) [31]. Systematically classifying various types of cyber attacks is an effective tool which can provide a solid basis for investigating relevant security solutions through different counter measuring techniques for attack mitigation [28]. In this chapter, we focus mainly on the impact of different types of cyber attacks on an important critical infrastructure, an agent-based smart grid. A few innovative countering measure techniques for the attacks have also been suggested in this chapter.

A new emerging technology called smart grid can be considered as an interconnection of a large number of autonomous nodes as agents which has a capability to reduce the energy losses and increase the reliability of the system. In smart grid infrastructure, networking and intelligent information and communication (ICT) functions are embedded into every aspect of the system component ranging from power generation, transmission, and distribution and consumer appliances. A communication framework is superimposed on this physical structure to enable the autonomy and build an interconnected grid topology. The pervasive embedding of ICT functionalities into typical agent-based smart grid architecture is shown in Fig. 16.3, in which such a system will expose it to a wide range of security threats, which will also render the secure operation of the system an extremely challenging task [10]. A smart grid security investigation report from the U.S. Government Accountability Office has cast a concern over the adequacy of the current security measures over smart grid's cyber infrastructure [32]. In shortly the goal of this chapter is to provide an idea of developing a framework for agent-based smart grid cyber security assessment through the following steps: (1) systematically identifying the different types of cyber attacks/threats that emerging on the smart grid; (2) analysing the potentially impact of cyber attack on the smart grid infrastructure which may cause damage to both cyber and physical domain and lastly (3) focusing attack resilient counter measuring techniques to mitigate the risk of cyber threats to ensure the secure operation of smart grid. A typical CPS infrastructure for smart electric power grid is shown in Fig. 16.16.

### 16.7.1 Attack Impacts on Smart Grid Dynamic States

Several types of cyber attack such as, integrity and availability (DoS/DDoS or replay) have a severe impact on the agent-based smart grid dynamic states like angle, frequency and voltage corruption that obtain from a time-stamped value comes from PMU. Integrity attack can alter the present value of the system's dynamic states i.e., the input to the dynamic node agent can be a different value from the true value, which will lead to the unstable device operation resulting a loss of service to the customers. On the other hand, it can also alter the load

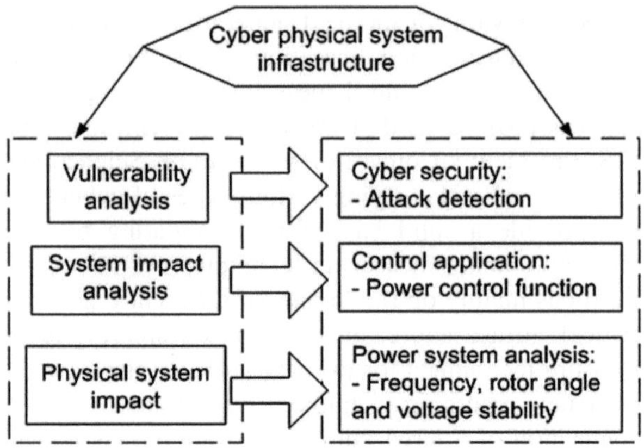


**Fig. 16.16** A typical CPS-based smart power grid infrastructure

threading and can corrupt the information on the supervisory control and data acquisition (SCADA) system, which may lead to a suboptimal system operation as load may not be adjusted as desired and make wrong trigger decision results in damaging physical grid, respectively, which may affect the power market economy as well as customer's usage. Availability attack may provide a time delay for which system dynamic state information will not be available or available after a time delay or delayed dynamic state can be taken as a present state which may make the unstable system operation that will lead to a loss of service as well as financial loss could happen.

## 16.7.2 Attack Impacts on Smart Grid Components

Besides affecting the smart grid dynamic states, cyber attack may also hamper its physical grid components. Being a distributed critical infrastructure, smart electric power grid consists of the following four major parts [10]: (i) power generation; (ii) transmission; (iii) distribution and, (iv) consumption. This section demonstrates about the different types of physical grid controllers which can be treated as agents under generation, transmission and distribution which are using communication infrastructure among them are vulnerable to various types of cyber attacks. The generation control loops are used for controlling the output power and the terminal voltage through both, local i.e., automatic voltage regulator (AVR) and governor control (GC), and remote wide-area i.e., automatic generation control (AGC) control schemes. The transmission systems normally operate at a HV generally used for power flow and their components controller includes switching and reactive power support devices like VAR compensation. The distribution system is normally responsible for delivering power to the end users or customers. With the emerging smart grid infrastructure, additional control loops enable direct control of

**Table 16.2** Impact of cyber attack on smart grid physical components

| Major parts | Physical grid control components | Cyber vulnerabilities |
|---|---|---|
| Generation | AVR, GC and AGC | Impacts and consequences |
| | Using modbus protocol to commutate with the computers in the plant control center via ethernet link | Corrupt the digital control logic or setting to disrupt the normal operation. Directly impacts on the system frequency, stability and economic operation |
| Transmission | State estimation and VAR compensation | Impacts and consequences |
| | Using wide-area communication through phasor measurement units (PMUs) and coordinated FACTS devices can communicate with other devices to determine the operating set point | Loss of critical information exchange results disruption of normal operation and incorrect operational data results unnecessary VAR compensation and can make unstable operating conditions |
| Distribution | Load shedding and AMI | Impacts and consequences |
| | Incorporates modern relays which use IP support communication protocols IEC61850 and deploys smart metering | Change of relay control logic could result unscheduled tripping that leading to load segment unserved. Tempering a meter data results economic loss |

load at the end user level and can deploy advanced metering infrastructure (AMI) to provide a real-time metering [32]. The possible cyber vulnerabilities for the above controllers are described in Table 16.2.

## *16.7.3 Attack Impacts on Smart Grid Appliances*

An agent-based smart grid becomes increasingly dependent on ICT for control and monitoring functions and this pervasive embedding of ICT can introduce a greater exposure to various cyber attacks [28]. It will introduce a number of power system applications which will heavily depend on cyber architectures for communication and control mechanism [33]. A typical agent-based smart grid infrastructure incorporates along with some applications like advanced metering infrastructure (AMI), distribution management systems (DMS), energy management systems (EMS), wide area measurement, protection, and control (WAMPAC) and power markets which are also being vulnerable to various types of cyber threats. Table 16.3 describes various cyber attacks impact on the above smart grid applications. A more detail of the cyber attack impact analysis on smart grid appliances can be found in [33].

**Table 16.3** Impact of cyber attack on smart grid applications

| Smart grid applications | | Cyber vulnerabilities |
|---|---|---|
| AMI | Smart meter provides a two way communication between the customer and utilities for control over user consumption, real-time pricing and also supports the integration of distributed renewable energy resources | Confidentiality attack may results wrong end user billing and can corrupt the privacy data. Integrity attack may affect meter's operation and control along with the communication of both pricing and status information |
| DMS | DMS incorporates applications of forecasting, state estimation, fault management, Volt/VAR control, restoration, consumption and load sharing | Integrity and availability attack can send false data or commands that may affect control and communication resources |
| EMS | EMS provides real-time communications for control and monitoring along with the applications of AGC, FACTS and state estimations | Integrity and availability attack may affect control operations |
| WAMPAC | WAMPAC incorporates PMUs that provides real-time grid state measurements | Availability and integrity attacks may potentially jam the communication network and may results in wrong state measurements |
| Power market | Commodity-based energy markets provide a balance between supply and demand of energy | Confidentiality, integrity and availability attacks may results various parties to place energy bid data |

## *16.7.4 Attack Impacts on Smart Grid Protection Scheme*

The easy availability of synchronized measurements using Global Positioning System (GPS) technology and the improvement in communication technology makes the smart grid protection of transmission lines quite faster and become more decentralized [34]. Since the agent-based protection scheme is largely depend on the communication infrastructure and ICT functionalities, hence this scheme can be vulnerable to various types of cyber attack like aurora attack, integrity attack, availability attack which could cause an unintentional or delayed tripping of the CBs by tampering the status or control signals used by protection device agents. These attacks may happen on the communication link between PMU-based protections relay agents as they use communicated information from the remote locations over a wide band channel.

The aurora attack may involves in an unintentional opening and closing of a CBs, resulting an out-of-synchronism condition that may damage rotating equipment connected to the power grid. That means the intent of an aurora attack is to intentionally open a breaker and close it out of synchronism to cause damage to the connected generators and motors [35]. If an attacker knows the accurate power system information i.e., number of breakers installed, placement of the breakers, any knowledge of the communication link of the protection relays if the communication link is not encrypted, then they can compromise with the communication channel and hence can control the protection relay operation by manipulating the signal or information used by the protection device agents.

Integrity attack may cause altering the information which the agent-based protection relays use to communicate with other relays e.g., tripping signal information; hence the other agent-based relays may have a chance to get the wrong information of tripping which could cause wrong protection device operation that may cause loss of service to the customers and a cascading tripping of the other CBs in a smart grid results a wide-spread blackout on it. On the other hand, DoS/DDoS attack may cause a delayed tripping of the CBs as it adding a time-delay signal to the original signal that send to the other relays. This type of cyber attack which could cause a protection relay to delay in operation that results an unstable condition in a smart grid.

## 16.8 Some Useful Innovative Counter Measuring Techniques for Cyber Attack Mitigation

Conventional cryptography based mechanisms can be an effective tool for mitigating the risk of confidentiality attack, integrity attack, and replay attack [31]. For DoS and DDoS attacks, network traffic analysis is an effective tool and popular detection mechanism [36]. The conventional cryptography based security mechanism generally depends on a password or PIN or token. As a matter of fact, sometimes this security solution cannot identify or address whether the password or PIN or token is provided by the genuine user or not [28]. In this case non-conventional security measurement e.g., a bio-cryptography based mechanism can be used which is more secured from being compromised and can be pre-stored in the server database. In worst case, a password or PIN or token can be re-issued if the existing technique has been compromised [37]. For agent-based protection relay security, these equipments can be protected by using breaker reclosing supervision with a time delay on reclosing, reclosing supervision by a backup protective relay, and rate of change of frequency, or wide-area synchronized phasor measurements [35]. An innovative physical process oriented countering measure technique, an abnormal state observer, can be a potential solution for a cyber attack on the system's dynamic states to mitigate its risk.

Other than the above methods some possible counter measuring techniques to mitigate the risk of cyber attacks are risk modelling methodology to assess the cyber vulnerabilities, risk mitigation algorithm to provide attack resilient control, intelligent power system control and domain-specific anomaly detection, and intrusion detection, coordinated attack defence to provide cyber risk modelling and mitigation, and planning and reliability studies, AMI security for remote attestation; trust management for dynamic trust distribution, cyber attack attribution, data sets and validation, etc., [32].

## 16.9 Conclusion

In this chapter, an intelligent multi-agent framework has been developed which is vulnerable to both physical and cyber attack. For smart grid protection, MAS has provide a power framework, in which agents dynamically adapt the changes in the system condition i.e., dynamically adapt increasing wind power penetration level by analysing and understanding a disturbance and hence choose an appropriate control action for coordinating the system protection devices by controlling the CBs by opening and reclosing them in order to enhance the transient stability of the smart grid. The developed MAS has meet the smart grid protection requirements and the preliminary simulation results give us a hope that implementation of the MAS with the grid protection relays may be a potential solution for smart grid protection which helps to improve the power system's transient stability.

On the other hand, as agent-based smart grid is a major CPS system whose security is a big concern, MAS established a structured framework where the security issues are effectively investigated. A few possible innovative countering measure techniques are also described in this paper which may help to mitigate the risk of various types of cyber threats mentioned in this chapter. In fact, a more detail investigations are required for designing a good taxonomy of cyber attack which may help to provide useful guidelines to mitigate the risk of cyber vulnerabilities.

## References

1. Rahman MS, Hossain MJ, Pota HR (2012) Agent based power system transient stability enhancement. In: IEEE international conference on power system technology (POWERCON)
2. Russell S, Norvig P (1995) Artificial intelligence: a modern approach. Prentice-Hill, Englewood Cliffs
3. Mohamed M, Karady GG, Yousuf AM et al (2007) New strategy agents to improve power system transient stability. World Acad Sci Eng Technol 3:678–683
4. Karady GG, Daoud A, Mohamed M (2002) On-line transient stability enhancement using multi-agent technique. In: IEEE PES winter meeting, pp 893–899
5. Karady GG, Mohamed M (2002) Improving transient stability using fast valving based on tracking rotor-angle and active power. In: IEEE PES summer meeting, pp 1576–1581
6. Hadidi R, Jeyasurya B (2011) A real-time multi-agent wide-area stabilizing control framework for power system transient stability enhancement. In: IEEE PES general meeting, pp 1–8
7. Abood AA, Abdalla AN, Avakian SK et al (2008) The application of multi-agent technology on transient stability assessment of Iraqi super grid network. Am J Appl Sci 5(11):1494–1498
8. Dou C, Mao C, Bo Z, Zhang X (2010) A multi-agent model based decentralized coordinated control for large power system transient stability improvement. In: 45th international universities power engineering conference (UPEC)
9. Giovanni R, Hopkinson K, Denis V, Thorp JS et al (2006) A primary and backup cooperative protection system based on wide area agents. IEEE Trans Power Delivery 21(3):1222–1230
10. Mo Y, Kim TH, Brancik K, Dickinson D, Lee H, Perrig A, Sinopoli B et al (2012) Cyber-physical security of a smart grid infrastructure. Proc IEEE 100(1):195–209
11. Panda S, Padhay NP, Patel RN et al (2007) Genetically optimized tcsc controller for transient stability improvement. Int J Inf Eng 1:19–25

12. Kerckhoffs A et al (1883) La cryptographie militairie. J Sci Militaires 9:5–38
13. Ahshan R, Iqbal MT, Mann George KI, Quaicoe JE et al (2013) Modelling and analysis of a micro-grid system powered by renewable energy sources. Open Renew Energy J 6:7–22
14. Kundur P (1994) Power system stability and control. McGraw-Hill, USA
15. Mahmud MA, Hossain MJ, Pota HR (2011) Nonlinear observer design for interconnected power systems. In: Australian control conference (AUCC), pp 161–166
16. Hossain MJ, Pota HR, Mahmud MA (2011) Decentralized STATCOM/ESS control for wind generators. In: Keyhani A, Marwali M (eds) Smart power grids. Springer, Berlin Heidelberg, pp 401–437
17. Ackermann T (2005) Wind power in power systems. Wiley, England
18. Multi-agent system. http://en.wikipedia.org/wiki/Multi-agent_system
19. Wooldridge M, Weiss G (1999) Intelligent agents. In: Wooldridge M, Weiss G (eds) Multi-agent systems. MIT press, Cambridge, pp 3–51
20. McArthur SDJ, Davidson EM, Catterson MV, Dimeas AL, Hatziargyriou ND, Ponci F, Funabashi T et al (2007) Multi-agent systems for power engineering applications-part I: concepts, approached, and technical challenges. IEEE Trans Power Syst 22(4):1743–1752
21. Panasetsky DA, Voropai NI (2009) A multi-agent approach to coordination of different emergency control devices against voltage collapse. In: IEEE power tech conference
22. McArthur SDJ, Davidson EM, Hossack JA, McDonald JR (2004) Automating power system fault diagnosis through multi-agent system technology. In: 37th Hawaii international conference on system science
23. Wooldridge M (2002) An introduction to multiagent systems. Wiley, New York
24. Panait L, Luke S et al (2005) Cooperative multi-agent learning: the state of the art. Auton Agent Multi-Agent Syst 11(3):387–434
25. Foner LN (1997) Entertaining agents: a sociological case study. In: 1st international conference on autonomous agents
26. Costea C, Horgos M (2008) An agent-based approach to power system control. In: 11th international conference on optimization of electrical and electronic equipment, pp 179–184
27. Zhang Z, McCalley JD, Vishwanathan V, Honavar V (2004) Multi-agent system solutions for distributed computing communications and data integration needs in power industry. In: IEEE PES general meeting, pp 45–49
28. Hu J, Pota HR, Gu S (2013) Taxonomy of attacks for agent-based smart grids. IEEE Trans Parallel Distrib Syst. doi:10.1109/TPDS.2013.301 [Available online: 5 December 2013]
29. Cardenas A, Amin S, Sastry S (2008) Secure control: towards survival cyber-physical systems. In: 28th international conference on distributed computing systems workshops (ICDCS), pp 495–500
30. President's Council of Advisers on Science and Technology (2007) Leadership under challenge: information technology R&D in a competitive world. An assessment of the federal networking and information technology R&D program
31. Sridhar S, Hahn A, Govindarasu M et al (2012) Cyber-physical system security for the electric power grid. Proc IEEE 100(1):210–224
32. Govindarasu M, Hahn A, Sauer P (2012) Cyber-physical systems security for smart grid. The future grid to enable sustainable energy systems. PSERC publication
33. Phadke AG, Thorp JS (2008) Synchronized phasor measurements and their applications. Power electronics and power systems. Springer, Virginia, USA
34. Salmon D, Zeller M, Guzman A, Mynam V, Donolo M (2009) Mitigating the aurora vulnerability with existing technology. In: 36th annual western protection relay conference
35. Simpson MG (2010) Plant systematics: an overview. Plant systematics, 2nd edn. Academic press
36. Hansman S, Hunt R et al (2005) A taxonomy of network and computer attacks. Comput Secur 24(1):31–43
37. Akhmatov V, Knudsen H, Bruntt M, Nielsen A, Pedersen JK, Poulsen NK (2000) A dynamic stability limit of grid-connected induction generator. In: International conference on power and energy systems, pp 235–244

# Chapter 17
# Vulnerabilities of Smart Grid State Estimation Against False Data Injection Attack

**Adnan Anwar and Abdun Naser Mahmood**

**Abstract** In recent years, Information Security has become a notable issue in the energy sector. After the invention of 'The Stuxnet worm' [1] in 2010, data integrity, privacy and confidentiality has received significant importance in the real-time operation of the control centres. New methods and frameworks are being developed to protect the National Critical Infrastructures like- energy sector. In the recent literatures, it has been shown that the key real-time operational tools (e.g., State Estimator) of any Energy Management System (EMS) are vulnerable to Cyber Attacks. In this chapter, one such cyber attack named 'False Data Injection Attack' is discussed. A literature review with a case study is considered to explain the characteristics and significance of such data integrity attacks.

**Keywords** State estimation · False data injection attack · Smart grid · Cyber security · Data integrity attack

## 17.1 Introduction

Power system State Estimation has been widely used at the utility control centres to know the system status during the power system operation. In order to ensure the stability and reliability of the power system, network operator monitors and controls the system states which are obtained from the state estimation processor. Generally, State Estimator provides an estimation of the data for all measured and

A. Anwar (✉) · A. N. Mahmood
School of Engineering and Information Technology, The University of New South Wales, Canberra 2600, Australia
e-mail: adnan.anwar@adfa.edu.au

A. N. Mahmood
e-mail: a.mahmood@adfa.edu.au

J. Hossain and A. Mahmud (eds.), *Renewable Energy Integration*, Green Energy and Technology, DOI: 10.1007/978-981-4585-27-9_17,

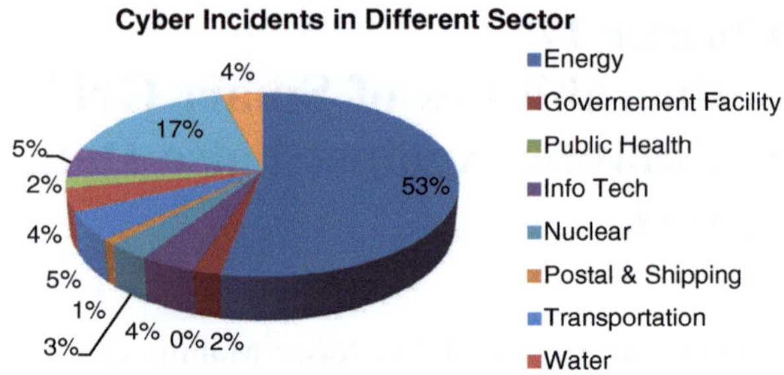


**Fig. 17.1** Cyber incidents in different sector [2]

unmeasured quantities. This advanced tool also filters out the measurement errors and noises and suppresses bad data. With the development of the power system research and engineering, the modern State Estimation programs have advanced capabilities which have enhanced the computational performances, as well as, accuracy. However, the challenges of accurate and efficient State Estimation programs have increased more because of the recent cyber attacks in the energy system infrastructure.

In a recent report of the 'Industrial Control Systems Cyber Emergency Response Team (ICS-CERT)', it has been mentioned that 198 cyber incidents happened in the financial year 2012 among which 41 % happened within 'Energy sector'. In the first half of the financial year 2013, 200 incidents happened across all sectors of the critical infrastructure among which the highest attack (111 in incidents) happened in the energy sector (53 %) as shown in Fig. 17.1 [2].

From the last few years, distributed energy resources and storage devices are widely used which have changed the power flow patterns of the grid [3]. These renewable sources have intermittent nature and most of the time they are not dispatchable. Therefore, demand response has been a crucial issue in a smart grid environment. To face the challenges, Advanced Metering Infrastructures (AMI) which is equipped with smart meters, may play a significant role. It is obvious that the use of smart meters and advanced communication network has helped the utility operators to implement the SCADA controls more easily; however, the communication system of the cyber-physical smart grid has been more vulnerable in terms of cyber attack which may affect the communication network. These types of cyber related crimes may have devastating impact on the physical power grid including operational failures and loss of synchronization of different critical equipments of a power grid. Moreover, a large scale blackout may occur due to a cyber attack in a smart grid.

In a smart grid environment, energy system control sector needs advanced communication means among different parts of the network which increases the use of commercial off-the-shelf technologies. As a result, the cyber security issues arise. State Estimator, which is one key operational tool in the Energy Management System, is also very vulnerable to cyber attacks. Due to any pre-planned cyber attack in any State Estimation programs, bad data detectors may not be able to identify the possible threats which are attacked by any intruder. As a result,

State Estimation programs will provide wrong information to the system operator. Based on the wrong estimation, operator may take misleading operational decision resulting a vital problem in the stable operation of power system. Therefore, advanced intrusion detection algorithms are desirable.

The organization of this chapter is as follows: A brief overview of traditional State Estimation is discussed in Sect. 17.2. The problem formulation with solution methodologies including Bad Data Detection techniques are also discussed in that section. The overview of Sate Estimation in Distribution System and Smart Grid is discussed in Sects. 17.3 and 17.4 respectively. A case study is illustrated to describe the vulnerabilities of the Smart Grid State Estimation in Sect. 17.5. A recent review on False Data Injection attack on State Estimator is discussed in Sect. 17.6. Finally, the chapter is concluded with brief remarks. This chapter intends to be a comprehensive reference in the field of cyber security of smart grid infrastructure.

## 17.2 Power System State Estimation

State Estimation is one of the most traditional power system analysis tools for reliable monitoring and control of Energy Management System. One early revolutionary work based on power system static state estimation was proposed in [4] and till then a significant number of research works have been conducted on this imperative issue. Although the traditional State Estimation has a long history in power transmission level, this powerful network analysis tool needs more attention in the low-voltage power distribution level [5], especially, when the grid adopts more communication infrastructures (i.e., Advanced Metering Infrastructure, Phasor Measurement Units (PMUs), etc.) and Distributed Energy Resources (DERs). As mentioned earlier, the evolution of the Smart Grid State Estimation and the vulnerabilities of this estimation tool in terms of Data Integrity Attack will be discussed in the following section; this section will provide a brief overview of the significance of the traditional Static State Estimation and different methodologies and techniques involved with it. In this section, the importance of Bad Data identification and well-established procedures of detecting bad data are also discussed.

Basically, State Estimation is a procedure which is used to determine the most approximate solution of the system states by analyzing the measured sensor values and the equivalent calculated values. In power system theories and applications, the term 'state estimator' implies a computer program for calculating the system states based on the measured data at different nodes of the network and the laws of electric power networks which explains the behavior of the physical network model. The purpose of the state estimation is to estimate the unmeasured variables, improve overall efficiency and to detect the bad measurement.

Generally, the states in a power system are the complex voltage magnitude and the angles of each bus. If the state vector is **x**, then the state vector for an 'n' bus system will be:

$$x = [\delta_2 \delta_3 \ldots\ldots \delta_n V_1 V_2 V_3 \ldots\ldots V_n]^T \quad (17.1)$$

where, $\delta_i$ indicates the phase angles and $V_i$ the voltage magnitude at the *i*-th bus. It is interesting to note that, the dimension of the state vector is $(2n - 1) \times 1$ as the phase angle at the reference bus is considered to be known which is generally assumed 0 rad. Although, the bus voltage magnitudes and angles are used in practice, current magnitudes-angles and power flows are also considered as state variables in some cases. At the first stage of the state estimation, measurement data are obtained from the Remote Terminal Units (RTUs) which are equipped with sensors. These measurements include voltage magnitudes, bus injections and both real and reactive power flows through different components of the network. However, the measurement data may be noisy and corrupted which increase the risk of direct use of these data. If the system states are known, then it is expected that from the laws of electric network (say, Kirchhoff's Current Law or Kirchhoff's Voltage Law) it is possible to calculate the power flow pattern of the network. However, it is not possible to directly measure the system states which motivate to develop and improve the methodologies related to State Estimations. As a result, at the second stage of state estimation, functions of state variables are used to calculate the expected values of the measurement data. Finally, any established method is employed to calculate the state variables from the measurement values and the calculated values. One such widely adopted method is Weighted Least Square (WLS) Method. After estimating the system states, the Bad Data Detection program is perform to identify the corrupted data. All of the steps of the state estimation are described briefly in following sub sections.

### *17.2.1 System Model of Measurement Data*

Consider a measurement vector z for an n-bus system, where, $z \in R^{M\times 1}, M > (2n - 1) \times 1$. Therefore, z should be:

$$z = \begin{bmatrix} z_1 \\ z_2 \\ \vdots \\ z_m \end{bmatrix}$$

It is assumed that the measurement vector should contain some error with the exact measurement function value. Therefore, **z** can be written as:

$$z = \begin{bmatrix} z_1 \\ z_2 \\ \vdots \\ z_m \end{bmatrix} = \begin{bmatrix} h_{1(x_1,x_2,..,x_3)} \\ h_{2(x_1,x_2,..,x_3)} \\ \vdots \\ h_{m(x_1,x_2,..,x_3)} \end{bmatrix} + \begin{bmatrix} e_1 \\ e_2 \\ \vdots \\ e_m \end{bmatrix} = h(x) + e \quad (17.2)$$

where, $\boldsymbol{h}(x) = \begin{bmatrix} h_{1(x_1,x_2,..,x_3)} \\ h_{2(x_1,x_2,..,x_3)} \\ \vdots \\ h_{m(x_1,x_2,..,x_3)} \end{bmatrix}$, $\boldsymbol{e} = \begin{bmatrix} e_1 \\ e_2 \\ \vdots \\ e_m \end{bmatrix}$, and, $\boldsymbol{x} = \begin{bmatrix} x_1 \\ x_2 \\ \vdots \\ x_m \end{bmatrix}$

Here, $\boldsymbol{h}(x)$ is the calculated function values for the state variables. $\boldsymbol{x}$ is the *vector* of state variables and $\boldsymbol{e}$ is the vector of measurement errors.

Generally, $\boldsymbol{e}$ is a zero-mean Gaussian noise vector where measurement errors are independent. Therefore, $\mathbf{E}(\mathbf{e_i}) = 0$, where $\mathrm{i} = 1, 2, \ldots, \mathrm{m}$. And $\mathbf{E}(\mathbf{e_i}\boldsymbol{e}_j) = 0$ and $Cov(\boldsymbol{e}) = \mathrm{E}[\mathrm{ee}^{\mathrm{T}}] = \mathrm{R} = \mathrm{diag}\ (\sigma_1^2, \sigma_2^2, \ldots, \sigma_{\mathrm{m}}^2)$.

### *17.2.2 Calculation of Measurement Function*

$h(\mathrm{x})$ is the vector of calculated functions. Generally, $\boldsymbol{h}(.)$ is a set of nonlinear functions of the state variables for AC approximations of the load flow equations whereas it would be a set of linear functions if the load flow equations are formulated considering DC approximations.

For a $\pi$-model of any network, the measurement function value can be calculated as follows [6]:

(a) Real and reactive power injection at bus i:

$$P_i = v_i \sum_{j \in n_i} v_j \left( G_{ij} \cos \delta_{ij} + B_{ij} \sin \delta_{ij} \right) \quad (17.3)$$

$$Q_i = v_i \sum_{j \in n_i} v_j (G_{ij} \sin \delta_{ij} - B_{ij} \cos \delta_{ij}) \quad (17.4)$$

(b) Real and reactive power flow from bus i to bus j are:

$$P_{ij} = v_i^2 \left( g_{si} + g_{ij} \right) - v_i v_j \left( g_{ij} \cos \delta_{ij} + b_{ij} \sin \delta_{ij} \right) \quad (17.5)$$

$$Q_{ij} = -v_i^2 \left( b_{si} + b_{ij} \right) - v_i v_j (g_{ij} \cos \delta_{ij} - b_{ij} \sin \delta_{ij}) \quad (17.6)$$

(c) Line current flow magnitude

$$I_{ij} = \sqrt{\frac{P_{ij}^2 + Q_{ij}^2}{V_i}} \tag{17.7}$$

where, the symbols have their usual meaning.

To calculate the $\boldsymbol{h}(x)$ values, any other functions can be used based on the formulation of the network model. For example, a multi-phase power flow model is proposed for state estimation in [7].

### 17.2.3 State Estimation: Formulation and Methodologies

As discussed, state estimation depends on the following equation,

$$z = \boldsymbol{h}(x) + \boldsymbol{e} \tag{17.8}$$

Therefore, state estimation can be formulated as an error minimization problem which is in fact a convex optimization problem described below:

$$\mathrm{x}' = \arg\min \sum_{i=1}^{m} W(z_i - \boldsymbol{h}_i(\boldsymbol{x}))^2 \tag{17.9}$$

where, W is the weighting matrix which can represents $W = R^{-1}$. To solve the value$\mathrm{x}'$, an iterative approach may be adopted. Some popular techniques are Gauss–Newton method and Newton–Raphson method [8]. Evolutionary algorithm, i.e., swarm intelligence based approaches (Particle Swarm Optimization) have also been used to solve this critical operational problem [9].

### 17.2.4 Bad Data Detection

Generally, it is assumed that the measured data will contain some errors. However, sometimes, measured data is so faulty that it affects the state estimation and inconsistent result occurs. As a result, Bad Data Detection becomes very important to obtain a successful state estimation. Different methodologies are used to detect and identify bad data. Such a widely adopted procedure is 'Largest Normalized Residual (LNR)' method [10]. Once the system states (x′) are estimated, then the residual is calculated as following

$$\boldsymbol{r} = z - \boldsymbol{h}(x') \tag{17.10}$$

At least one bad data exists if the value of the residual is less than a predefined threshold, which can be written as follows:

Bad data exists if $||r|| < \tau$

Some other techniques which are also used in literature are 'The $\boldsymbol{J}(x')$ Performance Index', 'Hypotheses Testing', 'Dormant and Perfect Measurement', 'Identification test', etc. [8].

## 17.3 State Estimation for Distribution Networks

State Estimation of transmission system is a well-established area for real-time monitoring and control of a complex power network. However, traditional techniques and methodologies for transmission style State Estimation do not fit for a low voltage power distribution network. Generally, balanced approximation of the power system is considered in most of the traditional State Estimation techniques, e.g., [6, 8]. Although this assumption of positive sequence network modeling is valid for high voltage transmission network, but does not work well for low voltage distribution system [11]. In reality, power lines are transposed and loads are not balanced in a distribution network. Moreover, there are three, two, and single-phase lines and transformers are both delta and wye connected. As a result, rather than a positive sequence modeling of the network, it is essential to have full multi-phase modeling for accurate simulation of distribution network as mentioned in [12]. Considering a-b-c phase modeling, some early researches on Distribution State Estimation are proposed in [11, 13, 14], where these issues are clearly pointed out. Other than this multi-phase property and untransposed phase conductors, distribution network exhibits some other characteristics as below [15, 16]:

(a) Feeders are mostly radial in nature
(b) Distributed loads with a small geographical area
(c) High R/X ratio
(d) Presence of Distributed Generation and no conventional generation
(e) Very low redundancy of measurement units.

Due to the distinct features of radial low voltage distribution feeders, Distribution State Estimation is different from the traditional one. Moreover, analysis procedure of this real-time operational tool is very challenging because of the following properties [17]:

(a) Limitation of measurement devices
(b) The pseudo-measurement of load data is obtained from the historical load data which may have very limited accuracy
(c) Significant number of current measurement devices are used.

These challenges are increased a lot in a smart grid environment which will be discussed in the next section.

## 17.4 Smart Grid State Estimation

Smart Grid State Estimation needs to face the new requirements and challenges of the future renewable energy based sustainable self-healing intelligent smarter grid. Different new aspects will have significant impact on the Smart Grid State Estimation. Three major aspects have been identified in [5] which will be discussed here briefly:

(a) Development of Advanced Measurement Technologies: Generally, measurement data of a power system is obtained through the SCADA network. Traditionally, RTUs are used for this purpose. RTU is a microprocessor-controlled electronic device that is responsible to measure network traffic through sensors and to transmit the telemetry data to the Distribution Management System for further processing. These measurements are non-synchronized and obtained too infrequently to understand the system operational characteristics. Especially capturing system dynamics is too difficult [5]. In recent years, PMUs have been adopted widely for better real-time monitoring and control of smart grid. PMUs have several advantages over traditional measurement devices, such as:

- It captures data more frequently, e.g., $20 \sim 60$ times per second [17].
- Measurement data are synchronized as they are sampled according to the Global Positioning System (GPS).
- Current measurement is also possible to those nodes where PMU is placed.

(b) New regulatory and pricing issues: In a Smart Grid concept, new regulatory issues are arising. In a consumer-centric electricity market, end users are capable to produce electricity and sell those to the Distribution Network Operator (DNO). There arises the need of dynamic pricing and new regulatory issues. As a result, DNO must have a clear knowledge about the whole distribution network, especially regarding the power flows through all the phases of the utility distribution grid. To obtain an accurate power flows, the role of Smart Grid State Estimation is vital.

(c) Demand response and Distributed Energy Resources: In order to fulfil the ever growing load demand, co-generation, distributed generation and storages are being employed in the grid. These devices are making the grid active from its traditional passive manner which will introduce a bi-directional power flow [18]. In order to understand the flow pattern, distribution network needs advanced modelling and analysis capabilities which motives to develop advanced State Estimation tools for Smart Grid.

Realizing the needs for developing accurate and fast State Estimation algorithms, a significant number of research works are going on throughout the world [19–24]. A multi-level State Estimation framework for Smart Grid is proposed in [19] where authors propose a new paradigm based on multi-level communication and computation architecture. At the lowest level, a local State Estimation (LSE) is proposed to deal with the distribution substation and its downstream radial feeders. Computed state variable values are then transmitted through the aid of communication infrastructure to its upper level which is Transmission System Operator (TSO) level. Rather than calculating only from 'raw' data, this time TSO-level SE will get a chance to update, smooth and modify the data from LSE by comparing with the raw measurement data. At the final stage, Regional State Estimation (RSE) will synchronize and purify the data obtained from TSO level SE. Numerical simulation is also carried out to explain and evaluate the working procedure of this multi-level schema in [19].

Signal Processing based approaches have also been used to solve this critical real-time operating problem. One such method is Belief Propagation based method to solve Distribution State Estimation [20]. One major challenge of Distribution State Estimation is limited measurement devices. The method proposed in [20] solves the problem of sparse measurement by addressing Belief Propagation based method for real-time Distribution State Estimation. One more advantage of this method is that it can deal with the renewable energy based distributed power generation sources. The performance of the proposed method is compared with the Electric Power Research Institute's distribution system analysis tool openDSS.

In [21], authors propose a method for Distribution State Estimation with the deployment of PMUs and Smart Meters. Authors suggest that rather than mixing the traditional meter measurements and the measurements obtained from Advanced Metering Infrastructure (AMI) and PMU, a two-step approach can be adopted. The advantage is that the current configurations of EMS software need not to be changed. According to that, a traditional State Estimation is performed at the first step and finally PMU measurements are considered to update and modify the pre-processing data.

Considering renewable energy resources, a method for State Estimation based on evolutionary algorithm is proposed in [25]. The proposed method can consider different practical issues including unbalanced power flows, VAR compensators, Voltage Regulators (VRs), tap changing transformers, etc.

## 17.5 Vulnerabilities of Smart Grid Sate Estimation: A Case study

Recent literature shows that a significant improvement of Smart Grid State Estimation is noticeable in terms of 'accuracy' and 'efficiency'. However, this operational tool is very prone to cyber vulnerabilities as discussed in [26]. As the State Estimation is highly dependent on the measurement data, any intruder can inject 'False Data' in such a way that the system is unable to detect it. Figure 17.2 shows such kind of scenario when State Estimation is under attack. This type of malicious modification of measurement data is known as 'False Data Injection Attack' [26]

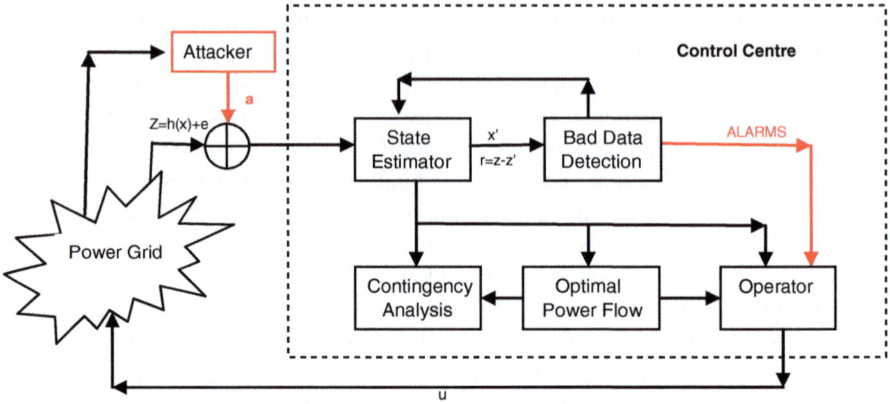


**Fig. 17.2** State estimation under attack [36]

or 'Data Integrity Attack' [27]. In the following sub-section, a step-by-step procedure of a false data injection attack is explained with example.

In terms of computational complexity, simplified DC approximation of a power network has more advantages over AC model of a power system. Instead of solving 'N' nonlinear equations, one need to solve a set of linear equations in DC approximation where the bus voltage is considered to be known and equivalent to 1 pu. As DC approximation does not need any iterative method, it is faster and reduces the computational burden in the State Estimation process.

For a DC State Estimation, the problem can be defined as:

$$z = \boldsymbol{H}x + \boldsymbol{e} \tag{17.11}$$

where, z is the vector of measurement data and $z \in R^N$, H is the Jacobian matrix and $\boldsymbol{e}$ is the error term. When the meter error follows normal distribution with zero mean, the solution becomes as follows [26]:

$$x' = (H^T WH)^{-1} H^T Wz \tag{17.12}$$

where, W is a diagonal matrix as follows:

$W = \begin{bmatrix} \sigma_1^{-2} & \cdots & . \\ \vdots & \ddots & \vdots \\ . & \cdots & \sigma_m^{-2} \end{bmatrix}$ and $\sigma_i^{-2}$ is the variance of *i*-th meter.

In order to avoid the bad measurement, the measurement residual $z - \boldsymbol{H}x$ should be below than the threshold value $\tau$. Generally, it can be said that there is at least one bad data if $||z - \boldsymbol{H}x|| > \tau$, otherwise, Bad Data does not exist. However, this assumption is not valid all time. Here, an example is shown to describe how to introduce False Data Injection Attack into the State Estimation. The theoretical concept is adopted from [26].

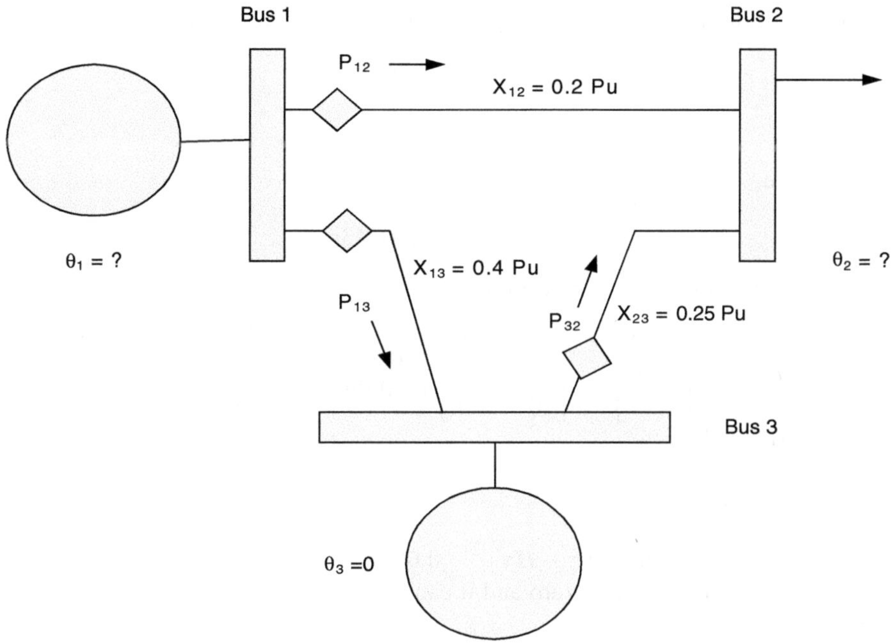


**Fig. 17.3** Three-bus test system

In this case, a three bus test system is considered as shown in Fig. 17.3, where three measurement devices are connected to measure the power through the lines 1–2, 1–3 and 3–2. The measurement powers are $P_{12} = 0.62$ pu, $P_{13} = 0.06$ pu, $P_{32} = 0.37$ pu and $\sigma = 0.01$. Here, $\theta_1$ and $\theta_2$ are the state variables and $\theta_3 = 0$ which is the reference angle. As, the problem solves DC state estimation, the voltages are considered 1 pu.

Following the DC power flow equations:

$$h_1(x) = P_{12} = \frac{(\theta_1 - \theta_2)}{X_{12}} = \frac{(\theta_1 - \theta_2)}{0.2} = 5(\theta_1 - \theta_2) \tag{17.13}$$

$$h_2(x) = P_{13} = \frac{(\theta_1 - \theta_3)}{X_{13}} = \frac{(\theta_1 - \theta_3)}{0.4} = 2.5\theta_1 \tag{17.14}$$

$$h_3(x) = P_{32} = \frac{(\theta_3 - \theta_2)}{X_{32}} = \frac{(\theta_3 - \theta_2)}{0.25} = -4\theta_2 \tag{17.15}$$

So, the H matrix becomes,

$$H = \begin{bmatrix} 5 & -5 \\ 2.5 & 0 \\ 0 & -4 \end{bmatrix}$$

Following the Eq. (17.12), the values of the state variables become,

$$\theta_1 = 0.0286$$

$$\theta_2 = -0.0943$$

Therefore,

$$\theta = [0.0286 - 0.0943]^T$$

Now, the residual matrix $\boldsymbol{r}$ becomes,

$$r = (z - Hx') = \begin{bmatrix} 0.62 \\ 0.06 \\ 0.37 \end{bmatrix} - \begin{bmatrix} 5 & -5 \\ 2.5 & 0 \\ 0 & -4 \end{bmatrix} \begin{bmatrix} 0.0286 \\ -0.0943 \end{bmatrix} = \begin{bmatrix} 0.62 \\ 0.06 \\ 0.37 \end{bmatrix} - \begin{bmatrix} 0.614 \\ 0.0714 \\ 0.3771 \end{bmatrix}$$
$$= \begin{bmatrix} 0.0057 \\ -0.0114 \\ -0.0071 \end{bmatrix}$$

So, the squared error is, $||z - \boldsymbol{H}x'||^2 = 0.00021429$.

This value is very close to zero and it can be said that a good assumption of the state variables are made.

Now, an attack scenario is introduced. It has been assumed that the measurement data is corrupted by the malicious modification of the measured data and therefore $z$ becomes $z_a$ where $z_a = z + a$ and $a$ is the attack vector. Here, $a = (a_1, \ldots, a_m)^T$ and m is the rank of $z$. It is expected that due to the change of the measured vector the values of the state variables will be altered. Considering that effect, the new state variables will be $x'_{false}$, where $x'_{false} = x' + c$. Here, c is a vector of non-zero values with a length n. According to [26], the residual of the base case ($||z - \boldsymbol{H}x'||$) and modified case ($\left|\left|z_a - \boldsymbol{H}x'_{false}\right|\right|$) would be same if $a = Hc$ that means $a$ is a linear combination of column vectors of H. At this stage four scenarios are considered as described below:

***Scenario 1***: This is the base case where no malicious modification of the measured data is made and therefore, it is assumed that the residual lies within the threshold and there is no probability of a False alarm. In the base case, the values of the measured data are $P_{Base\text{-}12} = 0.62$ pu, $P_{Base\text{-}13} = 0.06$ pu, $P_{Base\text{-}32} = 0.37$ pu.

***Scenario 2***: In this scenario, measured data are modified arbitrarily. Say, the corrupted measurements are $P_{False1\text{-}12} = 0.63$ pu, $P_{False1\text{-}13} = 0.05$ pu, $P_{False1\text{-}32} = 0.35$ pu.

***Scenario 3***: In this scenario, measured data are modified following the Attack Definition proposed in [26]. Considering $c$ as a vector of non-zero arbitrary chosen values of length n:

$$c = (c_1, \ldots, c_n)^T = [0.0050.001]^T$$

Then, the attack vector $a$ would be

$$a = Hc = \begin{bmatrix} 5 & -5 \\ 2.5 & 0 \\ 0 & -4 \end{bmatrix} * \begin{bmatrix} 0.005 \\ 0.001 \end{bmatrix} = \begin{bmatrix} 0.02 \\ 0.0125 \\ -0.004 \end{bmatrix}$$

So, corrupted measurement would be $z_a = z + a$, therefore,
$P_{\text{False2-12}} = 0.6400$ pu, $P_{\text{False2-13}} = 0.0725$ pu, $P_{\text{False2-32}} = 0.3660$ pu.

***Scenario 4:*** The attack formulation in this scenario is the same as it is discussed in the previous scenario, however, the value of c is different which is $c = (c_1, \ldots, c_n)^T = [0.010.04]^T)$

Therefore,

$$a = Hc = \begin{bmatrix} 5 & -5 \\ 2.5 & 0 \\ 0 & -4 \end{bmatrix} * \begin{bmatrix} 0.01 \\ 0.04 \end{bmatrix} = \begin{bmatrix} -0.15 \\ 0.0250 \\ -0.16 \end{bmatrix}$$

So, corrupted measurement would be $z_a = z + a$, therefore,
$P_{\text{False2-12}} = 0.4700$ pu, $P_{\text{False2-13}} = 0.0850$ pu, $P_{\text{False2-32}} = 0.2100$ pu.

Now, the results obtained from the previous scenarios are discussed. The base case is already discussed in the previous section. According to that, the state variables obtained in the base case are $\theta = [0.0286 - 0.0943]^T$. Using these values, the squared error becomes, $||z - \boldsymbol{H}x'||^2 = 0.00021429$. Considering this error value as normal operating limit (that means, it is less than the threshold $\tau$), we will evaluate how other corrupted measurement data in scenario 2, 3 and 4 perform during the bad data detection in the State Estimation process.

Now, the second scenario is considered where the vector of corrupted measurement data is

$$z_a = \begin{bmatrix} 0.63 \\ 0.05 \\ 0.35 \end{bmatrix}$$

Considering that, the value of $\theta$ becomes,

$$\theta_{Scenario2} = [0.0313 - 0.0919]^T$$

So, the residual matrix $r$ becomes,

$$r_{Scenario2} = (z - Hx') = \begin{bmatrix} 0.63 \\ 0.05 \\ 0.35 \end{bmatrix} - \begin{bmatrix} 5 & -5 \\ 2.5 & 0 \\ 0 & -4 \end{bmatrix} \begin{bmatrix} 0.0313 \\ -0.0919 \end{bmatrix} = \begin{bmatrix} 0.0141 \\ -0.0282 \\ -0.0176 \end{bmatrix}$$

Using these values, the squared error becomes, $||z - \boldsymbol{H}x'||^2 = 0.0013$ which is greater than the squared error in the base case. As a result, the False Data Injection attack may not be overlooked and be detected in the Bad Data Detection test. So, the intruder may not be successful to plan an attack.

At this stage of the discussion, Scenario 3 and Scenario 4 are considered where measurement data are also corrupted like the Scenario 2 but the attack vectors are created following the method described in [26]. Using those corrupted data State Estimation is performed and the obtained state variables are as follows:

$$\theta_{Scenario3} = [0.0336 - 0.0933]^T$$

and

$$\theta_{Scenario4} = [0.0386 - 0.0543]^T$$

So, the residuals are

$$r_{Scenario3} = (z - Hx') = \begin{bmatrix} 0.6400 \\ 0.0725 \\ 0.3660 \end{bmatrix} - \begin{bmatrix} 5 & -5 \\ 2.5 & 0 \\ 0 & -4 \end{bmatrix} \begin{bmatrix} 0.0336 \\ -0.0933 \end{bmatrix} = \begin{bmatrix} 0.0057 \\ -0.0114 \\ -0.0071 \end{bmatrix}$$

$$r_{Scenario4} = (z - Hx') = \begin{bmatrix} 0.4700 \\ 0.0850 \\ 0.2100 \end{bmatrix} - \begin{bmatrix} 5 & -5 \\ 2.5 & 0 \\ 0 & -4 \end{bmatrix} \begin{bmatrix} 0.0386 \\ -0.0543 \end{bmatrix} = \begin{bmatrix} 0.0057 \\ -0.0114 \\ -0.0071 \end{bmatrix}.$$

So, the squared error for both the case is $||z - \boldsymbol{H}x'||^2 = 0.00021429$.

In scenario 3 and scenario 4, the three measurement devices show different measurement data as they are attacked by the intruder and therefore, the system operator obtains two different set of state variables for these two different scenarios. But it is interesting to note that the residual values and squared errors calculated from both of these scenarios are the same and that is equal to the base case. Therefore, it is expected that the error value is below than the threshold and it will pass the Bad Data Detection technique although attack has been launched. So, the results can be summarized in Table 17.1.

From the Table 17.1, it can be seen that no bad data is detected in Scenario 1, Scenario 3 and Scenario 4. Although there is no False Data Injection attack in Scenario 1 but measurement data is manipulated in the remaining two scenarios. However, bad data detection technique fails to detect that. It is also interesting to note that squared error term for the above discussed cases is 0.00021429 but state variables vary a lot. As a result, the system operator may take misleading decisions.

## 17.6 Vulnerabilities of Smart Grid State Estimation: A Review

The new class of False Data Injection attack is first proposed in [26]. In that literature, authors show that the DC State Estimation is very vulnerable to malicious modification of the measurement data. With theorem and proof, some

**Table 17.1** Comparison of different attack scenarios

| Case | False data injection attack | State variables | | Squared error | Bad data detection |
|---|---|---|---|---|---|
| | | $\theta_1$ | $\theta_2$ | | |
| Scenario 1 | No | 0.0286 | −0.0943 | 0.00021429 | Not detected |
| Scenario 2 | Yes | 0.0313 | −0.0919 | 0.0013 | Detected |
| Scenario 3 | Yes | 0.0336 | −0.0933 | 0.00021429 | Not detected |
| Scenario 4 | Yes | 0.0386 | −0.0543 | 0.00021429 | Not detected |

heuristic approaches are proposed to attack the DC state estimation. Both random attacks and targeted attacks are considered. Two limitations of the proposed methodology are:

(1) The attacker needs the system 'Configuration' information prior to the attack,
(2) The proposed methodology is developed only for DC State Estimation.

A technique to detect false data injection is proposed in [28], where authors have focused on detecting a set of sensors whose measurements need to be protected in order to capture the false data injection in a DC State Estimator. The work presented in that paper considers the proposed attack model of [26]. The relation between the change of topology and the attack scenario is not considered in this research work [28].

The impact of False Data injection attack on the energy market is discussed in [29] which show that a successful attack can introduce a financial disaster. In that research, a convex optimization problem is formulated to find profitable attack. Although False Data Injection Attack is discussed based on financial issues, the work does not provide any intrusion detection or prevention technique to mitigate the problem.

A protection strategy against the False Data Injection Attack is proposed in [30]. In this research, the authors propose an effective algorithm to identify and protect the key measurements easily. A strategic plan for placement of PMU units is also described here. This work focuses system operator's point of view to utilize limited resources against the False Data Injection Attack. However, this paper does not consider accurate nonlinear AC State Estimation to define and protect attacks in power grid.

Generally, the bad data detection technique relies on the residual errors of the State Estimation procedure. However, it has been proven that this type of methodology for Bad Data Detection is vulnerable to False Data Injection Attacks [26]. To detect bad data, a Generalized Likelihood Ratio Test (GLRT) is proposed in [31]. This paper also considers the False Data Injection Attack from an intruder's point of view where attacker knows the information of mean-square error and GLRT of the system operator. This paper also limits the research idea to DC State Estimation.

Impact of cyber attack on the State Estimation considering a non-linear model is analyzed in [32]. Two widely used Bad Data Detection techniques are considered for comparison. This work concludes that False Data Injection Attack has a better probability to remain undetected if the attacker has a more accurate model of the system.

Another defense strategy against False Data Injection attack is proposed in [33]. The proposed detection framework has two stages. At the first stage, a linear unknown parameter solver is used and finally, a CUSUM algorithm is used to detect the intrusion maintain a certain low level of detection error rate.

Vulnerabilities of AC State Estimation due to the False Data Injection Attack are discussed in [34]. This work extends the hidden False Data Injection Attack model of [26] from a DC approximation to a non-linear AC model. Here, authors propose a Graph-theory based approach to determine critical measurement components which are vulnerable to cyber-attacks.

From the literature review, some decisions may be taken:

- From the attacker's point of view:

(a) Detail system model should be considered during the attack creation
(b) Different techniques exist to detect the Bad Data. It is important to note that attack vectors should be able to hide against most of the Bad Data Detection techniques.
(c) Attack should be introduced with limited knowledge of system and resources.

- From the system operator's point of view:

(a) System operator should be aware about the possible attack scenarios.
(b) Strategic protection and defense model should be introduced.

A brief description of different types of cyber attacks considering smart grid is given in [35].

## 17.7 Concluding Remarks

The role of State Estimation is crucial to operate the system in a stable condition. In recent time, Smart Grid State Estimation is very vulnerable to False Data Injection Attacks. In this chapter, the overview of State Estimation in both transmission level and distribution level is discussed. The evolution of State Estimation in the Smart Grid and its requirements are also explained. The review of False Data Injection attack is explained with a case study. It is expected that utilities, industries and academics should be more concerned to develop the countermeasures and protection strategies against this type of attacks.

## References

1. McMillan R (2010) Siemens: Stuxnet worm hit industrial systems. COMPUTERWorld. 14 Sept 2010
2. The Industrial Control Systems Cyber Emergency Response Team (ICS-CERT) (2013) Incident response activity (April-June 2013). http://ics-cert.us-rt.gov/sites/default/files/ICS-CERT_Monitor_April-June2013.pdf. Accessed 30 Aug 2013
3. Anwar A, Pota HR (2012) Optimum allocation and sizing of DG unit for efficiency enhancement of distribution system. In: 2012 IEEE international Power Engineering and Optimization Conference (PEOCO), Melaka, Malaysia, pp 165, 170, 6–7 June 2012
4. Schweppe FC, Wildes J (1970) Power system static-state estimation, part I: exact model. IEEE Trans Power Apparatus Syst PAS-89(1):120, 125 (Jan 1970)
5. Huang Y-F, Werner S, Huang J, Kashyap N, Gupta V (2012) State estimation in electric power grids: meeting new challenges presented by the requirements of the future grid. IEEE Signal Process Mag 29(5):33, 43 (Sept 2012)
6. Abur A, Expósito AG (2004) Power system state estimation: theory and implementation. CRC Press.
7. Sakis Meliopoulos AP, Zhang F (1996) Multiphase power flow and state estimation for power distribution systems. IEEE Trans Power Syst 11(2):939, 946 (May 1996)
8. Monticelli A (2000) Electric power system state estimation. In: Proceedings of the IEEE, vol 88, no 2, pp 262, 282 (Feb 2000)
9. Naka S, Genji T, Yura T, Fukuyama Y (2003) A hybrid particle swarm optimization for distribution state estimation. IEEE Trans Power Syst 18(1):60, 68 (Feb 2003)
10. Handschin E, Schweppe FC, Kohlas J, Fiechter A (1975) Bad data analysis for power system state estimation. IEEE Trans Power Apparatus Syst 94(2):329, 337 (Mar 1975)
11. Lu CN, Teng JH, Liu W-HE (1995) Distribution system state estimation. IEEE Trans Power Syst 10(1), 229, 240 (Feb 1995)
12. Kersting WH (2011) The whys of distribution system analysis. IEEE Ind Appl Mag 17(5):59, 65 (Sept–Oct 2011)
13. Baran ME, Kelley AW (1994) State estimation for real-time monitoring of distribution systems. IEEE Trans Power Syst 9(3):1601, 1609 (Aug 1994)
14. Lin W-M, Teng J-H (1996) State estimation for distribution systems with zero-injection constraints. IEEE Trans Power Syst 11(1):518, 524 (Feb 1996)
15. Haughton DA, Heydt GT (2013) A linear state estimation formulation for smart distribution systems. IEEE Trans Power Syst 28(2):1187, 1195 (May 2013)
16. Anwar A, Pota HR (2012) Optimum capacity allocation of DG units based on unbalanced three-phase optimal power flow. In: IEEE power and energy society general meeting, pp 1, 8, 22–26 July 2012
17. Novosel D, Vu K (2006) Benefits of PMU technology for various applications. In: 7-th CIGRE symposium on power system management, Cavtat, Croatia, pp 1–13, 5–8 Nov 2006
18. Anwar A, Pota HR (2011) Loss reduction of power distribution network using optimum size and location of distributed generation. In: 21st Australasian Universities Power Engineering Conference (AUPEC), pp 1, 6, 25–28 Sept 2011
19. Gomez-Exposito A, Abur A, de la Villa Jaen A, Gomez-Quiles C (2011) A multilevel state estimation paradigm for smart grids. In: Proceedings of the IEEE, vol 99, no 6, pp 952, 976 (June 2011)
20. Hu Y, Kuh A, Tao Y, Kavcic A (2011) A belief propagation based power distribution system state estimator. IEEE Comput Intell Mag 6(3):36, 46 (Aug 2011)
21. Liu J, Tang J, Ponci F, Monti A, Muscas C, Pegoraro PA (2012) Trade-Offs in PMU deployment for state estimation in active distribution grids. IEEE Trans Smart Grid 3(2):915, 924 (June 2012)
22. Park S, Lee E, Yu W, Lee H, Shin J (2013) State estimation for supervisory monitoring of substations. IEEE Trans Smart Grid 4(1):406, 410 (March 2013)

23. Xie L, Choi D-H, Kar S, Poor HV (2012) Fully distributed state estimation for wide-area monitoring systems. IEEE Trans Smart Grid 3(3):1154, 1169 (Sept 2012)
24. Zonouz S, Rogers KM, Berthier R, Bobba RB, Sanders WH, Overbye TJ (2012) SCPSE: Security-oriented cyber-physical state estimation for power grid critical infrastructures. IEEE Trans Smart Grid 3(4):1790, 1799 (Dec 2012)
25. Niknam T, Firouzi BB (2009) A practical algorithm for distribution state estimation including renewable energy sources. Renew Energy 34(11):2309–2316 (Nov 2009)
26. Liu Y, Reiter MK, Ning P (2009) False data injection attacks against state estimation in electric power grids. In: Proceedings of the 16th ACM conference on computer and communications security, 2009
27. Xie L, Mo Y, Sinopoli B (2011) Integrity data attacks in power market operations. IEEE Trans Smart Grid 2(4):659, 666 (Dec 2011)
28. Bobba RB, Rogers KM, Wang Q, Khurana H (2010) Detecting false data injection attacks on DC state estimation. In: Proceedings of the first workshop on secure control systems
29. Xie L, Mo Y, Sinopoli B (2010) False data injection attacks in electricity markets. In: 2010 First IEEE international conference on smart grid communications (SmartGridComm), pp 226, 231, 4–6 Oct 2010
30. Kim TT, Poor HV (2011) Strategic protection against data injection attacks on power grids. IEEE Trans Smart Grid 2(2):326, 333 (June 2011)
31. Kosut O, Jia L, Thomas RJ, Tong L (2010) On malicious data attacks on power system state estimation. In: 45th international universities power engineering conference (UPEC), pp 1, 6, 31 Aug 2010–Sept 3 2010
32. Teixeira A, Amin S, Sandberg H, Johansson KH, Sastry SS (2010) Cyber security analysis of state estimators in electric power systems. In: 49th IEEE conference on decision and control (CDC), pp 5991, 5998, 15–17 Dec 2010
33. Huang Y, Li H, Campbell KA, Han Z (2011) Defending false data injection attack on smart grid network using adaptive CUSUM test. In: 45th annual conference on information sciences and systems (CISS), pp 1, 6, 23–25 Mar 2011
34. Hug G, Giampapa JA (2012) Vulnerability assessment of AC state estimation with respect to false data injection cyber-attacks. IEEE Trans Smart Grid 3(3):1362, 1370 (Sept 2012)
35. Anwar A, Mahmood AN (2014) Cyber security of smart grid infrastructure. In: The state of the art in intrusion prevention and detection, CRC Press, USA (Jan 2014)
36. Teixeira A, Amin S, Sandberg H, Johansson KH, Sastry SS (2010) Cyber security analysis of state estimators in electric power systems. In: 49th IEEE conference on decision and control (CDC), pp 5991, 5998, 15–17 Dec 2010

# Chapter 18
# Impediments and Model for Network Centrality Analysis of a Renewable Integrated Electricity Grid

**A. B. M. Nasiruzzaman, Most. Nahida Akter and H. R. Pota**

**Abstract** Inclusion of renewable energy changes the power flow direction of the trans- mission grid, resulting in a bidirectional flow model of the power transmission systems. The changing nature of the grid demands for new and improved techniques to analyze the vulnerability of the power grid. In this chapter, a method for identifying critical nodes for smart and bulk power transmission grid environment is presented. A new model based on bidirectional power flow is considered. Three different models of power system based on complex network framework are analyzed. Applicability of these methods in smart grid environment is evaluated. The consequence of removing critical nodes found from the analysis is discussed. Four measures of impact based on topological and electrical characteristics are tested. The efficacy of bidirectional model is studied through rank similarity analysis.

**Keywords** Closeness centrality · Forward unidirectional graph · Backward unidirectional graph · Bidirectional graph · Path length · Connectivity loss · Load loss · Rank similarity

A. B. M. Nasiruzzaman (✉) · H. R. Pota
School of Engineering and Information Technology, The University of New South Wales, Canberra, ACT 2612, Australia
e-mail: nasiruzzaman@ieee.org

H. R. Pota
e-mail: H.Pota@adfa.edu.au

M. N. Akter
Department of Electrical and Electronic Engineering, Rajshahi University of Engineering and Technology, Kazla, Rajshahi 6201, Bangladesh
e-mail: reachnahida@gmail.com

J. Hossain and A. Mahmud (eds.), *Renewable Energy Integration*, Green Energy and Technology, DOI: 10.1007/978-981-4585-27-9_18,

## 18.1 Introduction

Utilities around the world are integrating smart and new technologies towards making the existing electrical power transmission grid much smarter [1]. The scope of smart grid includes various generation options, primarily in the distribution side—near consumers. Engagement of customers with the energy management systems is the most lucrative part of smart grid from the point of view of regulating energy usage. Excess of generation after local use can be transmitted long distance to meet the energy shortage of the destination area.

This introduces a new concept of power flowing from customer end towards the grid. The bidirectional power flow changes the whole power flow pattern of the existing grid [2]. Analytical methods, technical strategies, control system and protecting devices need to be changed along with, to mention a few. Metering and protecting equipments will experience flows coming from the reverse side. Proper operation of the equipments used earlier can be ensured either by changing the instruments themselves or by incorporating new measurement techniques [3].

Recent years have seen several very large scale blackouts initiating from small disturbances. In August 1996, a cascading outage occurred in the Western power grids of North America in USA and Mexico [4]. More than four million people suffered the consequences. Most affected areas were out of electricity for about 4 days. Another large scale blackout which affected around 55 million people happened in August 2003 [5]. Several northeast and mid-western states of USA and some provinces of Canada were affected.

The move towards the smart grid started after the blackouts happened all around the world [6]. From the frequent events of large scale-blackouts it is clear that the existing dynamics security assessment and monitoring system has not been working well [7]. The motivation of complex network framework based analysis approach comes from the necessity of new, alternative and improved methodology to assess the risk involved with cascading events in power system.

Degree centrality, betweenness centrality and closeness centrality measures are commonly used in social network research to find a person with most influence [8]. The person who has most number of links is the most central according to degree centrality. Betweenness centrality measures the importance of a person as an intermediary. The person who comes across a path of communication between two other persons most of the times is considered as central in between centrality. A person is said to be closeness central if he or she is closest to all other persons relative to other persons in the network of interest.

Connectivity of the network is hampered, when nodes with higher degrees are taken out from the system. Removing a node takes out with it many links, which degrades the performance of network. Betweenness central node is important because it has the most ability to control communication between other nodes. The node which has least distance from all other nodes is closeness central. This node is the most independent one since it can communicate with other nodes without the need of intermediate nodes.

Power grid topology has been analyzed by various researchers recently to explore its strength and weakness using complex network framework. The strength of the grid is found to be, from a pure topological analysis of USA power grid, small-world property [9]. This implies that various nodes within the system can be reached easily, which will make the communication that comes along with the smart grid easy and effective. The scale freeness of the topology of the grid is shown to be a weakness of the grid since it makes the system very much vulnerable to targeted attack [10]. This targeted attack can trigger cascading failure which will lead to blackout.

The research on power grid from a system point of view has been triggered after the publications of the preliminary topology based analysis results. Since results from pure topological approach is quite misleading [11], several researchers have a mix of both topological and electrical characteristics based complex network analysis of power system to find reasonably improved results [12, 13].

Motivated by the topological result that found the power grid robust against random failure but vulnerable to targeted attacks [10], critical node and link analysis of power grid have been carried out to explore the criticality of the power grid. If critical components can be spotted out which can initiate cascading effect, special preventive actions could be exercised so that to prevent large scale blackouts from happening.

Network efficiency, a topological measure of performance change after the inclusion or removal of nodes or lines from a grid, is analyzed in [14]. A weighted line betweenness based approach is utilized to find out critical lines responsible for spreading of large scale blackouts from small initial shock [15]. Vulnerable regions of power system are identified employing complex network theory based qualitative simulation in [16]. Transmission line reactance is incorporated to compute a new vulnerability index to identify critical lines [17].

A link is explored between power system reliability and small world effect [18]. Maximum flow based centrality approach is used to find out critical lines which removes the shortcoming of the assumption of power flowing through the shortest paths between source and load nodes [19]. This method has slow convergence but can be useful when used in conjunction of planning issues. A DC power flow model is used and hidden failure of protective equipment is considered to model the structural vulnerability of power grid [20]. Electrical parameters are incorporated extensively to improve the centrality indices for power system [21].

An extended topological approach proposed in [22] takes into consideration traditional topological metrics as well as operational behavior of power grids like real power flow allocation and line flow limits. Power Transfer Distribution Factor (PTDF) is used to simulate cascading event in an attempt to identify correlated lines [23].

All these analysis are carried out for electric grids where power flow is directed from generating nodes to load nodes. But since with the inclusion of distributed generations the power flow pattern is going to change, new methodologies have to be proposed which takes into account bidirectional power flow. Since communication is an important factor in smart grid, identifying those nodes in the system would be very much useful which are important for communication.

In this chapter, a method based on complex network theory has been proposed to identify critical components in smart grid. This method is a modification of closeness centrality which takes into account power flow distribution among various power lines during steady state. This is a reasonable extension of previous work carried out by researchers since it captures the power flow in smart grid environment. Rank similarity analysis result is carried out to verify that proposed index is useful although there is a slight change in network. The impact of removing critical components is identified using well known impact metrics like path length, connectivity loss and load loss.

The organization of the rest of the chapter is as follows. Sect. 18.2 provides a model for the analysis of smart power grid under complex network framework. A new model based on bidirectional power flow is considered and a method is discussed to find critical nodes in the power grid. The critical node identification procedure is illustrated in Sect. 18.3. The effect of removal of critical nodes on various topological and electrical measures is addressed in Sect. 18.4. Effect on the rank of critical nodes for different models, when the network is changed slightly is observed in Sect. 18.5. Conclusion is drawn and future research direction is provided in Sect. 18.6.

## 18.2 System Model and Methodology

The first step of analyzing power grid under complex network framework is to model the system as a directed graph [7]. Vertices in the graph represent generating stations, substations, loads etc. Edges of links represent transmission lines that connect various generating stations, substations and load points. In this model, only transmission system is considered. The overall distribution system is regarded as a lumped load at the distribution substation terminal.

Power flow analysis is conducted for the given test system during nominal condition. Newton–Raphson method is used to solve the simultaneous nonlinear algebraic power flow equations [24]. The direction of real power flowing through the lines is taken as the direction of edges in the modeled graph. From this point this graph will be known as forward unidirectional flow graph, which can be defined as:

**Definition 18.1** (*Forward Unidirectional Graph*) A nominal unidirectional graph model of a power system can be obtained from the normal operating states of the system. It can be represented by $\Gamma = (\varsigma, \mathrm{E}, \Omega)$ comprising of a set $\varsigma$, whose elements are called vertices or nodes, a set $\mathcal{E}$ of ordered pairs of vertices, called edges or lines and a set $\Omega$, whose elements are weights of edge set elements. There exists a one-to-one correspondence between set $E$ and set $\Omega$. An element $e = (x, y)$ of the edge set $E$, is considered to be directed from $x$ to $y$, where $y$ is called the head and $x$

is called the tail of the edge. In this model, transmission line impedances in pu is considered as weights of the edges between nodes. There exists a one-to-one correspondence between set $E$ and set $\Omega$.

In order to consider the bidirectional flow in smart grid, a backward unidirectional flow graph is also modeled, which is presented in a formal definition as follows:

**Definition 18.2** (*Backward Unidirectional Graph*) A backward unidirectional graph model of a power system can be obtained from the reversed operating states of the system. It can be represented by $\mathrm{G} = (\mathrm{V}, \mathrm{E}, \mathrm{W})$ comprising of a set V, whose elements are called vertices or nodes, a set E of ordered pairs of vertices, called edges or lines and a set W, whose elements are weights of edge set elements. There exists a one-to-one correspondence between set E and set W. An element $e = (x, y)$ of the edge set E, is considered to be directed from $x$ to $y$, where $y$ is called the head and $x$ is called the tail of the edge. In this model, transmission line impedances in pu is considered as weights of the edges between nodes. There exists a one-to-one correspondence between set E and set W.

As we can find out from the definition, the direction of edges in the backward unidirectional flow graph is exactly opposite to the nominal unidirectional flow graph. Now, the combination of the forward and backward unidirectional graph is considered to be the bidirectional graph, which is used to model the power flow pattern of the future smart power grid. The bidirectional graph can be defined as:

**Definition 18.3** (*Bidirectional Graph*) A bidirectional graph model of a power system can be obtained from the superposition of nominal unidirectional and backward unidirectional graph models. It can be represented by G = (V,E,W) comprising of a set V, whose elements are called vertices or nodes, a set E of ordered pairs of vertices, called edges or lines and a set W, whose elements are weights of edge set elements. There exists a one-to-one correspondence between set E and set W. An element $e = (x, y)$ of the edge set E, is considered to be directed from $x$ to $y$ , where $y$ is called the head and $x$ is called the tail of the edge. In this model, transmission line impedances in pu is considered as weights of the edges between nodes. There exists a one-to-one correspondence between set E and set W.

To illustrate uni- and bi-directional graph models in a power system, a simple example of 14 bus system [25] is used in this chapter. Figure 18.1 depicts the system with 14 bus bars, and 20 links connecting them, while Figs. 18.2 and 18.3 represent the forward and backward unidirectional graph model of Fig. 18.1. We can model the system as a graph which contains 14 nodes/vertices which correspond to the slack, voltage-controlled, and load bus bars of the original system. The transmission lines can be represented by the 20 links/edges which connects various nodes. The system data is given in Table 18.1.

Assume that, $k$ represent the intermediate bus within the shortest path originating from bus $s$ and ends at bus $t$. Let, $P_{st}$ represents the maximum power

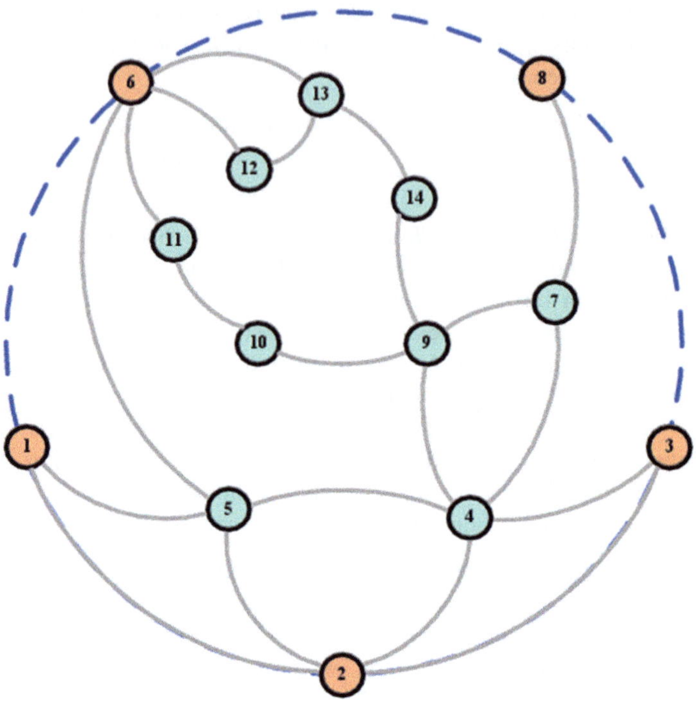


**Fig. 18.1** Topology of the IEEE 14 bus test system

flowing in the shortest electrical path between buses $s$ and $t$, and $P_{st}(k)$ is the maximum of inflow and outflow at bus $k$ within the shortest electrical path between buses $s$ and $t$. Then, let their fraction is represented by $r_{st}(k)$ as in:

$$r_{st}(k) = \frac{P_{st}(k)}{P_{st}} \tag{18.1}$$

where, the ratio $r_{st}(k)$ is an index of the degree to which buses $s$ and $t$ need bus $k$ to transmit power between them along the shortest electrical path. If a double sum is taken of (18.1) over all intermediate buses $k$ and all destination buses $t$ for the source buses $s$,

$$C_C^E(s) = \sum_{k=1}^{n} \sum_{t=1}^{n} \frac{P_{st}(k)}{P_{st}}, s \neq t \neq k \in V \tag{18.2}$$

a centrality measure for bus $s$ within the grid is obtained. This measure (18.2) adds up the real power of the lines originating at bus $s$ and terminating at all other buses. This quantity takes high values if the difference between numerator and denominator term is low. This fact represents that very few amount of power is lost in the shortest path. Such buses might have more direct influence on other buses since very few amount of power is lost.

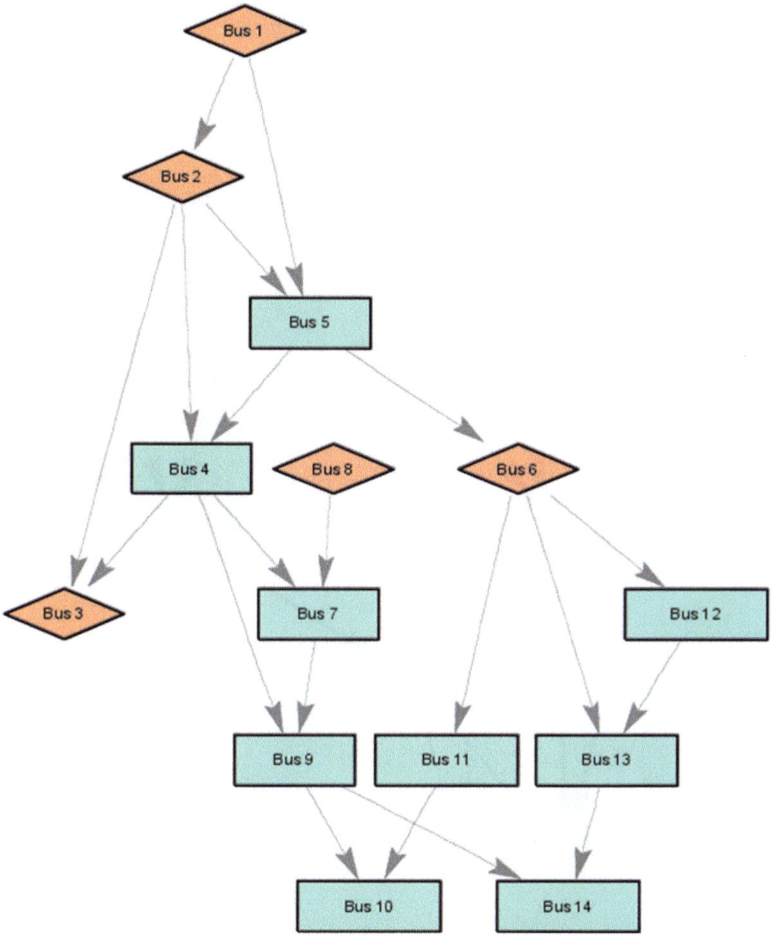


**Fig. 18.2** Forward unidirectional graph model of the IEEE 14 bus test system

Table 18.2 lists top ten critical nodes in IEEE 30 bus test system [24, 25] found from nominal and backward unidirectional as well as bidirectional model.

## 18.3 Measure of Pair Dependence of Various Buses

The concept of pair dependence of various buses is presented in [26], which is described here to maintain the flow of this chapter.

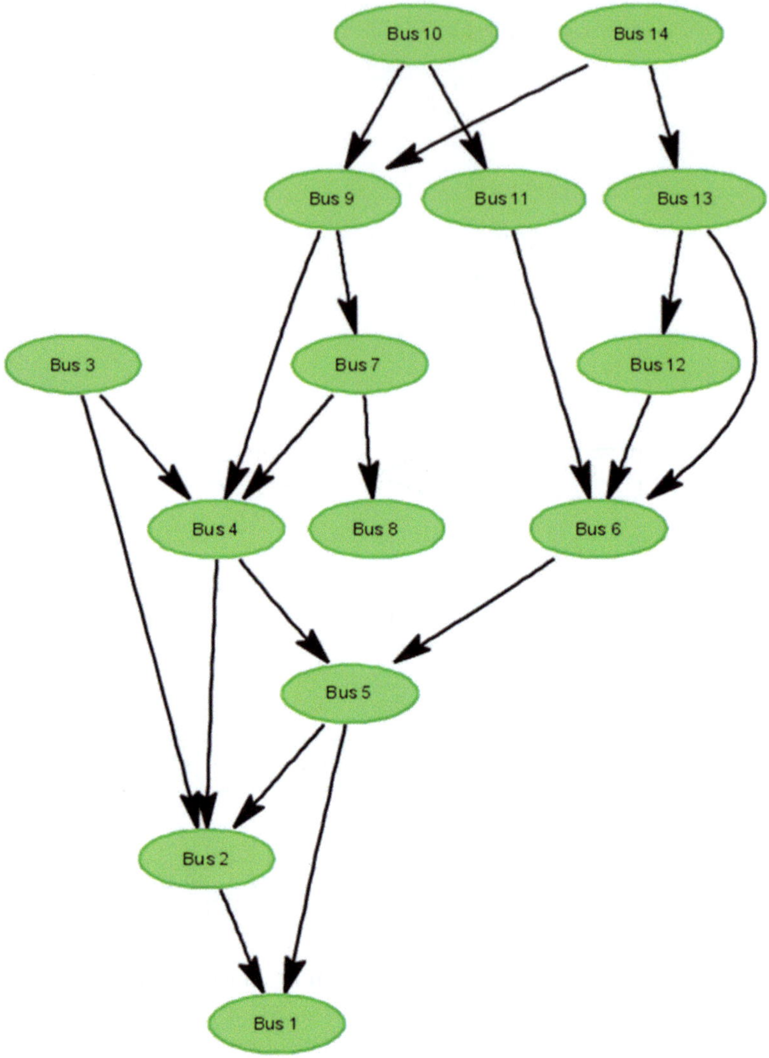


**Fig. 18.3** Backward unidirectional graph model of the IEEE 14 bus test system

### *18.3.1 Shortest Path*

The concept of shortest path is used by the researchers of power system who use complex network framework for network vulnerability analysis [17]. In order to assess the vulnerability of a power grid researchers used dynamic power system model where the concept of network flow is introduced [20]. The flow between two nodes *s* and *t* takes on shortest path between them. If there are two or more

**Table 18.1** System data for network in Fig. 18.1

| Branch number | From bus | To bus | From bus $P_{inj}$ (MW) | To bus $P_{inj}$ (MW) | Loss $P$ (MW) |
|---|---|---|---|---|---|
| 1 | 1 | 2 | 156.88 | −152.59 | 4.30 |
| 2 | 1 | 5 | 75.51 | −72.75 | 2.76 |
| 3 | 2 | 3 | 73.24 | −70.91 | 2.32 |
| 4 | 2 | 4 | 56.13 | −54.45 | 1.68 |
| 5 | 2 | 5 | 41.52 | −40.61 | 0.90 |
| 6 | 3 | 4 | −23.29 | 23.66 | 0.37 |
| 7 | 4 | 5 | −61.16 | 61.67 | 0.51 |
| 8 | 4 | 7 | 28.07 | −28.07 | 0.00 |
| 9 | 4 | 9 | 16.08 | −16.08 | 0.00 |
| 10 | 5 | 6 | 44.09 | −44.09 | 0.00 |
| 11 | 6 | 11 | 7.35 | −7.30 | 0.06 |
| 12 | 6 | 12 | 7.79 | −7.71 | 0.07 |
| 13 | 6 | 13 | 17.75 | −17.54 | 0.21 |
| 14 | 7 | 8 | 0.00 | 0.00 | 0.00 |
| 15 | 7 | 9 | 28.07 | −28.07 | 0.00 |
| 16 | 9 | 10 | 5.23 | −5.21 | 0.01 |
| 17 | 9 | 14 | 9.43 | −9.31 | 0.12 |
| 18 | 10 | 11 | −3.79 | 3.80 | 0.01 |
| 19 | 12 | 13 | 1.61 | −1.61 | 0.01 |
| 20 | 13 | 14 | 5.64 | −5.59 | 0.05 |

**Table 18.2** Top ten nodes of IEEE 30 bus test system in unidirectional & bidirectional power flow models

| Unidirectional nominal | Unidirectional backward | Bidirectional nominal |
|---|---|---|
| 1 | 24 | 1 |
| 3 | 19 | 3 |
| 2 | 26 | 2 |
| 4 | 18 | 4 |
| 6 | 23 | 6 |
| 13 | 21 | 24 |
| 12 | 25 | 19 |
| 9 | 29 | 13 |
| 14 | 30 | 12 |
| 28 | 17 | 14 |

paths between two buses then the path that has less weight is regarded as the shortest path between those two buses.

Only the physical connection is considered in traditional modeling approach by complex network researchers. The weight of the line between nodes reflects simply the topology of the network. If there is a connection between node $s$ and node $t$ then the weight of the corresponding line is taken as 1, otherwise it is 0 in traditional approach [20]. In case of a power system the main parameter of a

Table 18.3 Various possible connections between buses 1 and 4 of the system of Fig. 18.1

| Connection | Weight (pu) |
|---|---|
| 1–2–4 | 1.21 |
| 1–2–3–4 | 2.01 |
| 1–2–5–4 | 1.04 |
| 1–3–4 | 0.72 |
| 1–3–2–4 | 1.50 |
| 1–3–2–5–4 | 1.33 |

transmission line which has significant effect in the power flow in the line between buses is its impedance which is not considered in this model.

Several researchers have considered the reactance of the line [15], neglecting the line resistance which is very small for transmission systems. But, in order to generalize the model for both the transmission and the distribution system, the impedance, (i.e., both the reactance and resistance) needs to be taken into consideration [17].

In this chapter, we have used absolute measure of impedance,$|Z|$, as weight of the line. If we want to find shortest electrical path between buses 1 and 4, several paths are possible as given in Table 18.3. We can clearly see that the shortest path between buses 1 and 4 is $1-3-4$ whose weight is 0.72 pu.

Finding the shortest path set for a network is a problem of graph theory and several efficient algorithms are available.

### 18.3.2 Bus Dependency Matrix

In the context of complex network theory, when a pair of buses in the power system is connected via a transmission line without any other buses in between (intermediaries), they are said to be adjacent. A bus $s$ adjacent to bus $k$, another bus $t$ adjacent to bus $k$, creates a transmission path between buses $s$ and $t$ via bus $k$. The shortest electrical path linking a pair of buses is called a geodesic.

Let, $P_{st}$ is the maximum power flowing in the shortest electrical path between buses $s$ and $t$, and $P_{st}(k)$ is the maximum of inflow and outflow at bus k within the shortest electrical path between buses $s$ and $t$. Then, let their fraction is represented by $r_{st}(k)$ as in:

$$r_{st}(k) = \frac{P_{st}(k)}{P_{st}} \tag{18.3}$$

where, the ratio $r_{st}(k)$ is an index of the degree to which buses $s$ and $t$ needs bus $k$ to transmit power between them along the shortest electrical path.

The pair dependency of nodes in a network is defined in [27]. The concept of pair dependency in [27] is used here in case of electrical power grid. The dependency of bus pairs can be regarded as the degree to which a bus s must depend upon another bus $k$ to transmit its power along the shortest electrical path or geodesic to and from all other reachable buses $t$'s in the network. For a power grid with $n$ number of buses the dependency of bus $s$ upon bus $k$ to transmit power on any other buses in the network can be represented as follows:

$$d_{sk} = \sum_{t=1s\neq t\neq k\in V}^{n} r_{st}(k) = \sum_{t=1s\neq t\neq k\in V}^{n} \frac{P_{st}(k)}{P_{st}} \tag{18.4}$$

The dependency of bus pairs for the whole system can be calculated and the result can be summarized in a matrix **D** as follows:

$$\mathbf{D} = \begin{bmatrix} d_{11} & d_{12} & \cdots & d_{1n} \\ d_{21} & d_{22} & \cdots & d_{2n} \\ \vdots & \vdots & \ddots & \vdots \\ d_{n1} & d_{n2} & \cdots & d_{nn} \end{bmatrix} \tag{18.5}$$

Each element of **D** is an index of degree to which a bus designated by row number must depend upon another bus designated by column number to transmit its power along the shortest electrical path or geodesic to and from all other reachable buses in the network. Thus this matrix captures the information of importance of a bus as an intermediary with respect to other buses in the network. So we can call the matrix **D** as bus dependency matrix.

### *18.3.3 Steps to Find Bus Dependency Matrix from System Data*

The procedural steps to find bus dependency matrix from the system data is as follows:

1. Model the system as a graph.
2. Find a shortest path set for the graph using Johnson's algorithm [28].
3. Find flow in various lines of the system solving load flow problem.
4. Find the maximum power flowing in the shortest electrical path between buses $s$ and $t$, $P_{st}$, for the shortest path set.
5. Find$P_{st}(k)$, the maximum of inflow and outflow at bus $k$ within the shortest electrical path between buses $s$ and $t$.
6. Evaluate bus dependency matrix **D** from $P_{st}$ and $P_{st}(k)$.

### *18.3.4 Several Observations About Bus Dependency Matrix*

Several observations about the bus dependency matrix are enumerated as follows:

- The $(s,t)$-th element of the matrix represents the dependency of bus $s$ on bus $t$.
- Diagonal elements of the bus dependency matrix are zero.

- This matrix is non-symmetric.
- The row sum of the matrix could be used as an electrical closeness centrality measure.
- The column sum of the matrix is electrical betweenness centrality measure.

## 18.4 Measures of Impact

At first, the nominal network is solved and nodes are removed from the system one by one in the descending order of centrality measure. In order to measure the impact of removing critical nodes from the system various measures are being used. In this chapter, four measures are considered. The first two of them, path length and connectivity loss are purely topological. The last two measures are percentage of load lost due to the removal of critical nodes and number of overloaded lines.

### *18.4.1 Path Length*

The path length is used by researchers as a measure of network connectedness. It is the average length of the shortest paths between any two nodes in the network [29]. It is found that if a node is removed from a system, it generally increases the distance between other nodes. So, the increase in network characteristic path length is considered as a measure of impact analysis of removing critical nodes from the system.

A simple IEEE 30 bus test system is used to simulate the consequence of node removal on path length and the result is depicted in Fig. 18.4. It is seen that, if node with high centrality is removed found from nominal unidirectional graph model, the path length increases slightly. A mix result of increase and decrease in path length is found if backward unidirectional flow model is used. In case of bidirectional flow model the maximum impact is found.

### *18.4.2 Connectivity Loss*

This is a purely topological measure of impact a power grid encounters when some nodes are removed from the system. In this measure we calculate how much connectivity is lost in terms of how many generators a transmission or distribution node can access due to effect of removing a node from the system. The less is the number of generators a node is connected with, the less is the redundancy and the more is the vulnerability of the node. It is given as (18.6) originally proposed in [30].

Fig. 18.4 Change in path length in IEEE 30 bus test system for removal of critical nodes based on three different measures

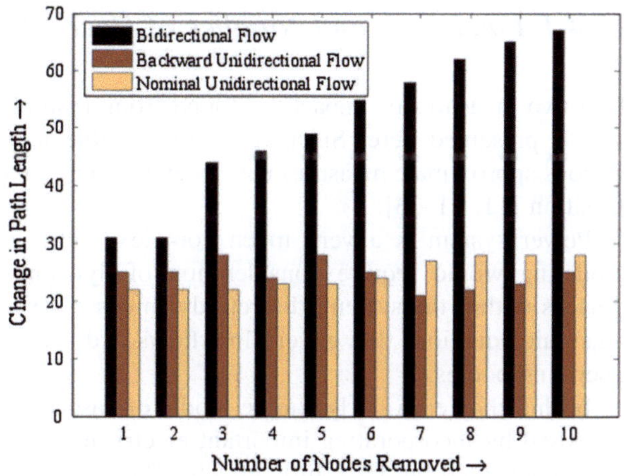


Fig. 18.5 Connectivity loss of IEEE 30 bus test system as a function of removal of critical nodes from three different points of view

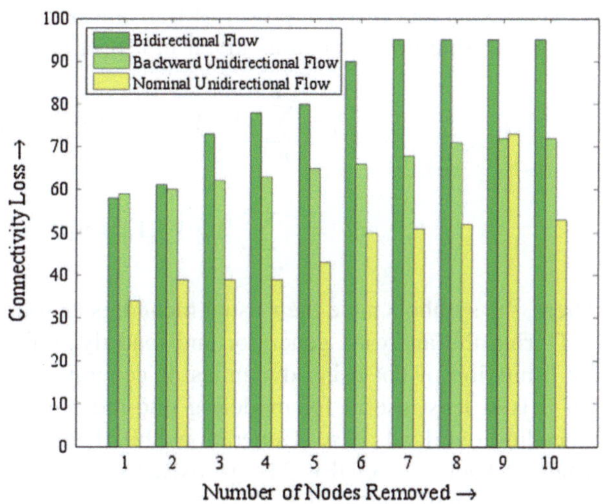


$$C = 1 - < \frac{N_g^i}{N_g} >_i \tag{18.6}$$

where, the averaging is done over each intermediate node, i.e., substations. $N_g$ is the total number of generators and $N_g^i$ is the number of generators that a node $i$ can reach. Impact on connectivity loss for three different models is presented in Fig. 18.5.

It is found that connectivity is lost to a great extent in all three cases, although the effect is highest in case of bidirectional flow model. Initially nominal and bidirectional method had similar impact, but the impact becomes more prominent in case of bidirectional flow model after removal of three nodes only.

### *18.4.3 Load Loss and Number of Overloaded Lines*

Last two measures of impact are found from a simple model of cascading failure that is presented here. Since it is not possible to exactly model the blackout, various approximate measures have been taken by several researchers to mimic the situation [11, 31–33].

Power system is a very much complex interconnected system whose exact modeling would require consideration of dynamics of rotating machines and devices within the system, discrete dynamics of switchgear elements, non-linear algebraic equations that govern line flows and social dynamics of governing and operating bodies.

In this chapter, a fairly simple model of cascading failure of the power grid is proposed by incorporating important electrical features ignoring those which are too complicated but have little effects. The detail of the model is described here.

At first AC power flow is used to calculate the steady state condition of the network. Real and reactive power of transmission lines are found from numerical solution of line flow equations given in (18.7) and (18.8)

$$P_i = \sum_{j=1}^{n} |V_i||V_j||Y_{ij}| \cos(\theta_{ij} - \delta_i + \delta_j) \tag{18.7}$$

$$Q_i = -\sum_{j=1}^{n} |V_i||V_j||Y_{ij}| \sin(\theta_{ij} - \delta_i + \delta_j) \tag{18.8}$$

where, the symbols have their usual meanings as found in power system literature.

During the analysis, generator and load dynamics are not included. Although the limitation of not using dynamics of generators and loads are well understood but it is at least useful for modeling one mechanism of cascading failure that is cascading overload. Also, Generation Shift Factors (GSF) and Line Outage Distribution Factors (LODF) [34] are used to recalculate flows in lines after disturbance. This helps achieving fast results without using actual load flow after each disturbance. The speed and accuracy of the result and comparison with actual load flow is out of the scope of this chapter and will be addressed in another research article in future.

The transmission lines are removed if overloaded. The number of lines tripped is taken as a measure of impact which is demonstrated in Fig. 18.6. It is clear that, the number of overloaded lines in nominal and backward unidirectional flow methods is almost same. The bidirectional flow model gives highest impact and a large number of lines are overloaded for removing only seven nodes.

Also, time delayed over current relays are used in every line so if there is a lot of overload it trips fast and if there is a little bit of overload it trips slowly. Another thing that is added to the model is ramping up of generators. As the system separates into sub grids, generators are allowed to ramp up or ramp down to rebalance a little bit.

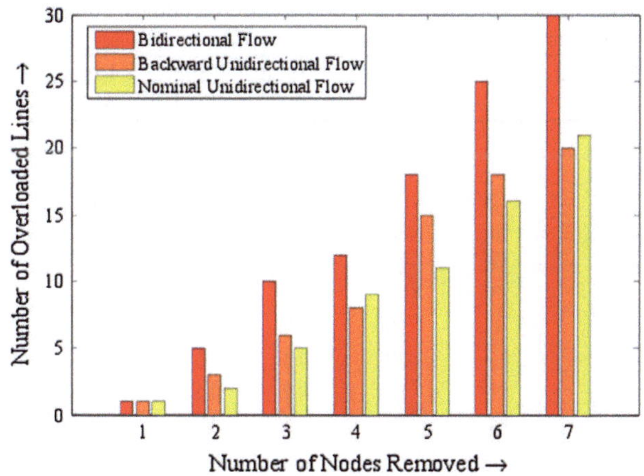


**Fig. 18.6** Number of overloaded lines increases drastically in bidirectional flow based algorithm

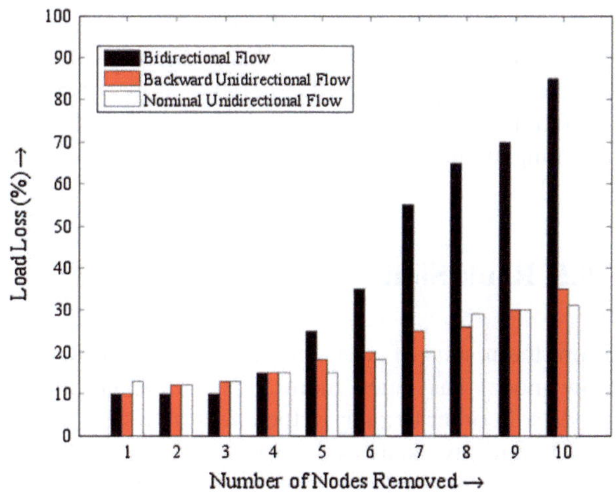


**Fig. 18.7** Three different effects on load loss due to loss of functionality of important nodes in IEEE 30 bus system

So, if a component failure disturbs the supply–demand balance, through generator set-point adjustment this balance is achieved. But if there is not enough ramping ability, then the ultimate choice is to trip lowest possible system load. The total amount of load lost during the successive removal of nodes is used as a measure of impact.

Figure 18.7 shows load loss as a percentage of total system loads. Up to six node removal the load loss is nearly equal and does not increase much for both unidirectional models. After five node removal, more than 50 % load of the system need to be shedded to ensure secure and reliable operation of the remaining system.

This introduces a new concept of power flowing from customer end towards the grid. The bidirectional power flow changes the whole power flow pattern of the

**Table 18.4** Top ten critical nodes in bidirectional power flow model for IEEE 30 bus system under various changed topological conditions

| Nominal case | Line 24–25 | Line 29–27 | Line 6–2 | Line 17–10 | Line 4–3 | Line 10–6 | Line 18–15 | Line 30–29 | Line 15–14 |
|---|---|---|---|---|---|---|---|---|---|
| 1 | 1 | 1 | 1 | 1 | 1 | 1 | 1 | 1 | 1 |
| 3 | 3 | 3 | 2 | 3 | 2 | 2 | 3 | 3 | 3 |
| 2 | 2 | 2 | 3 | 2 | 4 | 4 | 2 | 2 | 2 |
| 4 | 4 | 4 | 6 | 4 | 6 | 6 | 4 | 4 | 4 |
| 6 | 24 | 24 | 4 | 13 | 24 | 24 | 6 | 6 | 6 |
| 24 | 13 | 6 | 24 | 12 | 19 | 19 | 24 | 24 | 24 |
| 19 | 6 | 19 | 19 | 24 | 13 | 13 | 19 | 19 | 19 |
| 13 | 12 | 29 | 13 | 6 | 12 | 12 | 18 | 13 | 9 |
| 12 | 19 | 13 | 12 | 16 | 14 | 14 | 9 | 12 | 26 |
| 14 | 14 | 12 | 14 | 19 | 9 | 9 | 26 | 14 | 13 |
| 9 | 9 | 14 | 9 | 17 | 26 | 26 | 23 | 9 | 18 |

existing grid [18]. Analytical methods, technical strategies, control system and protecting devices need to be changed along with, to mention a few. Metering and protecting equipments will experience flows coming from the reverse side. Proper operation of the equipments used earlier can be ensured either by changing the instruments themselves or by incorporating new measurement techniques [27].

## 18.5 Rank Similarity of Critical Nodes

From the results of Sect. 18.4 it is clear that, the nodes found from bidirectional flow model has much more impact than nominal and backward unidirectional models. In order to analyze the effect of system change on ranks of critical nodes a rank similarity analysis is performed. A structural change like change in the direction of power flow is incorporated in the model and critical nodes are found out for the modified system. This change in network corresponds to a situation when there is a pushback of power from low voltage network via transmission system to meet energy needs in other area.

Table 18.4 compares the changes in top ten critical nodes in IEEE 30 bus test system. This analysis is carried out for bidirectional power flow model. Top row of Table 18.4 corresponds to the topological state of the system. The first column gives the top ten critical nodes from the bidirectional model. The rest of the columns list change in critical nodes for changed topology. As for example, the third column represents the top ten critical nodes when the nominal direction of flow is changed through line 29–27. It is clear that; changed topology does not affect much the node criticality.

On the other hand, slightly more change is observed in criticality for the unidirectional model as shown in Fig. 18.8. When power flow pattern through the grid is unidirectional, nominal unidirectional method is effective. But, in order to model

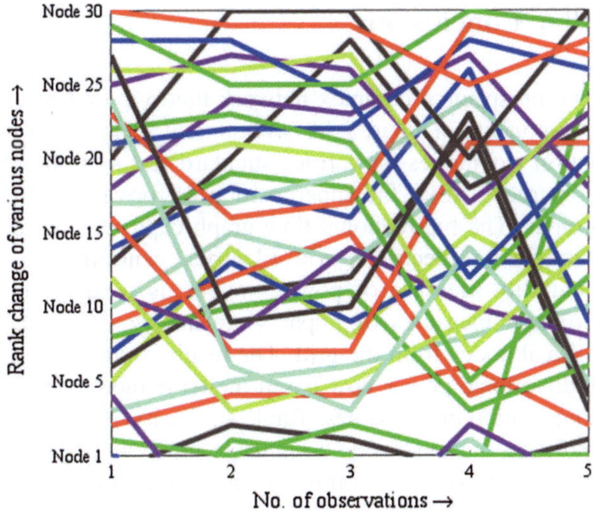


**Fig. 18.8** Variation of ranks of nodes in unidirectional model of IEEE 30 bus test system when the network is modified slightly

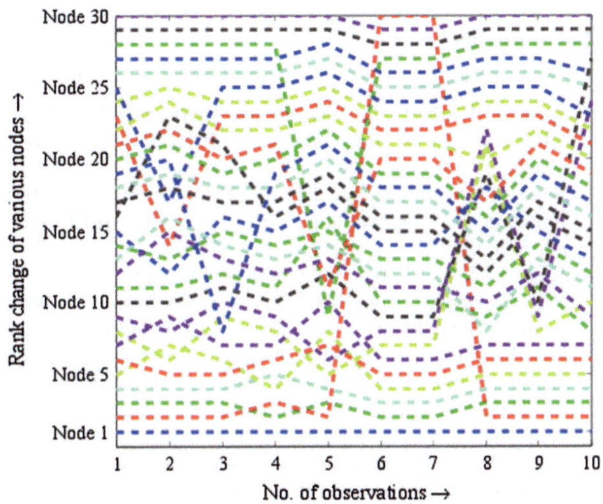


**Fig. 18.9** Rank similarity of nodes in bidirectional power flow model is better than that of unidirectional one

the situation in the future smart grid, bidirectional model gives better result in terms of rank similarity as given in Fig. 18.9.

This introduces a new concept of power flowing from customer end towards the grid. The bidirectional power flow changes the whole power flow pattern of the existing grid [18]. Analytical methods, technical strategies, control system and protecting devices need to be changed along with, to mention a few. Metering and protecting equipments will experience flows coming from the reverse side. Proper operation of the equipments used earlier can be ensured either by changing the instruments themselves or by incorporating new measurement techniques [27].

## 18.6 Conclusions

The prospect of complex network theory based research in analyzing the critical components in smart grid environment is analyzed here with Monte-Carlo simulation techniques on various standard test systems. A bidirectional flow graph is constructed from the superposition of forward and backward unidirectional flow graphs. The bidirectional flow graph captures the true power flow scenario of the future smart electricity grid. Electrical centrality measure, motivated by closeness centrality measure of power system, is used to find critical components. Four different measures of impacts are analyzed to quantify the effect of removing critical nodes from the grid. The results found from different measures show that, bidirectional power flow based model is more effective in smart grid environment than unidirectional ones. Rank similarity analysis shows that, critical nodes of bidirectional models do not change much with system topology change as a result of reverse power flow through transmission network in smart grid environment.

## References

1. Farhangi H (2010) The path of the smart grid. IEEE Power Energ Mag 8(1):18–28
2. Glover JD, Sarma MS, Overbye T (2011) Power system analysis and design. Fifth edition, Cengage Learning
3. Sood VK, Fischer D, Eklund JM, Brown T (2009) Developing a communication infrastructure for the smart grid. In: 2009 IEEE electrical power energy conference (EPEC), pp 1–7
4. Kosterev DN, Taylor CW, Mittelstadt WA (1999) Model validation for the august 10, 1996 WSCC system outage. IEEE Trans Power Syst 14(3):967–979
5. Andersson G, Donalek P, Farmer R, Hatziargyriou N, Kamwa I, Kundur P, Martins N, Paserba J, Pourbeik P, Sanchez-Gasca J, Schulz R, Stankovic A, Taylor C, Vittal V (2005) Causes of the 2003 major grid blackouts in North America and Europe, and recommended means to improve system dynamic performance. IEEE Trans Power Syst 20(4):1922–1928
6. Kaplan SM, Sissine F, Abel A, Wellinghoff J, Kelly SG, Hoecker JJ (2009) Smart grid: modernizing electric power transmission and distribution; energy independence, storage and security; energy independence and security act of 2007 (EISA); improving electrical grid efficiency, communication, reliability, and resiliency; integrating new and renewable energy sources. Government series, TheCapitol.Net
7. Chen G, Dong ZY, Hill DJ, Zhang GH (2009) An improved model for structural vulnerability analysis of power networks. Physica A 388(19):4259–4266
8. Freeman LC (1979) Centrality in social networks: I. conceptual clarification. Social Networks 1:215–239
9. Watts DJ, Strogatz SH (1998) Collective dynamics of 'small-world' networks. Nature 393(6684):440–442
10. Reka A, Barabási AL (2002) Statistical mechanics of complex networks. Rev Mod Phys 74:47–97
11. Hines P, Cotilla-Sanchez E, Blumsack S (2010) Do topological models provide good information about electricity infrastructure vulnerability? Chaos Interdisc J Nonlinear Sci 20(3):033122

12. Dwivedi A, Yu X (2011) A maximum flow based complex network approach for power system vulnerability analysis. IEEE Trans Industr Inf 99:1
13. Chen G, Dong ZY, Hill DJ, Xue YS (2011) Exploring reliable strategies for defending power systems against targeted attacks. IEEE Trans Power Syst 26(3):1000–1009
14. Sun k (2005) Complex networks theory: a new method of research in power grid. In: 2005 IEEE/PES transmission and distribution conference and exhibition: Asia and Pacific, pp 1–6
15. Chen X, Sun K, Cao Y, Wang S (2007) Identification of vulnerable lines in power grid based on complex network theory. In: 2007 IEEE power engineering society general meeting, pp 1–6
16. Zhao H, Zhang C, Ren H (2008) Power transmission network vulnerable region identifying based on complex network theory. In: Third international conference on electric utility deregulation and restructuring and power technologies, 2008. DRPT, pp 1082–1085
17. Dwivedi A, Yu X, Sokolowski P (2009) Identifying vulnerable lines in a power network using complex network theory. In: IEEE international symposiumon industrial electronics, 2009. ISIE, pp 18–23
18. Xu S, Zhou H, Li C, Yang X (2009) Vulnerability assessment of power grid based on complex network theory. In: Asia-Pacific power and energy engineering conference, 2009. APPEEC, pp 1–4
19. Dwivedi A, Yu X, Sokolowski P (2010) Analyzing power network vulnerability with maximum flow based centrality approach. In: 2010 8th IEEE international conference on industrial informatics (INDIN), pp 336–341
20. Chen G, Dong ZY, Hill DJ, Zhang GH, Hua KQ (2010) Attack structural vulnerability of power grids: a hybrid approach based on complex networks. Physica A 389(3):595–603
21. Wang Z, Scaglione A, Thomas RJ (2010) Electrical centrality measures for electric power grid vulnerability analysis. In: 2010 49th IEEE conference on decision and control (CDC), pp 5792–5797
22. Bompard E, Napoli R, Xue F (2010) Extended topological approach for the assessment of structural vulnerability in transmission networks. IET Gener Transm Distrib 4(6):716–724
23. Bompard E, Di W, Xue F (2011) Structural vulnerability of power systems: a topological approach. Electr Power Syst Res 81(7):1334–1340
24. Saadat H (2002) Power systems analysis. McGraw-Hill series in electrical and computer engineering, 2nd edn. McGraw-Hill Custom Publishing, 712 p, July 15 2002, ISBN-10:0072848693, ISBN-13:978-0072848694
25. IEEE power system test case archieve (2013) http://www.ee.washington.edu/research/pstca. Accessed 11 Nov 2013
26. Nasiruzzaman ABM, Pota HR (2013) Bus dependency matrix of electrical power systems. Int J Electr Power Energ Syst (Article in Press), http://dx.doi.org/10.1016/j.ijepes.2013.10.031
27. Freeman LC (1980) The gatekeeper, pair-dependency and structural centrality. Qual Quant 14(4):585–592
28. Johnson DB (1977) Efficient algorithms for shortest paths in sparse networks. J ACM 24:1–13
29. Albert R, Jeong H, Barabasi AL (2000) Error and attack tolerance of complex networks. Nature 406(6794):378–382
30. Albert R, Albert I, Nakarado GL (2004) Structural vulnerability of the North American power grid. Phys Rev E 69:025103
31. Carreras BA, Newman DE, Dobson I, Poole AB (2004) Evidence for self-organized criticality in a time series of electric power system blackouts. IEEE Trans Circuits Syst I (Reg Papers) 51(9):1733–1740
32. Dobson I, Carreras BA, Lynch VE, Newman DE (2007) Complex systems analysis of series of blackouts: cascading failure, critical points, and self-organization. Chaos 17(2):026103
33. Mei S, He F, Zhang X, Wu S, Wang G (2009) An improved OPA model and blackout risk assessment. IEEE Trans Power Syst 24(2):814–823
34. Wood AJ, Wollenberg BF (2006) Power generation operation and control, Wiley, New York

[illegible]

MIX
Papier aus verantwortungsvollen Quellen
Paper from responsible sources
FSC® C105338
FSC
www.fsc.org

If you have any concerns about our products,
you can contact us on
**ProductSafety@springernature.com**

In case Publisher is established outside the EU,
the EU authorized representative is:
**Springer Nature Customer Service Center GmbH**
**Europaplatz 3, 69115 Heidelberg, Germany**

Printed by Libri Plureos GmbH
in Hamburg, Germany